ALGORITHMS AND PROTOCOLS FOR WIRELESS AND MOBILE AD HOC NETWORKS

**WILEY SERIES ON PARALLEL
AND DISTRIBUTED COMPUTING**

Editor: Albert Y. Zomaya

A complete list of titles in this series appears at the end of this volume.

ALGORITHMS AND PROTOCOLS FOR WIRELESS AND MOBILE AD HOC NETWORKS

Edited by

Azzedine Boukerche, PhD
University of Ottawa
Ottawa, Canada

WILEY

A John Wiley & Sons, Inc., Publication

Copyright © 2009 by John Wiley & Sons, Inc. All rights reserved

Published by John Wiley & Sons, Inc., Hoboken, New Jersey
Published simultaneously in Canada

No part of this publication may be reproduced, stored in a retrieval system, or transmitted in any form or by any means, electronic, mechanical, photocopying, recording, scanning, or otherwise, except as permitted under Section 107 or 108 of the 1976 United States Copyright Act, without either the prior written permission of the Publisher, or authorization through payment of the appropriate per-copy fee to the Copyright Clearance Center, Inc., 222 Rosewood Drive, Danvers, MA 01923, (978) 750-8400, fax (978) 750-4470, or on the web at www.copyright.com. Requests to the Publisher for permission should be addressed to the Permissions Department, John Wiley & Sons, Inc., 111 River Street, Hoboken, NJ 07030, (201) 748-6011, fax (201) 748-6008, or online at http://www.wiley.com/go/permission.

Limit of Liability/Disclaimer of Warranty: While the publisher and author have used their best efforts in preparing this book, they make no representations or warranties with respect to the accuracy or completeness of the contents of this book and specifically disclaim any implied warranties of merchantability or fitness for a particular purpose. No warranty may be created or extended by sales representatives or written sales materials. The advice and strategies contained herein may not be suitable for your situation. You should consult with a professional where appropriate. Neither the publisher nor author shall be liable for any loss of profit or any other commercial damages, including but not limited to special, incidental, consequential, or other damages.

For general information on our other products and services or for technical support, please contact our Customer Care Department within the United States at (800) 762-2974, outside the United States at (317) 572-3993 or fax (317) 572-4002.

Wiley also publishes its books in a variety of electronic formats. Some content that appears in print may not be available in electronic formats. For more information about Wiley products, visit our web site at www.wiley.com.

Library of Congress Cataloging-in-Publication Data:

Algorithms and protocols for wireless, mobile ad hoc networks / edited by Azzedine Boukerche.
    p. cm.
  Includes bibliographical references and index.
  ISBN 978-0-470-38358-2 (cloth)
1. Ad hoc networks (Computer networks) 2.   Wireless LANs. 3.   Mobile computing. 4.   Computer algorithms. I.   Boukerche, Azzedine.
  TK5105.77.A44 2008
  621.384–dc22

                                                        2008016868

Printed in the United States of America

*This book is dedicated to my parents and my family who have always been there with me.
Love you all.*

Azzedine Boukerche

CONTENTS

Preface	ix
Contributors	xiii
About the Editor	xvii

1. **Algorithms for Mobile Ad Hoc Networks** — 1
   Azzedine Boukerche, Daniel Câmara, Antonio A.F. Loureiro, and Carlos M.S. Figueiredo

2. **Establishing a Communication Infrastructure in Ad Hoc Networks** — 21
   Michel Barbeau, Evangelos Kranakis, and Ioannis Lambadaris

3. **Robustness Control for Network-Wide Broadcast in Multihop Wireless Networks** — 51
   Paul Rogers and Nael B. Abu-Ghazaleh

4. **Encoding for Efficient Data Distribution in Multihop Ad Hoc Networks** — 87
   Luciana Pelusi, Andrea Passarella, and Marco Conti

5. **A Taxonomy of Routing Protocols for Mobile Ad Hoc Networks** — 129
   Azzedine Boukerche, Mohammad Z. Ahmad, Damla Turgut, and Begumhan Turgut

6. **Adaptive Backbone Multicast Routing for Mobile Ad Hoc Networks** — 165
   Chaiporn Jaikaeo and Chien-Chung Shen

7. **Effect of Interference on Routing in Multihop Wireless Networks** — 187
   Vinay Kolar and Nael B. Abu-Ghazaleh

8. **Routing Protocols in Intermittently Connected Mobile Ad Hoc Networks and Delay-Tolerant Networks** — 219
   Zhensheng Zhang

9. **Transport Layer Protocols for Mobile Ad Hoc Networks** — 251
   Lap Kong Law, Srikanth V. Krishnamurthy, and Michalis Faloutsos

10. **ACK-Thinning Techniques for TCP in MANETs** — 277
    Stylianos Papanastasiou, Mohamed Ould-Khaoua, and Lewis M. MacKenzie

11. **Power Control Protocols for Wireless Ad Hoc Networks** 315
   Junhua Zhu, Brahim Bensaou, and Farid Naït-Abdesselam

12. **Power Saving in Solar-Powered WLAN Mesh Networks** 353
   Amir A. Sayegh, Mohammed N. Smadi, and Terence D. Todd

13. **Reputation-and-Trust-Based Systems for Ad Hoc Networks** 375
   Avinash Srinivasan, Joshua Teitelbaum, Jie Wu, Mihaela Cardei, and Huigang Liang

14. **Vehicular Ad Hoc Networks: An Emerging Technology Toward Safe and Efficient Transportation** 405
   Maen M. Artimy, William Robertson, and William J. Phillips

15. **Performance Issues in Vehicular Ad Hoc Networks** 433
   Maria Kihl and Mihail L. Sichitiu

16. **Cluster Interconnection in 802.15.4 Beacon-Enabled Networks** 459
   Jelena Mišić and Ranjith Udayshankar

Index 481

PREFACE

With the recent technological advances in wireless communication and the increasing popularity of portable computing devices, wireless and mobile ad hoc networks are expected to play an increasingly important role in future civilian and military settings where wireless access to wired backbone is either ineffective or impossible. Mobile ad hoc networks (MANETs) are composed of a set of stations (nodes) communicating through wireless channels, without any fixed backbone support. Applications of MANETs include, but are not limited to, military operations, security, emergency, and rescue operations, among other applications where intense utilization of a communication networks is available for a very limited time. However, frequent topology changes caused by node mobility make routing in wireless ad hoc networks a challenging problem. In addition, limited capabilities of mobile stations require a control on node congestion due to message forwarding and limited battery consumption. Mobility of mobile hosts introduces also new challenging problems that were not encountered in the design and implementation of conventional wireless and wired networks. A critical and challenging problem of mobile ad hoc networking and computing is how to fully cope with the special characteristics of the wireless and mobile ad hoc environment, make balanced use of computation and communication resources, and take advantage of and support the user's mobility. Most of the available literature in this emerging technology concentrates on physical and networking aspects of the subject. However, in most of the these studies, they have neglected the description of the fundamental design of distributed algorithms and have not discussed how to apply them to wireless and mobile ad hoc and network setting environments. An important requirement for successful deployment of wireless and mobile ad hoc network-based applications is the careful evaluation of performance and investigation of alternatives algorithms, prior to their implementation. In light of this, the purpose of this book is to focus on several aspects of mobile ad hoc networking and computing and, in particular, algorithmic methods and distributed computing with mobile communication and computation capabilities.

This book is organized as follows. In Chapter 1, we address the design challenges of distributed algorithms and discuss several important algorithmic issues arising in wireless and mobile networks. Algorithms for MANETs must self-configure and must adjust to environment and data communication where they run, and goal changes posed from the user and application. Chapter 2 presents several techniques for enabling communication infrastructure and maintenance in ad hoc networks. Network-Wide Broadcast (NWB) algorithms provide a mechanism to deliver information to nodes in multihop and networks without depending on routing state. This makes them ideal for initial self-configuration or for operation under mobility where the routing state becomes stale. Chapter 3 presents a classification of existing NWB algorithms, focusing on their robustness characteristics representing different points of the space under different network densities, loss rates, and routing and mobility characteristics. The heterogeneity of portable devices to be interconnected, along with the large spectrum of communication requirements, has helped the design of a new set of multihop ad hoc network technologies, which are known as opportunistic networks, and they represent one of the most interesting scenarios for the application of the encoding techniques. Several studies have been devoted to possible applications of network coding over multihop networks. Chapter 4 describes the basic encoding techniques upon which network coding is based, and then it presents encoding techniques for efficient data distribution in multihop ad hoc networks.

Because of the frequent changes of mobile nodes, routing has always been one of the most challenging problem for MANET's designers. Chapter 5 provides a comprehensive taxonomy of existing ad hoc routing protocols, and it discusses the advantages and disadvantages of each of the routing protocols. Chapter 6 presents a set of well-known mobility-adaptive multicast routing protocols for MANETs. Traditional IP multicast and routing protocols are inappropriate for mobile ad hoc networks. This is mainly because multicast trees could easily break due to dynamic topologies of MANET. Chapter 7 discusses the evolution of multihop wireless network routing protocols from the perspective of accounting for link quality, and then it provides and overview of recent efforts to model and analytically derive near-optimal routing configurations. Chapter 8 provides an overview of the state of the art on routing protocols in intermittently connected mobile ad hoc networks and delay-tolerant networks.

The Transmission Control Protocol (TCP) is an efficient transport layer protocol designed for wired networks (such as the Internet). However, many studies have shown that it performs badly in the wireless and mobile ad hoc network (MANETs) environments. Chapter 9 provides a thorough understanding of the problems of traditional TCP over wired networks and presents the principles behind the proposed traditional TCP enhancements for mobile ad hoc networks. As a follow-up, Chapter 10 identifies the main challenges faced by TCP when used over a diverse ad hoc network environment, and it discusses recent ACK-thinning techniques that have been proposed for MANET environments.

Transmission power control in wireless ad hoc networks is concerned with the selection of transmit power for packet transmission at each node to achieve some desired performance targets. The transmit power level affects many aspects of the operation of wireless ad hoc network. Chapter 11 discusses some basic principles that can be used as a guideline for the design of efficient power control protocols and presents several well-known power control algorithms and protocols while highlighting their impacts on the layers of the protocol stack. In recent years, we have been witnessing a growing interest to WLAN mesh networks. This is mainly because they are viewed as a cost-effective way of deploying outdoor coverage in metro-area Wi-Fi hot zones. In a WLAN mesh, multihop relaying is used to reduce the infrastructure cost of providing wired network connections to each WLAN mesh node. However, before these multihop networks become a commodity for outdoor mesh deployments, we must ensure that there are continuous electrical power connections to mesh nodes. Chapter 12 discusses the design and resource allocation for solar-powered IEEE 802.11 WLAN mesh networks.

Wireless communication networks, in general, and mobile ad hoc networks (MANETs), in particular, have undergone tremendous technological advances over the last few years. However, due to the nature of MANETs, nodes are vulnerable to a variety of security threats. Unless we take these security issues seriously, mobile ad hoc networks will never be fully deployed and adopted by the regular users. Indeed, security and trust have been widely recognized as an important factor affecting consumer behavior. Chapter 13 presents a detailed understanding of reputation-and-trust-based systems for wireless network in general, as well as of mobile ad hoc networks in particular.

In these last few years, we have seen the development of a number of research studies investigating the use of ad hoc networks as a communication technology for vehicle-specific applications within the wider concept of Intelligent Transportation Systems (ITS). In this kind of network, vehicles have communication capability, which allows them to exchange messages with each other using vehicle-to-vehicle communication (V2V) and to exchange messages with a roadside network infrastructure using roadside-to-vehicle communication (R2V). Chapter 14 identifies the main features that distinguish vehicular ad hoc networks from traditional ad hoc networks, and it also presents a summary of the enabling technologies that are expected to support this emerging technology, while Chapter 15 presents the routing techniques that are suitable for vehicular ad hoc networks and highlights the transport and security issues for these networks. Last but not least, Chapter 16 discusses the main design and performance issues of cluster interconnection for beacon-enabled 802.15.4 clusters, an emerging technology for multihop wireless networks, and then highlights the pros and cons for each of the proposed approaches.

It is our belief that this is the first book that covers the basic and fundamental algorithms and protocols for wireless ad hoc and multihop networks, making their design and analysis accessible to all levels of readers.

Special thanks are due to all contributors for their support and patience, as well as to the reviewers for their hard work and timely reports, which make this book truly special. Last but not least, we wish to extend our thanks to Paul Petralia and Whitney Lesch from John Wiley & Sons for their support, guidance, and certainly their patience in finalizing this book.

AZZEDINE BOUKERCHE

University of Ottawa

CONTRIBUTORS

Nael B. Abu-Ghazaleh, Computer Science Department, Binghamton University, Binghamton, NY 13902

Mohammad Z. Ahmad, School of Electrical Engineering and Computer Science, University of Central Florida, Orlando, FL

Maen M. Artimy, Internetworking Atlantic Inc., Halifax, Nova Scotia, Canada, B3J 1L1

Michel Barbeau, School of Computer Science, Carleton University, 1125 Colonel By Drive, Ottawa, Ontario K1S 5B6, Canada

Brahim Bensaou, Department of Computer Science and Engineering, The Hong Kong University of Science and Technology, Kowloon, Hong Kong

Azzedine Boukerche, School of Information Technology and Engineering, University of Ottawa, Ottawa, Ontario K1N 6N5, Canada

Daniel Câmara, Department of Computer Science, Federal University of Minas Gerais, Below Horizonte, Brazil

Mihaela Cardei, Department of Computer Science and Engineering, Florida Atlantic University, Boca Raton, FL 33431

Marco Conti, IIT-CNR, Via G. Moruzzi 1, Pisa 56124, Italy

Michalis Faloutsos, Computer Science Department, University of California, Riverside, California, 92521 USA

Carlos M.S. Figueiredo, FUCAPI—Analysis, Research, and Technology Innovation Center, Belo Horizonte, Brazil

Chaiporn Jaikaeo, Department of Computer Engineering, Kasetsart University, Bangkok, Thailand

Maria Kihl, Department of Electrical and Information Technology, Lund University, Sweden

Vinay Kolar, Computer Science Department, Binghamton University, Binghamton, NY 13902, USA

Evangelos Kranakis, School of Computer Science, Carleton University, 1125 Colonel By Drive, Ottawa, Ontario K1S 5B6, Canada

Srikanth V. Krishnamurthy, Computer Science Department, University of California, Riverside, California, 92521 USA

Ioannis Lambadaris, Department of Systems and Computer Engineering, Carleton University, 1125 Colonel By Drive, Ottawa, Ontario K1S 5B6, Canada

Lap Kong Law, Trapeze Networks, Pleasanton, CA, 94588-4084

Antonio A.F. Loureiro, Department of Computer Science, Federal University of Minas Gerais, Below Horizonte, Brazil

Huigang Liang, Department of Management Information Systems, East Carolina University, Greenville, North Carolina 27858

Lewis M. MacKenzie, Department of Computing Science, University of Glasgow, Glasgow G12 8RZ, Scotland

Jelena Mišić, Department of Computer Science, University of Manitoba, Winnepeg, Manitoba, Canada

Farid Naït-Abdesselam, IRCiCA/LIFL—INRIA POPS, University of Lille 1, Lille, France

Mohamed Ould-Khaoua, Department of Computing Science, University of Glasgow, Glasgow G12 8RZ, Scotland

Stylianos Papanastasiou, Department of Computing Science, University of Glasgow, Glasgow G12 8RZ, Scotland

Andrea Passarella, IIT-CNR, Via G. Moruzzi 1, Pisa 56124, Italy

Luciana Pelusi, IIT-CNR, Via G. Moruzzi 1, Pisa 56124, Italy

William J. Phillips, Department of Engineering Mathematics and Internetworking, Dalhousie University, Halifax, Nova Scotia, Canada, B3H 4R2

William Robertson, Internetworking Program, Dalhousie University, Halifax, Nova Scotia, Canada, B3H 4R2

Paul Rogers, Computer Science Department, Binghamton University, Binghamton, NY 13902

Amir A. Sayegh, Department of Electrical and Computer Engineering, McMaster University, Hamilton, Ontario, Canada

Chien-Chung Shen, Department of Computer and Information Sciences, University of Delaware, Newark, DE 19716

Mihail L. Sichitiu, Department of Electrical and Computer Engineering, North Carolina State University, Raleigh, NC 27695

Mohammed N. Smadi, Department of Electrical and Computer Engineering, McMaster University, Hamilton, Ontario, Canada

Avinash Srinivasan, Department of Mathematics, Computer Science, and Statistics, Bloomsburg University, Bloomsburg, PA 17815

Joshua Teitelbaum, Microsoft, Redmond, Seattle 98052

Terence D. Todd, Department of Electrical and Computer Engineering, McMaster University, Hamilton, Ontario, Canada

Begumhan Turgut, Department of Computer Science, Rutgers University, Piscataway, NJ

Damla Turgut, School of Electrical Engineering and Computer Science, University of Central Florida, Orlando, FL 32816-2450

Ranjith Udayshankar, Department of Computer Science, University of Manitoba, Winnipeg, Manitoba, Canada

Jie Wu, Department of Computer Science and Engineering, Florida Atlantic University, Boca Raton, FL 33431

Zhensheng Zhang, San Diego Research Center, San Diego, CA 92121

Junhua Zhu, Department of Computer Science and Engineering, The Hong Kong University of Science and Technology, Kowloon, Hong Kong

ABOUT THE EDITOR

Azzedine Boukerche is a Professor and holds a Canada Research Chair position at the University of Ottawa. He is the Founding Director of Paradise Research Laboratory at the University of Ottawa. Prior to this, he held a Faculty position at the University of North Texas, and he was working as a Senior Scientist at the Simulation Sciences Division, Metron Corporation, located in San Diego. He was also employed as a faculty member at the School of Computer Science, McGill University, and he taught at Polytechnic of Montreal. He spent a year at the JPL/NASA-California Institute of Technology, where he contributed to a project centered around the specification and verification of the software used to control interplanetary spacecraft operated by JPL/NASA Laboratory. His current research interests include wireless ad hoc and sensor networks, wireless networks, mobile and pervasive computing, wireless multimedia, QoS service provisioning, large-scale distributed interactive simulation, parallel discrete event simulation, and performance evaluation and modeling of large-scale distributed and mobile systems. Dr. Boukerche has published several research papers in these areas. He was the recipient of and/or nominated for the Best Research Paper Award at IEEE/ACM PADS '97, IEEE/ACM PADS '99, ACM MSWiM 2001, ICC'08, and MobiWac'06, and he was the co-recipient of the 3rd National Award for Telecommunication Software 1999 for his work on distributed security systems on mobile phone operations.

Dr. A. Boukerche is a holder of an Ontario Early Research Excellence Award (previously known as Premier of Ontario Research Excellence Award), an Ontario Distinguished Researcher Award, and a Glinski Research Excellence Award. He is a Co-Founder of QShine International Conference on Quality of Service for Wireless/Wired Heterogeneous Networks (QShine 2004) and has served as a General Chair for the 8th ACM/IEEE Symposium on Modeling, Analysis, and Simulation of Wireless and Mobile Systems, the 9th ACM/IEEE Symposium on Distributed Simulation and Real-Time Application, and the 6th IEEE/ACM MASCOT '98 Symposium; he has also served as the Vice General Chair for the 3rd IEEE International Conference on Distributed Computing in Sensor Systems (DCOSS '07), Program Chair for IEEE Globecom 2007 and 2008 Ad Hoc, Sensor and Mesh Networking Symposium, and a Program Co-Chair for ICPP 2008, the 2nd ACM Workshop on QoS and Security for Wireless and Mobile Networks, ACM/IFIPS Europar 2002 Conference, IEEE/SCS Annual Simulation Symposium '02, ACM WWW '02, IEEE MWCN 2002, IEEE/ACM MASCOTS '02, IEEE Wireless Local Networks 03-04, IEEE WMAN 04-05, and ACM MSWiM 98-99.

Dr. A. Boukerche is an Associate Editor for *ACM/Springer Wireless Networks, IEEE Transactions on Vehicular Networks*, *IEEE Wireless Communication Magazine*, *IEEE Transactions on Parallel and Distributed Systems*, Elsevier's *Ad Hoc Networks, Wiley International Journal of Wireless Communication and Mobile Computing*, Wiley's *Security and Communication Network Journal*, Wiley's *Pervasive and Mobile Computing Journal*, Elsevier's *Journal of Parallel and Distributed Computing*, and *SCS Transactions on Simulation*. He also serves as a Steering Committee Chair for the ACM Modeling, Analysis and Simulation for Wireless and Mobile Systems Symposium, the ACM Symposium on Performance Evaluation of Wireless Ad Hoc, Sensor, and Ubiquitous Networks, and the IEEE/ACM Distributed Simulation and Real-Time Applications Symposium (DS-RT).

CHAPTER 1

Algorithms for Mobile Ad Hoc Networks

AZZEDINE BOUKERCHE

School of Information Technology and Engineering, University of Ottawa, Ottawa, Ontario K1N 6N5, Canada

DANIEL CÂMARA and ANTONIO A. F. LOUREIRO

Department of Computer Science, Federal University of Minas Gerais, Belo Horizonte, Brazil

CARLOS M.S. FIGUEIREDO

FUCAPI—Analysis, Research, and Technological Innovation Center, Belo Horizonte, Brazil

1.1 INTRODUCTION

In the fourth century B.C., the Greek writer Aeschylus wrote the play *Agamemnon*, which provides a detailed description of how fire signals were supposedly used to communicate the fall of Troy to Athens over a distance of more than 450 km. This very same idea is present in the third movie of the trilogy "The Lord of the Rings," where fire signals were used to call for help of an ally army. In both cases, as well as in others found mainly in the literature and movies, the problem with a fire signal is that there is only one meaning associated with it—in the examples above, the fall of Troy and the call for help, respectively. This limitation of using fire signals to relay a message was realized by Polybius, one of the most famous ancient Greek historians who lived 200 years after Aeschylus in the second century B.C. To overcome this limitation, Polybius proposed a very simple fire signal mechanism based on fire torches that could be used to relay different messages. He described the procedure a person should follow before they start transmitting a message to another one (i.e., how a connection could be established between a pair of communicating entities), and he also described how messages could be coded using fire torches. Since this was basically a visual communication system, other people could see the same message (broadcast) and the people responsible for relaying messages could be mobile.

Algorithms and Protocols for Wireless and Mobile Ad Hoc Networks, Edited by Azzedine Boukerche
Copyright © 2009 by John Wiley & Sons Inc.

Polybius can probably be considered the first data communication engineer for mobile ad hoc networks. What it is more amazing is that his ideas were used during the next 2000 years for relaying messages among people in scenarios similar to the ones described above.

A mobile ad hoc network (MANET)[1] is comprised of mobile hosts that can communicate with each other using wireless links. It is also possible to have access to some hosts in a fixed infrastructure, depending on the kind of mobile ad hoc network available. Some scenarios where an ad hoc network can be used are business associates sharing information during a meeting, emergency disaster relief personnel coordinating efforts after a natural disaster such as a hurricane, earthquake, or flooding, and military personnel relaying tactical and other types of information in a battlefield.

In this environment a route between two hosts may consist of hops through one or more nodes in the MANET. An important problem in a mobile ad hoc network is finding and maintaining routes since host mobility can cause topology changes. Several routing algorithms for MANETs have been proposed in the literature, and they differ in the way new routes are found and existing ones are modified.

Mobile ad hoc networks can be realized by different networks such as body area network (BAN), vehicular ad hoc network (VANET), wireless networks (varying from personal area network to wide area network), and wireless sensor network (WSN). Furthermore, MANETs can be realized by different wireless communication technologies such as Bluetooth, IEEE 802.11, and Ultra-Wide Band (UWB). However, each one of these networks combined with the communication technologies pose various challenges in the design of algorithms for them as discussed in the following.

1.2 DESIGN CHALLENGES

The design of algorithms for MANETs poses new and interesting research challenges, some of them particular to mobile ad hoc networks. Algorithms for a MANET must self-configure to adjust to environment and traffic where they run, and goal changes must be posed from the user and application.

Data communication in a MANET differs from that of wired networks in different aspects. The wireless communication medium does not have a foreseeable behavior as in a wired channel. On the contrary, the wireless communication medium has variable and unpredictable characteristics. The signal strength and propagation delay may vary with respect to time and environment where the mobile nodes are. Unlike a wired network, the wireless medium is a broadcast medium; that is, all nodes in the transmission range of a transmitting device can receive a message.

The bandwidth availability and computing resources (e.g., hardware and battery power) are restricted in mobile ad hoc networks. Algorithms and protocols need to save both bandwidth and energy and must take into account the low capacity and

---

[1] Ad hoc is a Latin expression that means *"for the particular end or case at hand without consideration of wider application"*. An ad hoc network means that the network is established for a particular, often extemporaneous service customized to applications for a limited period of time.

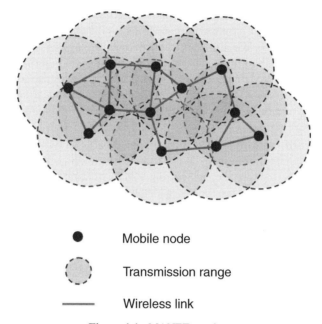

Figure 1.1. MANET topology.

limited processing power of wireless devices. This calls for lightweight solutions in terms of computational, communication, and storage resources.

An important challenge in the design of algorithms for a mobile ad hoc network is the fact that its topology is dynamic. Since the nodes are mobile, the network topology may change rapidly and unexpectedly, thereby affecting the availability of routing paths. Figure 1.1 depicts a snapshot of a MANET topology.

Given all theses differences, the design of algorithms for ad hoc networks are more complex than their wired counterpart.

1.3 MANETs: AN ALGORITHMIC PERSPECTIVE

1.3.1 Topology Formation

Neighbor Discovery. The performance of an ad hoc network depends on the interaction among communicating entities in a given neighborhood. Thus, in general, before a node starts communicating, it must discover the set of nodes that are within its direct communication range. Once this information is gathered, the node keeps it in an internal data structure so it can be used in different networking activities such as routing. The behavior of an ad hoc node depends on the behavior of its neighboring nodes because it must sense the medium before it starts transmitting packets to nodes in its interfering range, which can cause collisions at the other nodes.

Node discovery can be achieved with periodic transmission of beacon packets (active discovery) or with promiscuous snooping on the channel to detect the communication activity (passive discovery). In PRADA [1], a given source node sends periodically to its neighboring nodes a discovery packet, and in turn their neighbors reply with a location update packet (that might include, for instance, the node's geographical location). PRADA adjusts dynamically its communication range, called topology knowledge range, so it leads to a faster convergence of its neighboring nodes.

Packet Forwarding Algorithms. An important part of a routing protocol is the packet forwarding algorithm that chooses among neighboring nodes the one that is going to be used to forward the data packet. The forwarding algorithm implements a forwarding goal that may be, for instance, the shortest average hop distance from source to destination. In this case, the set of potential nodes may include only those in direct communication range from the current node or also the set of possible nodes in the route to the destination. The forwarding goal may also include some QoS parameters such as the amount of energy available at each node.

The following forwarding algorithms consider only nodes that are in direct communication range of the node that has a data packet to be forwarded, as depicted in Figure 1.2. The Most Forward within Radius (MFR) forwarding algorithm [2] chooses the node that maximizes the distance from node S to point p. In this case, as depicted in Figure 1.2, it is node 1. On the other hand, the Nearest Forward Progress (NFP) forwarding algorithm [3] chooses the node that minimizes the distance from node S to point q. In this case it is node 2. The Greedy Routing Scheme (GRS) [4] uses the nodes' geographical location to choose the one that is closest to the destination node D.

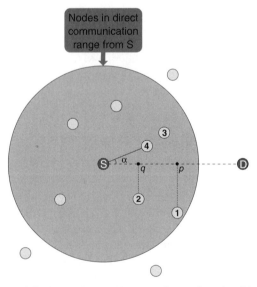

Figure 1.2. Strategies used by some forwarding algorithms.

In this case it is node 3. The compass-selected routing (COMPASS) algorithm [5] chooses the node that minimizes the angle α but considers the nodes that are closer to node D. In this case it is node 4. The random process forwarding algorithm [6], as the name suggests, chooses a random node that is in direct communication range from S.

The Partial Topology Knowledge Forwarding (PTKF) algorithm [1] chooses a node using a localized shortest path weighted routing where routes are calculated based on the local topological view and consider the transmission power needed to transmit in that link.

1.3.2 Topology Control

Topology control algorithms select the communication range of a node, and they construct and maintain a network topology based on different aspects such as node mobility, routing algorithm, and energy conservation [7]. Broadly speaking, topology control algorithms for ad hoc networks can be classified in hierarchical or clustering organization, as well as in power-based control organization [7, 8]. Furthermore, these algorithms can be either centralized, distributed, or localized.

Clustering Algorithms. The clustering process consists in defining a cluster-head node and the associated communication backbone, typically using a heuristic. The goal is to avoid redundant topology information so the network can work more efficiently. Clustering algorithms are often modeled as graph problems such as the minimum connected dominating set (MCDS) [9]. This problem asks for the minimum subset of nodes V' in the original graph $G = (V, E)$ such that V' form a dominating set of G and the resulting subgraph of the MCDS has the same number of connected components of G. It means that if G is a connected graph, so is the resulting subgraph. MCDS is an NP-complete problem [10], and thus we must look for approximate solutions [7]. In the case of the clustering algorithm, nodes in the dominating set represent the cluster heads and the other nodes are their neighbors. An inherit characteristic of an ad hoc network, which makes this problem much more difficult, is that its topology is dynamic.

The cluster heads can be elected using either deterministic or nondeterministic approaches. A deterministic solution is similar to a distributed synchronous algorithm in the sense that it runs in rounds. In this case there is just one round, and after finishing it the cluster heads are chosen. Suppose we have a node and its neighboring nodes—that is, its one-hop neighborhood. The lowest ID solution selects the node with the lowest identifier among them to create the minimal dominating set (MDS) [10]. The max degree solution selects the node with the highest degree among them [11, 12]. The MOBIC solution examines the variations of RSSI (received signal strength indicator) signal among them to select the cluster head [13].

A nondeterministic solution runs multiple incremental steps to avoid variations in the election process and to minimize conflicts among cluster heads in their one-hop neighborhood. Examples of this approach are CEDAR [11], SPAN [14], and solutions based on a spanning tree algorithm [9].

Power-Based Control Algorithms. A mobile node in a MANET must rely on a energy source (typically a battery) to execute all its tasks. Batteries need to be recharged to provide a continuous energy supply for a node. To extend the lifetime of nodes in an ad hoc network, we need algorithms to determine and adaptively adjust the transmission power of each node so as to meet a given minimization goal and, at the same time, maintain a given connectivity constraint. Some possible minimization goals are to control the maximum or average power and define a maximum or average connectivity degree. Some connectivity constraints are a simplex communication or a full-duplex communication (biconnected). Ramanathan and Hain [2] propose a topology control algorithm that dynamically adjusts its transmission power such that the maximum power used is minimized while keeping the network biconnected.

1.3.3 Routing

The main goal of an ad hoc network routing algorithm is to correctly and efficiently establish a route between a pair of nodes in the network so a message can be delivered according to the expected QoS parameters [15, 16]. The establishment of a route should be done with minimum overhead and bandwidth consumption. In the current wired networks, there are different link state [17] and distance vector [18] routing protocols, which were not designed to cope with constant topology changes of mobile ad hoc environments. Link-state protocols update their global state by broadcasting their local state to every other node, whereas distance-vector protocols exchange their local state to adjacent nodes only. Their direct application to a MANET may lead to undesired problems such as routing loops and excessive traffic due to the exchange of control messages during route establishment.

An ad hoc network has a dynamic nature that leads to constant changes in its network topology. As a consequence, the routing problem becomes more complex and challengeable, and it probably is the most addressed and studied problem in ad hoc networks. This reflects the large number of different routing algorithms for MANETs proposed in the literature [15].

Ideally, a routing algorithm for an ad hoc network should not only have the general characteristics of any routing protocol but also consider the specific characteristics of a mobile environment—in particular, bandwidth and energy limitations and mobility. Some of the characteristics are: fast route convergence; scalability; QoS support; power, bandwidth, and computing efficient with minimum overhead; reliability; and security. Furthermore, the behavior of an ad hoc routing protocol can be further complicated by the MAC protocol. This is the case of a data link protocol that uses a CSMA (Carrier Sense Multiple Access) mechanism that presents some problems such as hidden stations and exposed stations.

In general, routing algorithms for ad hoc networks may be divided into two broad classes: proactive protocols and reactive on-demand protocols, as discussed in the following.

Proactive Protocols. Proactive routing algorithms aim to keep consistent and up-to-date routing information between every pair of nodes in the network by proactively propagating route updates at fixed time intervals. Usually, each node maintains this information in tables; thus, protocols of this class are also called table-driven algorithms. Examples of proactive protocols are Destination-Sequenced Distance Vector (DSDV) [19], Optimized Link-State Routing (OLSR) [20], and Topology-Based Reverse Path Forwarding (TBRPF) Protocols [21].

The DSDV protocol is a distance vector protocol that incorporates extensions to make its operation suitable for MANETs. Every node maintains a routing table with one route entry for each destination in which the shortest path route (based on the number of hops) is recorded. To avoid routing loops, a destination sequence number is used. A node increments its sequence number whenever a change occurs in its neighborhood. When given a choice between alternative routes for the same destination, a node always selects the route with the greatest destination sequence number. This ensures utilization of the route with the most recent information.

The OLSR protocol is a variation version of the traditional link state protocol. An important aspect of OLSR is the introduction of multipoint relays (MPRs) to reduce the flooding of messages carrying the complete link-state information of the node and the size of link-state updates. Upon receiving an update message, the node determines the routes (sequence of hops) to its known nodes. Each node selects its MPRs from the set of its neighbors such that the set covers those nodes that are distant two hops away. The idea is that whenever a node broadcasts a message, only those nodes present in its MPR set are responsible for broadcasting the message.

The Topology-Based Reverse Path Forwarding is also a variation of the link-state protocol. Each node has a partial view of the network topology, but is sufficient to compute a shortest path source spanning tree rooted at the node. When a node receives source trees maintained at neighboring nodes, it can update its own shortest path tree. TBRPF exploits the fact that shortest path trees reported by neighbor nodes tend to have a large overlap. In this way, a node can still compute its shortest path tree even if it receives partial trees from its neighbors. In this way, each node reports part of its source tree, called reported tree (RT), to all of its neighbors to reduce the size of topology update messages, which can be either full or differential. Full updates are used to send to new neighbors the entire RT to ensure that the topology information is correctly propagated. Differential updates contain only changes to RT that have occurred since the last periodic update. To reduce further the number of control messages, topology updates can be combined with Hello messages so that fewer control packets are transmitted.

Reactive Protocols. Reactive on-demand routing algorithms establish a route to a given destination only when a node requests it by initiating a route discovery process. Once a route has been established, the node keeps it until the destination is no longer accessible, or the route expires. Examples of reactive protocols are Dynamic Source Routing (DSR) [22] and Ad Hoc On-Demand Distance Vector (AODV) [23].

The DSR protocol determines the complete route to the destination node, expressed as a list of nodes of the routing path, and embeds it in the data packet. Once a node receives a packet it simply forwards it to the next node in the path. DSR keeps a cache structure (table) to store the source routes learned by the node. The discovery process is only initiated by a source node whenever it does not have a valid route to a given destination node in its route cache. Entries in the route cache are continually updated as new routes are learned. Whenever a node wants to know a route to a destination, it broadcasts a route request (RREQ) message to its neighbors. A neighboring node receives this message, updates its own table, appends its identification to the message and forwards it, accumulating the traversed path in the RREQ message. A destination node responds to the source node with a route reply (RREP) message, containing the accumulated source route present in the RREQ. Nodes in DSR maintain multiple routes to a destination in the cache, which is helpful in case of a link failure.

The AODV protocol keeps a route table to store the next-hop routing information for destination nodes. Each routing table can be used for a period of time. If a route is not requested within that period, it expires and a new route needs to be found when needed. Each time a route is used, its lifetime is updated. When a source node has a packet to be sent to a given destination, it looks for a route in its route table. In case there is one, it uses it to transmit the packet. Otherwise, it initiates a route discovery procedure to find a route by broadcasting a route request (RREQ) message to its neighbors. Upon receiving a RREQ message, a node performs the following actions: checks for duplicate messages and discards the duplicate ones, creates a reverse route to the source node (the node from which it received the RREQ is the next hop to the source node), and checks whether it has an unexpired and more recent route to the destination (compared to the one at the source node). In case those two conditions hold, the node replies to the source node with a RREP message containing the last known route to the destination. Otherwise, it retransmits the RREQ message.

Some Comments. An important question is to determine the best routing protocol to be used in a MANET. This is not a simple issue, and the identification of the most appropriate algorithm depends on different factors such as QoS guarantees, scalability, and traffic and mobility pattern. Reactive protocols tend to be more efficient than proactive protocols in terms of control overhead and power consumption because routes are only created when required. On the other hand, proactive protocols need periodic route updates to keep information updated and valid. In addition, many available routes might never be needed, which increases the routing overhead. Proactive protocols tend to provide better quality of service than reactive protocols. In this class of protocols, routing information is kept updated; thus, a route to a given destination is available and up-to-date, which minimizes the end-to-end delay. Royer and Toh [15] present a comparison of these protocols in terms of their complexity, route update patterns, and capabilities.

The above classification is very broad, and there are other taxonomies to categorize routing protocols [24]. For instance, there are protocols that use a hybrid scheme to route messages; that is, they try to combine the advantages of some protocols, whereas there are protocols that use the node's geographical location to route messages.

It is interesting to observe that some IETF MANET Internet Drafts [mobile ad hoc networks (MANETs)] have reached a reasonable level of maturity, analysis, and implementation experience and became IETF standards. This includes the proactive protocols Optimized Link-State Routing (OLSR) [20] and Topology Dissemination-Based Reverse Path Forwarding (TBRPF) [21] and the reactive protocols Distributed Source Routing (DSR) [22] and Ad Hoc On-Demand Distance Vector (AODV) [23].

1.3.4 Multicasting and Broadcasting

An important aspect in the design of a routing protocol is the type of communication mode allowed between peer entities. Routing protocols for a MANET can be unicast, geocast, multicast, or broadcast. Unicast is the delivery of messages to a single destination. Geocast is the delivery of messages to a group of destinations identified by their geographical locations. Multicast is the delivery of messages to a group of destinations in such a way that it creates copies only when the links to the destinations split. Finally, broadcast is the delivery of a message to all nodes in the network.

Notice that, broadly speaking, there are two types of physical transmission technology that are largely used: broadcast links and point-to-point links. In a network with a single broadcast channel, all communicating elements share it during their transmissions. In a network that employs a wireless medium, which is the case of a mobile ad hoc network, broadcast is a basic operation mode whereby a message is received by all the source node's neighbors. In a MANET, the four communication modes that can be implemented by a routing protocol are realized by a wireless broadcast channel.

A multicast routing protocol is employed when a mobile node wants to send the same message or stream of data to a group of nodes that share a common interest. If there is a geographical area (location) associated with the nodes that will receive the message or stream of data, we use a geocast protocol. Thus, a geocast protocol is a special type of multicast protocol, such that nodes need their updated location information along the time to delivery a message. In a multicast communication, nodes may join or leave a multicast group as desired, whereas in a geocast communication, nodes can only join or leave the group by entering or leaving the defined geographical region.

In a MANET, a multicast communication can possibly bring benefits to the nodes such as bandwidth and energy savings. However, the maintenance of a multicast route, often based on a routing tree or mesh, is a difficult problem for mobile ad hoc multicasting routing protocols due to the dynamic nature of a MANET. In particular, the cost of keeping a routing tree connected for the purpose of multicast communication may be prohibited. In a multicast mesh, a message can be accepted from any router node, as opposed to a tree that only accepts packets routed by tree nodes. Thus, a multicast mesh is more suitable for a MANET because it supports a higher connectivity than a tree. The method used to build the routing infrastructure (tree or mesh) in a mobile ad hoc network distinguishes the different multicasting routing protocols.

Some of the route-tree-based multicast protocols for MANETs are AMRoute (Adhoc Multicast Routing Protocol) [25], DDM (Differential Destination Multicast) [26], and MAODV (Multicast Ad-hoc On-Demand Distance Vector routing) [27]. AMRoute uses an overlay approach based on bidirectional unicast tunnels to connect group members into the mesh. DDM is a stateless multicast protocol in the sense that no protocol state is maintained at any node except for the source node. Intermediate nodes cache the forwarding list present in the packet header. When a route change occurs, an upstream node only needs to pass to its downstream neighbors the difference to the forwarding nodes since the last packet. MAODV is the multicast version of the AODV protocol [23]. It uses a multicast route table (MRT) to support multicast routing. A node adds new entries into the MRT after it is included in the route for a multicast group. MAODV uses a multicast group leader to create an on-demand core-based tree structure.

Different from the previous route-tree-based multicast algorithms, LGT (Location-Guided Tree Construction Algorithm for Small Group Multicast) [28] uses the location information of the group members to build the multicast tree without the knowledge of the network topology. Two heuristics are proposed to build the multicast tree using location information: the Location-Guided k-rray tree (LGK) and the Location-Guided Steiner tree (LGS).

Some of the mesh-based multicast routing protocols for MANETs are CAMP (Core-Assisted Mesh Protocol) [29], FGMP (Forwarding Group Multicast Protocol) [30], and ODMRP (On-Demand Multicast Routing Protocol) [31]. CAMP generalizes the notion of core-based trees introduced for Internet multicasting. It uses core nodes for limiting the control traffic needed for the creation of a multicast mesh avoiding flooding. On the other hand, both FGMP and ODMRP use flooding to build the mesh. In the FGMP protocol, the receiver initiates the flooding process, whereas in the ODMRP the senders initiates it.

1.3.5 Transport Protocols

The Transmission Control Protocol (TCP) is by far the most used transport protocol in the Internet. It is the typical protocol for most network applications. TCP is a reliable connection-oriented stream transport protocol that has the following features: explicit and acknowledged connection initiation and termination; reliable, in-order, and not duplicated data delivery; flow control; congestion avoidance; and out-of-band indication of urgent data.

An important design issue of TCP is that it uses packet loss as an indication of network congestion, and it deals with this effectively by making corresponding transmission adjustment to its congestion window. In wired networks, error rates are quite low and the TCP's congestion avoidance mechanism works very well.

The mobile multihop ad hoc network introduces new challenges to the TCP protocol due to the frequent change in network topology, disconnections, variation in link capacity, and high error rate. In fact, issues present in the physical, MAC, and network layers can affect the performance of the TCP protocol. In a wireless mobile ad hoc network, packet losses are usually not caused by network congestion, but by error transmissions and frequent disconnections due to mobility, resulting in backoff mechanisms being inappropriately invoked. This reduces the network throughput and increases the delay for data transmission. The variation in link capacity, the presence of asymmetric links, and delayed acknowledgment of messages can seriously affect the

TCP's dynamic congestion window mechanism. In summary, standard TCP flow control and congestion control mechanisms do not work well in mobile ad hoc networks. The error control mechanisms of MAC protocols can affect the TCP performance. Timeouts in TCP and MAC can cause different perceptions for both protocols. For a MAC protocol as opposed to the TCP protocol, it is easier to distinguish a link failure from a congestion failure. Finally, the characteristics of the underlying routing protocol can impact the TCP performance. For instance, some reactive routing protocols (such as DSR) send back a path failure message to the source node whenever there is a broken link to the destination node. The source nodes starts a new route computation, thereby increasing the time to route packets and, probably, leading TCP to experience timeouts during each route computation time, especially if there is heavy traffic in the network.

There are at least two strategies to adapt the TCP protocol to a mobile ad hoc network [32]. The first one is to make TCP mobility-aware by adding to it mechanisms and support from the underlying protocols to diminish the impact caused by mobility. ELFN (explicit link failure notification) [33] provides the TCP sender with information about the link status and route failures. In this way, whenever there is some physical problem, the TCP avoids triggering the congestion avoidance mechanism (as if a congestion occurred) and consequently reducing the overall system performance. The second strategy uses a protocol stack such that the TCP keeps its original behavior whereas the underlying protocols incorporate the required mechanisms to mask out the negative effects of mobility on TCP. Notice that these protocols need to be designed to adjust to the principles of TCP. Atra [34] is a framework that aims to minimize the probability of a route failure, predict route failures in advance so the source node can recompute an alternate route before the existing one fails, and minimize the latency in conveying route failure information to the source node, when a prediction was not successfully predicted.

A third strategy is to design a new transport protocol for the mobile ad hoc network. The expected advantage of this approach is to have a protocol that fits better the ad hoc environment. On the other hand, its integration to an existing protocol stack may be more difficult. The ATP (Ad-Hoc Transport Protocol) [35] is a protocol designed to cope with the problems present in TCP arising from mobility. It is a rate-based transport protocol. ATP defines three entities: ATP sender, intermediate node, and ATP receiver. The ATP sender is responsible for connection management, reliability, congestion control, and initial rate estimation. The intermediate nodes help the ATP sender in its operations by providing network information regarding congestion control and initial rate estimation. The ATP receiver is responsible for collating the information provided by the intermediate nodes before sending the final feedback to the ATP sender for reliability, rate, and flow control.

1.3.6 Energy Conservation

Mobile devices in a MANET must operate under energy constraints since they typically rely on a battery, which has a finite capacity. For these mobile nodes, the most important system design criteria for optimization may be energy conservation [36].

Thus energy represents one of the greater constraints in designing algorithms for mobile devices [37]. It is interesting to notice that energy conservation is related to all network layers [38, 39], including MAC [40], routing [41], and application [42] protocols for MANETs.

Power-aware protocols are often based on the following techniques: active and standby modes switching, power setting, and retransmission avoidance. Mode switching between active and standby aims to avoid spending energy during system idle periods. Furthermore, power transmission must be set to the minimum level for the correct message reception at the destination. Retransmissions should be avoided since they waste energy by sending messages that will not be processed by the destination nodes. Power awareness is achieved using power management or power control mechanisms [40]. A power management mechanism alternates the state of a mobile device *wake* and *sleep* periods. Furthermore, the wireless data interface consumes nearly the same amount of energy in the receive, transmit, and idle states, whereas in the sleep state, a data interface cannot transmit or receive, and thus its power consumption is highly reduced [43]. However, it is not possible to have a mobile device most of the time in power-saving mode (sleep state), which will extend its battery lifetime but comprise the network lifetime, because ad hoc networks rely on cooperative efforts among participating nodes to deliver messages.

A possible strategy is to allow the network data interface to enter a power-saving mode while trying to achieve a minimum impact to the process of sending and receiving messages. In general, these algorithms depend on data collected from the physical and MAC layers. For instance, an algorithm can monitor the transmission error rates to avoid useless transmissions when the channel noise reduces the probability of a successful transmission [44, 45]. At the MAC layer, an algorithm can save energy by determining intervals during which the network data interface does not need to be listening [46]. This is the case, for instance, whenever a node transmits a message and the other nodes within the same interference and carrier sensing range must remain silent. During this period, these nodes can sleep with little or no impact on system behavior. Related to this strategy is to have a density control algorithm controlling the operational mode of mobile devices so only those needed to forward the data traffic are awake and the overall network lifetime is optimized [47]. The strategies of controlling the transmission power and the node density in a MANET must be performed very carefully. Reducing the transmission power and keeping the node density to a minimum level may lead to a smaller number of available data communication links among nodes and, hence, a lower connectivity that can increase the number of messages not transmitted.

1.3.7 Network Security

Mobile ad hoc networks are generally more prone to physical security threats than are fixed-wired networks [48, 49]. The broadcast nature of the wireless channels, the absence of a fixed infrastructure, the dynamic network topology, the collaborative multihop communication among nodes, and the self-organizing characteristic of the network increase the vulnerabilities of a mobile ad hoc network.

The starting point to provide a proper security solution for a mobile ad hoc network is to understand the possible forms an attack can happen. In a MANET, a security problem may happen at any network layer and include: data integrity attacks, by accessing, modifying, or injecting traffic; denial-of-service attacks; flow-disruption attacks, by delaying, dropping, or corrupting data passing through, but leaving routing traffic unmodified; passive eavesdropping; resource depletion attacks, by sending data with the objective of congesting a network or draining batteries; signaling attacks, by injecting erroneous routing information to divert network traffic, or making routing inefficient; and stolen device attacks.

Given the variety of possible attacks to a mobile ad hoc network, different solutions have been proposed to address them [49]. The first step is to protect the wireless network infrastructure against malicious attacks. Digital signatures can be used to authenticate a message and prevent attackers from injecting erroneous routing information and data traffic inside the network [50]. This scheme requires a certification authority function to manage the private–public keys and to distribute keys via certificates, which needs to be distributed over multiple nodes in the MANET [51]. A strategy is to use a threshold cryptographic model to distribute trust among the MANET nodes [52, 53]. This model tolerates a threshold t of corruptions/collusions in the network, whereas it allows any set of $t + 1$ nodes to make distributed decisions such as regarding admission of new nodes to the network. These proposals require that each node must receive a certificate and a secret share in a distributed manner. However, as long as each node is able to obtain an updated VSS information (Feldman's Verifiable Secret Sharing mechanism [54]), there is no need for node-specific certificates, and it is possible to create new secret shares in a distributed manner [55].

In many mobile ad hoc network applications, such as emergency disaster relief and information sharing in a meeting, it is important to guard against attacks such as malicious routing misdirection [56]. The problem is that ad hoc routing protocols were designed to trust all participants, are cooperative by nature, and depend on neighboring nodes to route packets. This naive trust model allows malicious nodes to attack a MANET by inserting erroneous routing updates, replaying old messages, changing routing updates, or advertising incorrect routing information. Furthermore, a mobile ad hoc environment makes the detection of these problems more difficult [57]. Some of the proposed solutions to the problem of secure routing in a MANET involve the use of a pre-deployed security infrastructure [58], concealing the network topology or structure as in the Zone Routing Protocol [59], and introducing mechanisms in the network to mitigate routing misbehavior such as the SAR (Security-Aware Ad Hoc Routing) technique [60] to add security attributes to the route discovery [61].

1.4 APPLICATIONS

Mobile ad hoc networks have been employed in scenarios where an infrastructure is unavailable, the cost to deploy a wired networking is not worth it, or there is no time to set up a fixed infrastructure. In all these cases, there is often a need for collaborative computing and communication among the mobile users who typically work as teams—for instance, medical personnel in a search and rescue mission, firefighters facing a hazardous emergency, policemen conducting surveillance of suspects, and soldiers engaging in a fight. When we consider all these usual driving applications managed by specialized people, we understand why there is a slow progress in deploying commercial ad hoc applications to ordinary people.

This situation may change with the deployment of opportunistic ad hoc networks [62]. These networks aim to enable user communication in an environment where disconnection and reconnection are common activities and link performance is dynamic. They are very suitable to support the situation where a network infrastructure has limited coverage and users have "islands of connectivity." By taking advantage of device mobility, information can be stored and forwarded over a wireless link when a connection "opportunity" arises, such as an appropriate network contact happens. In this view, the traditional MANET incorporates the special feature of connection opportunity.

A MANET can be used to provide access to crisis management applications, such as in a disaster recovery, where the entire communication infrastructure is destroyed and establishing communication quickly is crucial [63]. By using a mobile ad hoc network, an infrastructure could be set up in hours rather than days or weeks, as in the case of a wired networking.

One of many possible uses of a mobile ad hoc network is in noncritical and collaborative applications. One example is a business environment where the need for collaborative computing might be more important outside the office, such as in a business meeting at the client's office to discuss a project. Another viable example is to use a mobile ad hoc network for a radio dispatch system [64]. This system can be used, for instance, in a taxi dispatch system based on MANET.

When a user wants to use an existing application on the Internet in a mobile ad hoc network, it is important to investigate its performance. This is the case, for instance, of Gnutella, one of the most widely used peer-to-peer systems, which needs to be evaluated before putting it through typical ad hoc conditions such as node mobility and frequent network partitioning [65].

Another application area is communication and coordination in a battlefield using autonomous networking and computing [66]. Some military ad hoc network applications require unmanned, robotic components. Unmanned Airborne Vehicles (UAVs) can cooperate in maintaining a large ground mobile ad hoc network interconnected in spite of physical obstacles, propagation channel irregularities, and enemy jamming. The UAVs can help meet tight performance constraints on demand by proper positioning and antenna beaming.

A vehicular ad hoc network (VANET) is a mobile ad hoc network designed to provide communications among close vehicles and between vehicles and nearby fixed equipment. The main goal of a VANET is to provide safety and comfort for passengers. To this end, a special electronic device is placed inside each vehicle that will provide ad hoc network connectivity for the passengers and vehicle. Generally, applications in a VANET fall into two categories, namely safety applications and comfort applications [67, 68]. Safety applications aim to provide driver's information about future critical situations and, hence, have strict requirements on communication reliability and delay. Some of the safety applications envisioned for VANETs are intervehicle danger warning, intersection collision avoidance, and work zone safety warning. Comfort applications aim to improve the driving comfort and the efficiency of the transportation system and, hence, are more bandwidth-sensitive instead of delay-sensitive. Some of the comfort applications are on-board Internet access, high data rate content download (electronic map download/update), and driving through payment [68].

With numerous emerging applications, opportunistic ad hoc networks have the potential to allow a large number of devices to communicate end-to-end without requiring any pre-existing infrastructure and are very suitable to support pervasive networking scenarios. For instance, suppose we want to (a) communicate with a mobile user who is temporarily out of reach or (b) establish a public wireless mesh that includes not only fixed access points but also vehicles and pedestrians, or interconnect groups of roaming people in different locations via the Internet. It seems that finally mobile ad hoc networks and Internet are coming together to produce in the next few years viable commercial applications.

1.5 CONCLUDING REMARKS

A mobile ad hoc network is one of the most innovative and challenging areas of wireless networking and tends to become increasingly present in our daily life [69]. An ad hoc network is clearly a key step in the next-generation evolution of wireless data communication when we consider the different enabling networks and technologies. An ad hoc network inherits the traditional problems of wireless and mobile communications, including bandwidth optimization, power control, and transmission quality enhancement. In addition, MANETs pose new research problems due to the multihop nature and the lack of a fixed infrastructure. These problems are related to algorithms for different aspects such as network configuration, topology discovery and maintenance, and routing.

The problems in ad hoc networks face a very important and fundamental question that is the dynamic network topology. This has a serious impact on the design of algorithms for ad hoc networks since they are expected to work properly under different and unpredictable scenarios. Similar to other distributed problems, a designer can start reasoning about an algorithm for this type of network, initially considering a static version of the problem. In a static version, it is reasonable to assume that there is a global topological information of the network, the computation happens just once, and the proposed solution is a centralized algorithm. On the other hand, when we consider a dynamic solution for the same problem, it is reasonable to assume that there is only local information, the computation happens continuously along the time the network is operational, and the proposed solution is a distributed algorithm. Clearly, the dynamic solution is more useful for ad hoc networks. However, a detailed study of the static solution tends to provide valuable insight for the design of a distributed version, is useful to determine the upper bound on the performance of the algorithm, can even be applied to stationary ad hoc networks such as commercial mesh-based broadband wireless solutions, and is simple to understand.

REFERENCES

1. T. Melodia, D. Pompili, and I. F. Akyildiz. On the interdependence of topology control and geographical routing in ad hoc and sensor networks. *IEEE Journal on Selected Areas in Communications*, **23**(3):520–532, 2005.
2. R. Ramanathan and R. Hain. Topology control of multihop wireless networks using transmit power adjustment. In *IEEE INFOCOM*, Tel Aviv, Israel, March 2000, pp. 404–413.
3. H. Takagi and L. Kleinrock. Optimal transmission ranges for randomly distributed packet radio terminals. *IEEE Transactions on Communications*, **32**(3):246–257, 1984.
4. T. C. Hou and V. O. K. Li. Transmission range control in multihop packet radio terminals. *IEEE Transactions on Communications*, **34**(1):38–44, 1986.
5. G. G. Finn. Routing and addressing problems in large metropolitan scale internetworks. Technical Report RR–87–180, ISI Research Report, 1987.
6. R. Nelson and L. Kleinrock. The spatial capacity of a slotted aloha multihop packet radio network with capture. *IEEE Transactions on Communications*, **32**(6):684–694, 1984.
7. L. Bao and J. J. Garcia-Luna-Aceves. Topology management in ad hoc networks. In *Proceedings of the 4th ACM International Symposium on Mobile Ad Hoc Networking and Computing (MobiHoc '03)*, Annapolis, MD, ACM, New York 2003, pp. 40–48.
8. Y. Xu, S. Bien, Y. Mori, J. Heidemannn, and D. Estrin. Topology control protocols to conserve energy in wireless ad hoc networks. Technical Report Center for Embedded Networked Sensing Technical Report 6, UCLA, 2003.
9. S. Guha and S. Khuller. Approximation algorithms for connected dominating sets. *Algorithmica*, **20**(4):374–387, 1998.
10. M. R. Garey and D. S. Johnson. *A Guide to Theory of NP-Completeness*. Freeman, Oxford, UK, 1979.
11. R. Sivakumar, P. Sinha, and V. Bharghavan. Cedar: A core-extraction distributed ad hoc routing algorithm. *IEEE Journal on Selected Areas in Communications*, **17**(8):1454–14655, 1999.
12. L. Jia, R. Rajaraman, and T. Suel. An efficient distributed algorithm for constructing small dominating sets. In *Proceedings of the ACM Symposium on Principles of Distributed Computing (PODS '01)*, Newport, RI, 2001.
13. P. Basu, N. Khan, and T. D. C. Little. A mobility based metric for clustering in mobile ad hoc networks. In *Proceedings of the International Workshop on Wireless Networks and Mobile Computing (WNMC '01)*, Scottsdale, AZ, 2001, pp. 72–80.
14. B. Chen, K. Jamieson, H. Balakrishnan, and R. Morris. Span: An energy-efficient coordination algorithm for topology maintenance in ad hoc wireless networks. In *Proceedings of the 7th Annual International Conference on Mobile Computing and Networking (MobiCom '01)*, Rome, Italy, 2001, pp. 62–70.
15. E. M. Royer and C.-K. Toh. A review of current routing protocols for ad-hoc mobile wireless networks. *IEEE Personal Communications*, April 1999, pp. 40–45.
16. M. Ilyas, editor. *The Handbook of Ad Hoc Wireless Networks*. CRC Press, Boca Raton, FL, 2003.
17. J. M. McQuillan, I. Richer, and E. C. Rosen. The new routing algorithm for the ARPANet. *IEEE Transactions on Communications*, **28**(5):711–719, 1980.

18. C. Hedrick. Routing Information Protocol. Request for Comments 1058, June 1988. Available at http://www.ietf.org/rfc/rfc1058.txt.
19. C. E. Perkins and P. Bhagwat. Highly dynamic destination sequenced distance vector routing (DSDV) for mobile computers. *SIGCOMM '94—Computer Communications Review*, **24**(4):234–244, 1994.
20. T. Clausen and P. Jacquet. Optimized Link State Routing Protocol (OLSR). Request for Comments 3626, October 2003. Available at http://www.ietf.org/rfc/rfc3626.txt.
21. R. Ogier, F. Templin, and M. Lewis. Topology Dissemination Based on Reverse-Path Forwarding (TBRPF). Request for Comments 3684, February 2004. Available at http://www.ietf.org/rfc/rfc3684.txt.
22. D. Johnson, Y. Hu, and D. Maltz. The Dynamic Source Routing Protocol (DSR) for Mobile Ad Hoc Networks for IPv4. Request for Comments 4728, February 2007. Available at http://www.ietf.org/rfc/rfc4728.txt.
23. C. Perkins, E. Belding-Royer, and S. Das. Ad hoc On-Demand Distance Vector (AODV) Routing. Request for Comments 3561, February 2007. Available at http://www.ietf.org/rfc/rfc3561.txt.
24. M. Abolhasan, T. Wysocki, and E. Dutkiewicz. A review of routing protocols for mobile ad hoc networks. *Ad Hoc Networks*, **2**(1):1–22, 2004. Mobile Ad-hoc Networks (MANET). IETF Working Group. http://www.ietf.org/html. charters/manet-charter.html. MANET Internet drafts available at http://www.ietf. org/ids.by.wg/manet.html.
25. J. Xie, R. R. Talpade, A. McAuley, and M. Liu. AMRoute: Ad Hoc Multicast Routing Protocol. *Mobile Networks and Applications*, **7**(6):429–439, 2002.
26. L. Ji and M. S. Corson. Differential destination multicast—A MANET multicast routing protocol for small groups. In *Proceedings of the 20th Annual Joint Conference of the IEEE Computer and Communications Societies (INFOCOM '01)*, Anchorage, AK, April 2001, pp. 1192–1202.
27. E. M. Royer and C. E. Perkins. Multicast operation of the ad hoc on-demand distance vector routing protocol. In *Proceedings of the 5th Annual ACM/IEEE International Conference on Mobile Computing and Networking (MobiCom'99)*, Seattle, WA, August 1999, pp. 207–218.
28. K. Chen and K. Nahrstedt. Effective location-guided tree construction algorithms for small group multicast in MANET. In *Proceedings of the the ACM SIGMETRICS*, Marina del Rey, CA, June 2002, pp. 270–271.
29. E. L. Madruga and J. J. Garcia-Luna-Aceves. Scalable multicasting: The core-assisted mesh protocol. *Mobile Networks and Applications*, **6**(2):151–165, 2001.
30. C.-C. Chiang, M. Gerla, and L. Zhang. Forwarding group multicast protocol. *Journal of Cluster Computing*, **1**(2):187–196, 1998.
31. S.-J. Lee, M. Gerla, and C.-C. Chiang. On-demand multicast routing protocol. In *Proceedings of the Wireless Communications and Networking Conference (WCNC '99)*, New Orleans, LA, September 1999, pp. 1298–1302.
32. A. Al Hanbali, E. Altman, and P. Nain. A survey of TCP over ad hoc networks. *IEEE Communications Surveys & Tutorials*, **7**(3):22–36, 2005.
33. G. Holland and N. H. Vaidya. Impact of routing and link layers on TCP performance in mobile ad-hoc networks. In *Proceedings of the Wireless Communications and Networking Conference (WCNC '99)*, pages 1323–1327, New Orleans, LA, September 1999.

34. V. Anantharaman and R. Sivakumar. A microscopic analysis of TCP performance analysis over wireless ad hoc networks. In *Proceedings of the 21st Annual Joint Conference of the IEEE Computer and Communications Societies (INFOCOM '02)*, New York, June 2002, pp. 1180–1189.
35. K. Sundaresan, V. Anantharaman, H.-Y. Hsieh, and R. Sivakumar. ATP: A reliable transport protocol for ad-hoc networks. In *Proceedings of the 4th ACM International Symposium on Mobile Ad Hoc Networking and Computing (MobiHoc '03)*, New York, Annapolis, MD, 2003, pp. 64–75.
36. A. J. Goldsmith and S. B. Wicker. Design challenges for energy-constrained ad hoc wireless networks. *Wireless Communications*, **9**(4):8–27, 2002.
37. J. R. Lorch and A. J. Smith. Software strategies for portable computer energy management. *IEEE Personal Comunications*, **5**(3):60–73, 1998.
38. C. Jones, K. Sivalingam, P. Agarwal, and J. C. Chen. A survey of energy efficient network protocols for wireless and mobile networks. *Wireless Networks*, **7**(4):343–358, July 2001.
39. V. Srivastava and M. Motani. Cross-layer design: A survey and the road ahead. *IEEE Communications Magazine*, **43**(12):112–119, 2005.
40. S. Kumar, V. S. Raghavan, and J. Deng. Medium access control protocols for ad hoc wireless networks: A survey. *Ad Hoc Networks*, **4**(3):326–358, 2006.
41. F. Xie, L. Du, Y. Bai, and L. Chen. Energy aware reliable routing protocol for mobile ad hoc networks. In *Proceedings of the Wireless Communications and Networking Conference (WCNC '07)*, Hong Kong, China, March 2007, pp. 4313–4317.
42. R. Kravets and P. Krishnan. Application-driven power management for mobile communication. *Wireless Networks*, **6**(4):263–277, 2000.
43. J. L. Sobrinho and A. S. Krishnakumar. Quality-of-service in ad hoc carrier sense multiple access wireless networks. *IEEE Journal on Selected Areas in Communications*, **17**(8): 1353–1368, 1999.
44. M. Rulnick and N. Bambos. Mobile power management for wireless communication networks. *Wireless Networks*, **3**(1):3–14, 1997.
45. M. Zorzi and R. R. Rao. Energy constrained error control for wireless channels. In *Proceedings of the Global Telecommunications Conference (GLOBECOM '96)*, London, UK, November 1996, pp. 1411–1416.
46. S. Singh and C. S. Raghavendra. PAMAS—Power Aware Multi-Access Protocol with signalling for ad hoc networks. *ACM SIGCOMM Computer Communication Review*, **28**(3):5–26, 1998.
47. L. Ma, Q. Zhang, and X. Cheng. A power controlled interference aware routing protocol for dense multi-hop wireless networks. *Wireless Networks*, **6**:50–58, 2007.
48. P. Papadimitratos and Z. J. Haas. Secure data communication in mobile ad hoc networks. *IEEE Journal on Selected Areas in Communications*, **24**(2):343–356, 2006.
49. H. Yang, H. Luo, F. Ye, S. Lu, and L. Zhang. Security in mobile ad hoc networks: Challenges and solutions. *Wireless Communications*, **11**(1):38–47, 2004.
50. L. Zhou and Z. J. Haas. Securing ad hoc networks. *IEEE Network*, **13**(6):24–30, 1999.
51. A. M. Hegland, E. Winjum, S. F. Mjølsnes, C. Rong, Ø. Kure, and P. Spilling. A survey of key management in ad hoc networks. *IEEE Communications Surveys & Tutorials*, **8**(3): 48–66, 2006.

52. K. Jiejun, Z. Petros, H. Luo, S. Lu, and L. Zhang. Providing robust and ubiquitous security support for MANET. In *Proceedings of the IEEE 9th International Conference on Network Protocols (ICNP '01)*, Riverside, CA, November 2001, pp. 251–260.
53. M. Narasimha, G. Tsudik, and J. H. Yi. On the utility of distributed cryptography in P2P and MANETs. In *Proceedings of the IEEE 11th International Conference on Network Protocols (ICNP '03)*, Atlanta, GA, November 2003, pp. 336–345.
54. P. Feldman. A practical scheme for non-interactive verifiable secret sharing. In *Proceedings of the 28th Annual Symposium on Foundations of Computer Science (FOCS '87)*, Toronto, Canada, November 1987, pp. 427–437.
55. N. Saxena, G. Tsudik, and J.H. Yi. Efficient node admission for short-lived mobile ad hoc networks. In *Proceedings of the IEEE 13th International Conference on Network Protocols (ICNP'03)*, Boston, MA, November 2003, pp. 269–278.
56. H. Yih-Chun and A. Perrig. A survey of secure wireless ad hoc routing. *IEEE Security & Privacy Magazine*, 2(3):28–39, 2004.
57. N. Milanovic, M. Malek, A. Davidson, and V. Milutinovic. Routing and security in mobile ad hoc networks. *Computer*, 37(2):61–65, 2004.
58. K. Sanzgiri, D. LaFlamme, B. Dahill, B. N. Levine, C. Shields, and E. M. Belding-Royer. Authenticated routing for ad hoc networks. *IEEE Journal on Selected Areas in Communications*, 23(3):598–610, 2005.
59. Z. Haas and M. Pearlman. Zone routing protocol (ZRP): A framework for routing in hybrid ad hoc networks. In C. E. Perkins, editor, *Ad Hoc Networking*, Addison-Wesley, 2001, Reading, MA, pp. 221–253.
60. S. Yi, P. Naldurg, and R. Kravets. Security-aware ad hoc routing for wireless networks. In *Proceedings of the 2nd ACM International Symposium on Mobile Ad Hoc Networking and Computing (MobiHoc '01)*, Long Beach, CA, October 2001, pp. 299–302.
61. S. Gupte and M. Singhal. Secure routing in mobile wireless ad hoc networks. *Ad Hoc Networks*, 1(1):151–174, July 2003.
62. L. Pelusi, A. Passarella, and M. Conti. Opportunistic networking: Data forwarding in disconnected mobile ad hoc networks. *IEEE Communications Magazine*, 44(11):134–141, 2006.
63. Y. Shibata, H. Yuze, T. Hoshikawa, K. Takahata, and N. Sawano. Large Scale Distributed disaster information system based on MANET and overlay network. In *Proceedings of the 27th International Conference on Distributed Computing Systems Workshops (ICDCSW '07)*, Toronto, Canada, June 2007, p. 7.
64. E. Huang, W. Hu, J. Crowcroft, and I. Wassell. Towards commercial mobile ad hoc network applications: A radio dispatch system. In *Proceedings of the 6th ACM International Symposium on Mobile Ad Hoc Networking and Computing (MobiHoc '05)*, Urbana-Champaign, IL, May 2005, pp. 355–365.
65. M. Conti, E. Gregori, and G. Turi. A cross-layer optimization of Gnutella for mobile ad hoc networks. In *Proceedings of the 6th ACM International Symposium on Mobile Ad Hoc Networking and Computing (MobiHoc '05)*, Urbana-Champaign, IL, May 2005, pp. 343–354.
66. D. C. Reeve, N. J. Davies, and D. F. Waldo. Constructing predictable applications for military ad-hoc wireless networks. In *Proceedings of the Military Communications Conference (MILCOM '06)*, Washington, D.C., October 2006, p. 7.

67. M. Caliskan, D. Graupner, and M. Mauve. Decentralized disovery of free parking places. In *Proceedings of the Third ACM International Workshop on Vehicular Ad Hoc Networks (VANET '06)*, Los Angeles, CA, October 2006, pp. 30–36.
68. CAMP Vehicle Safety Communications Consortium. Vehicle Safety Communications Project: Task 3 Final Report—Identify Intelligent Vehicle Safety Applications Enabled by DSRC, March 2005. Available at http://ntlsearch.bts.gov/tris/record/tris/01002103.html.
69. I. Chlamtac, M. Conti, and J. J.-N. Liu. Mobile ad hoc networking: Imperatives and Challenges. *Ad Hoc Networks*, **1**(1):13–64, 2003.

CHAPTER 2

Establishing a Communication Infrastructure in Ad Hoc Networks

MICHEL BARBEAU[1] and EVANGELOS KRANAKIS[1]

School of Computer Science, Carleton University, Ottawa, Ontario, K1S 5B6, Canada

IOANNIS LAMBADARIS[1]

Department of Systems and Computer Engineering, Carleton University, Ottawa, Ontario, K1S 5B6, Canada

2.1 INTRODUCTION

Communication infrastructures must provide an efficient organizational framework for supporting the diverse elements of the underlying communication system. From small networks to complex networked systems, routing has always been the primary communication enabler. Routing protocols are often unable to maintain the required stability within a complex communication infrastructure when faced with mobility, dynamic faults, and devices with limited resources.

Emerging mobile ad hoc networks are a particular case in point offering an even bigger challenge to seamless infrastructure maintenance, especially if they are to combine protocol simplicity with performance. For example, the movement of nodes affects the network topology, causing frequent path breaks and service disruptions; the radio band is limited and error prone, resulting in data rates that may be significantly lower than in wired networks; and nodes must be able to manage their power resources because they are constrained by limited battery life. Since ad hoc networks cannot rely on a static underlying infrastructure, communication must be based on distributed, localized techniques that take advantage of the physical properties of the environment and require the minimum possible knowledge of the immediate surroundings.

---

[1]Research supported in part by Natural Sciences and Engineering Research Council of Canada (NSERC) and Mathematics of Information Technology and Complex Systems (MITACS).

Algorithms and Protocols for Wireless and Mobile Ad Hoc Networks, Edited by Azzedine Boukerche
Copyright © 2009 by John Wiley & Sons Inc.

2.1.1 Outline

The purpose of this chapter is to investigate techniques for enabling adequate infrastructure maintenance in ad hoc networks. In addition to clustering and hierarchical routing, which are mentioned in Section 2.2, methodologies are being explored for facilitating navigation of messages; that is, in Section 2.3, algorithms are being discussed for implanting beacons and anchors into the network infrastructure, while Section 2.4 outlines localization techniques that either establish the absolute (precise) location of a node in order to maintain navigational capabilities or provide a relative but consistent virtual coordinate system that enables routing. Finally, Section 2.5 provides details of two emerging protocols (ZigBee and MicrelNet) for setting up ad hoc network communication and discusses applications.

2.2 HIERARCHICAL ROUTING AND CLUSTERING

Flat-routed ad hoc networks require the discovery and maintenance of routes that may span a large number of unreliable wireless links (see Figure 2.1). Since some action must be taken every time a link breaks in order to maintain connectivity (especially in high node mobility), protocols on flat-routed networks have limited scalability. The alternative to flat routing are hierarchical routing protocols. Based on some form of establishment of hierarchy, like an addressing scheme, they can bypass intermediate hosts by establishing connection directly with more powerful hosts—for example, masters, beacons, or even base stations (see Figure 2.2).

Strategies have been devised that dynamically organize ad hoc networks into clusters and leverage the resulting cluster topology. Hierarchical routing schemes are utilized by clustering protocols. A clustered ad hoc network is dynamically organized into partitions, called *clusters*, whose objective is the maintenance of a relatively stable topology. Clustering in ad hoc networks can support hierarchical routing, make the route search process more efficient for reactive protocols, support hybrid-routing in which different routing strategies operate in different domains or levels of a hierarchy, and provide more control over access to transmission.

2.2.1 Clustering

Clustering in ad hoc networks can be defined as the *grouping* of nodes into *manageable* sets, also called *clusters*. A cluster consists of a single cluster head and many member nodes. Any node can become a cluster head if it has the necessary functionality, such

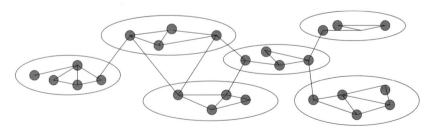

Figure 2.1. Flat routing infrastructure.

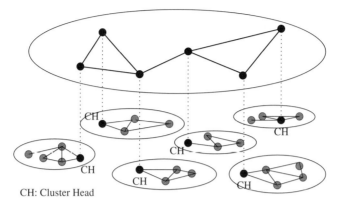

Figure 2.2. In hierarchical routing, depending on the available power range, intermediate hosts can be bypassed in order to communicate directly with master hosts (also called cluster heads and abbreviated CH) and the master hosts can communicate directly with themselves.

as processing and transmission power. Nodes register with the nearest cluster head and become members of that cluster.

Selection of cluster heads and partitioning of the nodes into clusters are essential aspects of mobile ad hoc networking. A clustering algorithm is basically a multi-leader election problem, and a cluster is similar to a *group* in distributed systems. Nodes within the cluster are considered members of the cluster head of that particular cluster. Sometimes, to be a member of a cluster head, a node has to lie within the transmission range of that cluster head. Nodes within the transmission range of more than one cluster head are usually called *gateway nodes*. A *distributed gateway* is a pair of neighboring nodes from different clusters physically located nearest to each other even though their clusters do not overlap (see Figure 2.3). Most

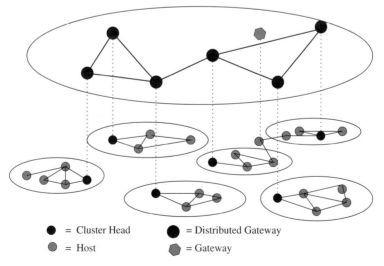

Figure 2.3. Partitioning of nodes in hierarchical routing.

cluster-head election algorithms require that each node be a member of one cluster head and that no two cluster heads be adjacent to each other. In a sense, clustering stabilizes the network topology, makes use of channel economy, eases location management, and provides a simple and feasible power control mechanism for an ad hoc network. Several techniques for hierarchical construction of ad hoc network infrastructures will be discussed in detail in the sequel. Additional information on topology control and distributed clustering can be found in Rajaraman [1] and Basagni [2], respectively.

2.3 ROUTING WITH VIRTUAL COORDINATES

Geographic awareness and on-demand topology formation are two well-known techniques for enabling scalability in ad hoc networks. The former (see Kranakis et al. [3], Bose et al. [4], Mauve et al. [5], and Al-Karaki and Kamal [6]) uses nodes' locations as addresses, while the latter (see Johnson et al. [7], Johnson and Maltz [8], Johnson [9] and Perkins [10]) floods the network only when needed. Despite the fact that geographic awareness is often limited by the need to find ways to route around voids, dead-ends, and obstacles as well as function in sparse environments with very low network densities, its advantages outweigh its limitations in that hosts need only keep their neighbors' state and support a fully general any-to-any communication pattern without explicit route establishment.

This section discusses how to establish location awareness. Subsection 2.3.1 discusses the general concept of beacons and how it can be used to build from scratch a coordinate system infrastructure. Subsection 2.3.2 discusses the limitations of anchor-based environments in creating unique naming and routing systems. Subsection 2.3.3 discusses a hybrid system that uses a combination of beacons with geographic information.

2.3.1 Beacon-Based Communication

A *radio beacon* is a (usually omnidirectional) transmitter carrying a constant signal on a specified radio frequency. It has been used in the past (in aviation, at sea) and are currently being used (in amateur radio to test signal propagation, as well as in satellites) in conjunction with direction finding equipment in order to identify the position of an object. A similar concept but with slightly different functionality are base stations as used in wireless networks. These are low-power, duplex, multichannel radios that are in a fixed location and they are being used by low-power, single-channel mobile phones and wireless routers. As such, they serve as as the liaison that links the mobile phone to the underlying land-line, public telephone infrastructure.

Building Virtual Positions. Beacon-supported protocols are based on selecting a few *beacon* nodes and constructing a basic communication tree from them to every other node. As a result, every node is aware of its distance (in hops) to every beacon

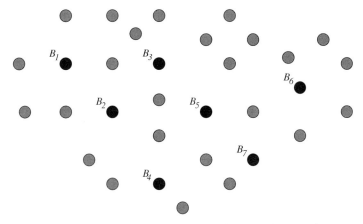

Figure 2.4. An ad hoc network with seven beacons (colored black). In such a network, every other node u (colored gray) has seven virtual coordinates $h_i(u)$, $i = 1, 2, \ldots, 7$, one for each virtual distance from a beacon.

and the resulting *vectors* can serve as coordinates. A beacon-based infrastructure consists of a set of, say, r nodes $B_1, B_2, \ldots, B_r$, which are specified to be the beacons. (Figure 2.4 depicts a simple beacon based ad hoc network with seven beacons.) For any other node u of the network, let $h_i(u)$ be the number of hops of the node u from the ith beacon B_i. Node u's *virtual position* is now determined by the vector of r virtual coordinates, an r-tuple $\vec{h}(u) = (h_1(u), h_2(u), \ldots, h_r(u))$. Since by following this procedure two different nodes may well have the same coordinates, it will be necessary to maintain an additional node identifier in order to break symmetry. To make routing decisions, nodes need to know not only their own virtual positions but also the virtual positions of their neighbors. Therefore they must periodically update their virtual coordinates using broadcast messages.

Routing with Beacons. Beacon vector routing (BVR) is based on employing these beacon nodes in order to construct spanning trees rooted at the beacons (one rooted tree per beacon) and spanning every other node of the ad hoc network. As a result, every node becomes aware of its distance (in hops) to every beacon, and the resulting *beacon vectors* can serve as coordinates. This *coordinate formation* protocol requires very little state, overhead, or preconfigured information (such as geographic location of nodes). After defining a distance metric over these coordinates, we can use a simple greedy distance-minimizing routing algorithm.

Routing requires the definition of a distance function between any pair $\{u, v\}$ of nodes. Such a metric must take into account the level of proximity of a node to (some of) the beacons. One possibility is to use the L_p-distance metric, $p > 1$, whereby

$$d_p(u, v) = \left(\sum_{i=1}^{r} w_i(u, v) |h_i(u) - h_i(v)|^p \right)^{1/p} \tag{2.1}$$

and $w_i(u, v)$ is an appropriately defined weight function between pairs of nodes indicating the weighted importance of beacon i for the nodes u, v. For example, $p = 1$ yields the well-known Manhattan distance, while $p = 2$ the Euclidean distance, in each case appropriately weighted with a function $w_i(\cdot, \cdot)$. In addition to the weight function $w_i(u, v)$, the distance function used in BVR uses a parameter $1 \leq k \leq r$ and defines the set $C(k, u)$ of the k beacons closest to the node u. The distance $d(k, u, v)$ between u and v is defined by Fonseca et al. [11] in such a way that

$$d(k, u, v) = \sum_{i \in C(k,v)} w_i(u, v)|h_i(u) - h_i(v)| \qquad (2.2)$$

It is also interesting to look at generalizations of Eqs. (2.2) and (2.1) so that

$$d(k, u, v) = \left(\sum_{i \in C(k,v)} w_i(u, v)|h_i(u) - h_i(v)|^p \right)^{1/p} \qquad (2.3)$$

To route to a target node t, a packet holds two fields: first, the identifier and virtual position $\vec{h}(t)$ of the target node and, second, a k-position vector indicating the minimum distance that the packet has seen so far using the set $C(i, t)$ of the $i \leq k$ beacons closest to t that it has seen so far. Thus, $C(i, t)$ is the set of nodes maintained in the memory of a host and can influence route selection by affecting distances between hosts as defined in Eqs. (2.2) and (2.3). If no neighbor is found that improves on this minimum distance, then the packet is forwarded toward the beacon closest to the destination. The beacon then forwards, first attempting a greedy approach and then failing the attempt using flooding.

2.3.2 Using Anchors

In a related model, first proposed by Wattenhofer et al. [12], the nodes of the network know the underlying topology (e.g., this could be a map of the network), while the beacons are called anchors and are linearly ordered $B_1 \prec B_2 \prec \cdots \prec B_k$. Given that each node in the network knows the underlying network topology, it is able to compute its hop distance to each one of the k anchors B_i. The beacon vectors (called pseudo-coordinates) are known to all the nodes of the network. Thus, the network can be thought of as being embedded in a pseudo-k-dimensional space while each node knows in addition to its own coordinate the coordinates of all its direct neighbors. When receiving a message containing the (pseudo-) coordinate of the destination, a node forwards the message to its geometrically closest (in the pseudo-geometric space) neighbor to the destination. The pseudo-geometric routing problem can be solved if there is a pseudo-geometric routing algorithm that guarantees message delivery.

To guarantee delivery, any routing algorithm must first resolve the *naming problem* in that all nodes must have unique identifiers, since the destination could not be uniquely identifiable otherwise, in general. Once this problem is solved, it is necessary to guarantee *reachability*—that is, i.e., that any destination must finally be reached from any source (this is also known as the routing problem).

Anchoring in Various Topologies. It is easy to see that in a ring on n nodes, using only one anchor A, there are $\lfloor (n-1)/2 \rfloor$ pairs of nodes with the same pairwise distance to A which cannot be distinguished. On the other hand, with two anchors A and A' (which are at distance $d \neq n/2$ from each other), or any three arbitrarily chosen anchors, each node has a unique coordinate. Therefore, one can choose a pair of anchors (e.g., two adjacent nodes) or any three arbitrarily chosen anchors, solve the naming problem in the ring. In addition, the pseudo-geometric routing problem on the ring can be solved (locally) with two anchors; furthermore, the chosen route between source and destination is a shortest path.

In a grid on n nodes, with one anchor alone at least $\sqrt{n}$ nodes will have the same coordinate in the grid. On the other hand, as depicted in Figure 2.5, if we choose one anchor A in the upper left corner and another anchor A' in the upper right corner of the grid, all nodes have different coordinates. Therefore, if chosen properly, two anchors are sufficient in order to solve the naming problem in the grid. However, if the anchors are chosen arbitrarily, then at least $(\sqrt{n}-1)^2 + 2$ anchors are needed to solve the naming problem. Concerning routing, if anchors A, A' are placed as depicted in Figure 2.5, a node can compute the position of each anchor, based on its coordinate and the coordinate of its neighbors. It is easy to see that the pseudo-geometric routing problem on the grid can be solved with two anchors. Furthermore, the chosen route between source and destination is a shortest path.

As shown in Wattenhofer et al. [12], in the case of trees it is not possible to solve the naming problem locally. If the anchors are chosen arbitrarily, up to $n-1$ nodes are required to solve the naming problem. However, in a tree it is always sufficient to choose all leaves as anchors. For unit disk graphs, at least $((\sqrt{n}+9)/12)^2$ anchors are

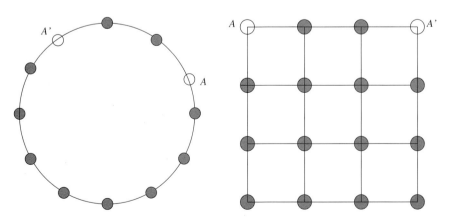

Figure 2.5. Any two anchors at distance $d \neq 2$ can solve the naming problem on a ring, while two anchors, in suitably chosen positions, can solve the naming problem on a grid.

needed, while if chosen arbitrarily, we need up to $n - 1$ anchors to solve the naming problem in unit disk graphs.

Wattenhofer et al. [12] first study the naming and pseudo-geometric routing problems as well as the minimum number of anchors needed to solve these two problems on a variety of graphs (lines, rings, grids, hypercubes, butterflies, trees, and unit disk graphs). Although these topologies may not be realistic, at least theoretically speaking, it is possible to give a lower bound on the number of anchors in certain types of networks. This can be important from a practical point of view because it gives the minimal amount of storage that is needed per node to solve either of the naming or routing problems.

2.3.3 Virtual Coordinates

Another approach due to Ratnasamy et al. [13] involves assigning virtual coordinates to each node and then applying standard geographic routing over these coordinates. Virtual coordinates can be constructed using only local connectivity information (which is available since nodes always know their neighbors). Assuming that each node knows its own routing table (consisting of all nodes in its distance two neighborhood), the following routing algorithm can be executed. First, the packet is forwarded to the node in the routing table closest (in virtual coordinates) to the destination, if (and only if) that node is closer to the destination than the current node. Second, if the node is closer to the destination than any other node in its routing table, then the packet is considered to have arrived to its destination. Third, if a packet is neither able to make greedy progress nor has reached a stopping point, then it performs an expanding ring search until a closer node is found (or a maximum TTL has been exceeded).

If such geometric coordinates are not known, then Ratnasamy et al. [13] base their algorithm on the so-called *perimeter* nodes (these are nodes lying on the outer boundary of the system). They distinguish and analyze three scenarios on the perimeter nodes: (a) Perimeter nodes know their location, (b) perimeter nodes know that they are perimeter nodes, but do not know their location, and (c) nodes know neither their location nor whether they are on the perimeter. In the sequel, we provide details of their construction for each of the three cases.

Perimeter Nodes Know Their Location. In the first case an iterative relaxation procedure is being used in order to build the coordinates of nonperimeter nodes using an analogy borrowed from the theory of graph embeddings (see Linial et al. [14]). Links are represented by forces pulling adjacent neighbors together. It is assumed that the force in the x-direction is proportional to the difference in the x-coordinates (similarly for the y-coordinates). Assuming its neighbors are held fixed, then a node's equilibrium position (the one where the sum of the forces is zero) is where its x-coordinate is the average of its neighbors' x-coordinates (similarly, for y-coordinates). Each nonperimeter node $u = (x_u, y_u)$ periodically updates its virtual coordinates using the iterative procedure

$$(x_u, y_u) = \left(\frac{\sum_{v \in N(u)} x_v}{|N(u)|}, \frac{\sum_{v \in N(u)} y_v}{|N(u)|} \right) \quad (2.4)$$

where $N(u)$ is the set of nodes which are neighbors of u. The relaxation equations imply that nonperimeter nodes having perimeter nodes among their neighbors will *tend to move toward perimeter nodes* that are closest to them in terms of the number of hops.

Perimeter Nodes Are Known. In this second case an algorithm is designed that does not use the fact that perimeter nodes know their exact geographic location. This is done by prefacing the previous relaxation method with a phase where perimeter nodes compute their own approximate virtual coordinates. The algorithm is in three steps. In the first step, each perimeter node broadcasts a *hello* message to the entire network so as to discover its relative position (distances in hops) to all other perimeter nodes in the network: Call $h(u, v)$ this distance measured in number of hops between perimeter nodes. In the second step, perimeter nodes broadcast their perimeter vector (i.e., vector of these distances to all other perimeter nodes) to the entire network. Finally, in the third step, every perimeter node uses a triangulation algorithm to compute the coordinates of all other perimeter nodes (including its own coordinates). Coordinates are chosen so as to minimize

$$\sum_{u, v \in \text{Perimeter Set}} (h(u, v) - d(u, v))^2$$

where $d(u, v)$ represents the Euclidean distance between the virtual coordinates of nodes u and v. Since at the end of the third step perimeter nodes know their own virtual coordinates, nonperimeter nodes can use the previous relaxation algorithm in order to compute their own virtual coordinates.

An important point raised by Ratnasamy et al. [13] is that message loss and node failure can cause perimeter nodes to have incomplete knowledge of the interperimeter distances and cause different perimeter nodes compute inconsistent coordinates. To address this problem, they use two designated bootstrapping beacons that flood the network with hello messages in order to *canonicalize* the computation and make all nodes performing it arrive at the same solution.

Without Location Information. In this third case, the assumption that perimeter nodes know they are on the perimeter is relaxed. A preparatory stage is added to the previous algorithm where a subset of nodes identify themselves as perimeter nodes. This is achieved by leveraging one of the bootstrap beacon nodes described before. Since these bootstrap nodes broadcast hello messages to the entire network, every node discovers its distance (in number of hops) to these bootstrap nodes. To decide if they are perimeter nodes, respective nodes use the *perimeter node criterion:*

A node decides that it is on the perimeter if it is the farthest away, among all its two-hop neighbors from the first bootstrap node. Furthermore, additional *coordinate projection to a circle* mechanism is added, allowing maintainance of a consistent virtual coordinate space even in the face of node mobility and failures.

2.4 (RELATIVE) LOCATION DETERMINATION

Although localization of network nodes is helpful in clustering, routing, and network map building, there are instances where either sensor nodes do not have GPS capability or satellite signals are blocked by obstacles or even a GPS receiver is affected by noise. In this section, methods are presented for localizing nodes that may not have GPS receiver capability. They all rely on the fact that the position of a nonlocalized node can be determined if the positions of localized neighbor nodes can be obtained. This information is relatively easy to obtain using advertisement messages and network flooding mechanisms. Firstly, there are methods using triangulation and distance estimates to localized neighbors. These distance estimates can be obtained by examining features of received radio signals (i.e., radio-location) or a logical method. Secondly, there are methods that are not based on distance estimates, but rather try to delimit the area in which a node yet to be localized should logically be contained.

A method for doing triangulation is discussed in Subsection 2.4.1. Estimation of distances using radio location is discussed in Subsection 2.4.2. Subsection 2.4.3 presents the logical distance estimation method used in the distance vector-hop algorithm. An area delimiting method, called the bounding box algorithm, is reviewed in Subsection 2.4.4. Finally, Subsection 2.4.5 explains how to build a consistent coordinate communication infrastructure using radio-location even when nodes cannot access directly GPS information.

2.4.1 Triangulation

The distance from node A to node B defines a circle around node B, and the position of A is on the circumference of this circle. In a two-dimensional model, the position of A is unambiguously determined as the intersection of three such circles. Each circle can be modeled by a quadratic equation, and the intersection point of these circles is theoretically the location of node A. The triangulation algorithm discussed in the sequel avoids quadratic equations and uses instead linear equations, which is a problem of lower difficulty that can be solved with linear algebra tools alone. This form of triangulation is used by GPS (see references 15 and 16) for their ad hoc positioning system.

Figure 2.6 depicts a node u that needs to determine its position. The positions of localized neighbors are used. The triangulation is an iterative procedure, which starts with an arbitrary estimate $(\hat{x}_u, \hat{y}_u)$ of the position of node u. The estimate is refined from one iteration to the next. The iterative procedure is repeated until the change from the previous estimate is below a given threshold. Firstly, the following procedure is applied for each localized neighbor v. (Figure 2.6 depicts an example using node

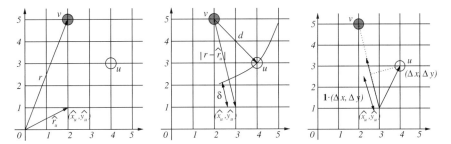

Figure 2.6. Triangulation, range, and projection.

v and node u.) Let r be a vector from the origin to the position of node v. Let $\hat{r}_u$ be a vector from the origin to the estimated position $(\hat{x}_u, \hat{y}_u)$ of node u. A unit vector is defined by

$$1_i = \frac{r_i - \hat{r}_u}{|r_i - \hat{r}_u|} \tag{2.5}$$

where the denominator in Eq. (2.5) is the length of the difference between vector r_i and vector $\hat{r}_u$ and the resulting unit vector defines the orientation of a vector from the estimated position $(\hat{x}_u, \hat{y}_u)$ to node v. If $|r - \hat{r}_u|$ is the distance between the estimated location $(\hat{x}_u, \hat{y}_u)$ and node v, then the distance correction is defined as the difference (see Figure 2.6)

$$\delta = |r_i - \hat{r}_u| - d \tag{2.6}$$

where d is the real distance between node u and landmark v. The goal is to iteratively reduce δ, for each neighbor v. The calculation used to achieve this goal is based on the observation that δ is approximately equal to the scalar product of the unit vector $\mathbf{1}$ and the difference between the real location (x_u, y_u) and estimated location $(\hat{x}_u, \hat{y}_u)$. This difference is denoted by $(\Delta x, \Delta y)$. It can also be viewed as a vector from $(\hat{x}_u, \hat{y}_u)$ to (x_u, y_u).

$$\delta \simeq \mathbf{1} \cdot (\Delta x, \Delta y)$$

The term δ can be viewed as the component of vector $(\Delta x, \Delta y)$ along the direction of vector $\mathbf{1}$ (see Figure 2.6). For each iteration of the procedure, the unknown $(\Delta_x, \Delta y)$ is resolved in the following the system of equations:

$$\begin{pmatrix} \Delta_1 \\ \vdots \\ \Delta_n \end{pmatrix} = \begin{pmatrix} 1_1[1] & 1_1[2] \\ \vdots & \vdots \\ 1_n[1] & 1_n[2] \end{pmatrix} (\Delta x, \Delta y)^T$$

where [.] is the vector projection operator. After each iteration, the $(\Delta x, \Delta y)$ is applied to improve the estimate $(\hat{x}_u, \hat{y}_u)$.

2.4.2 Distance Estimation Using Radio-Location

There are four radio-location techniques that can be used to estimate distances, namely, time of arrival (TOA), time difference of arrival (TDOA), angle of arrival (AOA) or signal strength.

To estimate the distance when using the TOA technique between node A and node B, node A sends a message to node B. The trip time of the signal is measured and multiplied by the propagation speed of signals (i.e., the speed of light) to yield the distance. Trip time measurement from A to B requires synchronized and accurate clocks at both locations. If round-trip time is measured instead (halved to obtain trip time), then this requirement is relaxed. No clock synchronization is required, but an accurate clock is needed at node A.

With the TDOA method, two nodes B_1 and B_2 simultaneously send a signal to node A. Times of arrivals t_1 and t_2 of signals respectively from B_1 and B_2 are measured by A. The time difference of arrival is calculated, i.e. $\delta_t = t_2 - t_1$. The time difference δ_t multiplied by the speed of light is mapped to the distance difference δ_d. The position (x_1, y_1) of B_1, position (x_2, y_2) of B_2 and δ_d define a hyperbola h with equation

$$\sqrt{(x-x_1)^2 + (y-y_1)^2} - \sqrt{(x-x_2)^2 + (y-y_2)^2} = \delta_d.$$

The positions of B_1 and B_2 are at the foci of the hyperbola and the position of node A is a solution of the hyperbola. The geometrical properties of the hyperbolas are such that all points located on the curve h are of equal time difference δ_t and equal distance difference δ_d. Two such hyperbolas can be defined by involving two different pairs of nodes (B_1, B_2) and (B_2, B_3) which produce two time differences of arrival δ_1 and δ_2. In a two-dimensional model, the observer of the times of arrival δ_1 and δ_2 (i.e., node A) is at the position corresponding to the intersection of the two hyperbolas. Note that there are cases in which the two hyperbolas intersect at two points. In these cases, a third independent measurement is required to resolve the ambiguity.

With the AOA technique, two nodes B_1 and B_2 determine the direction from which a signal from node A is coming. An imaginary line is traced from B_1 to A, and another imaginary line is traced from B_2 to A. The angle of arrival is defined as the angle that each of these lines makes with a line directed toward a common reference. The intersection of these two lines unambiguously determines the position of A. Note, however, that if A, B_1, and B_2 all lie on the same straight line, another independent measurement is required to resolve the ambiguity.

The signal-strength-based technique exploits the fact that a signal loses its strength as a function of distance. Giving the power of a transmitter and a model of free-space loss, a receiver can determine the distance traveled by a signal. If three different such signals can be received, a receiver can determine its position in a way similar to the TOA technique. The accuracy of this technique is generally not good due to transmission phenomena such as multipath fading and shadowing that cause important variation in signal strength.

Radio-location requires line-of-sight propagation between the nodes involved in a signal measurement. Line of sight means that a nonobstructed imaginary straight line can be drawn between the nodes. In other words, the accuracy is sensitive to radio propagation phenomena such as obstruction, reflection, and scattering. With all the distance-based techniques (i.e., TOA, TDOA, signal strength), three position-aware neighbors are required to determine the location of a position-unaware node.

2.4.3 Logical Distance Estimation

The Distance Vector-hop (DV-hop) algorithm uses triangulation to resolve the positions of nonlocalized nodes. In contrast to triangulation as used in the GPS system, an estimate of the distance is not computed using signal feature-based techniques. Instead, numbers of hops to localized nodes are used as a distance metric in a triangulation algorithm. The DV-hop algorithm has been introduced by Niculescu and Nath [16] and is presented in detail in the sequel.

In a connected network, it is assumed that among all the nodes there is a subset of localized nodes, called landmarks. The position of nonlocalized nodes is determined using the relative distance from the landmarks. An example is depicted in Figure 2.7. Filled circles represent landmarks, while unfilled circles represent nonlandmark nodes. In this example, there are three landmarks, numbered v_1, v_2, and v_3, positioned at coordinates (2, 5), (6, 2), and (5, 4). We demonstrate the localization of the nonlandmark node u. Using an advertisement message and controlled network flooding, each landmark advertises its position. When the advertisement is repeated from node to node, a hop count field, in the message, is incremented by one unit. Each landmark v_i collects the information in advertisement messages from

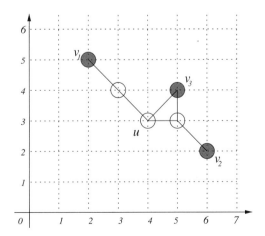

Figure 2.7. DV-hop network.

other landmarks and constructs a vector $[(x_1, y_1, h_1), \ldots, (x_n, y_n, h_n)]$, where x_j and y_j are the coordinates of landmark j and h_j is the distance in hops from v_i to v_j ($j = 1, \ldots, n$). Landmark v_i calculates a value, called the correction, by

$$c_{v_i} = \frac{\sum_{j=1,\ldots,n} \sqrt{(x_j - x_i)^2 + (y_j - y_i)^2}}{\sum_{v=j=1,\ldots,n} h_j}$$

whose numerator is the sum of Euclidian distances from v_i to every other landmark and the denominator is the sum of the number of hops from v_i to every other landmark. The ratio represents the average distance of a hop between landmark v_i and any other landmark. In a second round of advertisements and controlled network flooding, each landmark v_i communicates its correction value c_{v_i} to all other nodes. Corrections are collected by each nonlandmark node. For example, the correction for landmark v_1 is

$$c_{v_1} = \frac{\sqrt{(6-2)^2 + (2-5)^2} + \sqrt{(5-2)^2 + (4-5)^2}}{4+3} = 1.04$$

It is the average distance of one hop, according to landmark v_1. Each nonlandmark node ends up with a correction c, received from one of the landmarks. A distance estimate to every landmark is calculated. For every landmark $i = 1, \ldots, n$, the distance is defined as

$$d_i = c \times h_i$$

where h_i is the number of hops to landmark v_i. For the nonlandmark node u, if $c = c_1$, then $d_1 = 1.04 \times 2 = 2.08$, $d_2 = 1.04 \times 1 = 1.04$, and $d_3 = 1.04 \times 3 = 3.12$. This algorithm has been improved by Langendoen and Reijers [17].

2.4.4 Area Delimiting Methods

The algorithm presented in this section aims at locating the position of nodes, but without examining features of radio signals received from localized nodes. The algorithm, however, assumes that there are a number of nodes that, at the beginning, know their location, either by configuration or by using the GPS.

The bounding box algorithm is a method delimiting that area where a node yet to be localized is likely to be logically contained. The bounding box algorithm uses a grid model that divides the area in squares. Positions are given in reference to these squares, numbered using x and y indexes. Wireless communication ranges are also expressed as a number of squares. Hence, the coverage of a node can be expressed as a squared zone. The position of a node, whose location is initially unknown, can be bounded as the intersection of all the squared zones of its position aware neighbors. This information is obtained through advertisements. This procedure, called the bounding box algorithm, was proposed by Simic and Sastry [18]. We present the algorithm in details hereafter.

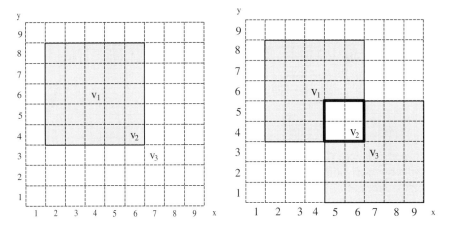

Figure 2.8. Zone in a bounding box network and intersection of zones in a bounding box network.

The region, containing all the nodes, is modeled as an m by m checkerboard of unit squares ($m \geq 1$). There are $n \geq 1$ nodes $v_1, \ldots, v_n$ distributed on the checkerboard. The location of a node v_i, when known, is defined by the coordinates (x_i, y_i) of a square on the checkerboard ($x_i, y_i \in 1, \ldots, m$). It is assumed that all nodes have the same communication range r. It is defined as a number of squares ($1 \leq r \leq m$).

An example is depicted in Figure 2.8. Node v_1 is located in a square whose coordinates are $(4, 6)$. In this example, the communication range r is two squares. Node v_1's coverage is an $(r + 1)^2$ grayed zone, hereafter called its zone. The node(s) within v_1's coverage (e.g., v_2 in the example) is/are its neighbor(s). Node v_i's zone is denoted by Z_i and is identified as a pair of ranges of coordinates delimiting the area

$$Z_i = [(x_i - r, x_i + r), (y_i - r, y_i + r)]$$

For instance, node v_1's zone is $Z_1 = [(2, 6), (4, 8)]$ and node v_3's zone is $Z_2 = [(4, 8), (2, 6)]$.

An intersection operation is defined over the zones. Given two zones Z_i and Z_j, their intersection is defined as

$$Z_i \cap Z_j = [(max(x_i, x_j) - r, min(x_i, x_j) + r), (max(y_i, y_j) - r, min(y_i, y_j) + r)]$$

which is the area where they overlap. Figure 2.8 depicts the intersection of Z_1 and Z_3. It is a zone defined as $[(5, 6), (4, 5)]$. Note that node v_2 has both nodes v_1 and v_3 as neighbors. The intersection of their zone, Z_1 and Z_3, defines a box bounding the location of node v_1.

The bounding box localization algorithm goes as follows. Among the n nodes, there are $k \leq n$ nodes that know their position, a square of the checkerboard. All nodes collaborate together to resolve the position of nonlocalized nodes. Nodes that know their location do advertise their position to their neighbors. Every nonlocalized node v_i eventually collects k advertisements from its neighbors $v_{i_1}, \ldots, v_{i_k}$ ($k \geq 1$). Using these positions and the range r, the corresponding zones $Z_{i_1}, \ldots, Z_{i_k}$ of neighbors are determined. The position of node v_i is defined as the bounding box resulting from the intersection of these zones, that is,

$$Z_{i_1} \cap \cdots \cap Z_{i_k}$$

This algorithm has a low message complexity. Every localized node needs to advertise its position once. There are at most n advertisement messages. The computation executed by every nonlocalized node (i.e., the intersection) is also of low complexity, on the order of n. The bounding box algorithm makes two assumptions. Firstly, the communication range is identical for all nodes. Secondly, every node covers a squared area. In practice, the communication range depends on several factors, including the transmission power, antenna gains, receiver sensitivity, and level of noise. The covered area is likely to be an irregular zone rather than a squared area.

2.4.5 Self-Positioning and Relative Coordinates

A distributed algorithm for the positioning of nodes without using any GPS information is described by Capkan et al. [19]. The *Self-Positioning Algorithm* (SPA) enables the nodes to find their positions within the network area using only range measurements between nodes in order to build a network coordinate system.

Local Coordinate System. In the first phase of the algorithm, each node builds a local coordinate system by becoming the origin $(0, 0)$ while the positions of its neighbors are computed relative to the position of this origin. Nodes can subsequently reach agreement on a common, global coordinate system for the entire network.

For each node u, let $N(u)$ be the set of its one-hop neighbors and let $D(u)$ be the set of all distances of u from all its hop-one neighbors $v \in N(u)$. Neighbors can be detected and distances can be measured using radio-location. Each node u sends its neighbor $N(u)$ and distance $D(u)$ sets to all its one-hop neighbors. At the end of this exchange, every node will know its two-hop neighbors as well as some of the distances between its one-hop and two-hop neighbors. The coordinates at u can be defined as follows. Node u selects two nodes p, q in its one-hop neighborhood $N(u)$ in such a way that u, p, q are not collinear and $d(p, q)$ is known (this means that we have either $p \in N(q)$ or $q \in N(p)$) and > 0. As depicted in Figure 2.9, the coordinates of u are $(0, 0)$, while $p = (d(u, p), 0)$ and $q = (d(u, q) \cos \gamma, d(u, q) \sin \gamma)$, where γ is the angle $\angle p, u, q$ which can be computed using the law of cosines

$$d(p, q)^2 = d(u, p)^2 + d(u, q)^2 - 2d(u, p)d(u, q) \cos \gamma$$

For any other node $v \in N(u)$, its coordinates (v_x, v_y) are computed as follows. First, the angles $\alpha_v = \angle(v, u, p)$ and $\beta_v = \angle(v, u, q)$ are computed from the known

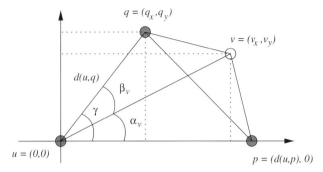

Figure 2.9. Node u builds its local coordinate system.

distances $d(v, p), d(u, v), d(v, q)$ by applying the law of cosines and subsequently the coordinates $v = (v_x, v_y)$ using the identities $v_x = d(u, v) \cos \alpha_v$ and

$$v_y = \begin{cases} d(u, v) \sin \alpha_v & \text{if } \beta_v = |\gamma - \alpha_v| \\ -d(u, v) \sin \alpha_v & \text{otherwise} \end{cases}$$

For nodes $v \in N(u)$ for which the distance of v to either p or q is not known, any pair of nodes whose coordinates have already been computed can be used. Notice that given u, among $\binom{|N(u)|}{2}$ possible pairs of nodes $p, q \in N(u)$ we can use p, q such that we have either $p \in N(q)$ or $q \in N(p)$.

From Local to Global Coordinate System. According to the previous computation, all nodes build a local coordinate system with each node at the origin $(0, 0)$. Two coordinate systems have the same direction if the their x and y axes have identical directions. It remains to describe how to adjust these local coordinate systems of the nodes so as to obtain a common coordinate system for all the nodes in the network. This will be a consequence of showing how to (a) elect a *universal origin* or center, (b) rotate directions so as to obtain identical directions, and (c) adjust the relative coordinates of the nodes in this new global network coordinate system.

The *local view set* V_u for node u is obtained first by selecting any two nodes $p, q \in N(u)$ as above and is the set of nodes $v \in N(u)$ such that node u can compute the location of the node v, in the local coordinate system of node u. For any two nodes, u and v, we adjust the direction of the coordinate system of the node v so as to have the same direction as the coordinate system of the node u. This involves rotating the coordinate system at node v by an appropriate angle in order to match the direction at node u. The required rotation angle is called the correction angle for the node v. For the given nodes u, v it is necessary that one be a neighbor of the other, that is, $u \in V_v$ and $v \in V_u$. Figure 2.10 depicts two different cases for the coordinate systems between nodes u and v. In the picture on the left the coordinate system at node v needs to be rotated, while in the picture on the right the coordinate system at

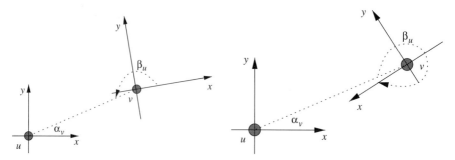

Figure 2.10. The two cases of coordinate system configurations for two neighboring nodes u and v. Rotation is required in both cases. For the case to the right, mirroring is also required.

node v needs to be rotated and subsequently mirrored. In both instances it is easy to see that the rotation depends on the angles α_u and β_v depicted. Two nodes u, v can distinguish the situation they are in by using a node $w \in V_u \cap V_v$. Figure 2.11 depicts the coordinate systems of two nodes u, v and a node w that is a neighbor to both. We consider the angles α_w and β_w formed by the vectors $\vec{uw}$ and $\vec{vw}$ in the coordinate systems of u and v, respectively. As illustrated in Figure 2.12, the position of the node v makes it possible to detect whether or not mirroring is necessary.

Final Positioning. After adjusting directions of local coordinate systems, the goal is that all the nodes in the network compute their position in the network coordinate system. The network coordinate system is in fact chosen as the local coordinate system of one of the nodes in the network, say u. In this case, all the nodes in the network have to adjust the directions of their coordinate systems to the direction of the coordinate system of the chosen node, and also every node has to compute its position in the coordinate system of this node. The procedure is as follows. Nodes in the local view V_u of u know their position by direct computation. In turn, a two-hop neighbor, say w, of u obtains its position as the vector sum of the position vector $\vec{uv}$ of a one-hop neighbor v of u and the position vector $\vec{vw}$. For additional details and experimental

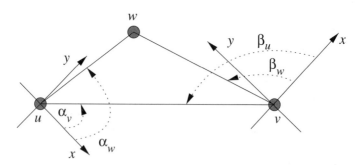

Figure 2.11. The coordinate systems of two nodes u, v and a node w that is a neighbor to both.

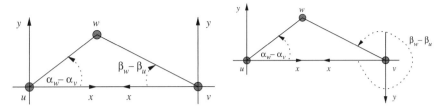

Figure 2.12. Determining whether or not mirroring is necessary.

results on this scheme the reader is advised to consult Capkun et al. [19]. In addition, Barbeau et al. [20] provides additional algorithms for improving localization in a sensor network.

2.5 PROTOCOLS AND APPLICATIONS

Practical networks relying on beacons for addressing and routing are based either on the emerging ZigBee protocol or on proprietary protocols. In what follows, the ZigBee protocol is introduced, along with ZigBee routing and data broadcasting. Furthermore, we give a brief introduction of an elementary wireless architecture/protocol called MicrelNet, which is offered from Micrel [21]. Finally, several applications are discussed in Subsection 2.5.3.

2.5.1 ZigBee/IEEE 802.15.4

The ZigBee wireless protocol architecture is targeted to low-data-rate, low-power-consumption, low-cost wireless personal networks (WPAN). The physical layer and link layer/access control was introduced in 2001 by IEEE (and is known as the IEEE 802.15.4 protocol). ZigBee [22] and IEEE eventually joined forces and a complete protocol developed, known as *ZigBee*.

The ZigBee protocol stack is shown in Figure 2.13. The physical layer of ZigBee specifies three possible bands of operation: (a) 868- to 868.6-MHz ISM band with one channel of 20 Kb/s, (b) 902- to 928-MHz ISM band with 10 channels, each one

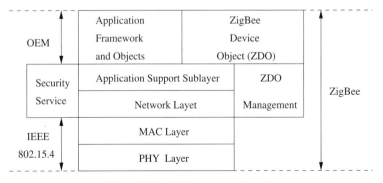

Figure 2.13. ZigBee protocol stack.

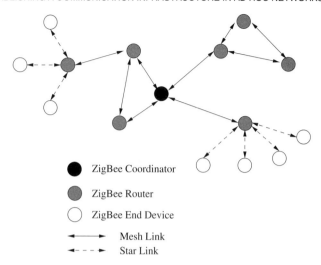

Figure 2.14. A ZigBee network.

with a rate of 40 Kb/s, and (c) 2.4- to 2.4835-GHz ISM band with 16 channels, each one at a rate of 250 Kb/s.

ZigBee specifies two types of device that can be network nodes: full-function device (FFD) and reduced-function device (RFD). Furthermore, two types for access medium are specified. The first one is a beacon mode in which intermittent communication based on polling happens between a beacon node and nonbeacon nodes. In this case there are no packet collisions. An FFD can function as a PAN coordinator or a router. The second method for medium access is random access based on Carrier Sense Multiple Access with Collision Avoidance (CSMA-CA).

An RFD is an end device that can communicate with a coordinator or router; however, it cannot forward traffic from other nodes. The end devices participate in the network formation by linking to the coordinator/routers in star-like link connections. Multihop connections can be established by relaying data from router to router using mesh-link connections. A representative ZigBee topology is shown in Figure 2.14.

The network layer of the ZigBee protocol implements a hierarchical routing strategy with table-driven optimizations applied where possible. The routing layer employs the Ad Hoc On-Demand Distance Vector (AODV) (see Parkins et al. [23] and Parkins and Belding-Royer [24]) and Motorola's Cluster-Tree algorithms WPANS [25]. The AODV algorithm will not be explained in detail since it falls outside the scope of this chapter.

The Cluster Tree Algorithm. This algorithm uses link-state packets to form either a single cluster network or a larger cluster tree network. Nodes select a cluster head (CH) and form a cluster in a self-organized manner. Self-developed clusters connect to each other using a designated device. We distinguish two cases: single-cluster and multiple-cluster networks.

Single-Cluster Network. On power up, a node scans the channels to search for *HELLO* messages from other nodes. If a *HELLO* message is not received within a time period, then the node assumes the role of a *CH* and broadcasts a *HELLO* message. If a reply is not received, the node goes back into the silent state and repeats the process. A *CH* can also be preconfigured based on special functionality, computing ability, or location information.

If a node receives a *HELLO* message from *CH*, then it replies with a *CONNECTION REQUEST* message to the *CH*. The *CH* replies with a *CONNECTIONS RESPONSE* that includes an address that the *CH* assigns to the node. The node finally replies with an *ACK* to the *CH* and completes its registration within the cluster.

A CH assumes a node id, which (for a single cluster) is ID_0 to indicate that the node is a cluster head. The ID_0 is also included as the most significant part of the addresses assigned to any other node in the cluster by the cluster head. Nodes that cannot directly contact the *CH* may reach it via other nodes already in contact following an analogous procedure. Further details can be found in WPANS [25].

Multiple-Cluster Network. In the case of a multiple-cluster network, a designator device (*DD*) is introduced. This device is responsible for assigning a unique cluster ID to each cluster head. The cluster ID combined with the node ID that each CH assigns to a node within a cluster will form an address and will be used for routing packets. The *DD* may also calculate the shortest route from a cluster to the *DD* and forward this information to all the nodes within the network.

The *DD* assigns cluster IDs (*CID*) using nodes that are found at the intersections of clusters. These nodes may happen to be ordinary nodes or cluster-head nodes. The details of the messages exchanged in order to form a multiple-cluster network can be found in WPANS [25]. An example of a multiple-cluster network is shown in Figure 2.15.

Routing inside a cluster is straightforward. Routing between clusters can be accomplished using the border nodes (nodes that are common within clusters), which can relay packets. Routing can be done in a hierarchical manner by appropriately

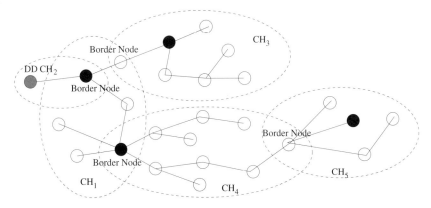

Figure 2.15. Clustering in a ZigBee network.

employing an overlay formed by border nodes and cluster heads (which act as beacons). Shortest paths can be computed using the designator device (DD), which can then distribute them to the entire network. This is a challenging research area, and efforts are ongoing.

Efficient Broadcasting in ZigBee Networks. In a ZigBee network, message broadcasting happens frequently. In order to minimize the transmission power and the processing involved in broadcasting, it is desirable that when a node has a message to broadcast, it is broadcasted to a particular subset of its neighbors and not to all its neighbors. Ding et al. [26] introduce an efficient message broadcast algorithm. The algorithm is based on the fact that in a ZigBee network, during network formation, nodes are assigned addresses by the coordinator device. The addresses are distributed hierarchically. In this case, it is possible for every device to know the address of its parent, the address space of the child nodes of its parent, and the address of its parents' neighbors. In the broadcasting protocol that is developed in Ding et al. [26], a device exploits the hierarchical addresses of its neighbors and their link qualities and makes a decision on whether to forward a broadcast message or not. As explained in Ding et al. [26], this is because the message broadcasting may have already been performed by the node's neighbors and a subsequennt broadcasting by the node may be redundant. This proposal may be possibly further elaborated for implementing a data broadcasting algorithm for ZigBee networks.

Localization in ZigBee Networks. Methodologies for determining the location of nodes in a wireless ad hoc/sensor network based on time of arrival (TOA), angle of arrival (AOA), and received-signal-strength (RSS) can be found in Patwari et al. [27]. Since ZigBee networks are using the IEEE 802.15.4 physical layer specification, the received signal strength (RSS) may be the most appropriate for location estimation. It can be measured via the link quality indicator (LQI), which reports the signal strength associated with a received packet to higher layers. Measurements are relatively simple to implement in hardware, and all commercial transceiver chipsets have an RSS indicator incorporated. However, such measurements may become very inaccurate. Major sources of error may be due to multipath signal arrivals and shadowing due to environmental factors. Errors to multipath signal arrivals can lead to frequency-selective shadowing, which in turn can be diminished by using spread spectrum transmission methodologies. Environmental shadowing (due to obstacles, etc.), however, may be very difficult to combat. Further sources of error may be due to the fact that the RSS functionality may vary from chip to chip because of manufacturing tolerances or because of the battery level changes that the nodes may encounter during their life. Calibration and synchronization procedures for RSS in these cases are proposed in Patwari et al. [27] and references therein.

2.5.2 The MicreInet Network Architecture

The RadioWire MicrelNet is an ad hoc network architecture developed and offered by [21]. The architecture employs the company's transceiver chipsets. The architecture

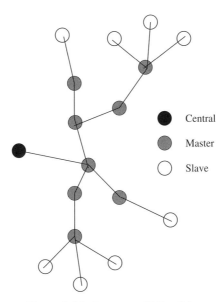

Figure 2.16. Structure of MicrelNet.

can implement a multilevel star-based topology. The MicrelNet architecture involves wireless nodes that can be one of the following types: central master, master, and slave. The network is shown in Figure 2.16.

Micrelnet has a tree structure. The central master is the root. Therefore, there is one central muster in a network. A master at level i (as well as the central master) can exchange information with a master at level $i - 1$, and it can connect to at most two masters at level $i + 1$. Slaves can connect to any master at any level and can exchange information only with it. According to Micrel [21] the architecture currently supports up to eight levels ($i = 1, \ldots, 8$).

Addressing and Routing in MicrelNet. Addressing in MicrelNet is hierarchic. Routing is done following the hierarchy of the address space.

Every node has a four-byte address programmed by a network administrator. The first byte refers to the network level to which the node belongs. At network level n, the n most significant bits of bytes two, three, and four are used for specifying a master address. The remaining bytes are all zeroes for a master. An example of addressing in MicrelNet is depicted in Figure 2.17. The figure shows two masters that are placed in level 3. Their level is indicated in the first byte. The addresses of these masters are specified by the three most significant bits of the second byte. The addresses in equivalent decimal notation are 3.192.0.0 and 3.224.0.0. A slave will have an address whose first two bytes will be the same with the corresponding two bytes in the address of its master. The last two bytes will contain the id of the slave node. Since the n most significant bits of the second byte denote the address of a master, the network may have at most eight levels. At any level a master node can have at most $2^{16} - 2$ slaves

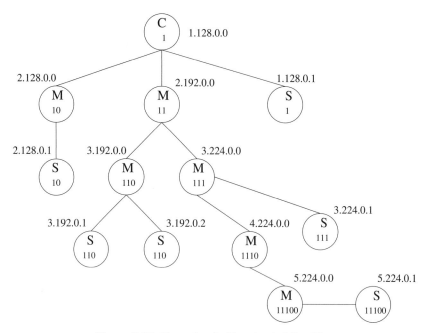

Figure 2.17. Example of addressing in MicrelNet.

(since the last two bytes cannot be either all 0s or all 1s). One exception to this rule is the case when a master has only one level below it, in which case it can have $2^{24} - 2$ slaves (i.e., bytes two, three, and four can all be used for addressing the slaves).

Under this address assignment scheme, each node can determine (a) if it is a central master, master, or slave and (b) if a slave can infer the address of its master by inspection of the n most significant bits of the second byte (n is the level given in the first byte of the address) filling with 0s as the rest of the address. A slave at level n always transmits data units to his master. A master checks the destination address of a received data unit. If the destination is in his own subnetwork, then it will forward to the appropriate slave; otherwise it will send the data to the master at level $n - 1$.

The MicrelNet architecture is simple from the organization, addressing, and routing point of view. Masters can be added as needed to extend the range and coverage of the network. For practical purposes a network accommodates a relatively large number of slaves. However, the addresses must be preassigned by the network user/administrator, and there is no provision for dynamic self-configuration at this time.

A number of open problems can be investigated. With respect to the localization problem, it may be assumed that masters have known locations (e.g., by means of GPS). The slaves can be localized with the aid of their masters. Finally, the dynamic configuration of the network (i.e., election of a primary master and masters at various network levels) is also an interesting problem that can be investigated. To this end, ideas from ZigBee protocol in the previous section using cluster formation and cluster-based routing may be readily available.

2.5.3 Applications to Real-time Tracking and Monitoring

The infrastructureless versatility of ad hoc networks makes them suitable for introducing more advanced features in support of multimedia data rates and integrating them with various network platforms. Specific applications being developed include networks for *personal* and *disaster* area communication, as well as sensor systems for military and measurement applications. The focus of this section is on providing examples for supporting real-time information collection.

Vehicle Tracking in a Transportation Network. Consider a bus route that starts at bus stop BS_0 and traverses a number of bus stops denoted by BS_i ($i = 1, \ldots, N$). A ring sensor network is deployed along the bus route. A sensor node is installed at every bus stop. Each sensor receives messages from its predecessor sensor and forwards messages to its successor sensor on the ring. There is clockwise flow of messages along the ring. Furthermore, such a sensor will be able to sense a bus that happens to be in close proximity. Figure 2.18 depicts a real-time bus tracking system.

Each bus is equipped with a radio transmitter that broadcasts a periodic beacon signal containing its id. When such a signal is received by a sensor node BS_i, a message is constructed consisting of the bus id, the id of BS_i, and a timestamp. This message is sent to the *successor* sensor BS_{i+1} and is subsequently relayed to all other bus station nodes until it arrives at BS_0 where it is stored in a local database. This database maintains the most recent messages regarding the location of each bus serving the particular route. The contents of the database can be made accessible to Web- based applications that may be consulted by commuters who would like to learn the next arrival time at a particular stop. The contents of the database may also be accessed using a telephone network-based application. To deal with a complex bus network, multiple

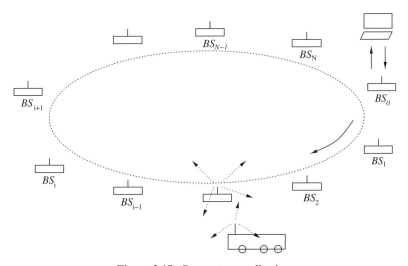

Figure 2.18. Bus system application.

overlapping rings can be used. Moreover, several databases can be used to collect information. The development of such a tracking network, assignment of the wireless sensor rings, and database synchronization may be a challenging research problem.

Smart Networked Toys. The toy industry presents an interesting area for sensor networks. Toys for children may be equipped with sensors enabling the establishment of communications with other toys in their neighborhood. Toys that happen to be in close proximity advertise their presence and form a network. They are programmed to follow an educational interaction schedule such as exchanging their names, introducing themselves and asking questions to children. They can be connected to the Internet through a home computer, and they can download appropriate software updates providing new and ongoing educational networking applications. This is expected to motivate children to learn by constantly using and exploring their toys.

Another application area is the remote control of models such as cars and airplanes (see Figure 2.19). Remote control toys are operating mostly in the 72- to 75-MHz radio range. The technology has not changed for the last 30 years. A sensor network can be used for the remote control of models. The frequency of operation can be defined at a 2.4-GHz ISM band. An active sensor node controls each model airplane and has the capability to interoperate with other sensors to achieve such goals as platoon flying and collision avoidance. Enhanced reliability regarding coordinated network access and reduction of packet loss may be implemented in such a network and is a requirement of crucial importance in the remote control hobby industry. Furthermore, various data from the models such as telemetry, speed, temperature, altitude, and pictures/video can be relayed to the operator.

Vehicle Platoon Formation. In the future, vehicles will have the capability to communicate together and form networks. Vehicle platoon formation is an interesting intervehicle communications and networking application (see Figure 2.20). The goal

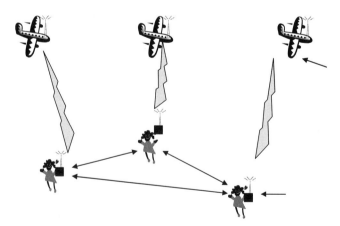

Figure 2.19. Smart-networked toy application.

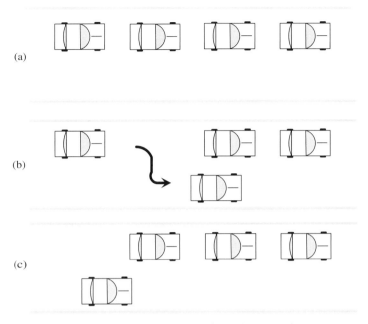

Figure 2.20. Vehicle platoon formation example.

is to improve driving comfort and safety and traffic flow. This is achieved by creating files of vehicles where the head vehicles lead the followers by communicating their parameters (in particular their speed). In the platoon, vehicles are controlled to ensure not only short distance but also safe vehicle-to-vehicle separation.

Vehicle platoon formation and maintenance can be done by the vehicles themselves, using ad hoc communications and networking and localization. Nearby vehicles first have to discover each other, establish communications links and routes (if required), exchanged positions and headings, and apply a platoon formation algorithm.

An example is depicted in Figure 2.20. In part (a), there is a platoon formation consisting of four vehicles in file. In part (b), the platoon is segmented because of the third vehicle leaving the platoon. In part (c), the fourth vehicle joins the second vehicle in file and the platoon is reconnected.

2.6 EXERCISES

1. Consider a heterogeneous network consisting of four elements: ad hoc hosts, base stations, public telephone network, and geosynchronous satellites. Provide *guidelines* for an efficient hierarchical protocol that will enable communication in a networked system connecting these elements. Justify your answers.

2. Elaborate on why the design of the protocols described in Section 2.5 have characteristics that are consistent with the principles of hierarchical routing and clustering, respectively, as well as with the principle of beacon vector routing in Section 2.3. In each case, isolate the specific design elements.

3. Consider the following "local" procedure for electing cluster heads in a unit disk graph. A node u is elected cluster head if its distance one neighborhood $N(u)$ is not a complete graph. Show that the resulting set C of cluster heads is (1) dominating, that is, every node is adjacent to a cluster head, and (2) connected, that is, every pair of nodes of C is connected via a path consisting only of nodes from C.

4. Consider a ring on n nodes. Show that using only one anchor A, there are $\lfloor (n-1)/2 \rfloor$ pairs of nodes with the same pairwise distance to A which cannot be distinguished, while with two anchors A, A' at distance $d \neq n/2$ from each other, or any arbitrary three anchors, each node has a unique coordinate. Solve the same problem on a $\sqrt{n} \times \sqrt{n}$ torus.

5. What can you say about the questions of exercise 4 on the random unit disk graph model? (See also Wattenhofer et al. [12].)

6. Give details of the mathematics involved in the four distance estimation techniques outlined in Subsection 2.4.2. In each case, depict the equations and explain how they can be solved. Elaborate why it is important to avoid using quadratic equations in distance estimation.

REFERENCES

1. R. Rajaraman. Topology control and routing in ad hoc networks: A survey, ACM SIGACT News, **33**(2): 60–73, 2002.
2. S. Basagni. Distributed clustering for ad hoc networks. In *ISPAN*, 1999, pp. 310–315.
3. E. Kranakis, H. Singh, and J. Urrutia. Compass routing in geometric graphs. In *Proceedings of 11th Canadian Conference on Computational Geometry, CCCG-99*, August 15–18, 1999, pp. 51–54.
4. P. Bose, P. Morin, I. Stojmenovic, and J. Urrutia. Routing with guaranteed delivery in ad hoc wireless networks. *Wireless Networks*, **7**:609–616, 2001.
5. M. Mauve, J. Widmer, and H. Hartenstein. A survey on position-based routing in mobile ad hoc networks. *IEEE Networks*, **November/December**: 2001, pp. 30–39.
6. J. Al-Karaki and A. Kamal. Routing techniques in wireless sensor networks: A survey. *IEEE Wireless Communications*, 2004, pp. 6–28.
7. D. B. Johnson, D. A. Maltz, and J. Broch. DSR: The dynamic source routing protocol for multi-hop wireless ad hoc networks. In C. E. Perkins, editor, *Ad Hoc Networking*, Addison-Wesley, Reading, MA, 2001, pp. 139–172.
8. D. B. Johnson and D. A. Maltz. Dynamic source routing in ad hoc wireless networks. In T. Imielinski and H. Korth, editors, *Mobile Computing*, Kluwer Academic Publishers, 1996, pp. 153–181.

9. D. B. Johnson. Scalable and robust internetwork routing for mobile hosts. In *Proceedings of the 14th International Conference on Distributed Computing Systems, IEEE Computer Society*, June 1994. Poznan, Poland, pp. 2–11.
10. C. Perkins. *Ad hoc Networking*. Addison-Wesley, Reading, MA, 2001.
11. R. Fonseca, S. Ratnasamy, J. Zhao, C.-T. Ee, D. Culler, S. Shenker, and I. Stoica. Beacon vector routing: Scalable point-to-point routing in wireless sensornets. In *Proceedings of NSDI '05, 2nd USENIX/ACM Symposium on Networked Systems Design & Implementation*, 2005.
12. M. Wattenhofer, R. Wattenhofer, and P. Widmayer. Geometric routing without geometry. In *Proceedings of International Colloquium on Structural Information and Communication Complexity (SIROCCO)*, Vol. 3499, Springer Lecture Notes in Computer Science, 2005, pp. 307–322.
13. S. Ratnasamy, C. Papadimitriou, S. Shenker, I. Stoica, and A. Rao. Geographic routing without location information. In *Proceedings of MOBICOM*, 2003, pp. 96–108.
14. N. Linial, L. Lovasz, and A. Wigderson. Rubber bands, convex embeddings and graph connectivity. *Combinatorica*, **8**(1):91–102, 1988.
15. B. W. Parkinson and J. J. Spiker, Jr., editors *Global Positioning System: Theory and Application (Volume I, Progress in Astronautics and Aeronautics)*. American Institute of Aeronautics and Astronautics, 1996.
16. D. Niculescu and B. Nath. Ad hoc positioning system (aps). In *IEEE Global Telecommunications Conference, 2001 (GLOBECOM 01)*, 2001, pp. 2926–2931.
17. K. Langendoen and N. Reijers. Distributed localization in wireless sensor networks: A quantitative comparison. *Computer Networks*, **43**(4):499–518, 2003.
18. S. Simic and S. Sastry. Distributed localization in wireless ad hoc networks. Technical Report UCB/ERL M02/26, UC Berkeley, 2002.
19. S. Capkun, M. Hamdi, and J.-P. Hubeaux. GPS-free positioning in mobile ad hoc networks. *Cluster Computing*, **5**:157–167, 2002.
20. M. Barbeau, E. Kranakis, D. Krizanc, and P. Morin. Improving distance based geographic location techniques in sensor networks. In *Proceedings of ADHOC-NOW 04*, Vol. 3158, Springer Lecture Notes in Computer Science, 2004, pp. 197–210.
21. Micrel. http://www.micrel.com, Micrel Corporation, 2006.
22. ZigBee. http://www.zigbee.org, Zigbee Alliance, 2001.
23. C. E. Perkins, E. M. Belding-Royer, and S. Das. Ad hoc on demand distance vector (AODV) routing. IETF RFC 3561, July 2003.
24. C. E. Perkins and E. M. Belding-Royer. Ad-hoc on-demand distance vector routing. In *WMCSA*, IEEE Computer Society, New York, 1999, pp. 90–100.
25. WPANS. Cluster tree network: IEEE P802.15 working group for WPANS, April 2001.
26. G. Ding, Z. Sahinoglu, P. Orlik, J. Zhang, and B. Bhargava. Reliable broadcasting in zigbee networks. In *IEEE Communications Society Conference on Sensor and Ad Hoc Communications and Networks (SECON)*, September 2005, pp. 510–520.
27. N. Patwari, J. Ash, S. Kyperountas, A. O. Hero, R. M. Moses, and N. S. Correal. Locating the nodes: Cooperative localization in wireless sensor networks. *IEEE Signal Processing Magazine. Special issue on Signal Processing in Positioning and Navigation*, **22**(4):54–69, 2005.

CHAPTER 3

Robustness Control for Network-Wide Broadcast in Multihop Wireless Networks

PAUL ROGERS[1] and NAEL B. ABU-GHAZALEH[1]
Computer Science Department, Binghamton University, Binghamton, NY 13902

3.1 INTRODUCTION

Mobile ad hoc networks (MANETs), mesh networks, and multihop sensor networks are instances of multihop wireless networks where nodes cooperate to forward traffic among each other. Such networks are important whenever infrastructure is unavailable (or expensive) and quick deployment is desired. They play an important role in many existing and emerging applications in the military, industry, research, and civilian domains.

Network-Wide Broadcast (NWB) algorithms provide a mechanism to deliver information to nodes in a multihop network without depending on routing state. This makes them ideal for initial self-configuration or for operation under mobility where the routing state becomes stale. Therefore, NWB is a heavily used primitive at the core of most multihop routing [1, 2] and group communication protocols [3, 4]. A common approach to performing NWB is flooding: a process where every node that receives a packet for the first time rebroadcasts it. Flooding has been shown to be wasteful, especially in dense networks—a problem called *the broadcast storm* [5].

Ni et al. [5], who identified the broadcast storm problem, also proposed several solutions to it based on nodes locally determining whether their rebroadcast is likely to be needed. An alternative, topology-sensitive approach to the problem attempts to construct a virtual backbone that is tasked with disseminating the broadcast. Such a virtual backbone must form Connected Dominating Sets (CDS) comprised of a connected subset of the nodes that together cover all the nodes in the network; this

---

[1]This work is partially supported by AFRL grant FA8750-05-1-0130 and NSF grant CNS-0454298.

Algorithms and Protocols for Wireless and Mobile Ad Hoc Networks, Edited by Azzedine Boukerche
Copyright © 2009 by John Wiley & Sons Inc.

set can then be tasked with rebroadcasting the NWB packet while all other nodes just receive it, significantly reducing the number of retransmissions relative to flooding. We survey these approaches in more detail in Section 3.3.

While the focus of NWB algorithms has mostly been on reducing the overhead, a different problem occurs that affects the robustness of most NWB protocols. This problem affects NWB protocols that rely on MAC level broadcast operations. This includes virtually all existing NWB algorithms, both flooding-based protocols [5] and virtual backbone approaches (e.g., references 1, 6, and 7). More specifically, because MAC broadcasts are unreliable, it is possible for rebroadcasts to be lost due to interference or transmission errors. The loss rate can be considerable if high interference exists or if link quality is bad [8, 9]. These losses may lead to the NWB, reaching only a subset of the nodes. A particularly damaging example of this problem occurs when the initial transmission of the NWB packet is lost at all receivers (e.g., due to a collision with another packet). In this case the NWB will not reach any nodes.

In Section 3.5 we present a classification of existing NWB algorithms, focusing on their robustness characteristics. In addition, we analyze the robustness properties of selected algorithms representing different points of the space under different network densities, loss rates, and mobility characteristics; the algorithms are described in Section 3.4. We show that under high loss rates and in networks that are sparse, coverage for existing NWB algorithms, including flooding and DCB, is poor. In addition, all the evaluated algorithms that reduce redundancy relative to flooding also reduce robustness. More importantly, algorithms that have static rebroadcast responsibilities are more vulnerable to losses than algorithms that take rebroadcast decisions locally. Moreover, algorithms that are topology-sensitive perform worse as mobility increases.

The results from the evaluation of NWB algorithms argue for the need of a robustness control component of NWB. Thus, NWB algorithms may be viewed as two components: (1) *redundancy control*, which is the component that attempts to reduce redundancy while maintaining coverage, and (2) *robustness control*, which is the component that attempts to recover from lost rebroadcasts and maintain coverage in the face of losses. The vast majority of existing NWB target redundancy control, without considering robustness control.

The next contribution of the paper is to present a classification of the solution space for robustness control in Section 3.6. More specifically, robustness control may be implemented at the network layer or at the MAC layer. Robustness control may be accomplished by statically introducing redundancy or by dynamically detecting losses and reacting to them. For the dynamic loss-sensitive approaches, we discriminate between (a) approaches that use explicit feedback to detect losses and (b) those that predict losses based on locally available information. Finally, we discriminate between (a) approaches that are sensitive to the state of the network and (b) those that are not. This section also presents a classification of existing algorithms that have robustness control components.

In order to evaluate the different approaches to robustness control, we implement representative protocols from different points of the solution space. We study these protocols for a number of different scenarios in Section 3.7. Finally, in Section 3.8 we present some concluding remarks.

3.2 BACKGROUND—MAC BROADCAST

Wireless communication properties make wireless transmissions unreliable for two primary reasons: (1) channel fading and (2) collisions. Channel fading can result in substantial signal power fluctuations at different time scales [10]. This leads to a large number of transmission errors, especially when communicating nodes are far from each other [8, 9].

Neither interference nor transmission occurs uniformly across the links of a network. More specifically, interference results in losses in areas closest to the interfering traffic, while transmission errors are affected by the surroundings (which determine the signal fading effects and the Doppler effect, as well as the path loss exponent which defines how fast the signal attenuates with distance) and increase with the distance between the sender and the receiver.

In a wireless medium, the channel is nonuniformly shared, giving rise to the well-known hidden terminal problem [11]. A hidden terminal is an interfering node out of carrier sense range of the sender, but in interference range with the receiver; such a transmission is not detected by the sender (it is hidden to it), causing a potential collision at the receiver. Techniques such as Carrier Sense Multiple Access and Collision Avoidance reduce collisions but do not eliminate them [12]. Finally, concurrent transmissions can cause collisions, especially when correlated transmissions occur within an NWB [5].

To improve reliability in the face of losses, MAC protocols like IEEE 802.11 [13] use retransmission to recover from losses of packets that are not acknowledged. Such an approach cannot be used with broadcast packets because there are a number of receivers. Accordingly, if a broadcast packet is lost due to a collision, the loss is not detected by the sender and no retransmission is carried out. This makes MAC level broadcast operations much more susceptible to losses. As a result, NWB operations that rely on MAC level broadcast suffer from losses that can lead to significant loss of coverage: *the NWB robustness problem*.

3.2.1 Broadcast Robustness

This section outlines the reasons that cause NWB algorithms to lack robustness. Wireless communication channels are unreliable for two primary reasons: (1) Transmission errors: signal propagation effects and interaction with the surrounding environment can cause deep fades in the received signal power (20 dB, or $100\times$ changes in signal power are common [14]). These fades can occur in small scales in time and space. This effect leads to a large number of transmission errors, especially when the two communicating nodes are far from each other; and (2) collisions: the wireless channel is non-uniformly shared, giving rise to the well-known hidden terminal problems [11].

Transmission errors and collisions makes MAC level broadcast operations susceptible to losses. The lack of a higher-level reliability mechanism (e.g., as is present in unicast operations via acknowledgments and retransmission) exacerbates the problem. As a result, NWB operations which rely on MAC level broadcast suffer from loss of coverage.

In addition to losses in the network from various application traffic flows, losses can occur with self-interference. In self-interference, a NWB operation results in two nodes broadcasting at the same time, resulting in collisions at receiving nodes. Since any node can issue a broadcast at any time, this can be problematic in an NWB operation. As a result, many NWB algorithms attempt to cut down on the number of broadcasts needed for an NWB.

Ni et al. [5] studied the effect of all nodes in the network performing a broadcast. They called this phenomenon the broadcast storm problem. Rebroadcasts carried out by nodes in the network were found to be redundant, because many neighbors had the broadcasted information already. Furthermore, the additional rebroadcasts caused contention in the network to go up, because the broadcasts interfered with each other. Lastly, since there is no backoff mechanism or reliability mechanism built into MAC level broadcasts, the number of collisions in the network went up.

3.3 OVERVIEW AND CLASSIFICATION OF NWB ALGORITHMS

In this section, we overview NWB approaches and classify them with an emphasis on reliability. The primary characteristics we are interested in are: (1) overhead, (2) resilience to mobility, and (3) robustness to losses.

3.3.1 Topology-Ambivalent (Flooding) Approaches

Flooding is an approach to NWB that operates as follows: Every node rebroadcasts a flood packet the first time it receives it (Figure 3.1 shows an example). Because no topology knowledge is needed, flooding and algorithms derived from it are *resilient to mobility* and do not require an ongoing overhead to discover neighbors/topology. Flooding is thought to be resilient to MAC losses due to the high redundancy generally present. However, this may not be true: In low-density areas of networks, the available redundancy is low. Moreover, the amount of available redundancy is fixed: It may be too high for some cases and too low for others.

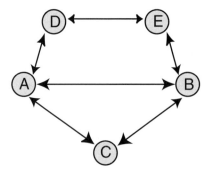

Figure 3.1. Flooding example.

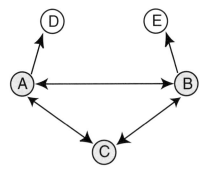

Figure 3.2. Optimized flooding example.

Flooding is a brute force approach that has high overhead, especially in dense networks [5]. Specifically, it leads to: (1) high redundancy in node coverage with many useless rebroadcasts; (2) high number of collisions: with all nodes rebroadcasting, collisions due to concurrent retransmission can occur; and (3) high contention for the use of the medium, leading to delays in delivering the NWB. Eliminating the redundancy inherent in flooding and alleviating the above effects are the primary motivations behind most NWB algorithms.

Using a geometric argument and ideal unit-disk transmission assumptions, Ni et al. [5] show that at most 61% of the rebroadcast coverage area does not overlap with the original transmission; on average this value is 41%. Furthermore, as additional rebroadcasts are overhead, the average nonoverlapping area drops rapidly. As a result of this observation, they suggest having nodes determine locally whether their rebroadcast is likely to be redundant, and eliminate it if it is so (an example is shown in Figure 3.2, where nodes D and E do not rebroadcast the packet, with no loss of coverage).

In the same paper, several criteria for deciding whether to rebroadcast or not are investigated, including:

1. *Probability-Based.* Probabilistically, a node decides whether or not to rebroadcast. This approach blindly eliminates rebroadcasts: It may not eliminate redundant broadcasts, or it may eliminate critical ones.
2. *Counter-Based.* If a certain number of previous rebroadcasts were heard, an additional one is probably not needed based on the geometric argument they developed. Thus, if the threshold of overhead transmissions is reached, the packet is not rebroadcast.
3. *Distance-Based.* Assuming that the distance of the node from whom we receive a rebroadcast can be determined, the following criterion can be used to decide whether to eliminate a packet or not. If the distance exceeds a certain threshold, the nonoverlapping area is significant and the packet can be rebroadcast; otherwise the packet is not rebroadcast.

4. *Location-Based.* In this approach, a rebroadcasting node includes its location on the packet. As a node receives rebroadcast packets, it can compute the additional area that would be covered by its own rebroadcast. If this area exceeds a threshold, the packet is rebroadcast; otherwise it is dropped.

The discrimination between needed and redundant algorithms improves in the order of listing of the criteria. However, the amount of needed information also increases. Furthermore, in particular, the distance-based and location-based algorithms are susceptible to the fact that their ideal premises do not hold in realistic environments. In general, these approaches are successful in reducing the overhead of NWBs relative to flooding. Like flooding, they are resilient to mobility. Because each node locally determines whether its rebroadcast is likely to be needed, the approach dynamically adapts to transmission losses.

The effect of the threshold in the above algorithms is investigated. However, it is likely that a single threshold will not work for all networks, or for all areas of a network. As a result, Tseng et al. [15] investigate the effect of adaptively adjusting the rebroadcast thresholds based on observed behavior. For example, in their adaptive counter scheme (where a rebroadcast is suppressed if a certain number of rebroadcasts of the same packet are overhead), the counter value is initialized to 1, and adjusted up based on observed rebroadcasts. The approach significantly improves on fixed-threshold performance.

Scott and Yasinsac [16] propose a solution to dynamically adjust the probability of transmitting a broadcast message. This work is built on the gossiping work of Haas et al. [17]. The solution attempts to (a) measure node density dynamically with a ping mechanism and (b) use that data to transmit the broadcast packet only if significant area and nodes will be covered.

3.3.2 Topology-Sensitive Approaches

These approaches build a virtual backbone that covers all nodes in the network. Ideally, the backbone forms a *Connected Dominating Set (CDS) of the topology graph*. A dominating set in a graph is a set of nodes where every node in the graph is a member of the dominating set or a direct neighbor of a member. In other words, all nodes in the network are one-hop reachable from the CDS. Thus, if only the dominating nodes rebroadcast NWB packets, all vertices of the graph are covered. An example CDS is shown in Figure 3.3; in this case, only nodes *A* and *B* transmit and all nodes are covered by the two transmissions.

CDS Construction. The Minimum Connected Dominating Set (MCDS) problem has been shown to be NP-complete for unit disk graphs,[2] by mapping the NP-hard set cover problem to it [18]. Das et al. [19] propose a centralized heuristic algorithm

---

[2] These are graphs where the neighbors of a vertex are the vertices contained by a disk of radius 1 around the vertex. Unit-disk graphs can represent the topology wireless networks with ideal omnidirectional antennas.

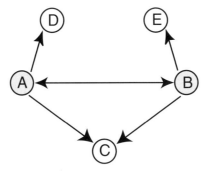

Figure 3.3. CDS example.

for deriving an approximate MCDS. However, a centralized approach is susceptible to mobility; changes in topology require global recalculation of the CDS.

More efficient algorithms, based on minimum spanning trees [6], graph-coloring [7], and clustering [5, 20, 21], have been proposed. However, these are also susceptible to mobility because dynamic changes in topology can require significant recomputation of the CDS. A localized CDS computation approach was proposed by Wu and Li [22]. In this approach, nodes track two-hop neighbor information. A node selects itself to be part of the CDS using a simple rule: If it has two neighbors that are not directly connected, it is a member. This rule results in a large CDS. As a result, Wu and Li suggest two rules for pruning the CDS whereby a node whose neighbors are fully covered by another node or a combination of two nodes is pruned from the CDS. Because it only requires information that is localized, this algorithm is better equipped to maintain the CDS in the presence of mobility.

Properties of Topology-Aware Algorithms. In order to build a virtual backbone, nodes exchange information about neighbors; this is an ongoing process. As a result, topology-sensitive approaches *degrade (in terms of overhead and/or coverage) with increasing mobility* since the neighborhood information becomes stale more quickly.

In terms of vulnerability to losses, an important distinction between topology-sensitive approaches is whether the forwarding responsibilities are statically or dynamically determined. In static approaches, the topology information is used to statically determine forwarding responsibilities. As a result, this approach becomes especially vulnerable to losses: If a loss of a packet to a node with forwarding responsibilities occurs, the remainder of the backbone reachable through it and the nodes they cover will not receive the NWB. In dynamic approaches [23, 24], each node locally determines whether it needs to be part of the CDS based on already heard transmissions. This makes them significantly more resilient to losses. For example, in the Scalable Broadcast Algorithm [24], each node tracks its two-hop neighbors. This enables a node A to determine which of its neighbors a node B covered by checking the two-hop neighbor information without requiring B to transmit its neighbor list.

In summary, NWB approaches assess forwarding responsibilities either in a topology-sensitive way or based on local estimates of the rebroadcast importance. Topology-sensitive protocols can provide more optimal NWBs from an overhead perspective, but require ongoing overhead to exchange neighborhood information and are more susceptible to mobility. Within the topology-sensitive model, another important classification is whether the forwarding responsibilities are assigned statically or dynamically at each forwarding node. Static responsibilities allow more optimized forwarding, but are more susceptible to loss of coverage due to transmission losses.

All the approaches described thus far focus on reducing the redundancy of NWB. However, the robustness of NWB has recently started to receive attention. However, we delay discussion of these approaches to Section 3.6 where the robustness control solution space is described.

3.4 REPRESENTATIVE ALGORITHMS

In this section, we overview representative algorithms from different points in the solution space. These algorithms will later be used in characterizing the performance of the different approaches. In addition to flooding, the following NWB algorithms were evaluated in order to provide representatives of the different classes of solutions.

3.4.1 Location-Based Algorithm (LBA)

In this algorithm, a node includes its location in the rebroadcasted packet. Upon reception of a broadcasted packet, the receiving node notes its current location in comparison with the location of the sending node, obtained from the packet header [5]. This information is then used to calculate the additional area that would be covered with an additional broadcast. Based on this information, a node decides whether its rebroadcast provides sufficient coverage to be worth sending. LBA is a dynamic flooding-based approach (not topology-sensitive). Flooding typically leads to higher overhead in dense networks. By including location information in each packet, the overhead is not raised quite as high as a traditional flood. We describe LBA in more detail by outlining its operation using pseudo-code.

```
When Node A receives a broadcast P from Node B:
     Distance := Distance(B, A);
     If Distance > DISTANCE_THRESHOLD Then
          Rebroadcast(P);
End If;
```

3.4.2 Ad Hoc Broadcast Protocol (AHBP)

In the Ad Hoc Broadcast protocol [25], nodes track two-hop neighbor information via 'HELLO' packets and use this information to explicitly select a set of one-hop neighbors to rebroadcast the packet such that all the two-hop neighbors are covered.

Thus, it is a static CDS-based algorithm. Only nodes marked as Broadcast Relay Gateway (BRG) nodes perform a broadcast. The node selection algorithm of AHBP is as follows:

1. Locate all two-hop nodes that can only be reached by a single one-hop node. Mark those one-hop nodes as BRG members.
2. Calculate the set of nodes that will be covered by the BRG nodes.
3. For all remaining one-hop neighbors, find the neighbor that will cover the most uncovered two-hop neighbors, and mark it as a BRG node.
4. Repeat steps 2 and 3 until all two-hop neighbors are covered.

Within a NWB packet, the set of BRG neighbors is kept in the packet header. As a result, upstream neighbors calculate which downstream neighbors will rebroadcast the packet, making this a static CDS-based approach. Static CDS algorithms can suffer greatly if a node with forwarding responsibilities does not receive a packet. In such a situation, the propagation of the NWB is halted, and the NWB is often worthless.

When a BRG node receives a broadcast, it uses its current two-hop neighbor information to calculate which one-hop neighbors received the transmission. These neighbors are then removed from the neighbor set used in the BRG selection algorithm described above.

Since AHBP is CDS based, its overhead is significantly smaller than flooding-based algorithms. However, since it is topology-sensitive, it is susceptible to mobility. Furthermore, since it is static, its robustness is low.

3.4.3 Scalable Broadcast Algorithm (SBA)

The Scalable Broadcast Algorithm (SBA) [24] is essentially a dynamic version of AHBP where nodes locally determine if all their two-hop neighbors have been covered based on overhead broadcasts. The two-hop neighbor information is obtained via 'HELLO' packets.

If node A sends a packet and it is received at node B, node B calculates if it has neighbors that are not covered by node B. If there are uncovered neighbors, it will retransmit the packet. The algorithm is described in more detail using pseudo-code.

```
Let T1 be the SBA Timer
Let L be the list of nodes that sent a particular
  broadcast
Let Nx be the neighbor list for node X
Let U be the list of uncovered nodes

When a source sends a packet, it is always broadcast.

When a node A receives a broadcast from node S:
```

```
If ({Na} > {Ns}) Then
  If T1<P> Then   // If a timer exists for this packet
    T1<P>.add(P.source);  // Add the originator of this
       broadcast
  Else
    T1 := new Timer(P, {Na}-{Ns});
    T1.start();
  End If;
End If;


T1.expire() {
  If ({U} > {N{L}}) Then
    Rebroadcast(P);
  End If;
}
```

Dynamic CDS algorithms adapt to losses far better than their static CDS counterparts. This resilience comes at a cost of higher forwarding costs than the static CDS approaches, but is more resilient to poor network conditions.

3.4.4 Double-Covered Broadcast (DCB)

DCB [26] is a static topology-sensitive CDS-based algorithm with fixed built in redundancy. Specifically, the CDS is built such that it provides double-coverage for every node. The intuition is that even if a single transmission is not received, the other is likely to enable recovery from this loss. Neighbor information is tracked via 'HELLO' packets. A node wishing to send a NWB will follow the following algorithm:

1. Select a subset of one-hop neighbors that cover all two-hop nodes. Each of these neighbors must be covered by at least two transmissions, one from the sending node and one from the one-hop neighbor set. The nodes selected serve as forwarding nodes.
2. Each forwarding node follows the previous step and broadcasts the packet.
3. The transmission of the forwarding nodes serves as an acknowledgment for the upstream sending nodes. If a sender does not detect rebroadcasts from all forwarding nodes, it recalculates the forwarding node set and repeats the process.

While this algorithm has built in redundancy, this redundancy is fixed: It may be insufficient in some situations (causing poor reliability), or unneeded in other situations, causing unnecessary addition retransmissions. Furthermore, when recalculating the forwarding node set, the latency of the NWB is increased, because a node must wait for a series of 'HELLO' messages to be sent.

3.5 CHARACTERIZING NWB UNRELIABILITY

In this section, the effect of the MAC broadcast unreliability on the performance of NWB algorithms is characterized to demonstrate their lack of robustness. NWB coverage is affected primarily by two factors: the redundancy of the NWB operation and the probability of MAC transmission loss. A dense network allows some NWB algorithms to use redundant paths providing loss-tolerance without losing coverage. This available redundancy is lower in sparse networks, or for algorithms that control redundancy to improve overhead, making NWBs less robust under such conditions.

3.5.1 Preliminaries and Experimental Setup

The primary performance metrics of interest are: (1) *node coverage*: percentage of nodes that receive the NWB; and (2) *overhead*: number of retransmissions. Raw overhead is difficult to interpret independently of the network size and the coverage success. Therefore, *normalized overhead*, defined as the number of retransmissions per receiving node, was tracked. For flooding, normalized overhead is always of interest regardless of the topology or the coverage success: every receiving node retransmits the packet once. Optimized NWBs target lowering the overhead while maintaining coverage that should result in a normalized overhead smaller than 1.

To simulate losses in the network, two approaches are used. First, packets are probabilistically dropped at receiving nodes; a similar approach to modeling losses was used in other works [26]. While this approach is somewhat unrepresentative (transmission errors are typically not uniform across all links), it provides a controllable approach to varying the loss rate. The second approach involves introducing fading to the network. Fading models allow for more realistic transmission behavior, but vary depending on the distance between nodes. As a result, the best way to judge the effect of fading is by comparing NWB behavior in networks of different densities. Previous work has been done using constant bit rate (CBR) traffic flows in order to increase the level of contention present in the network. These flows often interfere with each other, preventing a consistent level of contention in the network. As a result, data gathered from these studies are not as controlled the data presented below.

We evaluated the algorithms described in Section 3.4[3]. All experiments in this paper use the Network Simulator NS-2 [27]. NS-2 is a discrete event simulator with detailed models for the network protocol stack, including the IEEE 802.11 MAC protocol. NS-2 includes simplified models for wireless propagation. In all scenarios, nodes are randomly deployed with a uniform distribution in a fixed area of 1000 by

---

[3]The authors wish to thank Tracy Camp's group at Colorado Mines for the supplying the code for LBA, SBA, and AHBP. We also wish to thank W. Lou for the NS-2 implementation of DCB.

1000 m. The 802.11 MAC implementation in NS-2 was used for experimentation with the default parameters (range of 250 m). In each experiment, one NWB is generated per node. The NWBs are timed to ensure that successive operations do not interfere. Each data point represents an average of 20 scenarios with different random seeds. While the confidence intervals are not presented in the graphs to avoid clutter, they were tracked; the average is narrowly bound (95% confidence intervals within 1% of the mean).

3.5.2 Grid Topologies

Grid topologies consist of nodes that are evenly deployed in a two-dimensional grid throughout the simulation area. This configuration may be representative of static preplanned networks such as mesh network [28]. Grids also provide a level of control to experiments, due to the uniform density present throughout the network. The coverage of flooding for different density grids inside of a 1000-m by 1000-m area were studied.

Figure 3.4 plots the coverage achieved by flooding in different densities, using the controlled drop mechanism described earlier. As expected, grids that are more dense achieve a higher level of coverage. This is because denser grids provide more redundancy for the flood. In a sparse grid, the loss of a few packets can result in loss of coverage to significant portions of the network.

3.5.3 Random Topologies

This section describes studies done with nodes that are randomly deployed in a 1000-m by 1000-m area.

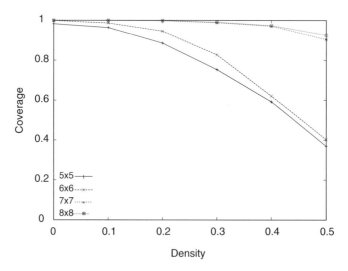

Figure 3.4. Coverage of Flooding (Grid Deployment, Controlled Drop).

CHARACTERIZING NWB UNRELIABILITY   **63**

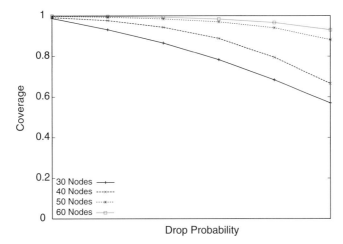

Figure 3.5. Coverage (random deployment, controlled drop, flooding).

Controlled Drop. Figure 3.5 charts the coverage obtained for flooding in different densities of networks. Similar to the density study done for grid scenarios (Figure 3.4), as the density increases, the overall coverage increases. This is because the redundancy built into the network has increased.

Figures 3.6 and 3.7 plot the coverage and overhead for various NWB protocols in a sparse 30-node environment. As the loss rate is increased, the overall coverage of the protocols decreases. This is particularly true for CDS-based approaches, where the loss of one packet in the CDS tree can stop the progression of the NWB.

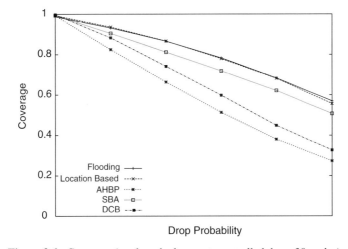

Figure 3.6. Coverage (random deployment, controlled drop, 30 nodes).

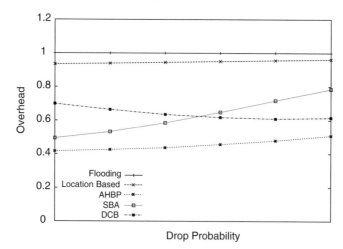

Figure 3.7. Overhead (random deployment, controlled drop, 30 nodes).

The coverage of flooding and LBA is very similar. However, LBA optimizes the number of rebroadcasts needed based on location, resulting in a slightly lower overhead. SBA, being a dynamic CDS algorithm, adapts to losses in the network, resulting in coverage near flooding. Furthermore, as the loss rate increases in the network, the number of broadcasts performed by SBA increases as well. This demonstrates the adaptive nature of the SBA protocol. For the static CDS approaches, DCB has a higher coverage level than AHBP. This behavior is expected, as each node is covered twice, as presented earlier.

The controlled drop experiments were repeated in a denser, 60-node environment (Figures 3.8 and 3.9). As the network is more dense, the protocols are generally more tolerant to losses in the network, resulting in higher coverage. This is especially true of flooding and LBA. The CDS-based approaches make a sparse network out of a dense network, and therefore they lose some of the built-in redundancy of a dense network. At high loss levels, the static approaches of AHBP and DCB drop off when compared to the dynamic approach of SBA and the flooding algorithms.

In terms of overhead, since more nodes are covered per broadcast, the overall normalized overhead tends to be less than that of a sparser network. The gain of using LBA over flooding is larger in a dense network, since fewer nodes rebroadcast per region of the network.

Losses Due to Self-Interference. As nodes propagate an NWB packet, there is a chance that the reception of that packet will coincide with the reception of another packet. If both packets are for the same NWB, those losses are considered to be self-interference. To measure levels of self-interference, the studies from above were repeated with the Two-Ray Ground Propagation Model, with no controlled drop. All resulting losses were due to collisions with other NWB data packets.

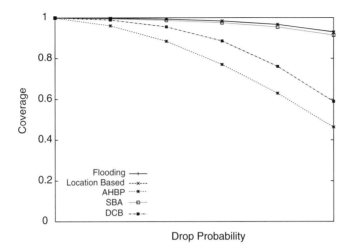

Figure 3.8. Coverage (random deployment, controlled drop, 60 nodes).

Examining Table 3.1, it can be seen that the flooding-based approaches have a higher level of self-interference than the neighbor knowledge protocols. This is as expected, since the neighbor knowledge protocols attempt to build a smaller set of rebroadcasting nodes. In addition, the dynamic approach of SBA has a higher level of self-interference, due to its higher overhead. AHBP, SBA, and DCB also all try to randomize when a packet is retransmitted by introducing some jitter. The jitter algorithm used in DCB is significantly more randomized than AHBP and SBA, because very little self-contention exists in the protocol.

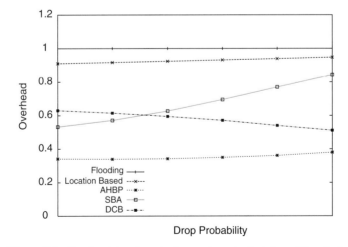

Figure 3.9. Overhead (random deployment, controlled drop, 60 nodes).

TABLE 3.1. Average Number of Packet Drops per NWB Due to Collisions

Protocol	30 Nodes	60 Nodes
Flooding	12.021	87.569
Location-based	7.24	69.228
AHBP	1.151	10.065
SBA	1.427	24.961
DCB	0.118	0.82

3.5.4 Cluster Topologies

Cluster topologies are networks that have groupings of nodes in specific areas, rather than being randomly scattered throughout the network. A cluster scenario was studied with four clusters equidistant from each other, connected by a cluster of four nodes. The resulting shape was that of a "plus."

Controlled Drop. Figure 3.10 plots the coverage achieved by the different protocols in the cluster scenarios. The results roughly mirror the results of the random topologies, with flooding achieving the highest level of coverage. Figure 3.11 plots the overhead of the different protocols. The results of the cluster experiments have lines that are not as smooth, because a loss of a NWB transmission has a high chance of preventing the rest of the network from receiving the broadcast. The results are somewhat similar to the results found in the 30-node random deployment environment (Figures 3.6 and 3.7). This is because a 30-node environment in a 1000-m^2 area is very sparse and loosely connected. The loss of a transmission along one of the links

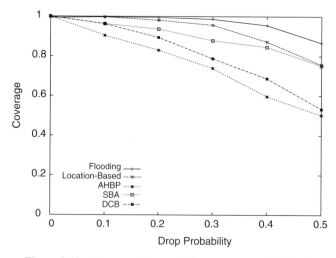

Figure 3.10. Coverage (cluster deployment, controlled drop).

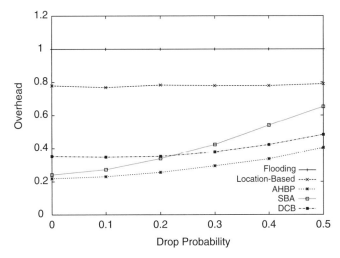

Figure 3.11. Overhead (cluster deployment, controlled drop).

has a high probability of terminating the NWB propagation. As the network density increases, this probability is lowered. By the same logic, if the clusters were made larger, with more nodes, the coverage results would increase proportionally.

3.5.5 Mobile Topologies

This section presents the coverage and overhead studies for various NWB protocols in a few different mobile environments. Topologies containing nodes within a 1000-m by 1000-m area were used for these studies.

Random Waypoint. The Random Waypoint Mobility Model is a model commonly used in MANET research. In this mobility model, nodes move in varying directions and speeds, with a pause time between each movement and direction change.

Figure 3.12 shows the tests using the random waypoint model with a pause time of 1 s between movements and a maximum speed of 15 m per second. This figure represents the coverage achieved for different NWB protocols. Similar to the static scenarios, flooding and LBA do better than the neighbor knowledge protocols. This behavior is noticed even more in a mobile environment, because the mobility causes neighbor knowledge to become stale. When Figure 3.12 is contrasted with the static 30-node results shown previously in Figure 3.6, the staleness of the neighbor knowledge data can easily be seen. Figure 3.13 shows the overhead of the protocols in a 30-node environment. The trends seen here are similar to those seen in the static scenarios.

Figures 3.14 and 3.15 show the coverage and overhead obtained in a 60-node environment, using the random waypoint model (1-s pause time, 15 m/s maximum

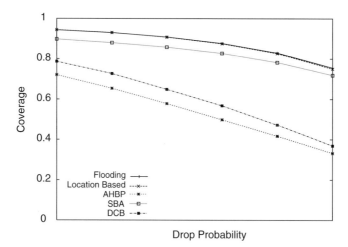

Figure 3.12. Coverage (RW, controlled drop, 30 nodes).

movement). These results again show that dense networks are more robust and resilient to losses. The overall coverage increases, and the overhead goes down, as more nodes are reached with the broadcasts that are sent.

3.5.6 Probabilistic Random Walk

The probabilistic version of the random walk model uses a set of probabilities to determine the next position of a node. This model also uses random directions and speeds, but makes use of a matrix in order to determine the node position for the next

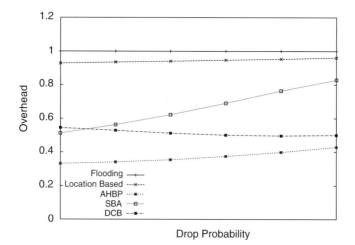

Figure 3.13. Overhead (RW, controlled drop, 30 nodes).

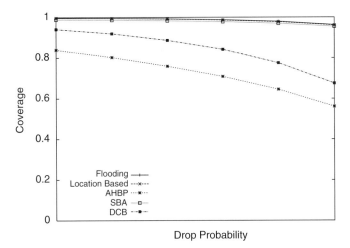

Figure 3.14. Coverage (RW, controlled drop, 60 nodes).

timeslot. Since the movements are probabilistic, the movements of nodes tend to be more realistic than purely random movements [29]. An interval of 0.5 was chosen for these studies.

This mobility model was chosen for study since there have been studies that have shown that the random waypoint mobility model is not always an accurate representation of mobility [30]. Their studies have shown that the general speed of the network slows over time. Bettstetter and Wagner [31] have also shown that the nodes tend to converge on the center of the network over time, when using the random waypoint model.

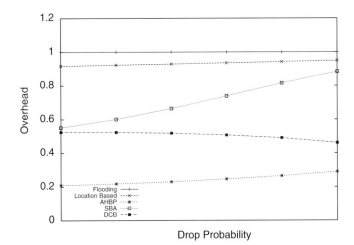

Figure 3.15. Overhead (RW, controlled drop, 60 nodes).

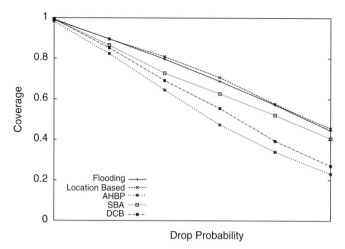

Figure 3.16. Coverage (PRW, controlled drop, 30 nodes).

Figure 3.16 shows the coverage obtained with the probabilistic random walk in a 30-node environment. The dropoff is much sharper, when compared with the random waypoint graphs presented earlier (Figure 3.12). This is due to the effects mentioned earlier—namely, that the random waypoint mobility model tends to slow down over time and converge toward the middle. As the network converges, and the neighbor information data is less likely to be out of date, the redundancy of the network is increased.

Figure 3.17 shows the overhead for the probabilistic random walk in a 30-node environment. Again, overall overhead is higher with this model, when compared with the random waypoint data (Figure 3.13).

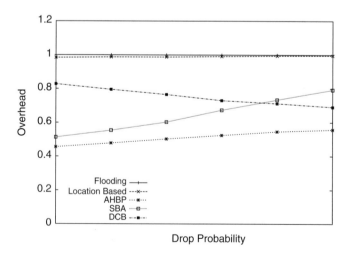

Figure 3.17. Overhead (PRW, controlled drop, 30 nodes).

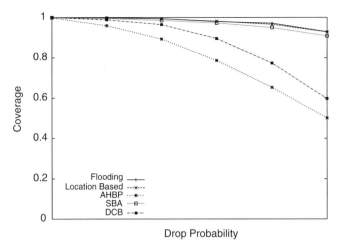

Figure 3.18. Coverage (PRW, controlled drop, 60 nodes).

As one would expect, the coverage obtained for a 60-node environment is higher than that of the 30-node environment, as can be seen when contrasting Figures 3.18 and 3.16.

3.5.7 Notes

The experiments presented in this section show that in sparse environments and environments under high load, the level of redundancy present in the NWB protocols is often not enough (see Figure 3.19). The protocols cannot adapt to losses and therefore are not robust. This weakness needs to be understood and addressed, which we will do in the next sections.

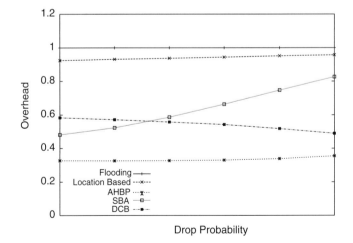

Figure 3.19. Overhead (PRW, controlled drop, 60 nodes).

3.6 ROBUSTNESS CONTROL—SOLUTION SPACE

Increasing NWB reliability requires increasing the probability of the reception of the NWB rebroadcast operations. This is especially true *in situations where their loss is likely and the redundancy in the network is low*. The goal of the solutions presented in this chapter is to improve the reliability of the NWB, rather than ensure complete reliability; guaranteed reliability requires too much overhead (for example, neighbor discovery and the use of unicast packets).

With solutions for robustness control, NWB algorithms can now be thought of in two parts: (1) *redundancy control*: this phase of the algorithm targets reducing the redundancy in the rebroadcasts; and (2) *robustness control*: this phase of the algorithm attempts to recover from losses occurring in the transmissions necessary to cover the nodes. The focus of most existing solutions is redundancy control; in contrast, our focus here is on robustness control.

There are several possible approaches for redundancy control which are classified in the remainder of this section. Please note that even though robustness control is discussed separately from the remainder of the NWB and redundancy control aspects of the algorithm, often the most appropriate choice is influenced by the structure of the underlying algorithm. Examples of such inter-dependence will be discussed as different points in the solution space are examined in more detail in the following three chapters.

Approaches to robustness control can be classified along multiple axes, including the following. First, we distinguish solutions according to how they trigger robustness control responses (typically additional rebroadcasts). *Fixed redundancy approaches* incorporate a fixed degree of redundancy in their coverage, beyond the minimum necessary, to leave a safety margin against broadcast losses. In contrast, *loss sensitive approaches*, trigger additional redundancy (typically in the form of additional rebroadcasts) only when losses are detected or predicted.

Among loss-sensitive approaches, there is a number of solutions possible based on the approaches for detecting/predicting losses and how redundancy is built in response. Broadly, losses detection/prediction can be accomplished by *explicit* or *implicit* feedback.

3.6.1 Explicit Feedback Algorithms

In explicit feedback algorithms, the receivers inform the senders of whether the broadcast was received successfully using an explicit transmission. Explicit feedback solutions face the following problem: If multiple receivers of a MAC broadcast have to provide acknowledgements, collisions occur. However, its possible to randomize/stagger the responses or require feedback from a subset of the receivers only.

With explicit feedback, the granularity of the feedback can be adjusted. The simplest scheme is to provide feedback on every NWB transmission. This has the cost of immediate latency in a NWB operation. Another approach that can be used to provide feedback about NWB successes is to inform the sending node periodically. This requires nodes to cache previously sent data, in case a failure situation arises.

Depending on the granularity and the subset of receivers responding, explicit feedback may also increase latency and load on the network (due to the acknowledgment traffic).

3.6.2 Implicit Feedback Algorithms

Schemes that use implicit feedback attempt to judge the success of a NWB transmission based on locally observed behavior and without requiring addition control packets to be exchanged. Briefly, this approach hypothesizes that the behavior of the neighbors in the case where they receive a broadcast will be different from their behavior in the case where they don't. Thus, by observing the behavior of nearby nodes, the loss probability can be predicted. A simple example of such implicit feedback is for a node to observe the channel and see if the NWB packet was retransmitted by neighboring nodes. If enough retransmissions are overhead, it is likely that the original transmission to the neighbor was received successfully.

Implicit feedback solutions can improve their decision by factoring in other criteria that can improve the loss prediction. For example, each node can measure the utilization of the channel; the higher the utilization, the higher the probability of a loss. However, this approach is imprecise because the local state at the sender may not match that at the receivers. An additional metric of interest is the expected number of rebroadcasts (a function of density and NWB algorithms), the knowledge of which can help refine the prediction made by the implicit feedback mechanism.

An interesting hybrid approach uses explicit feedback to estimate the probability of success of a broadcast and perhaps the criticality of it; this approach is termed *state-aware*. This approach is similar to the utilization and density estimation discussed above. However, because those estimates are created based on the imprecise source view, they may not be valid at the destination where losses occur. Furthermore, link utilization is only one of the factors that influence the quality of the wireless channel; other factors cannot be assessed without feedback from the receiver [32].

State-awareness can be combined with implicit feedback or fixed redundancy algorithms. Most NWB solutions view the link quality as boolean: Links either exist or they do not. The quality of links differs due to many variables, including the distance between the nodes, the power of transmissions between the nodes, and the overall traffic nearby [32]. NWB solutions can leverage this data in order to provide NWB robustness—for example, by building in more redundancy for nodes that are covered only by weak transmissions in a fixed redundancy NWB algorithm.

3.6.3 Classifying Existing Solutions

Full reliability for NWBs is an expensive operation. Using the 802.11 Medium Access Control (MAC) layer, the only mechanism for a reliable transmission is with unicast packets, which have an optional Request To Send/Clear To Send (RTS/CTS) handshake, followed by the data transmission, followed by an acknowledgment (ACK) packet. If the ACK is not received by the sending node, the process is retried, up to 7 times.

To address NWB robustness, two general approaches are used: (1) *MAC broadcast changes*: changes to the MAC layer are introduced to make NWB operations more reliable; and (2) *NWB algorithm changes*: NWB are done with a smaller set of nodes to cut down on overhead.

Tang and Gerla [33] proposed a full reliability MAC protocol to deliver NWB data. This protocol requires all nodes to maintain a neighbor list, a queue of previously sent broadcast packets, and a queue of previously received broadcast packets. Nodes maintain one-hop neighbor lists by periodically sending out 'HELLO' packets. When performing a NWB, a source node selects a neighbor and performs a RTS/CTS handshake. When sending the RTS, the packet also contains a piece of data indicating which NWB sequence number is about to be transmitted. The neighboring node replies with a CTS message, containing the last successfully received NWB id. The source node is then responsible for sending all NWB packets between the two values. This process continues in a round-robin fashion throughout the network.

This solution is an example of an explicit feedback algorithm. Their solution has a level of overhead, due to the extra packets that must be exchanged with each node, as well as a high amount of latency, if packets are dropped. The solutions of Hyper-flooding [34] and Hyper-gossiping [35] are more examples of explicit feedback algorithms. When nodes come into range, 'HELLO' packets are sent, indicating that they are a new node and that the NWB data should be resent. Lou and Wu's Double-Covered Broadcast (DCB) [26] solution is one that is an implicit feedback solution. Sending nodes keep track of which neighbors rebroadcast the NWB packet. If that retransmission is not heard, it is assumed that the first transmission failed. Pleisch et al. [36] propose the Mistrial protocol as an efficient flooding mechanism. This approach has nodes send out compensation packets periodically. The compensation packets contain versions of previous data packets, and nodes can deconstruct any missing packets. Since the extra compensation packets are sent without loss notification, it is a fixed redundancy algorithm.

We have also investigated a number of solutions from different points in the solution space. The Selective Additional Rebroadcast (SAR) protocol is a network layer implicit feedback solution. It requires no protocol transmissions to determine whether the packet was received successfully. This solution makes use of packet retransmissions, the contention of the channel, and past SAR successes in order to gauge whether or not to rebroadcast a packet. This work differs from Hyper-gossiping, because it is adapted to many NWB protocols, uses more advanced retransmission criteria, and does not incur the overhead of data caching. This work is different from the additional rebroadcast solution presented in this chapter for a number of reasons. The SAR approach uses more sophisticated mechanisms for gossiping; and it also builds off of other protocols besides flooding, such as AHBP and SBA. Those protocols have a lower overhead than flooding. Also, the SAR approach does not attempt to cache data for new nodes that come into the network at a later time. The SAR approach is different from DCB in that it attempts to strengthen the chance of a packet reception along the CDS tree, rather than increasing the size of the tree. This causes retransmission to only occur when needed.

Directed Broadcast (DB) is an explicit feedback solution. It requires nodes involved in the DB transmissions to send an ACK packet upon successful packet reception. This feedback is required from only a single neighbor. By using a unicast transmission between nodes, the reliability of unicast is leveraged, while keeping the efficiency of a broadcast. This approach differs from the approach used by Tang and Gerla, described earlier. Transmissions make use of the CDS tree built from neighbor-knowledge protocols, and no significant latency is introduced, because the feedback comes from receiving nodes immediately.

Shadowing-aware CDS is a solution that makes use of the state of the links in a MANET. These links are then used to create a CDS tree that can be used to gauge the chance of a packet reception. Current NWB robustness solutions do not make use of this information.

3.6.4 Summary

This section overviewed the potential solution space for increasing NWB robustness, along with the intuitive tradeoffs that result from these approaches. Existing solutions were classified in terms of this solution space. In the next section, we evaluate selected solutions from this space.

3.7 NWB ROBUSTNESS SOLUTIONS

This section presents an evaluation of two approaches for robustness control selected from different points in the solution space. Both mechanisms are independent of the redundancy control NWB aspect of the algorithm and can be used to control its robustness. Selective Additional Rebroadcast is an implicit feedback approach, while Directed Broadcast is an explicit feedback one.

3.7.1 Selective Additional Rebroadcast

In Selective Additional Rebroadcast (SAR), the base NWB algorithms are modified such that a node may transmit a broadcasted packet a second time [37] based on implicit feedback obtained from the behavior of the network observed by the node. For example, the node can use metrics such as the rebroadcast behavior, the neighboring topology, and the nearby network load to determine if a packet loss is likely and if the broadcast was critical. By doing this, coverage robustness is increased, because there is less of a chance that a critical rebroadcasts are lost.

Several criteria can be explored to decide when to rebroadcast a packet an additional time. SAR is designed to provide NWB robustness implicitly. No explicit feedback is used to predict whether an NWB transmission has been lost. This prediction can be based on recently observed behavior such as recent loss rates and the utilization of the channel. In addition, the prediction metrics can take into account the subsequent behavior of nearby nodes (which may differ based on whether the rebroadcast was received or not). We study the following retransmission policies:

1. *Probabilistic Solution*: A packet is rebroadcast with a fixed probability. In general, this solution is problematic because it does not adapt to the density or loss rates in the network. Therefore, it may result in large increases in the overhead when it is not needed (e.g., in dense areas and/or when interference is low). This solution is included because it provides a reference point that increases redundancy without using any prediction intelligence.

2. *Counter-Based Solution*: If the node does not hear n other nodes rebroadcast the packet within a certain amount of time, it will rebroadcast it again. This solution is attractive because it naturally adapts to the interference level and density of the network. In a dense/low-interference area, a number of rebroadcasts is likely to be received after a node rebroadcasts a packet. But in sparse/high-interference areas, this is not the case, and the algorithm rebroadcasts a packet to enhance reliability. First, the behavior of a fixed value of n across the network is studied, followed by the adaptation of the value to reflect the neighboring topology.

3. *Adaptive SAR*: Rather than having a fixed criteria that applies throughout the network, the appropriate threshold for rebroadcast can be adaptively set at the individual nodes. For example, a node can disable SAR if it determines that the SAR broadcasts have not helped the NWB recently. By doing this, the node can cut down on overhead when the broadcasts are not needed, and re-enable SAR later, in case new nodes have moved into the surrounding area. Alternatively, a node may assess its local neighborhood to compute the expected number of rebroadcasts and set its rebroadcast threshold to match those.

Furthermore, additional improvements to these metrics exist. Examples of these include:

1. *Medium Access Control (MAC) Utilization*: A node can take into account how busy the medium is, and it can use that to gauge how likely the loss of a NWB is. This is potentially unrepresentative, since the node is testing how busy the MAC layer is at the source, rather than at the destination.

2. *Density*: A node can use the number of neighbors it has to determine whether it is in a dense or sparse area of the network. Furthermore, it may be able to use this information to determine whether it is a leaf node, where a rebroadcast is wasteful, since no new nodes will be covered. Depending on the NWB algorithm being used, the density can be further defined as two different values: (1) *the physical* density: this consists of the number of neighboring nodes; and (2) *the NWB density*: this is the number of neighboring nodes that perform a NWB transmission. If not all neighboring nodes transmit, it is not useful to track all physical neighbors.

To explain how SAR works in more detail, in the following, we present pseudo-code for SAR integrated with the SBA algorithm. The reader may wish to refer to the SBA implementation in Section 3.4 to compare against the version with SAR.

```
Let T1 be the SBA Timer
Let T2 be the SAR timer
Let L be the list of nodes that sent a particular
  broadcast
Let Nx be the neighbor list for node X
Let U be the list of uncovered nodes

When a source sends a packet, it is always broadcast.

When a node A receives a broadcast from node S:

If ({Na} > {Ns}) Then
  If T1<P> Then  // If a timer exists for this packet
    T1<P>.add(P.source);  // Add the originator of this
      broadcast
  Else If T2<P> Then
    T2<P>.add(P.source);
  Else
    T1 := new Timer(P, {Na}-{Ns});
    T1.start();
  End If;
End If;


T1.expire() {
  If ({U} > {N{L}}) Then
    Rebroadcast(P);
    T2 := new Timer(P, {U} - {N{L}});
    T2.start();
  End If;
}

T2.expire() {
  If ({U} - {N{L}}) Then
  If Solution==Probabilistic
      R := Random(0, 1);
      If R > PROBABILITY_THRESHOLD Then
        Rebroadcast(P);
      End If;
    Else If Solution==Counter
      If size(N) < COUNTER_THRESHOLD Then
        Rebroadcast(P);
      End If;
    Else If Solution==Adaptive
```

```
            If size(N) == 0 Then
              W := W + 1;
            Else
              W := 0;
            End If;
            If W < INEFFECTIVE_THRESHOLD
              If size(N) != size(BRGs)
                Rebroadcast(P);
              End If;
            End If;
          End If Solution==MacBased
            metric := 1 - (size(N)/size(BRGs));
            metric := metric + macUtilization;
            If metric < MAC_THRESHOLD
              Rebroadcast(P);
            End If;
          End If;
        End If;
}
```

Using a purely probabilistic metric to use SAR, coverage is slightly increased, while overhead is increased by the percentage used as the probability value P (Figures 3.20 and 3.21). This approach is a *blind* method of controlling robustness. A node rebroadcasts solely based on a probability, and it takes no external observations into account. As a result, nodes perform the SAR broadcast when not necessary, and a critical rebroadcast may not occur.

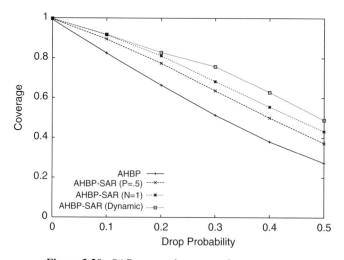

Figure 3.20. SAR approaches comparison, coverage.

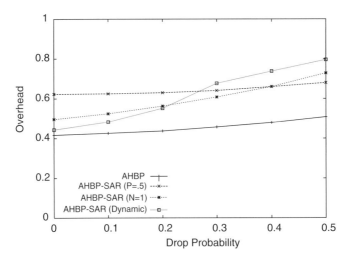

Figure 3.21. SAR approaches comparison, overhead.

Using the counter-based metric for SAR, coverage values are increased more than the probabilistic-based approaches; and the overhead fluctuates, based on network conditions. This approach does make use of observed network behavior; and the results are better, with higher coverages and lower overhead. A counter-based approach is *not blind*, but it also *not adaptive*. For example, nodes near the edge of the network (leaf nodes) may not have their counter value met, resulting in them performing the additional SAR broadcast. This broadcast is wasteful, since no new nodes will be covered.

The effect of the problem of leaf nodes broadcasting unnecessarily can be seen at a 0 loss rate in the counter-based solution. In this case, an oracle implementation of SAR would not increase the redundancy because no losses occur. However, in the simple counter implementation, there is an increase in overhead with SAR even at 0 loss rate.

To address this problem in an algorithm such as AHBP, a node can use its neighbor knowledge to see how strongly connected it is. A node that is a leaf node will not have a high neighbor count. This knowledge can be used to not cause additional SAR broadcasts if the node has heard a number of broadcasts equal to its CDS neighbor count. Furthermore, for static algorithms, such as AHBP, the neighbors that are to rebroadcast are known at the upstream sending node. The sending node can ensure that it has heard a transmission from each rebroadcasting CDS node, and it can cancel the additional SAR broadcast if the broadcast was propagated successfully.

An additional metric to be used is that of the MAC utilization. A node in a section of the network with more traffic will have a higher MAC utilization value, and the chances of a successful broadcast are smaller.

Figures 3.22 and 3.23 compare the base AHBP protocol, the counter-based AHBP-SAR protocol, AHBP-SAR with leaf node pruning, and AHBP-SAR with MAC utilization. It can be seen that the leaf node pruning results in coverage at the same level

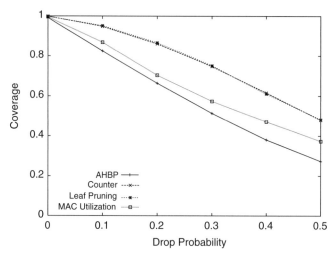

Figure 3.22. Coverage—Additional SAR Metrics.

as the counter-based solution, with a lower overhead. This is due to the suppression of broadcasts from leaf nodes that do not cover additional areas of the network. The MAC utilization version adapts to the increasing loss level in the network. Applied properly, this metric can also be valuable. Since each metric is valuable on its own, they are included in the criteria used for the adaptive SAR protocols.

The *adaptive* metrics cover cases where a counter-based metric does not perform well. They adjust according to more external network indicators, such as the observed network traffic load. Figures 3.20 and 3.21 show the coverage and overhead for the three approaches in a sparse 30-node environment, with modifications to the AHBP protocol. In these graphs, it can be seen that the probabilistic approach

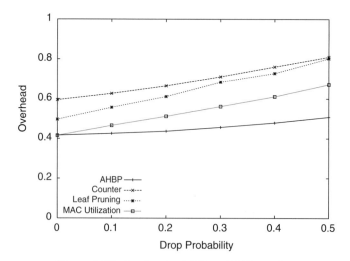

Figure 3.23. Overhead—Additional SAR metrics.

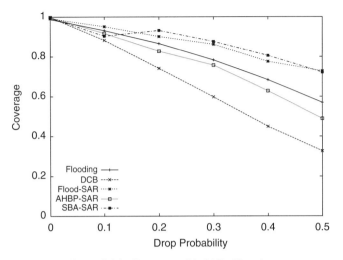

Figure 3.24. Coverage with SAR, 30 nodes.

has the poorest coverage, along with an overhead that does not adapt to the network conditions. The counter-based approach has coverage higher than the probabilistic approach, along with an overhead that tends to increase as the loss rate increases in the network. Lastly, the adaptive approach has the best overall coverage, along with an overhead that is increased as the loss rate increases in the network. This overhead starts out around the level of the base AHBP protocol, and it stays below the other approaches until the loss rate is sufficiently high. Even at very high loss rates, the overhead is below the level of flooding, which is constant at a value of 1.

Some sample results obtained with SAR can be seen in Figures 3.24 and 3.25. The overall coverage is increased greatly, with a small increase in overhead. Some

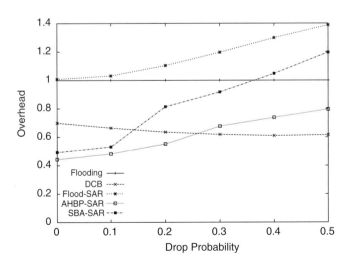

Figure 3.25. Overhead with SAR, 30 nodes.

broadcasts are critical, and the extra overhead incurred in order to ensure that the transmission is successful is worthwhile. For instance, in a network that is subdivided into two distinct sections which are only connected by a single link, it is extremely important that the NWB transmissions along the connecting link be successful. Furthermore, for NWB operations where the result is used many times, such as in routing, the extra overhead to discover an efficient route is rewarded every time a packet uses this superior route.

3.7.2 Directed Broadcast

Directed Broadcast (DB) introduces a new MAC level primitive that signals neighboring nodes that a packet being unicasted is really a public packet [38]. This allows nodes in a CDS tree to use reliable unicast transmissions to propagate NWB information. By keeping the unicast transmissions solely between the nodes in the CDS tree, the overall overhead is kept to a minimum. Furthermore, all nodes in the network are able to receive the packets.

As illustrated in Figure 3.26, DB provides substantial improvement in coverage, compared to the base protocols. AHBP with DB provides coverage near flooding, and SBA with DB provides coverage above flooding. The overhead of DB (Figure 3.27) is significantly lower than that of SAR. SBA with DB has overhead only slightly higher than the original protocol at lower loss levels. As losses increase, the overhead of SBA with DB is lower, due to the fact that less packets have been lost, and the protocol doesn't need to try to dynamically recover from those losses. The overhead of AHBP with DB remains well below flooding, except at very high loss levels, where it costs only slightly more than flooding.

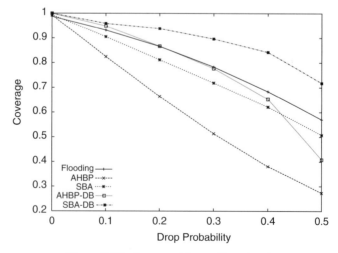

Figure 3.26. Coverage (directed broadcast).

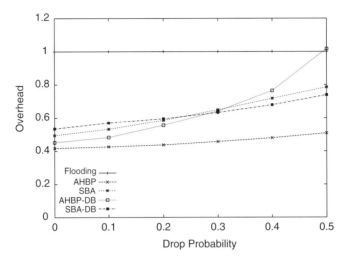

Figure 3.27. Overhead (directed broadcast).

3.8 CONCLUSIONS

Network-Wide Broadcasts (NWBs) are important operations in ad hoc networks that are used in several routing and group communication algorithms. Existing research has targeted efficient NWB to reduce the amount of redundancy inherent in flooding (the simplest NWB approach). As a result, the NWB becomes more susceptible to loss of coverage due to transmission losses that result from heavy interference or transmission errors. This problem arises because NWBs rely on an unreliable MAC level broadcast operation to reach multiple nodes with one transmission for a more efficient coverage of the nodes. In the presence of interference or transmission errors, this results in nodes not receiving the NWB.

We outline the NWB solution space and explain the properties of the different solutions in terms of overhead, resilience to mobility, and reliability. We demonstrate the problem using interference from CBR connections as well as using probabilistic dropping of packets (which allows us to control loss rate systematically). As the loss rate rises and as the density of the network goes down, the coverage achieved by an NWB operation drops. We showed that this is especially true for other NWB algorithms that also rely on MAC level broadcasts because they target reducing the redundancy present in floods, making them more susceptible to losses. Furthermore, we showed that approaches that statically determine the set of forwarding nodes perform much worse than ones that dynamically evaluate whether their rebroadcast is likely to be redundant. As a result, the recent DCB algorithm, while it improves coverage relative to other static approaches, remains substantially more vulnerable to losses than dynamic approaches.

3.9 EXERCISES

1. One criticism of robustness control is that the tradeoff is simply increasing the number of retransmissions, resulting in a higher chance of covering the network. Thus, all that redundancy control schemes do is increase this redundancy to increase coverage. Discuss whether or not this argument is correct (*Hint*: Is all increase in redundancy the same?)

2. Geocast is the problem of reaching nodes in a prespecified geographical area (assume that nodes know their own location).

    (a) Suggest an application that would need Geocast.
    (b) Suggest two solutions to implement Geocast. (*Hint*: Try to adapt solutions for NWB to Geocast.)
    (c) Comment on the robustness of the two solutions and suggest a different improvement for each to address robustness.

3. Repeat exercise 2 for Multicast (reaching nodes that are members of a multicast group). Only suggest one solution and one improvement.

4. Repeat exercise 2 for Anycast (reaching any one node among the members of an anycast group).

5. Of the NWB protocols covered, which are most resilient to mobility? Explain. What can be done to improve operation in the presence of mobility?

6. None of the protocols discussed considered the use of forward error correction—that is, encoding sufficient redundancy per packet or across packet to enable recovering from errors. Discuss the merits/demerits of such an approach. For what type of application would it work best?

REFERENCES

1. T. Clausen and P. Jacquet. Optimized link state routing protocol. Internet Draft, Internet Engineering Task Force, October 2003, http://www.ietf.org/internet-drafts/draft-ietf-manet-olsr-11.txt.
2. S. Das, C. Perkins, and E. Royer. Performance comaprison of two on-demand routing protocols for ad hoc networks. In *Proceedings of INFOCOM 2000*, March 2000.
3. M. Lewis, F. Templin, B. Bellur, and R. Ogier. Topology broadcast based on reverse-path forwarding (tbrpf). Internet Draft, Internet Engineering Task Force, June 2002, http://www.ietf.org/internet-drafts/draft-ietf-manet-tbrpf-06.txt.
4. E. Royer and C. Perkins. Multicast operation of the ad hoc on-demand distance vector routing protocol. In *Proceedings of the ACM International Conference on Mobile Computing and Networking (MobiCom'99)*, 1999, pp. 207–218.
5. S. Ni, Y. Tseng, Y. Chen, and J. Sheu. The broadcast storm problem in a mobile ad hoc network. In *Proceedings of ACM/IEEE International Conference of Mobile Computing and Networking (MOBICOM '99)*, September 1999.

6. K. Alzoubi, P.-J. Wan, and O. Frieder. Message-optimal connected dominating sets in mobile ad hoc networks. In *Proceedings of the 3rd ACM International Symposium on Mobile Ad Hoc Networking and Computing (MobiHoc 2002)*, 2002, pp. 157–164.
7. R. Gandhi, S. Parthasarathy, and A. Mishra. Minimizing broadcast latency and redundancy in ad hoc networks. In *Proceedings of the 4th ACM International Symposium on Mobile Ad Hoc Networking and Computing (MobiHoc 2003)*, 2003, pp. 222–232.
8. D. Aguayo, J. Bicket, S. Biswas, G. Judd, and R. Morris. Link-level measurements from an 802.11b mesh network. In *Proceedings of SIGCOMM 2004*, September 2004.
9. D. De Couto, D. Aguayo, B. Chambers, and R. Morris. Performance of multihop wireless networks: Shortest path is not enough. In *Proceedings of the First Workshop on Hot Topics in Networks (HotNetsI)*, October 2002.
10. T. Rappaport. *Wireless Communication: Principles and Practice*, 2nd edition, Prentice-Hall, Englewood Cliffs, NJ, 2001.
11. V. Bharghavan, A. Demers, S. Shenker, and L. Zhang. MACAW: A media access protocol for wireless lan's. In *Proceedings of SIGCOMM 1994*, 1994, pp. 212–225.
12. C. L. Fullmer and J. J. Garcia-Luna-Aceves. Solutions to hidden terminal problems in wireless networks. In *Proceedings of SIGCOMM 1997*, 1997, pp. 39–49.
13. B. Crow, I. Widjaja, J. Kim, and P. Sakai. IEEE 802.11 wireless local area networks. *IEEE Communications Magazine*, **September**: 116–126, 1997.
14. T. S. Rappaport, *Wireless Communications, Principles, and Practice*, Prentice-Hall, Englewood cliffs, NJ, 1996.
15. Y.-C. Tseng, S.-Y. Ni, and E.-Y. Shih. Adaptive approaches to relieving broadcast storms in a wireless multihop mobile ad hoc network. *IEEE Transactions on Computers*, **52**(5): 545–557, 2003.
16. D. Scott and A. Yasinsac. Dynamic probabilistic retransmission in ad hoc networks. In *Proceedings of the International Conference on Wireless Networks (ICWN '04)*, June 2004.
17. Z. Haas, J. Halpern, and L. Li. Gossip-based ad hoc routing. In *Proceedings of INFOCOM 2002*, June 2002.
18. B. N. Clark, C. J. Colbourn, and D. S. Johnson. Unit disk graphs. *Discrete Mathematics*, **86**:165–177, 1990.
19. B. Das, R. Sivakumar, and V. Bharghavan. Routing in ad-hoc networks using a spine. In *Proceedings of the International Conference on Computers and Communications Networks*, September 1997.
20. C.-R. Lin and M. Gerla. Adaptive clustering for mobile wireless networks. *IEEE Journal on Selected Areas in Communications*, **15**(7):pp. 1265–1275, 1997.
21. I. Stojmenovic, M. Seddigh, and J. Zunic. Dominating sets and neighbor elimination based broadcasting algorithms in wireless networks. In *Proceedings of the IEEE Hawaii International Conference on System Sciences (HICCS)*, January 2001.
22. J. Wu and H. Li. On calculating conncted dominating set for efficient routing in ad hoc wireless networks. In *Proceedings of the 3rd ACM International Workshop on Discrete Algorithms and Methods for Mobile Computing and Communication*, 1999, pp. 7–14.
23. H. Lim and C. Kim. Multicast tree construction and flooding in wireless adhoc networks. In *Proceedings of the ACM International Workshop on Modeling, Analysis and Simulation of Wireless and Mobile Systems (MSWIM)*, 2000, pp. 20–28.
24. W. Peng and X. Lu. On the reduction of broadcast redundancy in mobile ad hoc networks. In *Proceedings of MobiHoc 2000*, 2000.

25. W. Peng and X. Lu. AHBP: An efficient broadcast protocol for mobile ad hoc networks. *Journal of Science and Technology (Beijing, China)*, **6**:32–40, 2002.
26. W. Lou and J. Wu. Double-covered broadcast (dcb): A simple reliable broadcast algorithm in MANETS. In *Proceedings of INFOCOM 2004*, 2004, pp. 70–76.
27. UC Berkeley/LNBL/ISI. The ns-2 network simulator with the cmu mobility extensions 2002, http://www.isi.edu/nsnam/ns/.
28. Mesh Networks Inc. Mesh networks technology, 2004.
29. C. Chiang, *Wireless Network Multicasting*. Ph.D. thesis, University of California, Los Angeles, 1998.
30. J. Yoon, M. Liu, and B. Noble. Random waypoint considered harmful. In *Proceedings of the of INFOCOM 2003*, 2003. Available from http://www.eecs.umich.edu/ mingyan/pub/random-waypoint.pdf.
31. C. Bettstetter and C. Wagner. The spatial node distribution of the random waypoint mobility model. *Mobile Ad-Hoc Netzwerke, 1. Deutscher Workshop uber Mobile Ad-Hoc Netzwerke (WMAN 2002)*, 2002, pp. 41–58.
32. MIT, MIT Roofnet. 2002.
33. K. Tang and M. Gerla. MAC reliable broadcast in ad hoc networks. In *Proceedings of IEEE MILCOM 2001*, October 2001.
34. K. Viswanath and K. Obraczka. An Adaptive Approach to Group Communications in Multi-Hop Ad Hoc Networks. In *Proceedings of International Conference on Networking*, 2002.
35. M. Kabir. Region-Based Adaptation of Diffusion Protocols in MANETs. M.S. thesis, University of Stuttgart, Germany, November 2003.
36. S. Pleisch, M. Balakrishnan, K. Birman, and R. van Renesse. Mistral: Efficient flooding in mobile ad-hoc networks. In *MobiHoc '06: Proceedings of the Seventh ACM International Symposium on Mobile Ad Hoc Networking and Computing*, New York, ACM Press, New York, 2006, pp. 1–12.
37. P. Rogers and N. Abu-Ghazaleh. Towards Reliable Network Broadcast in Mobile Ad-Hoc Networks. Technical Report, SUNY—Binghamton, February 2004.
38. P. Rogers and N. Abu-Ghazaleh. Directed broadcast: A MAC level primitive for robust network broadcast. In *Proceedings of the IEEE International Conference on Wireless and Mobile Computing, Networking and Communications (WiMob 2005)*, August 2005.

CHAPTER 4

Encoding for Efficient Data Distribution in Multihop Ad Hoc Networks[1]

LUCIANA PELUSI, ANDREA PASSARELLA, and MARCO CONTI
IIT-CNR, Pisa, Italy

4.1 INTRODUCTION

The diffusion of pervasive, sensor-rich, network-interconnected devices embedded in the environment is rapidly changing the Internet. The nature of pervasive devices makes wireless networks the easiest solution for their interconnection. Furthermore, in a pervasive computing environment, the infrastructure-based wireless communication model is often not adequate: It takes time to set up the infrastructure network, while the costs associated with installing infrastructure can be quite high [1]. Therefore multihop ad hoc wireless technologies represent one of the most promising directions to further extend the current Internet and to diffuse traditional networking services even more widely [2, 3].

The heterogeneity of devices to be interconnected (from small sensors and actuators to multimedia PDAs), along with the large spectrum of communication requirements (from a few meters coverage and a few kilobits bandwidth to city-wide coverage and broadband communications), has produced a set of multihop ad hoc network technologies. On the one hand, we have sensor networks for communications among small-sized, low-cost, and low-energy-consuming devices (sensors) for which a high data rate is not necessary [4]. On the other hand, we have mesh networks that, by exploiting an infrastructure of wireless mesh routers, guarantee the exchange of multimedia information among devices located inside an urban area [5].

---

[1]This work was partially funded by the Information Society Technologies program of the European Commission under the HAGGLE and BIONETS FET-SAC projects.

Algorithms and Protocols for Wireless and Mobile Ad Hoc Networks, Edited by Azzedine Boukerche
Copyright © 2009 by John Wiley & Sons Inc.

The work in reference 6 provides a complete overview of the ad hoc networking techniques by presenting their principles and application scenarios and pointing out the open research issues. Among the ad hoc networking techniques, *opportunistic networks* [7] represent the most interesting scenario for the application of the encoding techniques analyzed in this chapter. Opportunistic networks represent the evolution of the multihop ad hoc network concept in which the end-to-end connectivity constraint is released. Indeed, typically, mobile ad hoc networks rely on the end-to-end principle; that is, a path must continuously exist between the sender and the receiver for successful communications. In reality, end-to-end connectivity is a strong requirement only for interactive services such as VoIP, gaming, and video streaming; many other applications can still correctly operate relaxing the end-to-end constraint—for example, data applications like messaging, e-mails, data sharing, and so on. In principle, they can operate even if a sender–receiver path never exists [7]. In opportunistic networks, a node stores the messages in its local memory until a suitable forwarding opportunity exists. In this way, packets are not discarded during network disconnections but are locally stored. The communication is still multihop with intermediate nodes acting as routers that forward the messages addressed to other nodes; but, in this case, forwarding is not *on-the-fly* since intermediate nodes store the messages when no forwarding opportunity exists (e.g., there are no other nodes in the transmission range, or neighboring nodes are considered not useful to reach the destination), and they exploit any contact opportunity with other mobile devices to forward information.

In all these scenarios, a big challenge is efficient data distribution. Applications like messaging, content distribution, and peer-to-peer applications are becoming more and more widespread in the today Internet. Thanks to their decentralized features, they are likely to play a major role in ad hoc environments such as opportunistic networks. One of their major requirements is networking techniques to efficiently convey data to possibly large sets of users. In the legacy wired Internet, researchers have proposed solutions based on IP multicast and, more successfully, on peer-to-peer overlay networks. While p2p solutions for wired networks are quite well established, designing similar systems to enable efficient data distribution over mesh, opportunistic and delay-tolerant networks is a big challenge still far to be satisfactorily addressed. In the field of sensor networks, content-distribution and p2p applications are not very likely. However, typical sensor network applications such as environmental monitoring require efficient data distribution techniques, too.

The very characteristics of wireless communications make legacy solutions for data distribution designed for the wired Internet not effective in ad hoc environments, even in static scenarios. Systems designed for ad hoc networks cannot assume that *bandwidth is for free*, as most legacy systems do. In addition, they have to efficiently cope with sudden changes of the wireless links characteristics. Dynamic topology reconfigurations, due either to user mobility or to energy management techniques that temporarily switch off some nodes, add further dimensions to the problem. These constraints are peculiar to wireless multihop ad hoc networks, and legacy systems are typically not able to cope with them. Data distribution systems, either targeted to p2p and content distribution paradigms or designed for sensor network applications, are thus today an exciting research area for multihop ad hoc networking environments. Encoding techniques are an important building block to address many issues that arise in these systems.

The main idea of encoding techniques applied to networking protocols is not to send plain data, but to combine (encode) blocks of data together in such a way that coded blocks can be interchangeable at receivers. In the simplest example, a source node willing to send k packets actually encodes the k packets into n encoded packets, with $n \gg k$. Encoding is performed so that a receiving node does not have to receive exactly the k original packets. Instead, *any* set of k packets out of the n encoded packets generated at the source is sufficient to decode the k original packets. Different receivers might get different sets of encoded packets and still be able to decode the same original information. These encoding algorithms, known as *erasure coding*, have been originally applied to implement efficient reliable multicast protocols for wired Internet [8]. Recently, Rizzo and Vicisano [8] have found interesting application scenarios in wireless ad hoc networks. For example, erasure code techniques have been used in reference 9 for the design of a reliable point-to-point protocol for data transmission over wireless sensor networks. Other applications of erasure code techniques over wireless sensor networks are reported in references 10 and 11. In reference 10 they have been exploited as a means to achieve minimum-energy data transmissions, while in reference 11 they have been used to support distributed data caching over the network and facilitate subsequent data retrieval. Erasure codes have also been used to support speech communications over well-connected mobile ad hoc networks [12] since they are able to reduce the total transfer delay experienced by data packets and make it adequate to real-time applications.

In the case of multihop ad hoc networks, erasure code techniques have been generalized by introducing several encoding phases of the same data. This generalization is known as *network coding*. With *network coding*, data packets are encoded both at source nodes and at intermediate hops in the path toward receivers. Assuming that the communication paradigm is one-to-many, the source node encodes locally generated data packets and sends them to a subset of current neighbors. Such intermediate nodes generate a new set of encoded data packets based on the received packets and further disseminate them toward the final destinations. Network coding has shown to be a very promising solution for data dissemination in multihop ad hoc networks [13]. Actually, it is able to provide very high reliability and exploit bandwidth very efficiently. It has been successfully applied to achieve optimal energy consumption in multicast [14], unicast [15], and broadcast [16] transmissions over mobile ad hoc networks. A lot of attention has also been devoted to possible applications of network coding over mesh networks [17, 18] as well as over opportunistic [19] and vehicular ad hoc networks [20].

In this chapter we provide a detailed overview of the research efforts on encoding techniques for multihop ad hoc networks. In Sections 4.2 and 4.3 we describe in detail the basic encoding techniques upon which network coding is based. Section 4.4 deals specifically with network coding, describing its operations and discussing its applications to multihop ad hoc networks. Finally, Section 4.5 draws conclusions and highlights open issues of this field.

4.2 CODING THEORY: MOTIVATION AND BASIC CONCEPTS

Reliable transmission of data over *noisy channels* has been a major concern to the communication engineering for a long time, and appropriate control systems as well as recovery mechanisms for corrupted data have been thoroughly studied. A real breakthrough in this field was achieved in 1948 when Claude Shannon [21] demonstrated that by efficiently *coding* messages at the sender before transmission and conversely decoding (possibly corrupted) messages that arrive at the receiver, it is possible to repair the effects of a noisy channel. The decoding system acts on the corrupted messages and strives to reconstruct the original messages. Encoding of data is equivalent to adding some *redundancy* to transmission, so the encoded data that are transmitted (i.e., the *code*) store the information that has originally been produced at the source more robustly. Indeed, redundancy is exploited at the receiver both to realize that the incoming code is corrupted and to correct it. In the first case the code is said to be *error detection code*, while in the latter it is called *error correcting code*.

The principle behind error detecting and correcting codes is also embedded in everyday language. Each language, in fact, is characterized by a vocabulary composed by a certain number of words. Words are sequences of letters (symbols) belonging to a particular alphabet. Although the total number of words included in a vocabulary is very high, it is far less than the total number of possible combinations of the alphabet letters. Thank to this property, if a misprint occurs in a long word, it can easily be recognized because the word (likely) changes into something that does not correspond to any other word of the vocabulary and rather resembles the correct word more than it resembles any other known word.

Adding redundancy to the information to transmit is equivalent to *mapping* the set of all the possible pieces of information that a source can generate, hereafter referred to as *words*, into a far bigger information set of so-called *code-words*. The mapping function is such that each word has a correspondent code-word associated to it, whereas a code-word may not correspond to any significant word produced at the source. The mapping process is referred to as *encoding process*. At the destination an inverse mapping is performed, named *decoding*, to reconstruct the original word transmitted by the source. Figure 4.1 shows the encoding, transmitting, and decoding processes for the data originated at a source.

The decoding algorithms adopted at the receiver can be of different types. A *complete* decoding algorithm decodes every possibly received code-word (even corrupted)

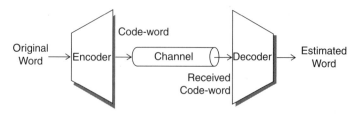

Figure 4.1. Encoding, transmitting, and decoding processes for the data originated at a source.

into a corresponding word. However, in some situations an *incomplete* decoding algorithm could be preferable, namely when a decoding error is very undesirable. When receiving new signals from the channel, a new code-word should be read. The receiver infers each single symbol (bit) of the code-word from the signals received from the channel, so for each received signal it has to decide whether it is much more similar to a 0-signal or to a 1-signal. Luckily, in most cases this decision is straightforward. Otherwise, it might be preferable to put a ? instead of guessing whether the symbol is 0 or 1. In the subsequent decoding stage the receiver has to reconstruct the original word from the received code-word possibly including some ?s. An incomplete decoding algorithm *corrects* only those code-words which contain few errors, whereas those with more errors cause decoding failures and give rise to so-called *erasures* in the sequence of the received words. The receiver can either ignore erasures or, if possible, ask for retransmission.

The success of an encoding/decoding system is strictly related to the characteristics of the channel over which transmissions occur and to the amount of redundancy which is introduced to transmission. Shannon formulated this relation in terms of *information rate* and *channel capacity*. He defined the *information rate* as the ratio between the number of significant words of a vocabulary and the total number of possible code-words that can be represented given a certain alphabet of symbols.[2] In addition, he defined the *channel capacity* as the total amount of information that a channel can transmit. Finally he demonstrated that in the case where the information rate is lower than the channel capacity, there exists an efficient system for encoding and decoding such that the probability of corruption due to transmission is lower than ε, for each arbitrarily small $\varepsilon > 0$. This means that it is possible to transmit any information with an arbitrarily small probability of error by using the right encoding technique.

The code system proposed by Shannon, known as *Shannon code*, relies on a *probabilistic* approach and uses a random generation for code-words to assign to the words produced at the source. This choice is motivated by the fact that it is quite unlikely that a random process generates code-words similar to one another. The *similarity* between code-words is to be intended in terms of the *Hamming distance*, which gives the total number of symbols that two code-words are differing in.[3] If for any couple of code-words generated by a coding system the code-words are sufficiently far from each other (differ in many symbols), then they are unlikely to be confused. Hence, whenever a code-word is sent over a channel and modified by noise, it is quite improbable that the corrupted code-word arriving at destination can be confused with another code-word. This would happen only if the noise reduces the Hamming distance between the two code-words so much that they could be considered the same.

---

[2] Given *code C*, the set of code-words corresponding to a certain vocabulary, and given *n* the number of bits over which code-words are represented, then the information rate (or just the rate) of the code *C* is defined as $R = \log_2 |C|/n$, where $|C|$ gives the total number of code-words of the set *C*. Please, note that code *C* is only a subset of all the elements that can be represented over *n* bits and that each element of it matches a word of the vocabulary. Hence the cardinality of code *C* is equal to the cardinality of the vocabulary.

[3] If $\underline{x} = (x_0, x_1, \ldots, x_{n-1})$ $\underline{y} = (y_0, y_1, \ldots, y_{n-1})$ and are two tuples, each one long *n* bits, then their Hamming distance is defined as follows:
$d(\underline{x}, \underline{y}) = |\{i | 0 \le i < n, x_i \ne y_i\}|$, where $|A|$ is the cardinality of the set "A · "

If the initial Hamming distance is quite long, then this should not happen and the reconstruction of the original source word should be easy and unique.

In the Shannon coding system a source word is a string of m symbols ($m > 0$) generated at the source site. These strings have to be encoded into strings of size n symbols (i.e., code-words), with n higher than m, and subsequently sent. Choosing n higher than m, words generated at the source are mapped over a bigger space with a higher dimension (so the Hamming distance between code-words is likely to be high). The choice of the correct lengths, m and n, of both words and code-words can be made once targeted a specific $\varepsilon > 0$. Then an appropriate *encoding/decoding system* can be selected such that the probability not to reconstruct the original word sent by the source is lower than ε.

Unfortunately, Shannon code cannot be implemented. Specifically, the decoding system cannot be implemented because it should evaluate the Hamming vicinity of the received corrupted code-word to any other code-word of the code and select the code-word with the minimum distance as the uncorrupted code-word (this search has a dramatic cost) and then apply the inverse mapping to reconstruct the original word.

Many codes have been conceived after the original Shannon code. They have such algebraic characteristics that not only can the decoding process be implemented but it can also be very fast. Anyway, the *delay* caused by both encoding and decoding computations as well as by the increase in the total transmission time due to the lengthening of the data to transmit has been the main limiting factor to the diffusion of encoding techniques and still remains their main drawback.

Error correcting codes have mainly been used, so far, in dedicated and critical communications systems (e.g., satellite systems) where they serve as a means to avoid expensive and long-lasting retransmissions. In these systems, encoding and decoding algorithms are typically implemented on dedicated hardware that allows overcoming the computational burden. Error correcting codes are also used in other special cases (e.g., in modems), over wireless or otherwise noisy links, in order to make the residual error rate comparable to that of dedicated, wired connections. Error detection codes have mainly been used, instead, in the context of general-purpose computer communications to discard the corrupted frames arriving from the communication channel actually, they are quite easier to implement than error correcting codes. Error detection codes [see, for example, the cyclic redundancy checksums (CRCs)] are managed at the lowest layers of the protocol stack (physical and data-link layers) that are typically HW-implemented in the network adaptor. While error correcting codes within streams of bits have seldom been implemented in these general-purpose systems, especially in SW, correction of errors can be implemented quite more efficiently if focusing on missing packets rather than on corrupted symbols (i.e., bits). In fact, SW-implemented codes have actually been successfully developed in the last few years. They are known as *erasure codes* because they deal with erasures—that is, missing packets in a stream. Erasure codes are implemented above the data-link layer where information is organized in *packets* rather than bit-streams (frames) and the channel noiseness is only perceived, after error processing and *detection* of the lower-layer protocols, through packet loss (i.e., erasures). Erasures originate from errors that cannot be corrected at the data-link layer (but those are not frequent with properly designed and working hardware), or, more frequently, from congestion in the network which causes otherwise valid packets to be dropped due to lack of buffer space. Erasures are easier to deal with than errors since the exact position of the missing packets is known. Recently, many efforts have been spent in the construction of *erasure codes*. They rely on the same concepts of the aforementioned error detecting and correcting codes but operate on packet-sized data objects rather than on bit-streams, and they can be implemented in software using general-purpose processors.

The motivation for recent deployment of high-layer erasure codes is slightly different from that of previous low-layer error codes. Erasure codes are attracting interest in the new emerging communication scenarios where existing protocols are becoming unsuitable, or at least inefficient, because, for example, they are based on frequent handshaking between the communicating peers or because they assume the existence of a continuous end-to-end connection between peers. Erasure codes have the potential to add to communication protocols (i) reliability of transmissions, (ii) scalability to the increasing number of hosts, (iii) adaptability to the topology changes of a network (e.g., due to mobility or even power-saving strategies), and (iv) robustness to node/link failures. These are very promising features, especially in scenarios like multicast transmissions, wireless networks (e.g., ad hoc, sensor, vehicular ad hoc, etc.), satellite networks, delay tolerant networks, and intermittently connected networks.

Software implementation for erasure codes is surely a challenging task due to the computational complexity. However, it has also been demonstrated that it can be done efficiently [8, 22, 23]. The rest of the chapter will focus on erasure codes and their more recent extension, network coding. In the following section some types of erasure code will be presented, from the oldest to the most recent ones, and some examples of applicative scenarios will be given for each of them, as well.

4.3 ERASURE CODES

Erasure codes are generally considered to be implementations of so-called *digital fountains*. Digital fountains are ideal, and unlimited data-streams are injected into the network by one or more source nodes. Nodes that are willing to receive information from those source nodes need to catch data from the stream until they have collected enough information to fulfill their needs. The way receivers catch data from the unlimited stream resembles the way people quench their thirst at a fountain. No matter which drops of water fall down on earth and which are drunk, the only important thing is that enough drops are caught. Source nodes generate digital fountains, sending out the data they are willing to send and adding redundancy to it via encoding. The redundancy added is practically unlimited because the generated data flow is infinite. Namely, suppose that k data packets are originated at a source node. After encoding, an unlimited stream of data packets is generated and injected into the network. The receiver node needs to receive *exactly* k data packets to be able to reconstruct the original source data. Moreover, *any* k data packets are suitable to the scope. Digital fountains have been introduced in reference 24 and so far have mainly been used in conjunction with multicast protocols to improve reliability of transmissions.

Multicast protocols suffer from the so-called feedback implosion that occurs when multiple receivers ask for retransmission of data to the server. Multiple retransmissions are due to the inevitably different loss patterns that affect different receiver nodes. Clearly, it is not affordable for sending nodes to manage multiple retransmissions of different data packets toward different receivers. If the sender generates a digital fountain instead, receiver nodes only have to receive an amount of data packets that corresponds to the exact number of original data packets. Indeed, *any* set of packets of size equal to the number of original data packets is appropriate. Hence, whenever receivers miss some packets, it is sufficient that they wait for the subsequent redundant packets to come. No retransmissions are needed toward different receivers, and no feedback channel is strictly necessary.

Content dissemination in environments with a vast number of receivers having heterogeneous characteristics can greatly benefit from this approach (see, for example, distribution of software, archived video, financial info, music, games, etc.). In these environments, receivers will wish to access data at times of their choosing, they all will have their own access speeds, and, in addition, their access times will probably overlap with one another. Besides addressing the retransmissions issue, in this case digital fountains allow receivers to join the transmitted stream at any point and without requiring any synchronization among them. Synchronization is also not needed between senders and receivers.

Other challenging target scenarios for digital fountains are satellite networks where channel characteristics (i.e., high latency and limited as well as intermittent capacity), make retransmission-based solutions totally unworkable. Finally, wireless networks are emerging as a very promising scenario for digital fountains because of the hardly unreliable communications and the common asymmetry between upstream and downstream channels. In addition, the recent deployment of opportunistic communication paradigms in the framework of the wireless networks characterized by intermittent connectivity appears to be quite an interesting application field, as well.

Although very promising, it should be noted that digital fountains are only an abstract concept. An unlimited stream of data is not feasible in practice, nor advisable, because it has the potential to flood and congest the network. During the last few years, multiple approximations of digital fountains have been proposed while trying to find a tradeoff between efficacy and computational costs (see references 25–30). All these approximations can be expressed in terms of (n, k)-*codes*, where k gives the number of source data blocks[4] which are encoded while n gives the number of data blocks that the encoder produces from the original k blocks and that are actually sent over the network. Figure 4.2 shows the encoding and decoding processes of an (n, k)-code. During transmission over the network, some encoded data blocks may get lost due to congestion or corruption and only a subset of encoded blocks of size k', $k \leq k' \leq n$, arrives at destination and enters the decoder. The decoder gives back the original k source data blocks.

---

[4]Here block is a generic piece of data. In the following sections, block will be used to refer to either a symbol or a packet or even a set of packets. Specific interpretations will be given when needed.

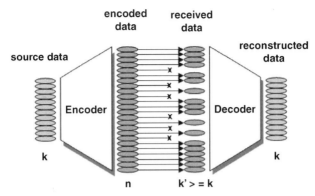

Figure 4.2. A graphical representation of the encoding and decoding processes of an (n, k)-code.

As will be shown in detail in the next sections, to be able to reconstruct the source data blocks, the decoder needs at least k blocks of encoded data. Some implementations actually need to get at least *more* than k blocks to decode the original data. In the optimal case, an (n, k)-code allows the receiver to recover from up to $n - k$ losses in a group of n encoded blocks.

In the remainder of this section, insights on four types of erasure codes will be given: Reed–Solomon codes, Tornado codes, Luby-Transform codes, and Raptor codes. The order followed in the presentation is chronological because these codes have been conceived in a temporal sequence, each one being an improvement of the previous one.

4.3.1 Reed–Solomon Codes

Reed–Solomon codes have been introduced in reference 25 as implementation of digital fountains. They are currently used to correct errors in many systems including wireless or mobile communications (e.g., cellular telephones, microwave links, etc.), satellite communications, digital television, and high-speed modems such as ADSL and xDSL. They are also used in storage devices (e.g., tape, Compact Disk, DVD, barcodes, etc.).

Assume that a source data block is a word and let a sequence of k words be represented by a vector, say $\mathbf{x}$, of k elements. Encoding is represented by an encoding function $f(\cdot)$ that is applied to $\mathbf{x}$ and produces an encoded vector of n code-words. When the encoding function is *linear*, the code is said to be linear too. Reed–Solomon codes are a special case of linear codes as will be better explained later, after the following brief introduction to general linear codes.

Linear Codes. Linear codes can be represented by a matrix G (encoding matrix), and encoding can be represented by a matrix–vector multiplication. Hence, a vector of code-words $\mathbf{y}$ corresponding to a vector of words $\mathbf{x}$ is simply given by the product $G\mathbf{x}$.

Remembering the definition of an (n, k)-code, the encoding process looks as follows:

$$y = \begin{pmatrix} y_0 \\ y_1 \\ \ldots \\ \ldots \\ \ldots \\ y_{n-1} \end{pmatrix} = G \cdot x = \begin{pmatrix} g_{0,0} & g_{0,1} & \cdots & \cdots & g_{0,k-1} \\ g_{1,0} & g_{1,1} & \cdots & \cdots & g_{1,k-1} \\ \ldots & \ldots & & & \ldots \\ \ldots & \ldots & & & \ldots \\ \ldots & \ldots & & & \ldots \\ g_{n-1,0} & g_{n-1,1} & \cdots & \cdots & g_{n-1,k-1} \end{pmatrix} \begin{pmatrix} x_0 \\ x_1 \\ \ldots \\ \ldots \\ \ldots \\ x_{k-1} \end{pmatrix}$$

where $\underline{x} = (x_0, x_1, \ldots, x_{k-1})^T$ is the vector of k source words, $y = (y_0, y_1, \ldots, y_{n-1})^T$ is the vector of n code-words, and $G_{(n \times k)}$ is the encoding matrix. The encoding matrix must have rank k.

Suppose the destination node receives at least k (otherwise decoding is not possible) out of the n code-words produced by the encoder at the source, and let y' be a vector of k elements picked up among the received code-words. The encoding process that has generated this vector follows.

$$y' = \begin{pmatrix} y_{i,0} \\ y_{j,1} \\ \ldots \\ \ldots \\ y_{l,k-1} \end{pmatrix} = G \cdot x = \begin{pmatrix} g_{i,0} & g_{i,1} & \cdots & \cdots & g_{i,k-1} \\ g_{j,0} & g_{j,1} & \cdots & \cdots & g_{j,k-1} \\ \ldots & \ldots & & & \ldots \\ \ldots & \ldots & & & \ldots \\ g_{l,0} & g_{l,1} & \cdots & \cdots & g_{l,k-1} \end{pmatrix} \begin{pmatrix} x_0 \\ x_1 \\ \ldots \\ \ldots \\ x_{k-1} \end{pmatrix}.$$

The encoding matrix $G'_{(k \times k)}$ is a $k \times k$ matrix obtained by extracting from $G_{(n \times k)}$ those rows that correspond to the elements of the vector y'. So, for example, if the jth code-word of the original code-word vector (i.e., y_j) is inserted as the second element in the vector y' (i.e., $y_{j,1}$), then the jth row of the matrix $G_{(n \times k)}$ is picked up and inserted as the second row in the matrix $G'_{(k \times k)}$.

Clearly, decoding means finding the solution to the linear equation $G' \cdot x = y'$, as follows.

$$x = G'^{-1} y'.$$

Note that the destination must be sure to identify the row in $G_{(n \times k)}$ corresponding to any received element of y. Note also that the set of rows corresponding to y' have to be linearly independent.

Encoding Process of RS-Codes. As has been introduced above, Reed–Solomon codes are a special case of linear codes. Source words are seen as coefficients of a polynomial of degree $k - 1$ whereas code-words are seen as values of the polynomial worked out at n different points that can be chosen arbitrarily.

Let the polynomial be as follows.

$$p(x) = a_0 + a_1 x^1 + a_2 x^2 + a_3 x^3 + \cdots + a_{k-1} x^{k-1}$$

where $a_0, a_1, \ldots, a_{k-1}$ are the k words generated at the source for transmission and $p(x)$ is a single code-word obtained by evaluating the polynomial at the point x. The encoding process for a Reed–Solomon (n, k)-code is thus as follows,

$$\begin{pmatrix} p(x_0) \\ p(x_1) \\ p(x_2) \\ \ldots \\ \ldots \\ \ldots \\ p(x_{n-1}) \end{pmatrix} = \begin{pmatrix} 1 & x_0 & x_0^2 & x_0^3 & \ldots & \ldots & x_0^{k-1} \\ 1 & x_1 & x_1^2 & x_1^3 & \ldots & \ldots & x_1^{k-1} \\ 1 & x_2 & x_2^2 & x_2^3 & \ldots & \ldots & x_2^{k-1} \\ \ldots & \ldots & \ldots & \ldots & \ldots & \ldots & \ldots \\ \ldots & \ldots & \ldots & \ldots & \ldots & \ldots & \ldots \\ \ldots & \ldots & \ldots & \ldots & \ldots & \ldots & \ldots \\ 1 & x_{n-1} & x_{n-1}^2 & x_{n-1}^3 & \ldots & \ldots & x_{n-1}^{k-1} \end{pmatrix} \begin{pmatrix} a_0 \\ a_1 \\ a_2 \\ \ldots \\ \ldots \\ \ldots \\ a_{k-1} \end{pmatrix}$$

where $x_0, x_1, \ldots, x_{n-1}$ are the n points selected for evaluation of the polynomial. They can be chosen arbitrarily, for example, for simplicity of encoding, or alternatively they can be *all* the possible integer values that can be represented over the number of bits available.

The encoding matrix of Reed–Solomon codes has a special form characterized by a geometric progression in each row. Such matrices are named *Vandermonde* matrices.

Decoding Process of RS-Codes. The decoding process consists in reconstructing all the polynomial coefficients $a_0, a_1, \ldots, a_{k-1}$ in a unique way. As is commonly known, for this to be possible, it is sufficient to know the value of the polynomial at exactly k points; that is, receiving k code-words is sufficient. Hence, assuming that the identity (e.g., the sequence number) of the code-words that have arrived to destination is known, the coefficients of the polynomial can be derived at the destination site by solving the following system of equations.

$$\begin{pmatrix} y_{i,0} \\ y_{j,1} \\ \ldots \\ \ldots \\ y_{l,k-1} \end{pmatrix} = \begin{pmatrix} 1 & x_{i,0} & x_{i,0}^2 & \ldots & x_{i,0}^{k-1} \\ 1 & x_{j,1} & x_{j,1}^2 & \ldots & x_{j,1}^{k-1} \\ \ldots & \ldots & \ldots & \ldots & \ldots \\ \ldots & \ldots & \ldots & \ldots & \ldots \\ 1 & x_{l,k-1} & x_{l,k-1}^2 & \ldots & x_{l,k-1}^{k-1} \end{pmatrix} \begin{pmatrix} a_0 \\ a_1 \\ \ldots \\ \ldots \\ a_{k-1} \end{pmatrix}$$

The matrix utilized in the decoding process is a submatrix of the encoding matrix and is obtained by selecting the k rows that correspond to the arrived code-words (in the example the ith, jth, $\ldots$, and the lth rows have been selected). The system admits a solution if the matrix is nonsingular. The determinant of the above $k \times k$

Vandermonde matrix corresponds to the following expression.

$$\det(V) = \prod_{0 \leq l < t < k} (\hat{x}_t - \hat{x}_l),$$

given $\hat{\underline{x}} = (\hat{x}_0, \hat{x}_1, ..., \hat{x}_{k-1})^T = (x_{i,0}, x_{j,1}, ..., x_{l,k-1})^T$, the second column of the Vandermonde matrix. Hence, the determinant is non-null if and only if all the $\hat{x}_i - s$ are non-null and different from each other.

It should finally be noted that, to allow decoding Reed–Solomon codes, the encoding matrix has to be *known* at both the source and destination sites.

Arithmetic of RS-Codes. Reed–Solomon codes are based on a mathematics area known as *Galois fields* or *finite fields*. A Galois field $GF(p)$, also called *prime field*, is a field that includes p elements, from 0 to $p-1$, with p prime. It is *closed* under additions and multiplications modulo p. Operating on a prime field is relatively simple, since field elements can be thought of as integers modulo p; and sums and products are just ordinary sums and products, with results computed modulo p. Galois fields can be used for the representations of the elements involved in RS codes—that is, the blocks of source data, the blocks of encoded data, and the elements of the encoding matrix. However, from the implementation standpoint, this is not feasible or at least not efficient for two main reasons. Firstly, the number of bits necessary to represent an element in the Galois field is $\lceil \log_2 p \rceil > \log_2 p$. This introduces inefficiency for encoding and also for computation of operations because the operand sizes may not match the word size of the processor. In addition, operations modulo p are expensive because they need a division to be applied. It is much more efficient to work on a so-called *extension field* $GF(p^r)$ that includes $q = p^r$ elements, with p prime and $r > 1$. Field elements can be represented as polynomials of degree $r-1$ with coefficients in $GF(p)$. Operations in extension fields with $p = 2$ can be extremely fast and simple to implement. Modulo operations are not needed. Efficient implementations can be realized exploiting only XORs and table lookups. A thorough dissertation on the arithmetic of extension field can be found, for example, in references 31 and 32.

Systematic Codes. When the code-words include a verbatim copy of the source words, the code is said to be *systematic*. This corresponds to including the identity matrix I_k in the encoding matrix. The advantage of a systematic code is that it simplifies reconstruction of the source words in case very few losses are expected. In Figure 4.3 for example, two out of the k code-words arrived at destination are actual original words. Hence, the system of equations that must be solved to reconstruct the original words includes $k-2$ equations instead of k.

By using $x_i \in GF(q = p^r)$, it is possible to construct encoding matrices with up to $q-1$ rows corresponding to the total number of non-null x_i's of the field. The number of columns of the encoding matrix is instead k corresponding, as usual, to the total number of original source words, and the relation $q-1 \geq k$ must hold true. As a result, the maximum size of the encoding matrix is $(q-1) \times (q-1)$.

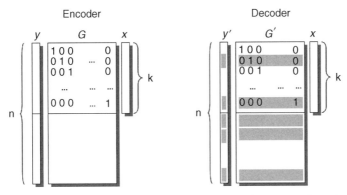

Figure 4.3. The encoding/decoding process in matrix form, for systematic code (the top k rows of G constitute the identity matrix I_k). y' and G' correspond to the grey areas of the vector and matrix on the right.

The encoding matrix can further be extended with the identity matrix I_k to generate systematic codes that help simplify the decoding process. Hence, the final maximum size of the encoding matrix is as follows.

$$((q-1) + k_{\max}) \cdot k_{\max} = ((q-1) + (q-1)) \cdot (q-1) = 2(q-1) \cdot (q-1)$$

where $k_{\max}$ is the maximum number of source words that can be encoded over $GF(q)$.

Applications. A major hurdle in implementing a digital fountain through standard Reed–Solomon codes (RS codes) is the unacceptably high computational time caused by both encoding and decoding. It is of order $O(n^2)$, where n corresponds to the number of generated code-words. The large decoding time for RS codes arises from the *dense* system of linear equations which is generated. However, efficient implementations of RS codes are actually feasible as has been shown in references 22 and 23. By means of thorough optimisations, the developed code performs encoding and decoding on common microprocessors at speeds up to several MB/s. The code has been tested on a wide range of systems, from high-end workstations to old machines and small portable systems (see, e.g., DECstation 2100, SUN IPX, Sun Ultra1, PC/FreeBSD) with CPU speeds ranging from 8 up to 255 MHz. This variety is motivated by the fact that erasure codes can be used in very different contexts and with very different speed requirements. Results effectively demonstrate the feasibility of *software FEC (forward error correction)* and advocate its use in practical, real applications. The same code has finally been utilized to implement a reliable multicast data distribution protocol [8]. Namely, the use case referred is a large file transfer from a server to multiple receivers. The file is already encoded (offline) and stored at the server ready for transmission. Receivers can join the multicast session whenever they need, generally at different time instants. The server needs a counter of the maximum number of encoded packets that have to be sent out to each receiver. This maximum number is

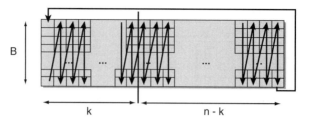

Figure 4.4. The arrangement of data and the transmission order at the server.

the sum of the number of packets containing original data and the number of packets containing redundant data. Each time a new receiver joins the session, the server simply reinitializes the counter to the maximum value. Then, it continues transmitting the file from exactly the same point it had arrived when the last receiver joined the session and in the same order already begun. The counter is decremented after sending a packet. Once arrived to the end of the file, the server loops through the file until all the requests are fulfilled—that is, until the counter goes to zero, meaning that all the receivers have received the appropriate number of packets. Since with RS codes reconstruction of source data is possible after having successfully received any k-set of distinct encoded packets (k being the size in packets of the original data set), each receiver leaves the session when it has received enough packets to decode the entire file. When a new receiver joins the session, transmission of data does not need to start again from the beginning of the file. The new receiver receives the same packets the other receivers are receiving, but it will stop receiving later than the others.

Figure 4.4 shows the scheme followed by the server to transmit the file content. The source file is subdivided into B blocks of data (on the vertical axis of the figure). The size of each block is such that it can be sent over k packets; thus it can be said that each block includes k packets. Each block is encoded separately, so redundancy is added to each single block independently from the others. Hence, assuming a systematic code, each encoded block includes k original packets plus $n-k$ redundant packets (on the horizontal axis of the figure). Finally, the server sends out packets by repeatedly looping through all the blocks. Namely, packets scheduled for transmission are picked from consecutive blocks, rather than sequentially from the same block. Such *interleaving* protects single blocks from the effects of a burst of losses, since it is preferable that losses are spread among all the blocks rather than concentrated in a unique block. Encoding is applied to single blocks of data rather than to the entire file to reduce the computational burden of both encoding and decoding. As has been mentioned above, in fact, the encoding and decoding costs are quadratic in the size of the encoded block ($O(n^2)$). Therefore, given that $n > k$, if k were the size of the entire file, the computational cost would be very high, whereas, by encoding over smaller chunks of file at a time, it is much more efficient (even though encoding must be repeated more times, once for each block).

The transmission scheme of Figure 4.4 is known as *data carousel* (see also reference 24). It limits both encoding and decoding times, allows multiple receivers to be served at the same time, and eliminates the need for several retransmissions towards

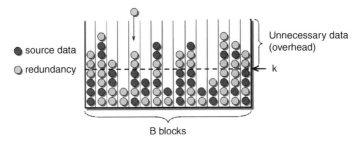

Figure 4.5. Waiting for the last blocks to fill.

different receivers. However, it has a drawback in the *receipt overhead*. In fact, since transmission loops through different blocks, after having completely received and decoded a certain number of blocks, a receiver can potentially receive and discard packets belonging to the blocks already decoded before receiving the missing packets necessary to decode the last blocks. The receipt overhead is measured in terms of unnecessary received packets (see Figure 4.5).

Starting from the achievements of references 8, 22 and 23, an implementation of RS codes has been provided also for wireless sensor networks. Kim et al. [9] propose a point-to-point data transmission protocol for WSNs that *combines* multiple strategies, namely, (i) erasure codes, (ii) retransmission of lost packets, and (iii) a route-fix technique. RS codes are used to add redundancy to transmission such that the receiver is able to reconstruct original data despite losses or congestion (up to a certain point). Nevertheless, as underlined in reference 9, retransmissions cannot be eliminated completely. Rather, it is necessary to provide integration between encoding and retransmission of data and to fine-tune this integration for better efficiency. Retransmissions are managed hop-by-hop such that the point of retransmission is moved progressively forward toward the destination node. Experimental results show that these two strategies together are very efficient and result in good reliability. Nevertheless, it has also been noted that both link-level retransmissions and erasure codes have poor performance in case of route failures. When a route is no longer available, consecutive losses occur and retransmissions are of no use because the route toward the destination does not exist any more. Neither erasure codes are helpful in this case because receivers cannot receive the minimum amount of encoded messages that is needed to reconstruct the original messages. A viable solution consists in finding an alternative route as soon as possible. Kim et al. [9] have also integrated route fixing into the framework with encoding and retransmission of data and have used the Beacon Vector Routing (BVR) protocol [33] (a particular case of geographic routing for WSNs) at the routing layer to provide the special support needed. Results show improvements in both reliability of transmissions and RTTs experienced by messages.

4.3.2 Tornado Codes

Tornado codes were introduced in 1997 to be fast, loss resilient codes, suitable to error-prone transmission channels. A preliminary version of these codes was presented in reference 26, and further improvements were later added in reference 27. The overall structure of Tornado codes is related to the low-density parity-check codes introduced by Gallagar [34].

Tornado codes cannot overtake the information-recovery capacity of RS codes, which is already optimal, but focus on faster encoding and decoding times. Namely, given the length of the output block of the encoder corresponding to an input block of length k ($k < n$), encoding and decoding times of Tornado codes are of order $O(n \ln(1/\varepsilon))$ for some $\varepsilon > 0$ (instead of $O(n^2)$ as in RS-codes). The price paid for much faster encoding and decoding times, with respect to RS codes, is that arrival of k packets is no longer sufficient for the receiver to reconstruct the source data.

Encoding Process. Suppose the source data to be encoded is a vector of k symbols in the finite field $GF(q)$ and let a symbol be a *bit*. The encoding process to yield the correspondent Tornado code C relies on a bipartite graph B. Let the overall code be referred to as $C(B)$. It consists of a *systematic code* and thus includes the k bits of the original source message, named *message bits*, as well as $\beta \cdot k$ redundant bits ($0 < \beta < 1$) named *check bits*. The bipartite graph consists of k left nodes, $\beta \cdot k$ right nodes, and some edges to connect left nodes to right nodes. Left nodes correspond to message bits and right nodes correspond to check bits. Check bits are obtained by XOR-ing over the source message bits according to the bipartite graph B. Namely, a check bit is worked out by XOR-ing over all the message bits it is connected to in the graph B (see Figure 4.6a). The bipartite graph is sparse, random, and irregular. Its construction is actually a bit complex, and it is principally oriented to the reduction of decoding costs (see below).

By making things a bit more complex than has been described so far, a Tornado code is generally obtained with a multi-bipartite graph—that is, a sequence (cascade) of successive bipartite graphs. In the first stage, a set of $\beta \cdot k$ redundant bits

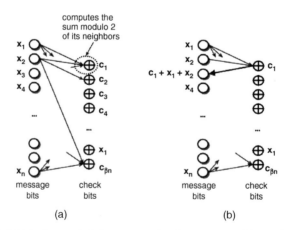

Figure 4.6. (a) A bipartite graph defines a mapping from message bits to check bits. (b) Bits x_1, x_2, and c are used to solve for x_3.

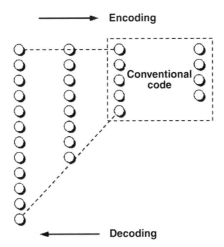

Figure 4.7. The code levels.

is generated from k message bits with a first bipartite graph, as explained above. Then, a second graph is added starting from the check bits of the previous graph. Thus the new graph has βk left bits and $\beta^2 k$ right bits. Then, a further graph can be added with $\beta^2 k$ left bits and $\beta^3 k$ right bits, and so on. More formally, it should be said that a set of codes $C(B_0), C(B_1), ..., C(B_m)$ is constructed by means of a set of bipartite graphs $B_0, B_1, ..., B_m$, where a generic graph B_i has $\beta^i k$ left nodes and $\beta^{i+1} k$ right nodes. The value m is chosen such that $\beta^{m+1} k$ is roughly $\sqrt{k}$. The check bits of the last graph are finally encoded via another erasure code, say C, which can be, for example, a Reed–Solomon code as proposed in reference 27. In Figure 4.7 the last level erasure code is a systematic code with left nodes as original data and right nodes as redundant data. The final code includes the original k message bits, the check bits produced by the bipartite graphs at the m different levels, and the last-level RS code. The entire code can be referred to as $C (B_0, B_1, B_2, ..., B_m, C)$. Given k the number of original message bits, the total number of check bits is as follows.

$$\sum_{i=1}^{m+1} \beta^i k + \beta^{m+2} k/(1-\beta) = k\beta(1-\beta).$$

Hence, the code $C (B_0, B_1, B_2, ..., B_m, C)$ has globally k message bits and $k\beta(1-\beta)$ check bits.

Decoding Process. Decoding relies on the assumption that the receiver knows the position of each received symbol within the graph and is performed going backward throughout the multi-bipartite graph. The RS code C is decoded first. If the decoding process tackles to reconstruct all the original $\beta^{m+1} k$ bits of the RS code, then the last level of check bits is complete whereas the next-to-last level as well as all the previous ones still need reconstruction. The decoding process proceeds one level at a time and requires finding each time a check bit on the right side such that only one adjacent message bit (among those that have generated it) is missing. The missing message bit can thus be obtained by XOR-ing among the check bit and its known message bits. As an example (see Figure 4.6b), suppose that c is a check bit whereas x_1, x_2, and x_3 are the message bits it depends on, i.e., $c = x_1 \oplus x_2 \oplus x_3$. Suppose also that, after transmission, c, x_1, and x_2 are known whereas x_3 is unknown. Then, the following holds true.

$$x_1 \oplus x_2 \oplus c = (x_1 \oplus x_2) \oplus (x_1 \oplus x_2) \oplus x_3 = x_3.$$

The missing message bits are yielded one at a time by exploiting the graph relations in which they are involved together with other known message bits, to produce other known check bits. Once decoded a single level, the previous one is then decoded. By repeating the same procedure for each bipartite graph, the original k-bit message can finally be reconstructed, as well.

Properties of Tornado Codes. This subsection provides some insights into interesting features of Tornado codes related to constraints and requirements of the encoding process, and it also discusses some advantages and drawbacks.

Property 1. The first property is about the feasibility of Tornado codes. This is strictly connected to the particular characteristics of the multi-bipartite graph in that successful decoding is only possible if an appropriate *degree distribution* is fulfilled during construction of the graph. The degree of a node belonging to the kth graph level gives the number of nodes of the $(k-1)$th level that this node is generated by. Luby et al. [27] make use of linear programming techniques to select the degrees to assign to the nodes of the same level and specifically define appropriate *sequences* of node degrees.

Property 2. The shape of the multi-bipartite graph also impacts the performance of Tornado codes. Both encoding and decoding times strictly depend on the number of edges which are present in the multi-bipartite graph. Clearly, the higher the number of edges, the higher the computational burden of the code. The aforementioned encoding approach based on linear programming converges in *linear* time and also allows linear time decoding, as has been formally demonstrated by Luby et al. [27]. Limiting encoding and decoding complexity to linear time is also the reason why the last bipartite graph in the cascading sequence of bipartite graphs should have $\sqrt{k}$ right nodes (where k is the number of nodes associated with the original message), and the total number of graphs in the sequence should consequently be $O(\log(k))$. $\sqrt{k}$ nodes are the ingress nodes of the last stage of encoding; and, given the quadratic complexity of the RS code that is used, by reducing the input size of the encoder, also the complexity is limited. The overall time complexity is about linear.

Property 3. *Code efficiency* of Tornado codes is worse than code efficiency of RS codes. This is because Tornado codes need slightly more than k encoded bits to arrive to destination for successful decoding of the original k message bits. Namely, 1.063 k encoded bits, at least, must be collected at destination. However, better results have been obtained with an improved version of Tornado codes. This new version uses a three-level graph and does not perform RS coding at the last level of the graph but rather repeats the same type of random bipartite graph as the previous two levels. Results show that 1.033 k encoded bits are needed with this version of Tornado codes to reconstruct the original k message bits.

Property 4. Some of the limits of Tornado codes are related to the *stretch factor*, which is defined as the ratio n/k. Tornado codes can admit only a small stretch factor so as to limit the computational burden. For example, a typical value is 4. However, a limited stretch factor limits in practice the applicability of Tornado codes. Another drawback related to the stretch factor is that, when using RS code or another kind of erasure code at the last stage, the stretch factor must be *predetermined* before encoding takes place. Hence, researchers need to estimate, in advance, the loss probability of the channel to determine the amount of redundancy to put into the data. For most Internet applications, the loss probability varies widely over time. Thus, in practice, the loss rate has to be overestimated, making the codes slower to decode and far less efficient in transmitting information. If the loss probability is underestimated instead, the code fails.

4.3.3 LT Codes

With Luby-Transform codes (LT codes for short) it is no longer necessary to tune the encoding process over the loss characteristics of the transmission channel. LT codes do not rely on a fixed rate to add redundancy to the data to send out; rather, they are said to be *rateless* codes. They were first devised in 1998 and can definitely be considered the first full realization of the concept of digital fountains [28]. Let a symbol be an l-bit string. Encoding is performed over k input symbols and has the potential to generate a limitless number of encoded symbols. Each encoded symbol is produced, on average, by $O(\ln(k/\delta))$ symbol operations, being $0 < \delta \leq 1$. The original k input symbols can be reconstructed with probability $(1 - \delta)$ from any set of $k + O(\sqrt{k}\ln^2(k/\delta))$ encoded symbols out of the total number of encoded symbols produced. The total number of operations needed to recover the original k symbols is about $O(k \ln(k/\delta))$. Hence, encoding and decoding times depend on the ingress data length independent of how many encoded symbols are generated and sent by the decoder. The same relations hold for the encoder and decoder memory usage. Therefore, given the reduced overall costs of LT codes with respect to the other aforementioned erasure codes, encoding can be performed over bigger chunks of data than those used in RS codes and Tornado codes, or even over the entire file at once. Luby-Transform codes are similar in construction and properties to Tornado codes because they also rely on a graph and the encoded symbols are obtained by XOR-ing over other initial (or previously encoded) symbols (see Figure 4.8). However, the graph construction for LT codes is quite different from

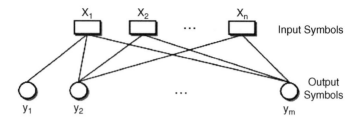

Figure 4.8. Bipartite graph illustrating input symbols and output symbols.

Tornado codes, and the resulting graph has logarithmic density. Moreover, LT codes are not systematic codes.

Encoding Process. Let the input file to be encoded have length N. The file is partitioned into $k = N/l$ input symbols; that is, each input symbol has length l. Let an input symbol be referred to as a packet. Similarly, each output symbol is referred to as an encoded packet. The encoding operation defines a bipartite graph connecting output symbols to input symbols. For each output symbol (encoded packet), encoding is performed by (i) randomly choosing the degree d from a given degree distribution, (ii) choosing, uniformly at random again, d distinct input symbols as neighbors[5] of the encoded symbol, and (iii) XOR-ing over the selected d input symbols taken as neighbors. The result of the bit-by-bit XOR among the selected symbols is the value of the encoded symbol. The degree distribution is computed easily independently of the data length (see below). To work properly, the decoder should be able to reconstruct the bipartite graph without errors; that is it needs to know which d input symbols have been used to generate any given output symbol, but not their values.

Decoding Process. The decoder needs to know the degree and set of neighbors of each encoded symbol. Hence, for example, the degree and list of neighbor indices may be given explicitly to the decoder for each encoded symbol, or, as another example, a key may be associated with each encoded symbol and then both the encoder and the decoder apply the same function to the key to produce the degree and set of neighbors of the encoded symbol. The encoder may randomly choose each key it uses to generate an encoded symbol, and keys may be sent to the decoder along with the encoded symbols.

The decoder receives K output symbols and the bipartite graph, from which it tries to recover the input symbols. Typically, K is slightly larger than k. The decoding process can be described as follows (see the Belief Propagation algorithm [35]).

---

[5]The neighbors of an encoded symbol are those symbols of the previous layer of the bipartite graph to which the encoded symbol is connected by edges. An encoded symbol is obtained by XOR-ing over all its neighbors of the previous graph layer.

1. Find an output symbol y_i that is connected to only one input symbol x_j. If there is no such output symbol, the decoding algorithm halts at this point and fails to recover all the input symbols.
2. Set $x_j = y_i$.
3. Add (i.e., perform a bit-by-bit XOR) x_j to all the output symbols that are connected to x_j.
4. Remove all the edges connected to the input symbol x_j.
5. Repeat until all $\{x_j\}$ are determined, or otherwise no more output symbols can be found that have exactly one neighbor. In the first case, decoding ends successfully; in the latter, it fails.

LT codes require a simple decoder, but it turns out that the *degree distribution* is a critical part of the design. The decoding process as described above will not even start if there is no output symbol of degree 1. This means that a good degree distribution is required for good decoding performances. Indeed, whether or not the decoding algorithm is successful depends solely on the degree distribution. The degree distribution used to produce LT codes is the *Soliton Distribution* or the more efficient *Robust Soliton Distribution*. Further details on the degree distribution analysis and the Soliton and Robust Soliton distributions can be found in Luby [28].

Applications. In 1998 a new company (Digital Fountain, Inc.) was launched by the inventor of LT codes (Michael Luby) [36]. The company aimed at exploiting the digital fountain's technology as a tool for downloading popular software packages, such as Adobe Acrobat; and, indeed, Adobe was an early investor, as were Cisco Systems and Sony Corporation. Obviously, Digital Fountain Inc. took a patent on LT codes, and this has been limiting utilization of LT codes in the research community as well as in other research and development centres. Digital Fountain's researchers have been working on a new generation of codes, and the group is also addressing other related research issues, such as application of the technology to video-on-demand and streaming media. Recently, the group has been focusing on congestion controls to ensure that the Digital Fountain packets behave in ways that do not cause network flow problems. Indeed, security issues are being addressed in collaboration with some universities to conceive, for example, a packet authentication system for verifying that data have come from a particular source. Launched to address the problems of streaming media and multicast, Digital Fountain—like many new technology companies—has found an unexpected niche. Without the requirement for TCP acknowledgments, data encoded with Luby transform codes travels faster than ordinary packets; this advantage becomes significant over long distances. Thus, the technology has proved particularly useful to companies that frequently transmit extremely large data files over long distances, such as movie studios and oil exploration companies.

Inspired to LT codes, the work in Dimakis et al. [11] proposes using digital fountains in wireless sensor networks. The sensor network is viewed as a big DataBase with k independent data-generating nodes and n data-storage nodes. Each storage node is assumed to have a limited storage capacity of only one single data packet. Data packets are small and generated independently from each other. It is indeed assumed that no correlation exists between them. The k source data packets are encoded and distributed after addition of redundancy to the n storage nodes. A collector/sink node wishing to retrieve information from the sensor network needs only to query any set of storage nodes of size slightly larger than k to reconstruct the entire data sensed at the source nodes. The solution proposed is inspired by the family of protocols named Sensor Protocols for Information via Negotiation (SPIN). These protocols assume that all the nodes have enough storage space to store all the information gathered in the network. Hence the information gathered by the sensors is disseminated throughout the entire network, and a user can query any node to get the required information immediately by only communicating with one node. The solution in Dimakis et al. [11] is different in that each sensor node has not enough memory to store all the data generated in the network and thus collects a combination of the total information gathered in the network. Therefore, one single interrogation is not sufficient, but slightly more than k are required.

4.3.4 Raptor Codes

Raptor codes are a recent extension of LT codes (2004) being conceived with the aim to improve the decoding probability of LT codes [29]. The decoding graph of LT codes, in fact, needs to be on the order of $k \log(k)$ in the number of edges (where k is the total number of input symbols) to make sure that all the input symbols can be recovered with high probability. Raptor codes focus on relaxing this condition such that the decoding process requires a smaller number of edges to be available. Raptor codes introduce a first level of encoding where the input symbols are precoded and a new set of symbols comprising some redundancy is generated. The new set of symbols is then given as input to a second level of encoding that performs LT coding (see Figure 4.9).

Precoding consists of a traditional erasure correcting code, say C, with a fixed stretch factor. The choice of code C depends on the specific application in mind. One possible choice is to use a Tornado code, but other choices are also possible such as

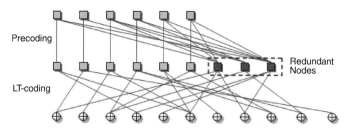

Figure 4.9. Raptor codes: The input symbols are appended by redundant symbols (dark squares) in the case of a systematic precode. An appropriate LT code is used to generate output symbols from the precoded input symbols.

an LT code with the appropriate number of output symbols. The second-level LT code needs appropriate tuning such that it is capable to recover all the input symbols (i.e., symbols given as input to pre-coding) even in the face of a fixed fraction of erasures.

Encoding Process. Let C be a linear code producing n output symbols from k input symbols, and let D be a degree distribution. A *Raptor code* is thus an LT code with distribution D on n symbols which are the coordinates of the code words in C. Code C is called *precode* of the Raptor code. The input symbols of a Raptor code are thus the k symbols used to construct the code word in C. The code word consists of n *intermediate symbols*. Then, the output symbols are the symbols generated by the LT code from the n intermediate symbols using the degree distribution D. Typically, C is a *systematic* code, though this is not strictly necessary. The encoding cost of a Raptor code takes into account the encoding cost of both precoding and LT coding. It thus includes the number of arithmetic operations sufficient for generating a code word in C from the k input symbols. The encoding cost is generally a per-symbol cost. Hence, it is worked out dividing the total cost to generate n output symbols by k.

Decoding Process. A reliable decoding algorithm of length m for a raptor code is an algorithm that can recover the initial k input symbols from any set of m output symbols and errs with probability at most $1/k^c$ for some constant $c > 1$. The decoding process will include in the first stage an LT decoding process that will recover a fraction of the n intermediate symbols that have been given as input to the LT encoder in the encoding phase. Then, the second stage will consist of a classical erasure decoding process that should be able to recover the k original input symbols of the entire Raptor encoder. The decoding cost of a Raptor code is the expected number of arithmetic operations sufficient to recover the k input symbols, divided by k. However, the overall efficiency of Raptor codes should not be measured only in terms of the number of operations per single input symbol but also in terms of *space* and *overhead*. Space refers to the memory consumption for storage of intermediate symbols. Overhead instead is a function of the decoding algorithm used and is defined as the number of output symbols that the decoder needs to collect in order to recover the input symbols with high probability.

Different types of Raptor codes can be obtained depending on the type of erasure code C which is used for precoding and the particular degree distribution used in the second level encoding. Hence, a complex precoding[6] can be combined with a simple degree distribution of the LT encoder, or vice versa a simple precoding algorithm can be combined with a complex degree distribution. In between these two extremes, other solutions are also possible.

Optimized Raptor codes applied to k original input symbols have the potential to generate an infinite stream of output symbols. Any subset of symbols of size $k(1 + \varepsilon)$, for some $\varepsilon > 0$, is then sufficient to recover the original k input symbols with high

---
[6]Sometimes complex erasure codes can be used when intermediate symbols (code words in C) can be calculated offline via preprocessing.

probability. The number of operations needed to produce each output symbol is of order $O(log(1/\varepsilon))$ whereas order $O(k \cdot log(1/\varepsilon))$ operations are needed to recover all the k original data symbols. Differently from LT codes, Raptor codes can be *systematic codes*.

Applications. Raptor codes are being used in commercial systems by Digital Fountain Inc. for fast and reliable delivery of data over heterogeneous networks. The Raptor implementation of Digital Fountain reaches speeds of several gigabits per second, on a 2.4-GHz Intel Xeon processor, while ensuring very stringent conditions on the error probability of the decoder.

4.4 NETWORK CODING

Network coding is recently emerging as a new promising research field in the networking framework [37]. Although sharing some basic concepts with erasure coding and digital fountains, network coding differs from the aforementioned encoding techniques in many interesting aspects. First of all, the target of network coding is something different from that of digital fountains. Network coding has been conceived to allow *better resource utilization* during transmission over the network, especially as regards *bandwidth* and *throughput*. In addition, network coding allows *optimal delays* and better traffic distribution throughout the network, thus leading to overall *load balancing*, which is very promising in scenarios where power saving is a major concern (see, e.g., wireless ad hoc and sensor networks). It adds *robustness* against node and connections failures (even permanent), as well as *flexibility* against changes in the network topology [38]. Therefore, network coding is not concerned with reliable transmissions, which is instead the major target of erasure coding, but focuses more generally on the overall optimization of network usage. Moreover, as far as robustness and flexibility are concerned, network coding seems to be quite suitable to new emerging scenarios of extreme and challenged networks like, for example, delay tolerant networks (DTNs), intermittently connected ad hoc and sensor networks, underwater acoustic sensor networks, and vehicular ad hoc networks. These scenarios are prone to link failures because nodes can move, crash, or simply go to sleep mode to save energy. It has been shown that network coding-based algorithms for data dissemination where all nodes are interested in knowing all the information produced are actually very efficient.

Apart from targets and performance, the other great difference between erasure and network coding is recursive coding at intermediate nodes. With network coding, both source nodes and intermediate/relay nodes encode the original and/or to-be-forwarded data; however, in generic relay systems, relay nodes simply repeat data toward the next identified hops (see Figure 4.10).

To illustrate the advantages of network coding, let us consider a classical example. Figure 4.11 shows the graph representation of a one-source two-sink network. The capacity of each link (edge) of the network is one bit per second. The value of the

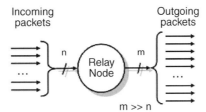

Figure 4.10. Relay nodes perform encoding over n ingress data packets and produce m data packets to send out. The number of encoded packets is greater than the number of incoming data packets.

max-flow from the source node to each of the destination nodes t_1 and t_2, according to the well-known max-flow min-cut theorem, is 2. So it should be possible to send 2 bits, b_1 and b_2, to t_1 and t_2 simultaneously.

However, this is only possible by using network coding. Figure 4.11 shows the network coding scheme that allows the max-flow. The symbol "+" denotes XOR operations. Instead of forwarding b_1 and b_2 in two separate transmissions, node 3 encodes them and sends just a single encoded packet, $b_1 + b_2$, which eventually reaches t_1 and t_2 via node 4. Because node t_1 and t_2 also get, respectively, b_1 and b_2 via different routes, they are able to recover the missing bit from the encoded packet received from node 3.

The above network scheme is well known and called *butterfly network*. It was presented in 2000 in Ahlswede et al. [37], when the term network coding was originally proposed. The same article also demonstrated that network coding is necessary to achieve the theoretical upper bound limit for throughput utilization (max-flow) in optimal multicast schemes that include two or more destination nodes. However, throughput benefits of network coding are not only limited to multicast flows [39] and can extend to other traffic patterns (e.g., unicast transmissions) [40]. Throughput gains are different, depending on the network graph. Specifically, they can be very significant in directed graphs, whereas in undirected graphs (e.g., a wired network where all links are half-duplex) the throughput gain is at most a factor of two [41, 42].

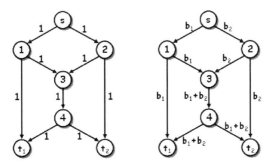

Figure 4.11. A one-source two-sink network with coding: link capacities (left) and network coding scheme (right).

In Sections 4.4.1 and 4.4.2, some insights on the encoding and decoding processes of network coding will be given. Indeed, a particular type of network coding will be described, which is called *random, linear network coding* [43]. Linear coding is one of the simplest coding schemes for network coding. It regards a block of data as a vector over a certain base field and allows a node to apply a linear transformation to this vector before forwarding it [44]. When the transformation is performed at each node independently from other nodes, along with using random coefficients, linear coding is said to be random network coding. Its efficacy has been extensively studied, mainly analytically, for different types of network corresponding to different graphs [45–47]. Cases that have been studied include: (i) one-source networks and networks with multiple sources either independent or linearly correlated; (ii) acyclic networks and cyclic networks; and (iii) delay-free and delayed networks. In all these environments, random network coding has brought interesting and valuable benefits. Finally, some recent applications of network coding will be described in Section 4.3.

4.4.1 Encoding Process

As mentioned before, encoding is performed at both the source and intermediate relay nodes which are visited by the data flowing over the network. Data are organized into packets. Assume that all the incoming/generated *packets* have the same length of L bits. If a packet is shorter, then it is padded with trailing 0s. The content of a packet can be interpreted as a sequence of *symbols*, where each symbol corresponds to a sequence of s consecutive bits. Hence, each 0-padded packet, L bits long, is a sequence of L/s, say l, symbols. Each symbol belongs to the field $GF(2^s)$ since it can be represented over s bits (Figure 4.12).

In Linear Network Coding the encoding process works out a linear combination over a set of incoming packets (at a relay node) or original packets (at the source) and produces new packets for sending out, which have the same size (L bits) of the packets that have been combined together. The arithmetic operations involved in linear combinations (addition and multiplication) are performed over the field $GF(2^s)$.

Assume having a set of n ingress packets $M^0, M^1, \ldots M^{n-1}$. Each packet is a vector of $l = L/s$ symbols. Ingress packets are *original packets* in case of source nodes, whereas in the case of relay nodes they may be original and *incoming packets* as well. Incoming packets are intended to come from one or even multiple nodes. However, suppose for simplicity that the incoming packets come all from the same source node.

Each node performing encoding generates n coefficients, one for each ingress packet: $g_0, g_1, \ldots g_{n-1}$. Coefficients are symbols, that is, $g_i \cdot GF(2^s)$. Then the node generates an outgoing packet X of size L bits as follows:

$$x = \sum_{i=0}^{n-1} g_i M^i$$

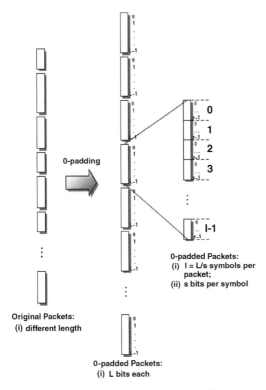

Figure 4.12. Scheme illustrating the encoding process.

With much detail the encoding process for a single outgoing packet X works as follows:

$$X = \begin{pmatrix} x_0 \\ x_1 \\ x_2 \\ \vdots \\ \vdots \\ x_{l-1} \end{pmatrix} = \sum_{i=0}^{n-1} g_i M^i = \sum_{i=0}^{n-1} g_i \cdot \begin{pmatrix} m_0^i \\ m_1^i \\ m_2^i \\ \vdots \\ \vdots \\ m_{l-1}^i \end{pmatrix} = g_0 \cdot \begin{pmatrix} m_0^0 \\ m_1^0 \\ m_2^0 \\ \vdots \\ \vdots \\ m_{l-1}^0 \end{pmatrix} + g_1 \cdot \begin{pmatrix} m_0^1 \\ m_1^1 \\ m_2^1 \\ \vdots \\ \vdots \\ m_{l-1}^1 \end{pmatrix}$$

$$+ g_2 \cdot \begin{pmatrix} m_0^2 \\ m_1^2 \\ m_2^2 \\ \vdots \\ \vdots \\ m_{l-1}^2 \end{pmatrix} + \cdots + g_{n-1} \cdot \begin{pmatrix} m_0^{n-1} \\ m_1^{n-1} \\ m_2^{n-1} \\ \vdots \\ \vdots \\ m_{l-1}^{n-1} \end{pmatrix}$$

where $x_i, m_i^j, g_j \in GF(2^s)$ x_i is the ith symbol of the egress vector (packet) X. m_i^j is the ith symbol of the jth ingress packet M^j. g^j is the coefficient that multiplies the jth ingress packet M^j. The same encoding process is then repeated more times. Specifically, from n ingress packets a node generates m outgoing, egress packets with $m \gg n$ so that at the receiving site decoding is possible after having received *any* n out of m packets. Hence, the encoding node actually generates m sets of coefficients, one for each egress packet to generate, and then performs encoding as follows (all the three formulas below are equivalent):

$$X^j = \sum_{i=0}^{n-1} g_i^j M^i, \qquad j = 0, \ldots, m-1,$$

$$\begin{pmatrix} X^0 \\ X^1 \\ X^2 \\ \ldots \\ \ldots \\ \ldots \\ X^{m-1} \end{pmatrix} = \begin{pmatrix} g_0^0 & g_1^0 & g_2^0 & \cdots & g_{n-1}^0 \\ g_0^1 & g_1^1 & g_2^1 & \cdots & g_{n-1}^1 \\ g_0^2 & g_1^2 & g_2^2 & \cdots & g_{n-1}^2 \\ \ldots & \ldots & \ldots & \ldots & \ldots \\ \ldots & \ldots & \ldots & \ldots & \ldots \\ \ldots & \ldots & \ldots & \ldots & \ldots \\ g_0^{m-1} & g_1^{m-1} & g_2^{m-1} & \cdots & g_{n-1}^{m-1} \end{pmatrix} \cdot \begin{pmatrix} M^0 \\ M^1 \\ \ldots \\ \ldots \\ M^{n-1} \end{pmatrix}$$

where X^i is the ith egress packet that is generated and g_i^j is the ith symbol of the jth encoding vector—that is, the jth set of coefficients which is generated to produce the jth encoded packet. All the cofficients g_i^j together form the $(m \times n)$-sized encoding matrix G. Each encoded data packet X^j is named *information vector*, and the corresponding vector of coefficients it is obtained from, g^j, is said to be an *encoding vector*. The encoding node produces and sends out packets by including both encoding and information vectors; thus an outgoing packet is actually a tuple $<g^j, X^j>$ for $j = 0, \ldots, m-1$. Each encoding vector includes n symbols, whereas each information vector includes l symbols, as many as in the ingress data items $M^i, i = 0, \ldots, n-1$:

$$\langle g^j, X^j \rangle = \langle (g_0^j, g_1^j, g_2^j, \ldots, g_{n-1}^j), (x_0^j, x_1^j, x_2^j, \ldots, x_{l-1}^j) \rangle$$

When an encoding vector includes all zeros but one single one in the ith position, it means that the ith ingress data item is not encoded.

$$g^i = e^i = (0, 0, \ldots, 0, 1, 0, \ldots, 0) \Rightarrow X^i = M^i$$

Recursive Encoding. Suppose that the following conditions hold.

1. An intermediate relay receives and stores a set of r-tuples, as follows:

$$\langle g^0, X^0 \rangle, \langle g^1, X^1 \rangle, \langle g^2, X^2 \rangle, \ldots, \langle g^{r-1}, X^{r-1} \rangle$$

2. All the information vectors received $(X^0, X^1, X^2, \ldots, X^{r-1})$ come from the same set of original data $(M^0, M^1, M^2, \ldots, M^{n-1})$ produced by a single source node somewhere.

Then, the intermediate node performs re-encoding on the set of information vectors that it has received: $X^0, X^1, X^2, \ldots, X^{r-1}$. It is not necessary that $r \gg n$ because the intermediate node does not need decoding at this stage. It simply re-encodes the pieces of data it has received. This results in adding some more redundancy into the network for some part of the original data. Re-encoding is performed by generating a certain number of encoding vectors, as many as the number of information vectors that the node wants to send out. Suppose, for example, that the node wants to generate k information vectors. It should be noted that there is not a mathematical constraint on the values allowed for r and k. Rather, the right tuning of these parameters depends on the networking scenario and is still a challenging open issue.

After having chosen the appropriate value for k, the relay node acts as follows:

a. It selects k new encoding vectors h^j, $j = 0, \ldots, k-1$. Each encoding vector includes as many symbols as the number of ingress information vectors to which the new encoding is applied. In this case each encoding vector is r symbols long:

$$h^j = (h_0^j, h_1^j, h_2^j, \ldots, h_{r-1}^j), \quad j = 0, \ldots, k-1$$

b. It produces k new information vectors as follows:

$$Y^j = \sum_{i=0}^{r-1} h_i^j X^i; \quad j = 0, \ldots, k-1$$

The new encoding matrix is therefore

$$H_{(k \times r)} = \begin{pmatrix} h_0^0 & h_1^0 & \cdots & h_{r-1}^0 \\ h_0^1 & h_1^1 & \cdots & h_{r-1}^1 \\ \cdots & \cdots & \cdots & \cdots \\ h_0^{k-2} & h_1^{k-2} & \cdots & h_{r-1}^{k-2} \\ h_0^{k-1} & h_1^{k-1} & \cdots & h_{r-1}^{k-1} \end{pmatrix}$$

whereas the overall re-encoding process can be represented as follows:

$$\begin{pmatrix} Y^0 \\ Y^1 \\ \vdots \\ Y^{k-2} \\ Y^{k-1} \end{pmatrix} = \begin{pmatrix} h_0^0 & h_1^0 & \cdots & h_{r-1}^0 \\ h_0^1 & h_1^1 & \cdots & h_{r-1}^1 \\ \vdots & \vdots & \cdots & \vdots \\ h_0^{k-2} & h_1^{k-2} & \cdots & h_{r-1}^{k-2} \\ h_0^{k-1} & h_1^{k-1} & \cdots & h_{r-1}^{k-1} \end{pmatrix} \cdot \begin{pmatrix} X^0 \\ X^1 \\ \vdots \\ X^{r-1} \end{pmatrix}$$

$$= \begin{pmatrix} h_0^0 & h_1^0 & \cdots & h_{r-1}^0 \\ h_0^1 & h_1^1 & \cdots & h_{r-1}^1 \\ \vdots & \vdots & \cdots & \vdots \\ h_0^{k-2} & h_1^{k-2} & \cdots & h_{r-1}^{k-2} \\ h_0^{k-1} & h_1^{k-1} & \cdots & h_{r-1}^{k-1} \end{pmatrix} \cdot \begin{pmatrix} g_0^0 & g_1^0 & g_2^0 & \cdots & g_{n-1}^0 \\ g_0^1 & g_1^1 & g_2^1 & \cdots & g_{n-1}^1 \\ \vdots & \vdots & \vdots & \cdots & \vdots \\ g_0^{r-1} & g_1^{r-1} & g_2^{r-1} & \cdots & g_{n-1}^{r-1} \end{pmatrix} \cdot \begin{pmatrix} M^0 \\ M^1 \\ M^2 \\ \vdots \\ M^{n-1} \end{pmatrix}$$

$$\stackrel{\Delta}{=} G'_{(k \times n)} \cdot \begin{pmatrix} M^0 \\ M^1 \\ M^2 \\ \vdots \\ M^{n-1} \end{pmatrix}$$

where $G'_{(k \times n)}$ is equal to $H_{(k \times r)} \cdot G_{(r \times n)}$ and is hereafter called re-encoding matrix.

c. It produces the k final encoding vectors to include in the outgoing tuples. They correspond to the rows of G' worked out as follows:

$$g'^j_i = \sum_{t=0}^{r-1} h^j_t g^t_i; \quad i = 0, \ldots, n-1; \quad j = 0, \ldots, k-1; \quad t = 0, \ldots, r-1$$

d. It sends out k-tuples as follows:

$$\langle g'^0, Y^0 \rangle, \langle g'^1, Y^1 \rangle, \langle g'^2, Y^2 \rangle, \ldots, \langle g'^{k-1}, Y^{k-1} \rangle$$

This process is equivalent to *encoding only once* the original data $M^0, M^1, \ldots M^n - 1$ with the new set of encoding coefficients. At the receiver site the number of incoming tuples to collect prior to decoding corresponds to the number of symbols included in a single encoding vector or, from another standpoint, to the number of original data packets (as can be seen, it is always n). Finally, it should be noted that the outgoing information vector is again l symbols long, as follows.

$$Y^i = (y^i_0, y^i_1, y^i_2, \ldots, y^i_{l-1})^T$$

The computational cost of performing re-encoding at each intermediate node depends on the following operations.

(i) The product between the new encoding matrix, $H_{(k \times r)}$, and the old encoding matrix, $G_{(r \times n)}$, to obtain the final encoding matrix to be included in the outgoing tuples $G'_{(k \times n)} = H_{(k \times r)} \cdot G_{(r \times n)}$.
(ii) The product between the new encoding matrix, $H_{(k \times r)}$, and the matrix of encoded data, $X_{(r \times l)}$, to obtain the outgoing information vectors $Y_{(k \times l)}$.

How to Choose Encoding Vectors? The choice of coefficients to include in encoding vectors is, as can easily be expected, quite critical. An inaccurate choice can produce tuples that are linearly dependent on each other. When the destination node receives a tuple that is linearly dependent on another tuple already received, it discards the new tuple because it does not contain innovative information and is thus of no use (see Section 4.4.2 on the decoding process for further details). This directly impacts on the decoding time, causing the destination node to wait for more packets than strictly necessary to be able to do the decoding. A simple algorithm to generate such coefficients is proposed in reference 48 and gives rise to the so-called *Random Linear Coding*. Each node in the network selects *uniformly at random* its coefficients over the field $GF(2^s)$ in a completely *independent* and *decentralized* manner. Simulation results indicate that even for small field sizes (e.g., with $s = 8$), the probability of selecting linearly dependent combinations is *negligible*.

Other approaches have been proposed in references 45 and 49. The first one refers to a centralized algorithm where a single entity decides which coefficients each node in the network has to assign. The latter describes deterministic decentralized algorithms that apply to restricted families of network configurations.

Generations. *Generation* is the name assigned to a set of vectors (packets) that are encoded together at a source node. Typically, in fact, a source node produces a continuous flow of packets but obviously it cannot apply encoding to them all at the same time because otherwise (i) it should wait for the entire packet flow to be generated to perform encoding, thus causing long delay to transmission, and (ii) it should generate far long encoding vectors to include in outgoing packets. This would cause severe overhead in transmission. Therefore, source vectors are grouped into generations and each generation has a corresponding matrix (that includes all the encoding vectors). To allow successful decoding at destination, it is necessary that only vectors (packets) belonging to the same generation are combined. This implies that packets should carry some information about the generation they belong to or, at least, this should be deducible in some other way so that intermediate nodes do not encode together packets of different generations, thus making them impossible to decode.

4.4.2 Decoding Process

In network coding, decoding is only needed at destination. Suppose n source data packets have been generated in the network. Let them be, $M^0, M^1, \ldots, M^{n-1}$ with

$M^i = (m_0^i, m_1^i, m_2^i, \ldots, m_{l-1}^i)$. Suppose that the receiver node has just received r data packets containing r-tuples:

$$\langle g^0, X^0 \rangle, \langle g^1, X^1 \rangle, \langle g^2, X^2 \rangle, \ldots, \langle g^{r-1}, X^{r-1} \rangle$$

Summing up, the encoding vectors g^0, g^1, g^{r-1} are such that $g^i = (g_0^i, g_1^i, g_2^i, \ldots, g_{n-1}^i)$ whereas the r encoded data packets $X^0, X^1, \ldots, X^{r-1}$ are such that $X^i = \sum_{j=0}^{n-1} g_j^i \cdot M^j$, and the following relations hold true.

$$x_z^i = \sum_{j=0}^{n-1} g_j^i \cdot m_z^j; \Rightarrow$$

$$\begin{pmatrix} x_0^i \\ x_1^i \\ x_2^i \\ \ldots \\ x_{l-1}^i \end{pmatrix} = g_0^i \cdot \begin{pmatrix} m_0^0 \\ m_1^0 \\ m_2^0 \\ \ldots \\ m_{l-1}^0 \end{pmatrix} + g_1^i \cdot \begin{pmatrix} m_0^1 \\ m_1^1 \\ m_2^1 \\ \ldots \\ m_{l-1}^1 \end{pmatrix} + g_2^i \cdot \begin{pmatrix} m_0^2 \\ m_1^2 \\ m_2^2 \\ \ldots \\ m_{l-1}^2 \end{pmatrix} + \ldots + g_{n-1}^i \cdot \begin{pmatrix} m_0^{n-1} \\ m_1^{n-1} \\ m_2^{n-1} \\ \ldots \\ m_{l-1}^{n-1} \end{pmatrix}$$

The above relations can also be written as follows:

$$X_{(r \times 1)} = G_{(r \times n)} \cdot M_{(n \times 1)}$$

$$\begin{pmatrix} X^0 \\ X^1 \\ X^2 \\ \ldots \\ \ldots \\ \ldots \\ X^{r-1} \end{pmatrix} = \begin{pmatrix} g_0^0 & g_1^0 & g_2^0 & \ldots & g_{n-1}^0 \\ g_0^1 & g_1^1 & g_2^1 & \ldots & g_{n-1}^1 \\ g_0^2 & g_1^2 & g_2^2 & \ldots & g_{n-1}^2 \\ \ldots & \ldots & \ldots & \ldots & \ldots \\ \ldots & \ldots & \ldots & \ldots & \ldots \\ \ldots & \ldots & \ldots & \ldots & \ldots \\ g_0^{r-1} & g_1^{r-1} & g_2^{r-1} & \ldots & g_{n-1}^{r-1} \end{pmatrix} \cdot \begin{pmatrix} M^0 \\ M^1 \\ \ldots \\ M^{n-1} \end{pmatrix}$$

To go back to the original messages $M^0, M^1, \ldots, M^{n-1}$, it is necessary that the following conditions hold true.

(i) $r \geq n$,
(ii) The encoding matrix $G_{(r \times n)}$ has rank n. This happens if and only if the encoding matrix includes n rows that are linearly independent.

Provided conditions (i) and (ii) are met, the original data $M^0, M^1, \ldots, M^{n-1}$ can be derived via standard linear system solution techniques.

A computational-efficient way of doing this online can be sketched as follows: If r packets have been received, a decoding matrix can be constructed as

$$\left(\begin{array}{cccccc|cccc} g_0^0 & g_1^0 & g_2^0 & g_3^0 & \cdots & g_{n-1}^0 & x_0^0 & x_1^0 & \cdots & x_{l-1}^0 \\ g_0^1 & g_1^1 & g_2^1 & g_3^1 & \cdots & g_{n-1}^1 & x_0^1 & x_1^1 & \cdots & x_{l-1}^1 \\ g_0^2 & g_1^2 & g_2^2 & g_3^2 & \cdots & g_{n-1}^2 & x_0^2 & x_1^2 & \cdots & x_{l-1}^2 \\ \cdots & \cdots & \cdots & \cdots & \cdots & \cdots & \cdots & \cdots & \cdots & \cdots \\ \cdots & \cdots & \cdots & \cdots & \cdots & \cdots & \cdots & \cdots & \cdots & \cdots \\ g_0^{r-2} & g_1^{r-2} & g_2^{r-2} & g_3^{r-2} & \cdots & g_{n-1}^{r-2} & x_0^{r-2} & x_1^{r-2} & \cdots & x_{l-1}^{r-2} \\ g_0^{r-1} & g_1^{r-1} & g_2^{r-1} & g_3^{r-1} & \cdots & g_{n-1}^{r-1} & x_0^{r-1} & x_1^{r-1} & \cdots & x_{l-1}^{r-1} \end{array} \right)$$

By Gaussian elimination it is possible to reduce the system to the following one in which the original messages can be read.

$$\left(\begin{array}{c|cccc} & m_0^0 & m_1^0 & \cdots & m_{l-1}^0 \\ & m_0^1 & m_1^1 & \cdots & m_{l-1}^1 \\ I_{n \times n} & m_0^2 & m_1^2 & \cdots & m_{l-1}^2 \\ & m_0^{n-1} & m_1^{n-1} & \cdots & m_{l-1}^{n-1} \\ \hline G'_{((r-n) \times n)} & & X'_{((r-n) \times l)} & & \end{array} \right)$$

To construct this matrix dynamically, as packets arrive to the receiver, they are added to the decoding matrix row by row (a row is $< (g_0^i, g_1^i, g_2^i, \ldots, g_{n-1}^i), (x_0^i, x_1^i, \ldots, x_{l-1}^i) >$). The ith information vector X^i is said to be *innovative* if it increases the rank of the matrix. If a packet is not innovative, it is reduced to a row of all 0s by Gaussian elimination and is ignored. While performing Gaussian elimination, as soon as the matrix contains a row of the form $< e_i, X>$, the node knows that the original packet M^i is equal to X (remember that we have defined e_i in Section 4.4.1 as a vector that includes all 0s but one single 1 in the ith position).

4.4.3 Applications

In reference 19, network coding was proposed to allow *efficient communication in extreme networks*—that is, in delay-tolerant networks or, more generally, intermittently connected networks. According to the forwarding scheme described, nodes do not simply forward packets they overhear but send out information that is coded over the content of several packets they have received. Simulation results show that this algorithm achieves the reliability and robustness of flooding at a small fraction of the overhead. Two interesting topics have been discussed, namely generations and redundancy.

The solution for generation recognition which has been proposed in reference 19 is as follows. Let $X_{(i)}^j$ be the *jth* information vector that originates at the *ith* source node S_i. Then a function $f(X_{(i)}^j)$ determines which generation the packet belongs to and $\Gamma_\gamma = \{X_{(i)}^j | f(X_{(i)}^j) = \gamma\}$ is the set of all the source vectors of a generation γ. Namely, the generation membership is determined through hashing over the *sender address* and the *packet identifier*. A new hash function is generated whenever the matrix becomes too big. The hash function is then used at a node to determine which generation to insert a given packet into, provided that the size of this generation does not exceed a certain max threshold. It has been demonstrated that managing generations through hashing works better than simply incrementing the generation index from time to time, especially in ad hoc networks.

Another interesting topic raised in reference 19 is on the forwarding strategy and, specifically, on how to decide when to send a new packet. The solution that has been proposed relies on a so-called *forwarding factor* $d > 0$ and establishes that whenever an innovative packet is received or generated at a node for a given generation, this has to forward a certain number of packets depending on the value of d and on whether the node is a source or an intermediate node. Specifically, in case of an intermediate node, it first generates $\lfloor d \rfloor$ vectors from the corresponding matrix and rebroadcasts them to the neighbors, and then it generates and sends a further vector with probability $d - \lfloor d \rfloor$. In the case of a source node, whenever it generates a new original packet it encodes and broadcasts to the neighbors $\max(1, \lfloor d \rfloor)$ vectors. It then produces and sends out a further vector if $d > 1$ with probability $d - \lfloor d \rfloor$. The delivery policy at source nodes is obviously a bit different, since at least one packet must be generated from each newly produced packet. In other words, a new packet is sent out by the source at least once.

With the network coding model described above, a node sends out, on average, $dG + 1$ packets where G is the maximum generation size ($G = m$ in the simplest case). Since receivers can decode all original vectors when they receive a number of innovative packets equal to the generation size, a good value for d strongly depends on the number of neighbors (i.e., the node density). Similar to probabilistic routing, a high forwarding factor results in a high decoding probability at the expense of a high network load.

In references 50 and 51, network coding is used in a peer-to-peer content distribution network named *Avalanche*. In a peer-to-peer content distribution network, a server splits a large file into a number of blocks, and then peer nodes try to retrieve the original file by downloading blocks from the server but also distributing downloaded blocks among them. When using network coding, the blocks sent out by the server are random linear combinations of all the original blocks. Similarly, peers send out random linear combinations of all the blocks available to them. Network coding in *Avalanche* minimizes download times with respect to the case where it is not used. Network coding gives the system more robustness in case the server leaves early or when peers only join for short periods.

Finally, some promising application fields for network coding are wireless ad hoc, mesh, and sensor networks, where it can contribute to achieving throughput gains and more secure systems. Network coding seems to facilitate protection from eavesdroppers, since information is more spread out and thus more difficult to overhear. It also simplifies the protection against modified packets in a network. In a network with no additional protection, an intermediate attacker may make arbitrary modifications to a packet to achieve a certain reaction at the attacked destination. However, in the case of network coding, an attacker cannot control the outcome of the decoding process at the destination, without knowing all other coded packets the destination will receive. Given that packets are routed along many different paths, this makes controlled man-in-the-middle-attacks more difficult.

4.5 CONCLUSIONS AND OPEN ISSUES

Encoding techniques are becoming popular in the evolving Internet scenario. The advent of wireless technologies, the diffusion of an ever-greater and heterogeneous number of portable devices, and the diffusion of multimedia and content-distribution applications as well as their adoption over wireless networks are posing and renewing a number of networking challenges, such as reliability, scalability, and efficient data distribution over wireless networks.

Encoding techniques have successfully been used to increase reliability and scalability—for example, in multicast scenarios, in parallel download from many sources, and over overlay networks. Moreover, they are commonly applied to multimedia streaming applications to manage reliability issues without requiring feedback channels and by minimizing synchronizations and interactions among multiple receivers and between senders and receivers.

Finally, it should be noted that the advantages of encoding are not only limited to transmission reliability and scalability but also regard security. It is much harder, for example, for the attackers to catch useful information from a network because they should have to collect a sufficiently high number of data packets that need to be decoded together and they should also know exactly how to do the decoding.

The main drawback of the encoding techniques analyzed in this chapter is the *computational burden* involved in both the encoding and decoding processes. Also the *delay* might be negatively impacted being increased by the time spent in encoding and decoding and the time spent to transmit the extra information produced during encoding (encoded data has higher size than the original). However, many optimizations have been studied and implementations of encoding techniques have also been provided even for wireless sensor networks (WSNs). Since WSNs are particularly scarce in resources, this means that resource usage has successfully been minimized.

Network coding is a sort of generalization of encoding techniques. Most of the work conducted on network coding is analytical, so far, since it has been introduced as a way to improve the use of network bandwidth. It has been shown that network coding allows achievement of the *max* possible *flow* on the networks under investigation, including many different topology patterns. Network coding has also the advantage to increasing *robustness* of the network against node failures, even permanent, and *adaptability* to variations of the network topology. By disseminating information throughout the network, network coding also provides *load balancing* of the network traffic, and, in addition, it adds *reliability* to transmissions as do the other encoding techniques.

The main drawback of network coding is the *computational burden* that is added not only to the source and destination nodes but also to the intermediate nodes that act as relays and provide re-encoding together with usual forwarding. Another challenge in network coding is the traffic overhead injected into the network. However, specific analyses do not exist yet, neither simulative nor experimental. It can easily be envisioned that the next efforts on network coding will be on both simulation and experimentation to better understand real strengths and limits of this new technique. Other interesting challenges to be addressed in the framework of network coding are how to minimize the traffic injected into the network while still retaining sufficient redundancy in the encoding process, as well as how to efficiently deal with dynamically generated content.

4.6 EXERCISES

1. Consider the encoding process of the Reed–Solomon codes. Given the extension field GF(2^8), which is the maximum possible size of the encoding matrix? Give a schematic representation of the matrix.

2. Consider the bipartite graph of 0 (Figure 4.13). Assume it is part of the multi-bipartite graph that describes a complete Tornado code and that the decoding process has already produced the second-level check bits $c_0, \ldots, c_5$ whose values follow.

$$c_0 = 1, \quad c_1 = 1, \quad c_2 = 1, \quad c_3 = 1, \quad c_4 = 1, \quad c_5 = 0.$$

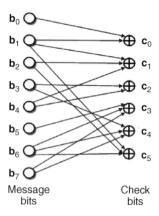

Figure 4.13. First-level bipartite graph of a Tornado code.

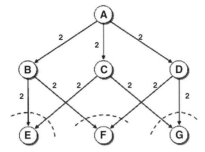

Figure 4.14. Node A is the source node of the graph. Receiver nodes are nodes E, F, and G. Each edge has 2-bit capacity. Each destination node has the potential to receive four bits at the same time, two for each incoming edge.

Assume also that the original message bits b_1 and b_4 are known because they have arrived incorrupted at destination.

$b_1 = 0$ and $b_4 = 0$

Complete the decoding process.

3. Consider the graph of Figure 4.14. The graph includes seven nodes. Node A is the only source, whereas there are three destination nodes: E, F, and G. As is shown, each edge of the graph has a capacity of two bits. Hence, each of the destination nodes has the potential to receive four bits at the same time (see the max-flow min-cut theorem). Suppose the source node A has to send the four bits b_1, b_2, b_3, and b_4. Show a possible transmission scheme of the four bits in the two cases *with* and *without* applying network coding. Then, compare the total number of transmissions needed in the two cases and evaluate the percent saving achieved when using network coding. Use simple *xor* operations for the encoding.

4. Consider the graph of Figure 4.15. Suppose Node A generates four messages destined to node D: M^1, M^2, M^3, and M^4. Suppose node B receives the messages

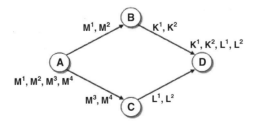

Figure 4.15. Node A is the source node of four messages $M_1, \ldots, M_4$ destined to node D. Intermediate nodes B and C receive (each) two out of the four source messages. They apply encoding over the received messages and forward the resultant encoded messages to node D.

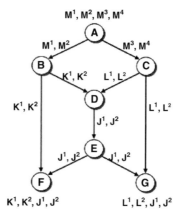

Figure 4.16. Node A is the source node of four messages $M_1, \ldots, M_4$ destined to both nodes F and G. Intermediate nodes B and C each receive two out of the four source messages. They apply encoding over the received messages and then forward the resultant encoded messages to node D. Node D performs encoding over the received messages, too, while node E simply forwards the received messages to nodes F and G.

M^1 and M^2 while node C receives the messages M^3 and M^4. Both nodes B and C apply a linear combination over the received messages and produce two encoded messages each. Node B produces messages K^1 and K^2, and node C produces messages L^1 and L^2. Finally, nodes B and C send out the encoded messages to the destination node D. Assume that the original messages M^1, M^2, M^3, and M^4 can be represented as four-element vectors with the following expressions:

$$M^1 = \begin{pmatrix} 1 \\ 1 \\ 2 \\ 0 \end{pmatrix}, \quad M^2 = \begin{pmatrix} 2 \\ 1 \\ 4 \\ 2 \end{pmatrix}, \quad M^3 = \begin{pmatrix} 3 \\ 1 \\ 1 \\ 3 \end{pmatrix}, \quad M^4 = \begin{pmatrix} 1 \\ 0 \\ 1 \\ 0 \end{pmatrix}$$

Select, at will, the coefficients of the possible linear combinations that are applied at nodes B and C and work out possible values for the vectors L^1, L^2, K^1, and K^2. Show the form of the *encoding vectors* that are sent together with the *information vectors* and finally show the algebraic steps of the decoding process applied at node D, and demonstrate that the decoding process produces back the original messages M^1, M^2, M^3, and M^4. For simplicity, work out additions and products in the field $(R, +, \cdot)$.

5. Consider the graph of Figure 4.16. Similarly to the case of exercise 4, node A generates four messages: M^1, M^2, M^3, and M^4. Node B receives the messages M^1 and M^2, while node C receives the messages M^3 and M^4. Both node B and node C apply a linear combination over the received messages and produce two *encoded messages* each. Node B produces the messages K^1 and K^2, and node C produces

the messages L^1 and L^2. Nodes B and C send out their encoded messages to node D.

This time, assume that there are two destinations, namely, node F and node G. After receiving the four encoded messages K^1, K^2, L^1, and L^2, node D applies a second-stage encoding and produces the messages J^1 and J^2 and then forwards them to node E. Node E sends out the same messages J^1 and J^2 it receives, without encoding. Finally the destination nodes F and G receive four messages each. Node F receives K^1, K^2, J^1, and J^2, while node G receives L^1, L^2, J^1, and J^2.

Starting from the same original and encoded vectors of exercise 4 (i.e., M^1, M^2, M^3, and M^4; and K^1, K^2, L^1, and L^2), think of a possible linear combination that node D can apply to generate the encoded information vectors J^1 and J^2. Give the exact expression of the *encoding vectors* (vectors of coefficients) which have to accompany the *information vectors* J^1 and J^2 sent by node D. Finally, verify that the decoding process produces the exact original messages both at node F and node G.

For simplicity, work out additions and products in the field $(R, +\cdot)$.

REFERENCES

1. M. Conti. Wireless communications and pervasive technologies. In D. Cook and S. K. Das, editors, *Environments: Technologies, Protocols and Applications*, John Wiley & Sons, New York, 2004, pp. 63–99.
2. I. Chlamtac, M. Conti, and J. Liu. Mobile, Ad hoc Networking: Imperatives and Challenges, *Ad Hoc Networks Journal*, **1**(1):13–64, 2003.
3. J. P. Macker and S. Corson, Mobile ad hoc networks (MANET): Routing technology for dynamic, wireless networking. In S. Basagni, M. Conti, S. Giordano, I. Stojmenovic, editors, *Mobile Ad Hoc Networking*, IEEE Press and John Wiley & Sons, New York, 2004.
4. I. F. Akyildiz, T. Melodia, K. R. Chowdhury. A survey on wireless multimedia sensor networks. *Computer Networks*, **51**(4):921–960, 2007.
5. R. Bruno, M. Conti, and E. Gregori. Mesh networks: Commodity multihop ad hoc networks. *IEEE Communications Magazine*, March:123–13, 2005.
6. M. Conti. Principles and applications of ad hoc and sensor networks. In H. Bidgoli, editor, *The Handbook of Computer Networks*, John Wiley & Sons, New York, 2007.
7. L. Pelusi, A. Passarella, and M. Conti. Opportunistic networking: Data forwarding in disconnected mobile ad hoc networks. *IEEE Communications Magazine*, **44**(11):134–141, 2006.
8. L. Rizzo and L. Vicisano. RMDP: An FEC-based reliable multicast protocol for wireless environments. *ACM SIGMOBILE Mobile Computing and Communications Review*, **2**(2):23–31, 1998.
9. S. Kim, R. Fonseca, and D. Culler. Reliable transfer on wireless sensor networks. In *Proceedings of the First IEEE Communications Society Conference on Sensor and Ad Hoc Communications and Networks (IEEE SECON 2004)*, October 4–7, 2004.
10. P. Karlsson, L. Öberg, and Y. Xu. An address coding scheme for wireless sensor networks. in *Proceedings of the 5th Scandinavian Workshop on Wireless Ad-Hoc Networks (ADHOC '05)*, Stockholm, May 3–4, 2005.

11. A. G. Dimakis, V. Prabhakaran, and K. Ramchandran. Ubiquitous access to distributed data in large-scale sensor networks through decentralized erasure codes. In *Proceedings of the Fourth International Conference on Information Processing in Sensor Networks (IPSN 2005)*, April 25–27, 2005, Sunset Village, UCLA, Los Angeles, CA.
12. H. Dong, I. D. Chakeres, A. Gersho, E. M. B.-R. U. Madhow, and J. D. Gibson. Speech coding for mobile ad hoc networks. In *Proceedings of the Asilomar Conference on Signals, Systems, and Computers*, Pacific Grove, CA, November 2003.
13. S. Deb, M. Effros, T. Ho, D. R. Karger, R. Koetter, D. S. Lun, M. Médard, and N. Ratnakar. Network coding for wireless applications: A brief tutorial. In *Proceedings of the International Workshop on Wireless Ad-hoc Networks (IWWAN)*, London, UK, May 23–26, 2005.
14. Y. Wu, P. A. Chou, and S.-Y. Kung. Minimum-energy multicast in mobile ad hoc networks using network coding. *IEEE Transactions on Communications*, **53**(11):1906–1918, 2005.
15. Y. Wu, P. A. Chou, and S.-Y. Kung. Information exchange in wireless networks with network coding and physical-layer broadcast. In *Proceedings of the 39th Annual Conference on Information Sciences and Systems (CISS)*, Baltimore, MD, March 16–18, 2005.
16. J. Widmer, C. Fragouli, and J.-Y. Le Boudec. Low-complexity energy-efficient broadcasting in wireless ad hoc networks using network coding. In *Proceedings of the 1st Workshop on Network Coding, Theory, and Applications*, April 2005.
17. S. Katti, H. Rahul, W. Hu, D. Katabi, M. Medard, and J. Crowcroft. XORs in the air: Practical wireless network coding. In *Proceedings of the ACM SIGCOMM 2006*, Pisa, Italy, September 11–15, 2006.
18. A. A. Hamra, C. Barakat, and T. Turletti. Network coding for wireless mesh networks: A case study. In *Proceedings of the 2nd IEEE International Conference on the World of Wireless, Mobile and Multimedia Networks (WoWMoM2006)*, Niagara-Falls/Buffalo, NY, June 26–29, 2006.
19. J. Widmer and J.-Y. Le Boudec, Network coding for efficient communication in extreme networks. In *Proceedings of the ACM SIGCOMM 2005 Workshop on Delay Tolerant Networks*, Philadelphia, PA, August 22–26, 2005.
20. U. Lee, J.-S. Park, J. Yeh, G. Pau, and M. Gerla. Codetorrent: Content distribution using network coding in VANETs. In *Proceedings of the 1st International ACM Workshop on Decentralized Resource Sharing in Mobile Computing and Networking (ACM MobiShare) in Conjunction with ACM Mobicom 2006*, Los Angeles, CA, September 25, 2006.
21. C. E. Shannon. A Mathematical Theory of Communication. In *The Bell System Technical Journal*, **27**:379–423, 623–656, 1948.
22. L. Rizzo. Effective erasure codes for reliable computer communication protocols. *In ACM Computer Communication Review*, **27**(2):24–36, 1997.
23. L. Rizzo. On the feasibility of software FEC. In *DEIT Technical Report LR-970131*, available as http://www.iet.unipi.it/luigi/softfec.ps, January 1997.
24. J. B. Byers, M. Luby, and M. Mitzenmacher. A digital fountain approach to asynchronous reliable multicast. *IEEE Journal on Selected Areas in Communications*, **20**(8):300–304, 2002.
25. I. S. Reed and G. Solomon, Polynomial codes over certain finite fields. *Journal of the Society for Industrial and Applied Mathematics*, 8:300–304, 1960.
26. M. Luby, M. Mitzenmacher, A. Shokrollahi, D. Spielman, and V. Stemann. Practical loss-resilient codes. In *Proceedings of the 29th Annual ACM Symposium on Theory of Computing*, 1997, pp. 150–159.

27. M. Luby, M. Mitzenmacher, A. Shokrollahi, and D. Spielman. Efficient erasure correcting codes. *IEEE Transactions on Information Theory*, **47**(2):569–584, 2001.
28. M. Luby. LT codes. In *Proceedings of the 43rd Annual IEEE Symposium on Foundations of Computer Science (FOCS)*, 2002, pp. 271–282.
29. A. Shokrollahi. Raptor codes. In *Proceedings of the IEEE International Symposium on Information Theory (ISIT 2004)*, Chicago Downtown Marriott, Chicago, IL USA, June 27–July 2, 2004. Preprint available at http://algo.epfl.ch/pubs/raptor.pdf.
30. A. Shokrollahi, S. Lassen, and M. Luby. Multi-stage code generator and decoder for communication systems. *U. S. Patent Application #20030058958*.
31. J. H. Lint. *Introduction to the Coding Theory*. Springer-Verlag, Berlin, 1982.
32. S. Lin. *An Introduction to Error-Correcting Codes*. Prentice-Hall, Inc., Englewood Cliffs, NJ, 1970.
33. R. Fonseca, S. Ratnasamy, J. Zhao, C. T. Ee, D. Culler, S. Shenker, and I. Stoica. Beacon vector routing: Scalable point-to-point routing in wireless sensornets. In *Proceedings of the 2nd Conference on Symposium on Networked Systems Design & Implementation*, Vol. 2, USENIX Association, Berkeley, CA, May 2005, pp. 329–342.
34. R. G. Gallagar. *Low Density Parity-Check Codes*. MIT Press, Cambridge, MA, 1963.
35. T. Richardson, M. A. Shokrollahi, and R. Urbanke. Design of capacity approaching irregular low-density parity-check codes. *IEEE Transactions on Information Theory*, **47**:619–637, 2000.
36. S. Robinson. Beyond Reed–Solomon: New codes for Internet multicasting drive Silicon Valley Start-up. *SIAM News*, 35 (4): 00–00, 2002.
37. R. Ahlswede, N. Cai, S.-Y. R. Li, and R. W. Yeung. Network information flow. *IEEE Transactions on Information Theory*, **46**:1204–1216, 2000. http://personal.ie.cuhk.edu.hk/~pwkwok4/Yeung/1.pdf.
38. R. Koetter and M. Médard, Beyond routing: An algebraic approach to network coding. In *Proceedings of INFOCOM*, 2002.
39. T. Noguchi, T. Matsuda, and M. Yamamoto. Performance evaluation of new multicast architecture with network coding. *IEICE Trans. Comm.*, June, 2003.
40. D. S. Lun, M. Médard, and R. Koetter. Network coding for efficient wireless unicast. In *Proceedings of the IEEE International Zurich Seminar on Communications*, ETH Zurich, Switzerland, February 22–24, 2006.
41. C. Chekuri, C. Fragouli, and E. Soljanin. On average throughput and alphabet size in network coding. *IEEE/ACM Transactions on Networking*, **14**:2410–2424, 2006. DOI= http://dx.doi.org/10.1109/TIT.2006.874433.
42. Z. Li and B. Li, Network coding in undirected networks. *CISS*, 2004.
43. C. Fragouli, J.-Y. Le Boudec, and J. Widmer. Network coding: An instant primer. In *ACM Computer Communication Review*, **36**(1):63–68, 2006.
44. S.-Y. R. Li, R. W. Yeung, and N. Cai. Linear network coding. *In IEEE Transactions on Information Theory*, **49**(2):371–381, 2003.
45. P. Sanders, S. Egner, and L. Tolhuizen. Polynomial time algorithms for network information flow. In *Proceedings of the 15th ACM Symposium on Parallel Algorithms and Architectures*, 2003.

46. T. Ho, R. Koetter, M. Médard, D. R. Karger, and M. Effros. The benefits of coding over routing in a randomized setting. In *Proceedings of the IEEE International Symposium on Information Theory*, June 2003.
47. T. Ho, M. Médard, J. Shi, M. Effros, and D. R. Karger. On randomized network coding. In *Proceedings of 41st Annual Allerton Conference on Communication, Control, and Computing*, October 2003.
48. T. Ho, R. Koetter, M. Médard, D. R. Karger, and M. Effros. The benefits of coding over routing in a randomized setting. In *Proceedings of the IEEE International Symposium on Information Theory (ISIT 2003)*, Yokohama, Japan, June 29–July 4, 2003.
49. C. Fragouli and E. Soljanin. Decentralized network coding. Information Theory Workshop, Oct. 2004.
50. C. Gkantsidis and P. Rodriguez. Network coding for large scale content distribution. In INFOCOM, Miami, FL, March 2005.
51. Avalanche: File swarming with network coding. http://research.microsoft.com/pablo/avalanche.aspx.

CHAPTER 5

A Taxonomy of Routing Protocols for Mobile Ad Hoc Networks

AZZEDINE BOUKERCHE

School of Information Technology and Engineering, University of Ottawa, Ottawa, Ontario K1N 6N5, Canada

MOHAMMAD Z. AHMAD and DAMLA TURGUT

School of Electrical Engineering and Computer Science, University of Central Florida Orlando, FL 32816–2450

BEGUMHAN TURGUT

Department of Computer Science, Rutgers University, Piscataway, NJ 08854–8019

5.1 INTRODUCTION

Ad hoc networks are wireless networks without a fixed infrastructure, which are usually assembled on a temporary basis to serve a specific deployment such as emergency rescue or battlefield communication. They are especially suitable for scenarios where the deployment of an infrastructure is either not feasible or is not cost effective. The differentiating feature of an ad hoc network is that the functionality normally assigned to infrastructure components, such as access points, switches, and routers, needs to be achieved by the regular nodes participating in the network. For most cases, there is an assumption that the participating nodes are mobile, do not have a guaranteed uptime, and have limited energy resources.

In infrastructure-based wireless networks, such as cellular networks or WiFi, the wireless connection goes only one-hop to the access point or the base station; the remainder of the routing happens in the wired domain. At most, the decision that needs to be made is which base station a mobile node should talk to, or how it should handle the transfer from one station to another during movement. Routing in the wired domain was long considered a mature field, where trusted and reliable solutions exist.

Algorithms and Protocols for Wireless and Mobile Ad Hoc Networks, Edited by Azzedine Boukerche
Copyright © 2009 by John Wiley & Sons Inc.

The infrastructure's topology, its bandwidth, and its routing and switching resources are provisioned to provide a good fit with the expected traffic.

In ad hoc networks, however, routing becomes a significant concern, because it needs to be handled by ordinary nodes that have neither specialized equipment nor a fixed, privileged position in the network. Thus, the introduction of ad hoc networks signaled a resurgent interest in routing through the challenges posed by the mobility of the nodes, their limited energy resources, their heterogeneity (which under some conditions can lead to asymmetric connections), and many other issues. These challenges were answered with a large number of routing algorithms, and ad hoc routing remains an active and dynamically evolving research area. Ad hoc routing algorithms are serving as a source of ideas and techniques to related technologies such as wireless sensor networks and mesh networks.

In this chapter, we survey the field of wireless ad hoc routing. While we attempted to include most of the influential algorithms, our survey cannot be exhaustive: The number of proposed algorithms and variations exceeds 1000. However, we strived to represent most of the research directions in ad hoc routing, thus giving the reader an introduction to the issues, the challenges, and opportunities offered by the field.

5.2 AD HOC NETWORKS APPLICATIONS

In this section, we present some applications of ad hoc networks [1]:

Conferencing. Mobile conferencing is without a doubt one of the most recognized applications. Establishment of an ad hoc network is essential for mobile users where they need to collaborate in a project outside the typical office environment.

Emergency Services. Responding to emergency situations such as disaster recovery is yet another naturally fitting application in the ad hoc networking domain. During the time of emergencies, several mobile users (policeman, firefighters, first response personnel) with different types of wireless devices need to not only communicate but also maintain the connectivity for long periods of time.

Home Networking. The wireless computers at home can also create an ad hoc network where each node can communicate with the others without taking their original point of attachment into consideration. This approach is alternative to assigning multiple IP addresses to each wireless device in order to be identified.

Embedded Computing Applications. Several ubiquitous computing [2] internetworking machines offer flexible and efficient ways of establishing communication methods with the help of ad hoc networking. Many of the mobile devices already have add-on inexpensive wireless components, such as PDAs with wireless ports and Bluetooth radio devices.

Sensor Dust. This application can be considered a combination of ad hoc and sensor networks. In hazardous or dangerous situations, it makes sense to distribute a group of sensors with wireless transceivers to obtain critical information about the unknown site by the creation of ad hoc networks of these sensors.

Automotive/PC Interaction. The interaction between many wireless devices (laptop, PDA, and so on) being used in the car for different purposes can create an ad hoc network in order to carry out tasks more efficiently. An example can be finding the best possible mechanic shop to fix a car problem in a new city on the way to a meeting.

Personal Area Networks and Bluetooth. A Personal Area Network (PAN) creates a network with many of the devices that are attached or carried by a single user. Even though the communication of devices within PAN does not concern mobility issues, the mobility becomes essential when different PANs need to interact with each other. Ad hoc networks provide flexible solutions for inter-PAN communications. For instance, Bluetooth provides a wireless technology built-in to many of the current PDAs; and up to eight PDAs, called *piconet*, can exchange information.

5.3 DESIGN ISSUES

The main reasoning behind the design of mobile ad hoc networks was to respond to the military needs for battlefield survivability [1]. The situations in the battlefield require soldiers to move from place to place without any constraints by wireline communications and communicate with each other without depending on any fixed infrastructure. Since it is almost impossible to have a fixed backbone network in certain territories such as desert, timely deployment of mobile nodes communicating via wireless medium becomes critical. Another deciding point is the consideration of the physics of electromagnetic propagation. The fact is that frequencies much higher than 100 MHz are limited by their propagation distance. Therefore, for a mobile host to communicate with another mobile host beyond its transmission range, multihop routing protocols become necessary. This means that messages are transferred from one host to another via other intermediate hosts.

Mobile ad hoc networks inherit all the issues/problems related to mobile computing and wireless networking and perhaps even more due to lack of infrastructure in these types of networks. There are several unique design challenges that need to be considered, including deployment, coverage, connectivity, and so on.

Deployment. Deployment of such networks are accomplished dynamically on the fly, and the lifetime of these networks are usually short-lived. The deployment of ad hoc networks simply eliminates the cost of laying cables and maintenance of an infrastructure. In order to have a partially functioning network immediately, an incremental deployment with minimal configuration is possible. The requirements vary between different types of deployments such as commercial, military, emergency-operations, and so on.

Coverage. Adequate coverage of the entire area in question has to be provided to enable effective communication within devices which may not be within direct transmission ranges of each other. In many cases, the coverage area is determined by the particular deployed application. For example, in the home networking example, some of the nodes can be fixed static entities while others are mobile.

Connectivity. Another unique feature of ad hoc networking is the lack of connectivity due to mobility. Because the nodes can move at all times, the connectivity graph is continually changing. If all these nodes lie within the transmission range of each other, the network is said to be *fully* connected, which results in a complete graph [3]. Since this situation does not generally hold in practice, routing is needed between any two nodes which are connected directly. The only way to connect any two nodes farther apart is via the intermediary nodes between them, which create the need for *multihop* routing. Ad hoc networks cannot precompute a static routing table; rather, they must dynamically adjust routing based on the mobility of the nodes. Due to the mobility and dynamic topology changes, the protocols are designed to keep the network structure *stable* as long as possible.

5.4 AD HOC NETWORKS ROUTING PROTOCOLS

There have been many existing routing protocols for ad hoc networks emphasizing different implementation scenarios. However, the basic goals have always been to devise a routing protocol that minimizes control overhead, packet loss ratio, and energy usage while maximizing the throughput. Because these types of network can be used in a variety of situations (disaster recovery, battlefields, conferences, and so on), they differ in terms of their requirements and complexities. The routing protocols in ad hoc networks can hence be divided into five categories based on their underlying architectural framework as follows and are shown in Figure 5.1.

- Source-initiated (reactive or on-demand), Section 5.4.1
- Table-driven (proactive), Section 5.4.2
- Hybrid, Section 5.4.3
- Location-aware (geographical), Section 5.4.4
- Multipath, Section 5.4.5

5.4.1 Source-Initiated Protocols

Source-initiated routing represents a class of routing protocols where the route is created only when the source requests a route to a destination. A route discovery procedure is invoked when the route is requested by the source and special route request packets are flooded to the network starting with the immediate neighbors. Once a route is formed or multiple routes are obtained to the destination, the route discovery process comes to an end. Route maintenance procedure maintains the active routes for the duration of their lifetimes.

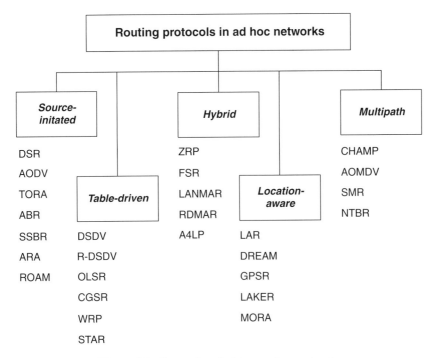

Figure 5.1. Categories of ad hoc routing protocols.

***Dynamic Source Routing (DSR)* [4].** One of the most widely referred routing algorithms is Dynamic Source Routing (DSR), which is an "on-demand" routing algorithm and it has *route discovery* and *route maintenance* phases.

Route discovery contains both *route request* message and *route reply* messages. In route discovery phase, when a node wishes to send a message, it first broadcasts a route request packet to its neighbors. Every node within a broadcast range adds their node id to the route request packet and rebroadcasts. Eventually, one of the broadcast messages will reach either to the destination or to a node that has a recent route to the destination. Since each node maintains a *route cache*, it first checks its cache for a route that matches the requested destination. Maintaining a route cache in every node reduces the overhead generated by a route discovery phase. If a route is found in the route cache, then the node will return a route reply message to the source node rather than forwarding the route request message further. The first packet that reaches the destination node will have a complete route. DSR assumes that the path obtained is the shortest since it takes into consideration the first packet to arrive at the destination node. A route reply packet is sent to the source that contains the complete route from source to destination. Thus, the source node knows its route to the destination node and can initiate the routing of the data packets. The source caches this route in its route cache.

In route maintenance phase, two types of packets are used, namely *route error* and *acknowledgements*. DSR ensures the validity of the existing routes based on the acknowledgments received from the neighboring nodes that data packets have been transmitted to the next hop. Acknowledgment packets also include *passive acknowledgments* as the node overhears the next hop neighbor is forwarding the packet along the route to the destination. A route error packet is generated when a node encounters a transmission problem, which means that a node has failed to receive an acknowledgment. This route error packet is sent to the source in order to initiate a new route discovery phase. Upon receiving the route error message, nodes remove the route entry that uses the broken link within their route caches.

***Ad Hoc On-Demand Distance Vector (AODV)* [5].** AODV routing protocol is developed as an improvement to the Destination-Sequenced Distance-Vector (DSDV)

ALGORITHM 1. AODV Routing Protocol [5]

```
// S is the source node; D is the destination node
// RT = Routing Table

S wants to communicate with D if RT of S contains a route to D
  S establishes communication with D
else
  S creates a RREQ packet and broadcasts it to its neighbors
  // RREQ contains the destination Address(DestAddr),
  // Sequence Number (Seq) and Broadcast ID (BID)
  for all nodes N receiving RREQ
    if (RREQ was previously processed)
      discard duplicate RREQ
    end if
    if (N is D)
      send back a RREP packet to the node sending the RREQ
    else if (N has a route to D with SeqId >= RREQ.Seq)
      send back a RREP packet
    else
      record the node from which RREQ was received
      broadcast RREQ
    end if
  end for
  while (node N receives RREP) and (N != S)
    forward RREP on the reverse path
    store information about the node sending RREP in the RT
  end for
  S receives RREP
  S updates its RT based on the node sending the RREP
  S establishes communication with D
end if
```

routing algorithm [6]. The aim of AODV is to reduce the number of broadcast messages sent throughout the network by discovering routes on-demand instead of keeping a complete up-to-date route information.

A source node seeking to send a data packet to a destination node checks its route table to see if it has a valid route to the destination node. If a route exists, it simply forwards the packets to the next hop along the way to the destination. On the other hand, if there is no route in the table, the source node begins a *route discovery* process. It broadcasts a *route request* (RREQ) packet to its immediate neighbors, and those nodes broadcast further to their neighbors until the request reaches either an intermediate node with a route to the destination or the destination node itself. This route request packet contains the IP address of the source node, current sequence number, the IP address of the destination node, and the sequence number known last. Figure 5.2 denotes the forward and reverse path formation in the AODV protocol. An intermediate node can reply to the route request packet only if they have a destination sequence number that is greater than or equal to the number contained in the route request packet header. When the intermediate nodes forward route request packets to their neighbors, they record in their route tables the address of the neighbor from which the first copy of the packet has come from. This recorded information is later used to construct the reverse path for the route reply (RREP) packet. If the same RREQ packets arrive later on, they are discarded. When the route reply packet arrives from the destination or the intermediate node, the nodes forward it along the established reverse

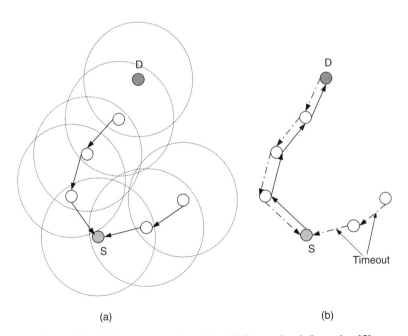

Figure 5.2. (a) Reverse path formation. (b) Forward path formation [5].

path and store the forward route entry in their route table by the use of symmetric links. Route maintenance is required if either the source or the intermediate node moves away. If a source node becomes unreachable, it simply reinitiates the route discovery process. If an intermediate node moves, it sends a *link failure notification* message to each of its upstream neighbors to ensure the deletion of that particular part of the route. Once the message reaches to source node, it then reinitiates the route discovery process.

Local movements do not have global effects, as was the case in DSDV. The stale routes are discarded; as a result, no additional route maintenance is required. AODV has a route aging mechanism; however, it does not find out how long a link might be alive for routing purposes. The latency is minimized due to avoidance of using multiple routes. Integration of multicast routing makes AODV different from other routing protocols. AODV combines unicast, multicast, and broadcast communications; currently, it uses only symmetric links between neighboring nodes. AODV provides both a route table for unicast routes and a multicast route table for multicast routes. The route table stores the destination and next-hop IP addresses and destination sequence number. Destination sequence numbers are used to ensure that all routes are loop free, and the most current route information is used whenever route discovery is executed. In multicast communications, each multicast group has its own sequence number that is maintained by the multicast group leader. AODV deletes invalid routes by the use of a special route error message called Route Error (RERR).

Temporally Ordered Routing Algorithm (TORA) **[7].** TORA is adaptive and scalable routing algorithm based on the concept of link reversal. It finds multiple routes from source to destination in a highly dynamic mobile networking environment. An important design concept of TORA is that control messages are localized to a small set of nodes nearby a topological change. Nodes maintain routing information about their immediate one-hop neighbors. The protocol has three basic functions: route creation, route maintenance, and route erasure.

Nodes use a "height" metric to establish a directed cyclic graph (DAG) rooted at the destination during the route creation and route maintenance phases. The link can be either an upstream or downstream based on the relative height metric of the adjacent nodes. TORA's metric contains five elements: the unique node ID, logical time of a link failure, the unique ID of a node that defined the new reference level, a reflection indicator bit, and a propagation ordering parameter. Establishment of DAG resembles the query/reply process discussed in Lightweight Mobile Routing (LMR) [8]. Route maintenance is necessary when any of the links in DAG is broken. Figure 5.3 denotes the control flow for the route maintenance in TORA.

The main strength of the protocol is the way it handles the link failures. TORA's reaction to link failures is *optimistic* that it will reverse the links to re-position the DAG for searching an alternate path. Effectively, each link reversal sequence searches for alternative routes to the destination. This search mechanism generally requires a *single-pass* of the distributed algorithm since the routing tables are modified simultaneously during the outward phase of the search mechanism. Other routing

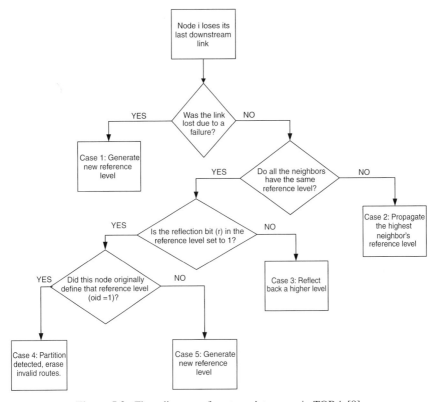

Figure 5.3. Flow diagram of route maintenance in TORA [9].

algorithms such as LMR uses two-pass whereas both DSR and AODV use three-pass procedure. TORA achieves its single-pass procedure with the assumption that all the nodes have synchronized clocks (via GPS) to create a temporal order of topological change of events. The "height" metric is dependent on the logical time of a link failure.

Associativity-Based Routing (ABR) **[10].** ABR uses the property of "associativity" to decide on which route to choose. In this algorithm, route stability is the most important factor in selecting a route. Routes are discovered by broadcasting a *broadcast query* request packet; with the assistance of these packets, the destination becomes aware of all possible routes between itself and the source. Based on these available routes, a path is selected using the associativity property of these routes.

The ABR algorithm maintains a "degree of associativity" by using a mechanism called *associativity ticks*. According to this, each node in the network maintains a tick value for each of its neighbors. Every periodic link layer HELLO message increases the tick value by one each time it is received from a neighbor. Once the tick value reaches a specified threshold value, it means that the route is *stable*. If the neighbor goes out of the range, then the tick value is reset to zero. Hence a tick level above the threshold value is an indicator of a rather stable association between these two nodes.

Once a destination has received the *broadcast query* packets, it has to decide which path to select by checking the tick-associativity of the nodes. The route with the highest degree of associativity is selected since it is considered the most stable of the available routes.

ABR is quite an effective algorithm in selecting routes because it focuses on the route stability to a great extent. However, some inherent drawbacks include memory requirements for the routing tables, excessive storage needs for storing the ticks, and additional computation to maintain the tick count along with greater power requirements.

***Signal Stability-Based Adaptive Routing (SSBR)* [11].** The SSBR differs when compared with the conventional routing algorithms. The main routing criteria is the signal and location stability. The basic routing framework is similar to any other standard on-demand routing protocol: The route request is broadcast throughout the network, the destination replies back with the route reply message, and then the sender sends data through the selected route. However, the signal strength (link quality) between neighboring nodes plays a major role in the route selection process in this protocol.

SSBR is comprised of two subprotocols: *Dynamic Routing Protocol (DRP)* and *Static Routing Protocol (SRP)*. The DRP interacts with the network interface device driver dealing with signal strengths through an API to determine the actual strength of a received signal. Using this signal information, the DRP maintains a *signal stability table* that categorizes each link with the neighboring nodes as strong or weak. This table is updated with every new packet received. For instance, if a HELLO packet is received, the signal strength is monitored and the signal stability table is upgraded, whereas for other packets such as route update packets, data packets, and so on, the packet is sent to the SRP for further processing. The SRP performs the routine tasks such as forwarding packets according to the existing routing table, replying to route requests, and so on.

The route request is given an option on the type of link it requests—that is, strong, weak or a combination of both. If the route request specifies only strong links, all the route request packets coming from a perceived weaker link are dropped. Thus, the final discovered path consists of only strong links. If there are multiple paths from source to destination using strong links, the destination can choose among them. The destination can simply choose the first route request it receives. If, however, no strong links are found, the protocol could fall back on other available weaker links.

Two enhancements to the selection process are proposed. In the first case, the link strength (strong or link) is added for each hop into the route request packet and then forwarded toward the destination. In this case, the destination does not select the first route request packet received, but waits for a period of time to choose the best route among all the route requests within a set time interval. The second improvement suggests that any intermediate node can make an unnecessary route reply for a route it already has a prior information about.

The SSBR algorithm uses a similar scheme as the ABR algorithm to determine the reliability of links before selecting a particular route. While the former uses signal strength, the latter uses ticks; however, the goal of selecting a route with greatest reliability remains the same.

The Ant-Colony-Based Routing Algorithm (ARA) [12].
This work presents a novel technique for ad hoc routing by using the concepts of *swarm intelligence* and the *ant colony* based meta heuristic. These ant colony algorithms focus on solving the complex problems by cooperation and without any direct communication. Ants communicate with each other using *stigmergy*, a means of indirect communication between individuals by modifying their operational environment.

Ants find the shortest path to a destination by making the use of *pheromones* (Figure 5.4). Initially, the ants randomly choose the path. On the way, they drop off pheromones. The ants chosen the shortest path reach the destination earlier than the others and travel back along the same path. They keep releasing additional pheromones on this path gradually increasing the pheromone concentration. The successive groups of ants coming from the source recognize the path with greater pheromone levels due to the higher concentration and hence start taking this route which ends up as the best route. This method of selecting the shortest path is used to determine the optimal route for ad hoc networks. This technique performs well with the dynamic link changes due to node mobility along with scalability issues.

Using the ant colony meta heuristic, the pheromone change of an edge $e(v_i, v_j)$ when moving from node v_i to v_j can be derived as follows: $\phi_{i,j} := \phi_{i,j} + \delta_\phi$. The

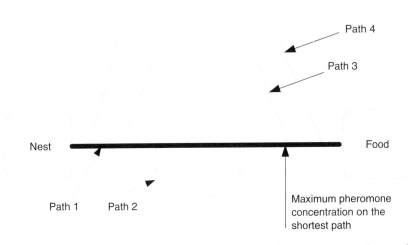

Figure 5.4. Pheromone concentration along the shortest path used by ants to discover food from their nest [12].

pheromone concentration also decreases exponentially with time according to the following equation: $\phi_{i,j} := (1-q).\phi_{i,j}$.

The algorithm has the following phases:

Route Discovery. This phase uses two types of control packet: the *forward ant* (FANT) and the *backward ant* (BANT). The FANT establishes the pheromone track to the source node, while the BANT establishes the pheromone track to the destination. When the route is required, the source broadcasts FANT packets to all its neighbors. FANTs are distinguished by unique sequence numbers. A node that receives a FANT for the first time creates a routing table record that contains the destination address, next hop, and pheromone value. The source address of the FANT is taken as the destination address, the previous node address is taken as the next hop, and the pheromone value is calculated based on the total number of hops required by the FANT to reach a particular node. When the FANT reaches the destination, the node updates its own information and sends the BANT back. Once the BANT reaches the source, the path is to be used. Figures 5.5 and 5.6 denote the forward and backward ants' route discovery phases.

Route Maintenance. No special route maintenance packets are required by ARA as it uses the transmitted data packets to maintain the route. The pheromone value of the path is increased by δ_ϕ each time a data packet is sent along this path and it also decreases according to the equation above.

Route Failure Handling. Due to node mobility, route failures which are determined by missing acknowledgments may occur frequently. If a route failure occurs, an existing alternate path is used; otherwise, the source initiates the route discovery phase.

The expected overhead of ARA is very low since the route maintenance is carried out by the data packets.

Routing On-Demand Acyclic Multipath (ROAM) [13].

The ROAM algorithm uses coordination among nodes along directed acyclic subgraphs that are defined only

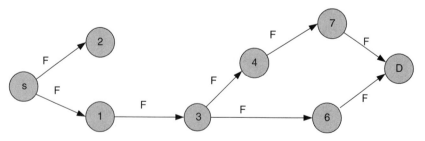

Figure 5.5. Forward ant route discovery phase. A forward ant (F) is sent from the sender (S) toward the destination node (D). The forward ant is relayed by other nodes, which initialize their routing table and the pheromone values [12].

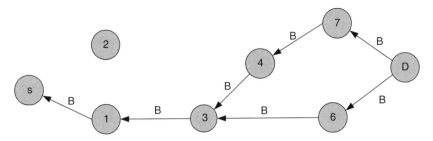

Figure 5.6. Backward ant route discovery phase. The backward ant (B) has the same task as the forward ant. It is sent by the destination node toward the source node [12].

on the routers' distances to the respective destinations. It is an extension of the DUAL [14] routing algorithm. The main motivation comes from the fact that conventional on-demand schemes tend to use flooding during route discovery repeatedly until a destination is obtained. If no route is found out initially, the source does not know whether to initiate another route discovery. This may be a problem when a malicious router indefinitely queries the network for a nonexistence route causing network congestion. Standard protocols have no mechanisms to protect against such type of attacks. In ROAM, either a search query results in the destination path or all the routers determine that the destination is unreachable.

Each router in ROAM maintains *distance*, *routing*, and *link cost* tables. While the distance table maintains the distances of nodes for each destination and neighbors from the respective node, the routing table contains a column vector containing the distance to each destination, the feasible distance, and the reported distance. The link cost table provides the link costs to each of the adjacent neighbors of the router. *Queries*, *replies*, and *updates* are the three types of control packets used in the routing protocol.

A router updates its routing table for a destination when it needs to (i) add an entry for a particular destination, (ii) modify its distance to the destination, and (iii) erase the entry for the destination.

The routers in ROAM are in either *active* or *passive* states. If a router has sent queries to all its neighbors and awaiting a reply, it is in an active state; otherwise, it is in a passive state. Selection of loop-free paths allows a router select a neighbor as its successor only if its is a *feasible successor*. This provides a shortest loop-free path to the destination and is determined by two different algorithms based on the fact that they are either passive or active. A diffusing search starts by a router when it requests a path to a destination, and this packet is propagated through routers that have no entry of the node. The first router that has an available route to the destination responds to the source with the distance to the node. At the end of this search, the source either has a finite distance to the destination or realizes that the destination is unreachable. Link costs are also updated based on the packets received.

The ROAM provides loop-free multipaths if the successors are selected using the passive and active successor algorithms. The very nature of this algorithm makes it suitable for wireless networks with limited mobility.

5.4.2 Table-Driven Protocols

Table-driven protocols always maintain up-to-date information of routes from each node to every other node throughout the network. Routing information is stored in a routing table in each of the nodes and route updates are propagated throughout the network to keep the routing information as recent as possible. Different protocols keep track of different routing state information; however, each has the common goal of reducing route maintenance overhead as much as possible. These types of protocols are not suitable for highly dynamic networks due to the extra control overhead generated to keep the routing tables consistent and fresh for each node in the network.

Destination-Sequenced Distance-Vector (DSDV) [6]. The Destination-Sequenced Distance-Vector (DSDV) is a table-driven routing protocol based on Bellman–Ford routing algorithm. Every mobile node maintains a routing table that contains all of the possible destinations in the network and each individual hop counts to reach those destinations. Each entry also stores a sequence number that is assigned by the destination. Sequence numbers are used to identify stale entries and avoidance of loops. In order to maintain routing table consistency, routing updates are periodically sent throughout the network. Two types of update can be employed; *full dump* and *incremental*. A *full dump* sends the entire routing table to the neighbors and can require multiple network protocol data units (NPDUs). *Incremental* updates are smaller updates that must fit in a packet and are used to transmit those entries from the routing table since the last full dump update. When a network is stable, incremental updates are sent and full dump are usually infrequent. On the other hand, full dumps will be more frequent in a fast moving network. The mobile nodes maintain another routing table to contain the information sent in the incremental routing packets. In addition to the routing table information, each route update packet contains a distinct sequence number that is assigned by the transmitter. The route labeled with the most recent (highest number) sequence number is used. The shortest route is chosen if any of the two routes have the same sequence number.

Analysis of a Randomized Congestion Control Scheme with DSDV Routing in Ad Hoc Wireless Networks [15]. In the randomized version of the DSDV (R-DSDV) protocol, the control messages are propagated based on a routing probability distribution instead of a periodic basis. Local nodes can tune their parameters to the traffic and route the traffic through other routes with lesser load. This implies implementing a congestion control scheme from the routing protocol's perspective.

The randomization of the algorithm is with respect to the routing table advertisement packets and the rate at which they are sent. In DSDV, whenever there is any change in the routing table, advertisement packets are propagated to update the state information at each node. R-DSDV sends these update messages only at a probability

$Pr_{n,adv}$ for a node n in the network. This can reduce the control packet overhead; however, there may be a corresponding delay in updating all the nodes. If there is a routing table update at node n, the node can sends a regular message with a probability $1 - Pr_{n,adv}$ or an update message with probability $Pr_{n,adv}$. Thus, the rate at which routing table advertisement sent is $\rho_n = F_{send} \times Pr_{n,adv}$, where F_{send} is the frequency at which a node is allowed to send a message. Piggybacking can be used with this scheme to transfer routing table updates with the regular messages.

Optimized Link State Routing (OLSR) [16]. The OLSR optimizes a pure link state because it reduces the size of information sent in each message and also reduces the total control overhead by minimizing the number of retransmissions flooding an entire network. It uses a multipoint relaying technique to flood the control messages in a network in an efficient manner.

The aim of using the multipoint relay is to reduce retransmissions within the same region. Each node selects a set of one-hop neighbors which are called the *multipoint relays (MPR)* for the node. The neighbors of the node which are not MPRs process the packets but do not forward them since only the MPRs forward the packets and the node forwards any of the broadcast messages to these MPR nodes.

The multipoint relay set must be chosen in an efficient manner to ensure that its range covers all the two-hop neighbors. This set must also be the minimum set to broadcast the least number of packets. The multipoint relay set of a node N should be such that every two-hops neighbor of N has a bidirectional link with the nodes in the MPR set of N. These bidirectional links can be determined by using periodic HELLO packets which contain information about all neighbors and their link status. Thus, a route is a sequence of hops from source to destination through multipoint relays within the network. Source does not know the complete routes only next hop information to forward the messages.

Cluster-Head Gateway Switch Routing (CGSR) [17]. The CGSR protocol (see Algorithm 2) is a clustering scheme that uses a distributed algorithm called the Least Cluster Change (LCC). By aggregating nodes into clusters controlled by cluster heads, a framework for developing additional features for channel access, bandwidth allocation, and routing is created. Nodes communicate with the cluster head, which, in turn, communicate with other cluster heads within the network (see Figure 5.7).

Selecting a cluster head is a very important task because frequently changing cluster heads will have an adverse effect on the resource allocation algorithms that depend on it. Thus cluster stability is of primary importance in this scheme. The LCC algorithm is stable in that a cluster head will change only under two conditions: when two cluster heads come within the range of each other or when a node gets disconnected from any other cluster.

CGSR is an effective way for channel allocation within different clusters by enhancing spatial reuse. The explicit requirement of CGSR on the link layer and MAC scheme is as follows: Each cluster is defined with unique CDMA code and

ALGORITHM 2. CGSR Routing Protocol [17]

// CMT = Cluster Member Table; RT = Routing Table

S wants to communicate with D S looks up the clusterhead of D
  in CMT
S looks up the next hop towards D in RT S forwards packet to
clusterhead of next hop while packet is not delivered
   if destination CH is within direct range of current CH
      current CH forwards packet to destination CH
      destination CH delivers packet to D
   else
      current CH forwards packets to Gateway Node
      gateway node sends packet to CH of next hop
   end if
end if

---

hence each cluster is required to utilize spatial reuse of codes. Within each cluster, TDMA is used with token passing.

Gateway nodes are defined as those nodes which are members of more than one cluster and therefore need to be communicating using different CDMA codes based on their respective cluster heads. The main fators affecting routing in these networks are token passing (in cluster heads) and code scheduling (in gateways). This uses a sequence number scheme as in DSDV [6] to reduce stale routing table entries and gain loop-free routes. A packet is routed through a collection of these cluster heads and gateways in this protocol.

Wireless Routing Protocol (WRP) [18]. WRP is one of the earliest works on routing algorithms and is similar to the distributed Bellman–Ford algorithm. It is a

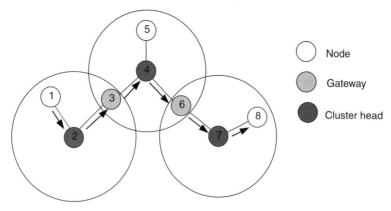

Figure 5.7. Cluster gateway switch routing [17].

table-driven protocol where routing tables are maintained for all destinations. The routing table contains an entry for each destination with the next hop and a cost metric. The route is chosen by selecting a neighbor node that would minimize the path cost. Link costs are also defined and maintained in a separate table, and various techniques are available to determine these link costs.

To maintain the routing tables, frequent routing update packets have to be sent. These are sent to all neighbors of a node and contain all the routes which the node is aware of. As the name suggests, these are just update messages and hence only the recent path changes are sent in these messages and not the whole routing table. To keep the links updated, empty HELLO packets are sent at periodic intervals only if no other update messages need to be sent. These empty HELLO packets are not required to be acknowledged specifically.

Source-Tree Adaptive Routing (STAR) **[19].** STAR is an efficient link state protocol. Each node maintains a source tree which contains preferred links to all possible destinations. Nearby soure trees exchange information to maintain up-to-date tables. A route selection algorithm is executed based on the propagated topology information to the neighbors. The routes are maintained in a routing table containing entries for the destination node and the next hop neighbor.

In this protocol, link state update (LSU) messages are used to update changes of the routes in the source trees. Since these packets do not experience timeout, no periodic messages are required. STAR protocol provides two distinct approaches: optimum routing (ORA) and least overhead routing (LORA). The ORA approach obtains the shortest path to the destination while LORA minimizes the packet overhead. STAR also requires a neighbor protocol to make sure that each node is aware of its active neighbors. The STAR protocol has been further developed as SOAR [20].

5.4.3 Hybrid Protocols

Hybrid routing schemes combine the power of on-demand and table-driven routing protocols. Static routing is generally used at the fringes of the network where route changes are not frequent while in the core of the network on-demand routing has more significance. These schemes create a bridge between the two major types of routing protocols, and the overall performance obtained can be further improved.

Zone Routing Protocol (ZRP) **[21].** ZRP is a well-known hybrid routing protocol that is most suitable for large-scale networks. Its name is derived from the use of "zones" that define the transmission radius for every participating node. This protocol uses a proactive mechanism of node discovery within a node's immediate neighborhood, while interzone communication is carried out by using reactive approaches.

ZRP utilizes the fact that node communication in ad hoc networks is mostly localized, thus the changes in the node topology within the vicinity of a node are of primary importance. ZRP makes use of this characteristic to define a framework for node communication with other existing protocols. Local neighborhoods, called

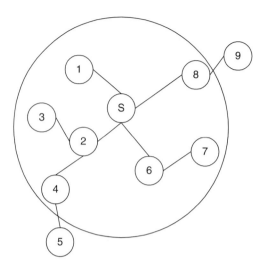

Figure 5.8. Example routing zone with $\rho = 2$.

zones, are defined for nodes. The size of a zone is based on ρ factor, which is defined as the number of hops to the perimeter of the zone. There may be various overlapping zones, which helps in route optimization. See Figure 5.8.

Neighbor discovery is accomplished by either the Intrazone Routing Protocol (IARP) or simple "Hello" packets. IARP is proactive approach and always maintains up-to-date routing tables. Since the scope of IARP is restricted within a zone, it is also referred to as a "limited scope proactive routing protocol." Route queries outside the zone are propagated by the route requests based on the perimeter of the zone (i.e., those with hop counts equal to ρ), instead of flooding the network.

The Interzone Routing Protocol (IERP) uses a reactive approach for communicating with nodes in different zones. Route queries are sent to peripheral nodes using the Bordercast Resolution Protocol (BRP). Since a node does not resend the query to the node in which it received the query originally, the control overhead is significantly reduced and redundant queries are also minimized.

ZRP provides a hybrid framework of protocols, which enables the use of any routing strategy according to various situations. It can be optimized to take full advantage of the strengths of any current protocols. The ZRP architecture can be seen in Figure 5.9.

Fisheye State Routing (FSR) [22]. FSR is a hierarchical routing protocol which aims at reducing control packet overhead by introducing the multilevel scopes. It is essentially a table-driven protocol that implements the "fisheye" technique proposed in Kleinrock and Stevens [23]. This technique is very effective to reduce the size of information required to represent graphical data. It uses the concept that the eye of a fish captures with greater detail the view nearer to the focal point while detail decreases as the distance from the focal point increases.

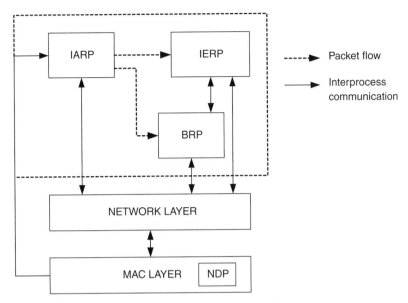

Figure 5.9. ZRP architecture [21].

FSR is similar to link state routing, because it maintains a routing table at each node. The only difference is in the maintenance of these tables. FSR introduces the *scopes* concept, which depends on the number of hops a packet traveled from its source. A higher frequency of update packets are generated for nodes within smaller scope whereas for farther-away nodes, updates are fewer in general. Each node maintains a local topology map of the shortest paths which is exchanged periodically between the nodes.

Fisheye state routing allows distinct exchange periods for different entries in the routing tables. These scopes are considered based on the distance between each node. The foremost benefit is the reduction of the message size, since the routing information of the far-away nodes is omitted. With an increase in size of the network, a "graded" frequency update plan can be adopted across scopes to minimize the overall overhead. This protocol scales well to large size of networks while keeping the control overhead low without compromising on the accuracy of route calculations. Routes to farther destinations may seem stale; however, they become increasingly accurate as a packet approaches its destination.

Landmark Ad Hoc Routing (LANMAR) **[24].** This protocol combines properties of link state and distance vector algorithms and builds subnets of groups of nodes which are likely to move together. A *landmark* node is elected in each subnet, similar to FSR [22]. The key difference between the FSR protocol and the LANMAR protocol is that the LANMAR routing table consists of only the nodes within the scope and landmark nodes, whereas FSR contains the entire nodes in the network its table.

During the packet forwarding process, the destination is checked to see if it is within the forwarding node's neighbor scope. If so, the packet is directly forwarded to the address obtained from the routing table. On the other hand, if the packet's destination node is much farther, the packet is first routed to its nearest landmark node. As the packet gets closer to its destination, it acquires more accurate routing information; thus in some cases it may bypass the landmark node and become routed directly to its destination. The link state update process is again similar to the FSR protocol. Nodes exchange topology updates with their one-hop neighbors. A distance vector, which is calculated based on the number of landmarks, is added to each update packet. As a result of this process, the routing tables entries with smaller sequence numbers are replaced with larger ones.

***Relative Distance Micro-discovery Ad Hoc Routing (RDMAR)* [25].** The RDMAR protocol is very similar to existing reactive protocols since it uses the two standard phases of route discovery and route maintenance. However, route discovery broadcast messages are limited by a maximum number of hops which is calculated using the relative distance between the source and destination. Each node also maintains a routing table that contains the next hop neighbor of each known destination, an estimated relative distance between all known source and destination nodes, a timestamp at which the current entry was made, a timeout field indicating the time at which a particular route is no longer active, and a flag specifying if a route still exists or not.

The estimated distances are measured by the source nodes using the last known distance between the respective nodes, the last time when the route was updated, and also the estimated speed of the destination node. Each node also maintains two other data structures: (a) a *data retransmission buffer* that queues data being transmitted until an explicit acknowledgment is received and (b) a *route request table* that stores all necessary information which pertains to the most recent route discovery.

Route discovery and route maintenance is carried out by broadcasting route request packets and expecting a route reply packet from the destination. Each node also occasionally probes for bidirectional links by sending a packet on the link where it has just received a packet. Route maintenance is performed when a route failure occurs and the node re-sends the data up to a maximum number of retries. This is why the intermediate nodes buffer data packets until they receive link level acknowledgments from the next hop node. When a link failure occurs at an intermediate node close to the destination, this node sets the "emergency" flag in its route request packets such that it increases the possibility of a faster recovery time. If, however, the route has completely failed, the intermediate node forwards a *failure notification* to the source node by unicasting it to all neighboring nodes involved. When a node receives a failure notification, it updates its routing tables accordingly.

***Scalable Location Update-Based Routing Protocol (SLURP)* [26].** The SLURP focuses on developing an architecture scalable to large-size networks. A location update mechanism maintains location information of the nodes in a decentralized fashion by mapping node IDs to specific geographic subregions of the network where any node located in this region is responsible for storing the current location information for all the nodes situated within that region. When a sender wishes to send a packet to a destination, it first queries nodes in the same geographic subregion of the destination to get a rough estimate of its position. It then uses a simple geographic routing protocol to send the data packets. Since the location update cost is dependent on the speed of the nodes, for high speeds, a larger number of location update messages are generated. By theoretical analysis, it is shown that the routing overhead scales as $O(v)$ where v is the average node speed and that it and also scales as $O(N^{3/2})$ where N is the number of nodes within the network. It can be noted that the routing packet overhead scales linearly with respect to node speeds and with $N^{3/2}$ with the present number of nodes within the network.

***A^4LP Routing Protocol* [27, 28].** A^4LP is specifically designed to work in networks with asymmetric links. The routes to In-, Out-, and In/Out-bound neighbors are maintained by periodic neighbor update and immediately available upon request, while the routes to other nodes in the network are obtained by a path discovery protocol. A^4LP proposes an advanced flooding technique: *m-limited forwarding*. Receivers can rebroadcast a packet only if it qualifies a certain *fitness* value specified by the sender. The flooding cost is reduced and shortest high-quality path is likely to be selected by using *m*-limited forwarding. Moreover, the metrics used to choose from multiple paths are based on the *power consumed per packet* and *transmission latency*. A^4LP, is also both *location-* and *power-aware* routing protocol supporting asymmetric links that may be suitable for heterogeneous MANET.

5.4.4 Location-Aware Protocols

Location-aware routing schemes in mobile ad hoc networks assume that the individual nodes are aware of the locations of all the nodes within the network. The best and easiest technique is the use of the Global Positioning System (GPS) to determine exact coordinates of these nodes in any geographical location. This location information is then utilized by the routing protocol to determine the routes.

***Location-Aided Routing (LAR)* [29, 30].** The LAR protocol suggests an approach that utilizes location information to minimize the search space for route discovery toward the destination node. The aim of this protocol is to reduce the routing overhead for the route discovery, and it uses the Global Positioning System (GPS) to obtain the location information of a node.

The intuition behind using location information to route packets is very simple and effective. Once the source node knows the location of the destination node and also has some information of its mobility characteristics such as the direction and speed of movement of the destination node, the source sends route requests to nodes only in the "expected zone" of the destination node. Since these route requests are flooded throughout the nodes in the expected zone only, the control packet overhead is considerably reduced. If the source node has no information about the speed and the direction of the destination node, the entire network is considered as the expected zone.

Before sending a packet, a source node determines the location of the destination node and defines its "request zone," the zone in which it initiates flooding with the route request packets. In some cases, the nodes outside the request zone may also be included. If the source node is not inside the destination node's expected zone, the request zone has to be increased to accommodate the source node. Also, a situation may occur where all neighboring nodes of the destination node may be located outside the request zone. In this case, the request zone has to be increased to include as many neighboring nodes as possible.

LAR defines two schemes to identify whether a node is within the request zone.

- *Scheme 1.* In this scheme, the source node simply includes the smallest rectangle containing the current location of the source node and the expected zone of the destination node based on its initial location and current speed. The speed factor may be varied to either include the current speed or the maximum obtainable speed within the network. This expected zone will be a circle centered at the initial location of the destination node with a radius dependent on its speed of the movement. The source node sends the route request packets with the coordinates of the entire rectangle. The nodes receiving these packets check to see whether their own locations are within the zone. If so, they forward the packet using the regular flooding algorithm; otherwise, the packets are simply dropped.
- *Scheme 2.* Here, the source node calculates the distance between itself and the destination node based on the GPS coordinates and includes these values within the route request packets. An intermediary node receiving this packet calculates its distance from the destination. If its distance from the destination is greater than that of the source, the intermediary node is not within the request zone and hence drops the packet. Otherwise, it forwards the packet to all its neighbors.

LAR essentially describes how location information such as GPS can be used to reduce the routing overhead in an ad hoc network and ensure maximum connectivity. See Figure 5.10.

Distance Routing Effect Algorithm for Mobility (DREAM) [31]. The
DREAM protocol (Algorithm 3) also uses the node location information from GPS systems for communication. DREAM is a part proactive and part reactive protocol where the source node sends the data packet "in the direction" of the destination node by selective flooding. The difference from the other location-based protocols is that only the data packets are forwarded to the next hop neighbor, not the control packets. Each node maintains a table with the location information of each node, and the periodic location updates are distributed among the nodes to keep this information as up-to-date as possible. Collectively updating location table entries indicates the proactive nature of the protocol, while the fact that all intermediate nodes in a

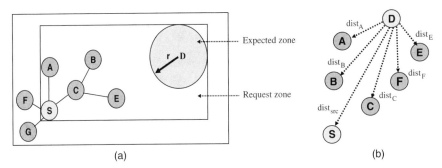

Figure 5.10. LAR routing protocol. The diagrams (a) and (b) present LAR1 and LAR2 schemes [29, 30].

route perform a lookup and forward the data packet in the general direction of the destination reflects DREAM's reactive properties.

DREAM is based on two classical observations: the *distance effect* and the *mobility effect*. The distance effect states that the greater the distance between two nodes, the slower they appear to move with respect to each other. Hence, the location information tables can be updated depending on the distance between the nodes without making any concessions on the routing accuracy. Two nodes situated farther apart view the other to be moving relatively slowly, requiring less frequent location updates compared with nodes closer to each other. The mobility effect determines how often the location information packets can be generated and forwarded. In an ideal scenario, whenever a node moves, it should update entire the network but not generate any packets if it remains idle. However, a node keeps generating location update packets at periodic intervals that can be a function of the node's mobility. Thus, the nodes with higher mobility generate more frequent location update messages. This allows

ALGORITHM 3. DREAM Routing Protocol: Send Procedure [31]

// LT = Location Table

Find destination node from packet header D
if (no LT entry for D or the information is not valid)
   invoke recovery mode
else
   find existing neighbors from the LT
   if no neighbors exist
     invoke recovery mode
   else
     set the timer when the packet is forwarded
     transmit the packet to all the neighbors
   end if
end if

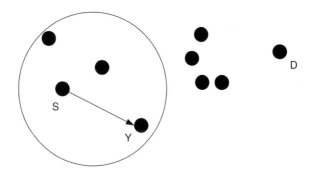

Figure 5.11. *Y* is *S*'s closest neighbor in greedy forwarding [32].

each node to send control packets based on their mobility and helps to reduce the overhead by a great extent.

Since DREAM does not need any route discovery procedure, it does not incur the delay seen in other reactive protocols. It is energy- and bandwidth-efficient because control message generation is optimized with respect to node mobility. In addition, it is also inherently loop-free, robust, and, more importantly, adaptive to mobility.

Greedy Perimeter Stateless Routing (GPSR) [32]. Similar to the other location-based protocols, GPSR also uses the location of the node to selectively forward the packets based on the distance. The forwarding is carried out on a *greedy* basis by selecting the node closest to the destination (Figure 5.11). This process continues until the destination is reached. However, in some scenarios, the best path may be through a node which is farther in geometric distance from the destination. In this case, a well-known right-hand rule is applied to move around the obstacle and resume the greedy forwarding as soon as possible.

Let us note that the location information is shared by beacons from the MAC layer. A node uses a simplistic beaconing algorithm to broadcast beacon packets containing the node ID and its *x* and *y* coordinates at periodic intervals, helping its neighbors to keep their routing tables updated. With greater mobility, the beaconing interval must be reduced to maintain up-to-date routing tables; however, this results in greater control overhead. To reduce this cost, the sender node's location information is piggybacked with the data packets.

Location Aided Knowledge Extraction Routing for Mobile Ad Hoc Networks (LAKER) [33]. This protocol minimizes the network overhead during the route discovery process by decreasing the zonal area in which route request packets are forwarded. During this process, LAKER extracts knowledge of the nodal density distribution of the network and remember a series of "important" locations on the path to the destination. These locations are named "guiding routes," and with the help of these guiding routes the route discovery process is narrowed down.

LAKER uses the same forwarding strategy as DSR and caches the *forwarding routes* and also creates its own *guiding routes*. While a forwarding route is a series of nodes from the source to the destination, the guiding route contains a series of locations along this route where there may be a cluster of nodes. Even though individual nodes may move around a bit, the basic cluster topology generally remains similar for an extended period of time. Thus, the information found in the first route discovery round is stored and used during the subsequent route discoveries. Since LAKER uses the guiding route caches in route discovery, the mobility model chosen becomes very important. The restricted random waypoint mobility model [34], which is an extension of Broch et al. [35], is used. Since LAKER is a descendant of DSR and LAR, it uses an on-demand request-reply mechanism for route discovery. The control packet format in LAKER contains characteristics of both DSR and LAR along with their forwarding route and guiding route metrics that aim to decrease the forwarding area of these route request packets as described earlier. Figure 5.12 shows that the request zone in LAKER is more specific in comparison with LAR; therefore, there is a greater probability of accuracy in determining the exact location of the destination node. Knowledge extraction is carried out by keeping track of the number of neighbors of each node until a certain threshold value is reached. The route reply packet forwards these discovered forwarding and guiding routes to the source.

Movement-Based Algorithm for Ad Hoc Networks (MORA) [36].
In addition to forwarding packets based on the location information, MORA also takes into account the direction of the movement of the neighboring nodes. The metric used for

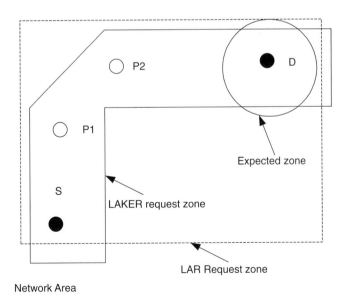

Figure 5.12. Request zone LAKER versus LAR [33].

making the forwarding decision is a combination of the number of hops that have an arbitrary weight assigned and a function independent of each node.

While calculating this function F, the primary goal remains to make full use of the directions of the neighboring nodes' movement in selecting the optimal path from source to destination. The function F should depend upon the distance of the node from the line joining the source and destination (sd) and the direction it is moving toward. The function should reach the maxima when the node is moving on sd and should decrease with an increase in the distance from this line. The MORA protocol has two versions:

1. *UMORA*. This is the Unabridged-MORA version because it is very similar to source routing on IP networks. Here, a short message (called a *probe*) is used to localize the position of the destination. The destination sends a probe along various different routes. Each node receiving this packet keeps updating its own weight function accordingly. After a fixed period of time, the source has all the paths to the destination and corresponding weight functions and selects the most suitable path based on this information.

2. *D-MORA*. This version is the Distributed MORA. It is a scalable algorithm and uses a single path from source to destination. A short probe message is also forwarded from the destination to the source. In every k hop, the node receiving the packet polls for information from the neighboring nodes. The packet is then forwarded to the node with the higher link weight. The path information is attached to the packet header and forwarded to the next node.

5.4.5 Multipath Protocols

Multipath routing protocols create multiple routes from source to destination. The main advantage of discovering multiple paths is that the bandwidth between links is used more effectively with greater delivery reliability. It also helps during times of the network congestion. Multiple paths are generated on demand or by using a proactive approach and are of great significance because routes generally get disconnected quickly due to node mobility.

CacHing and Multipath Routing Protocol (CHAMP) [37]. The CHAMP protocol uses data caching and shortest multipath routing. It also reduces packet drops in the presence of frequent route breakages. Every node maintains a small buffer for caching the forwarded packets. This technique is helpful in the case when a node close to the destination encounters a forwarding error and cannot transmit the packet. In such a situation, instead of the source retransmitting again, an upstream node that has a cached copy of the packet may retransmit it, thereby reducing end-to-end packet delay. In order to achieve this, multiple paths to the destination must be available.

In CHAMP each node maintains two caches: (a) a route cache containing forwarding information and (b) a route request cache that contains the recently received and processed route requests. Those entries that have not been used for a specific route lifetime are deleted from the route cache. A node also maintains a send buffer for waiting packets and a data cache for storing the recently forwarded data packets. A route discovery is initiated when there is no available route. The destination replies back with a corresponding route reply packet. There may be multiple routes of equal length established, each with a forwarding count value that starts with a zero from the source and is increased by one with every retransmission.

Ad Hoc On-Demand Multipath Distance Vector Routing (AOMDV) [38].

The AOMDV protocol (Algorithm 4) uses the basic AODV route construction process, with extensions to create multiple loop-free and link-disjoint paths. AOMDV mainly computes the multiple paths during route discovery process, and it consists of two main components: (a) a rule for route updates to find multiple paths at each node and (b) a distributed protocol to calculate the link-disjoint paths.

In this protocol, each route request and route reply packet arriving at a node is potentially using a different route from the source to the destination. All of these routes cannot be accepted, since they can lead to creation of loops (see Figure 5.13).

The proposed "advertised hop count" metric is used in such a scenario. The advertised hop count for a particular node is the maximum acceptable hop count for any path recorded at that node. A path with a greater hop count value is simply discarded, and only those paths with a hop count less than the advertised value is accepted. Values greater than this threshold means the route most probably has a loop.

The following proven property allows us to have disjoint routes [38]: *Let a node S flood a packet m in the network. The set of copies of m received at any node I (not equal S), each arriving via a different neighbor of S, defines a set of node-disjoint*

ALGORITHM 4. AOMDV Routing Protocol: Route Update Rules [38]

// seqnum(d,i)= Sequence number for destination d at node i
// advertised_hopcount(d,i) = Advertised hop count for d at i
// route_list(d,i) = Route list for d at i

if (seqnum(d,i) is less than seqnum(d,j))
  initialize seqnum(d,i) to seqnum(d,j)
  if (i is not equal d)
    initialize advertised_hopcount(d,i) to infinity
    initialize the route_list(d,i) to NULL
    insert j and advertised_hopcount(d,j) + 1 to route_list (d,i)
  else
    initialize advertised_hopcount(d,i) to 0
  end if
else if ((seqnum(d,i) equals seqnum(d,j)) and
    ((advertised_hopcount(d,i),i) is greater than advertised_hopcount(d,j),j))
  insert j and advertised_hopcount(d,j) + 1 to route_list(d,j)
end if

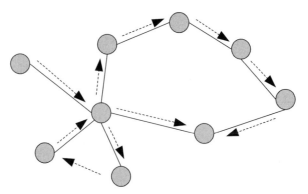

Figure 5.13. Example of a potential routing loop scenario with multiple path computation [38].

paths from I to S. This distributed protocol is used in the intermediate nodes where multiple copies of the same route request packet is not immediately discarded. Each packet is checked to find whether it provides a node-disjoint path to the source.

Split Multipath Routing (SMR) **[39].** The SMR protocol establishes and uses multiple routes of maximally disjoint paths from source to destination (Algorithm 5). Similar to any reactive multipath protocol, multiple routes are discovered on demand and the route with the shortest delay is selected.

When a node wants to send a packet to a destination for which a route is not known, it floods a RREQ packet into the network. Due to flooding, several duplicate RREQ messages reach the destination along various different paths. Source routing is used because the destination needs to select multiple disjoint paths to send the RREP packet.

Unlike the conventional routing protocols such as AODV and DSR, the intermediate nodes in SMR are not allowed to send back RREPs even if they have the route to the destination. This is because the destination can only make a decision on the validity of maximally disjoint multiple paths from all of its received RREQ packets. If the intermediate nodes reply back, it is almost impossible for the destination to keep track of the routes forwarded to the source. Intermediate nodes also use a different packet forwarding approach. Instead of dropping all duplicate RREQs, each node only forwards those RREQ packets that arrived using a different link from the first RREQ packet and that have a hop count lower than the first RREQ packet.

The destination considers the first received RREQ packet as the path with the shortest delay. It immediately sends an RREP back an minimize the route acquisition latency. To find the maximal disjoint path to the already replied route, it waits for additional time to determine all possible route instances. In some cases, there may be more than one maximal disjoint route and, if so, the shortest hop distance route is selected.

ALGORITHM 5. SMR Routing Protocol [39]

```
// Source(S) has data to send but has no route to
   Destination(D)

Transmit a RREQ containing source ID and sequence number
if (receiving node is not D)
   check sequence number for RREQ and incoming link
   if (duplicate packet and different incoming link)
      forward the packet to the neighbor nodes
   else
      drop the packet
   end if
else RREQ at destination
   perform route selection
end if

D receives first RREQ packet
D sends back RREP along the source route in RREQ
D waits for a threshold time duration
At the end of the time interval, D selects the maximally disjoint route from the route already
   chosen and the one with the shortest latency
D sends another RREP to S along this new route
S can use any or both routes to send packets
```

***Neighbor-Table-Based Multipath Routing (NTBR)* [40].** An initial theoretical analysis showing that nondisjoint multipath routing has a higher route reliability than the conventional disjoint multipath routing led to the development of a neighbor-table-based multipath protocol that does not require the disjoint routes. In NTBR, every node maintains a neighbor table that records routing information related to its k hop neighbors. k can be set to any value; however, the control overhead also increases with an increase of this value.

The NTBR protocol has route discovery and route maintenance mechanisms. It also maintains a route cache at each node in the network. The route caches are maintained by the neighbor tables which also serve to make an estimate of the lifetime of the wireless links. This information is used to keep track of the route lifetime. Every node transmits periodic beacon packets to its two-hop neighbors. The neighbor table is established based on the information from the beacon packet by using any of the following approaches:

- *Time-Driven*: In this approach, each node essentially waits for a predefined time-out interval before deciding whether a link is active or not. The node waits for a beacon packet from either its one-hop or two-hop neighbors and adds the information to its neighbor routing table. However, the problem with this approach is

that there is always a timeout between the actual topology change and the time in which this information is realized by the node.
- *Data-Driven*. This approach alleviates the problem arising from the time driven mechanism. In this scheme, one field of the beacon packet is used to inform whether a node is unreachable or not. The address of the unreachable node is added into the beacon packet, and all nodes receiving the packet update their neighbor routing tables accordingly.

The route cache contains all the routing information for a particular node, and it is updated by monitoring any packet passing through the network. To extract individual routes, the *route extraction reason* mechanism is used which simply prioritizes the routes extracted from different packets. The routes from route replies are assigned the highest priority, while the routes from route request packets or neighbor tables become second followed by the routes from data packets. These priorities are used during the route selection process. The route discovery and route maintenance are similar to the other routing algorithms.

5.5 CONCLUSIONS

In this chapter, we have discussed various routing protocols in ad hoc networks. The common goals of designing a routing algorithm is to reduce control packet overhead, maximize throughput, and minimize the end-to-end delay; however, they differ in ways of finding and/or maintaining the routes between source–destination pairs. We have divided the ad hoc routing protocols into five categories: (i) source-initiated (reactive or on demand), (ii) table-driven (proactive), (iii) hybrid, (iv) location-aware (geographical), and (v) multipath. We have then compared the protocols based on common characteristics (see Table 5.1).

5.6 EXERCISES

1. If N is the total number of nodes operating in the network, calculate the control packet overhead complexity and the memory complexity of the following protocols: DSDV, GSR, DREAM, and OLSR.
2. Route discovery and route maintenance are two important characteristics of reactive routing protocols. If N is the number of nodes in the network and D is the diameter of the network, calculate the time and communication complexity of both the route discovery and route maintenance operations for the following protocols: AODV, DSR, and TORA.
3. It has been shown that the net throughput in DSDV decreases at a greater rate than in AODV. What could be the main reason for this? What other network characteristics can be the deciding factors in determining the effectiveness of deploying one of these protocols in a real ad hoc network?

TABLE 5.1. Comparison of Routing Protocols

Protocol	Category	Metrics	Route Recovery	Route repository	Loop Free	Communication Overhead	Feature
DSR	Reactive	Shortest path, next available	New route, notify source	Route cache	Yes	High	Completely on demand
AODV	Reactive	Newest route, shortest path	Same as DSR, local repair	Routing table	Yes	High	Only keeps track of next hop in route
TORA	Reactive	Shortest path, next available	Reverse link	Routing table	Yes	High	Control packets localized to area of topology change
ABR	Reactive	Strongest associativity	Local broadcast	Routing table	Yes	Medium	High delays in route repair
SSBR	Reactive	Strongest signal strength	New route, notify source	Routing table	Yes	Medium	Uses a signal stability table
ARA	Reactive	Shortest path	Alternate route, backtrack	Routing table	Yes	Medium	Uses swarm intelligence concepts
ROAM	Reactive	Shortest path	Erase route, start new search	Routing table	Yes	Low	Removes count-to-infinity problem
DSDV	Proactive	Shortest path	Periodic broadcast	Routing table	Yes	High	Distributed algorithm
R-DSDV	Proactive	Shortest path	Randomized updates	Routing table	Yes	Low	Probablistic table updates
OLSR	Proactive	Shortest path	Periodic updates	Routing table	Not always	High	Uses MPRs as routers
CGSR	Proactive	Shortest path	Periodic updates	Routing table	Yes	Low	Clusterhead is critical node
WRP	Proactive	Shortest path	Periodic updates	Routing table	Yes	Low	Uses HELLO messages
STAR	Proactive	Shortest path	Specific updates	Routing table	Yes	Low	Updates at specific events

(*continued*)

TABLE 5.1. (Continued)

Protocol	Category	Metrics	Route Recovery	Route repository	Loop Free	Communication Overhead	Feature
ZRP	Hybrid	Shortest path	Start repair at failure point	Interzone, intrazone tables	Yes	Medium	Routing range defined in hops
FSR	Hybrid	Scope range	Notify source	Routing tables	Yes	Low	Updates are localized
LANMAR	Hybrid	Shortest path	Notify source	Routing table at landmark	Yes	Medium	Using landmarks increases scalability
RDMAR	Hybrid	Shortest path	New route, notify source	Routing table	Yes	High	Localized query flooding
SLURP	Hybrid	MFR for interzone, DSR for interzone	Notify source	Route cache at location	Yes	High	Eliminates global route discovery
A4LP	Hybrid	Power consumed	Notify source	Routing table	Yes	Medium	Uses asymmetric links
LAR	Hybrid	Hop count	Notify source	Route cache	No	Medium	RERR message on link break
DREAM	Geographical	Hop count	Any available method	Routing table	No	Low	Location table at each node
GPSR	Geographical	Shortest path	N/A	Route cache	Yes	High	Greedy and perimeter forwarding
LAKER	Geographical	Hop count	Notify source	Route cache	No	Low	Knowledge guided route discovery
MORA	Geographical	Weighted hop count	New route, notify source	Routing table	No	High	RREQ based on metric "m"
CHAMP	Multipath	Shortest path	Notify source	Route cache	Yes	High	Performs load balancing
AOMDV	Multipath	Advertised hop count	Local repair	Routing table	Yes	High	Multipath extensions to AODV
SMR	Multipath	Least delay	Notify source	Routing table	Yes	High	Two disjoint routes chosen
NTBR	Multipath	Link active	Notify source	Route cache	Yes	High	Nondisjoint routes selected

4. DSDV and AODV are both deployed on an experimental network under the same network characteristics for a comparative study. Which protocol will have higher bandwidth consumption, and why? Note that DSDV sends periodic route broadcasts while AODV carries out route maintenance using periodic "HELLO" packets. Considering this property, which protocol would accumulate a greater control overhead?
5. List the major points of distinction between the LAR and the DREAM routing protocols. LAR routes packets by using the GPS system to determine the location of the nodes in the network. However, does it take into account the presence of obstructions in the network area? Will such obstructions degrade the protocol's performance or is there any technique to route "around" these obstructions?
6. Describe the primary disadvantages of multipath approaches to routing in comparison with unipath protocols. Multipath reactive protocols do not allow intermediate nodes to reply to RREQ packets. Why? Is it beneficial for the network? Explain your reasoning.
7. Multipath routing brings in the traffic allocation problem. If a source has a set of paths to the destination, how should it allocate traffic to each of the different routes? Traffic allocation is the job of the transport layer, but there seems to be a need of some cross-layer interactions in determining the appropriate route and send the correct amount of traffic on most suitable routes. Stronger links should be allocated more traffic than other weaker links. Discuss probable solutions and issues in determining the appropriate traffic quantities in the presence of multiple paths.
8. Routing is carried out under a varied set of constraints and requirements. Candidate selection for a node has been shown to be an NP-Complete problem. It is almost impossible to optimize all parameters (for example, hop count, control packet overhead, and link quality, among others) when a route is selected. Hence, what should be the major design goal a researcher who proposes a new routing protocol for ad hoc networks?

REFERENCES

1. C. Perkins, editor. *Ad Hoc Networking*, Addison-Wesley, Reading, MA, 2001.
2. M. Weiser. Some computer science issues in ubiquitous computing. *ACM SIGMOBILE Mobile Computing and Communications Review*, **3**(3):12, (1999).
3. B. Bollobas. *Random Graphs*, Academic Press, New York, 1985.
4. D. Johnson and D. Maltz. Dynamic source routing in ad hoc wireless networks. In T. Imielinski and H. Korth, editors, *Mobile Computing*, Kluwer Academic Publishers, Norwell, MA 1996, pp. 153–181.
5. C. Perkins and E. Royer. Ad hoc on-demand distance vector routing. In *Proceedings of the Second IEEE Workshop on Mobile Computing Systems and Applications*, 1999, pp. 99–100.

6. C. Perkins and P. Bhagwat. Highly dynamic destination-sequenced distance-vector routing (DSDV) for mobile computers. In *Proceedings of the ACM SIGCOMM*, 1994, pp. 234–244.
7. V. Park and M. Corson. A highly adaptive distributed routing algorithm for mobile wireless networks. In *Proceedings of the IEEE INFOCOM*, 1997, pp. 1405–1413.
8. M. S. Corson and A. Ephremides. A distributed routing algorithm for mobile wireless networks. *ACM/Baltzer Wireless Networks Journal*, 1(1):61–81, (1995).
9. Temporally ordered routing algorithm (TORA), http://wiki.uni.lu/secan-lab/Temporally-Ordered+Routing+Algorithm.html.
10. C.-K. Toh. Associativity-based routing for ad hoc mobile networks. *Wireless Personal Communications Journal, Special Issue on Mobile Networking and Computing Systems*, 4(2):103–139, (1997).
11. R. Dube, C. Rais, K. Wang, and S. Tripathi. Signal stability-based adaptive routing (SSA) for ad hoc mobile networks. *IEEE Personal Communications Magazine*, 4:36–45, (1997).
12. M. Gunes, U. Sorges, and I. Bouazizi. ARA—the ant-colony based routing algorithm for MANETS. In *Proceedings of the IEEE ICPP Workshop on Ad Hoc Networks (IWAHN)*, 2002, pp. 79–85.
13. J. Raju and J. Garcia-Luna-Aceves. A new approach to on-demand loop-free multipath routing. In *Proceedings of the IEEE International Conference on Computer Communications and Networks (IC3N)*, 1999, pp. 522–527.
14. J. Garcia-Luna-Aceves. Loop-free routing using diffusing computations. *IEEE/ACM Transactions on Networking*, 1(1):130–141, (1993).
15. A. Boukerche, S. K. Das, and A. Fabbri. Analysis of a randomized congestion control scheme with DSDV routing in ad hoc wireless networks. *Journal of Parallel and Distributed Computing*, 61(7):967–995, (2001).
16. T. Clausen, P. Jacquet, A. Laouiti, P. Muhlethaler, A. Qayyum, and L. Viennot. Optimized link state routing protocol for ad hoc networks. In: *Proceedings of the IEEE INMIC*, 2001, pp. 62–68.
17. C. Chiang, H. Wu, W. Liu, and M. Gerla. Routing in clustered multihop, mobile wireless networks. In *Proceedings of IEEE SICON*, 1997, pp. 197–211.
18. S. Murthy and J. Garcia-Luna-Aceves. An efficient routing protocol for wireless networks. MONET 1(2):183–197, (1996).
19. J. Garcia-Luna-Aceves and M. Spohn. Source-tree routing in wireless networks. In: *Proceedings of the IEEE ICNP*, 1999, pp. 273–282.
20. S. Roy and J. Garcia-Luna-Aceves. Using minimal source trees for on-demand routing in ad hoc networks. In *Proceedings of the IEEE INFOCOM*, 2001, pp. 1172–1181.
21. P. Samar, M. Pearlman, S. Haas, Independent zone routing: an adaptive hybrid routing framework for ad hoc wireless networks, *IEEE/ACM Transactions on Networking*, 12(4): 595–608, 2004.
22. G. Pei, M. Gerla, and T.-W. Chen. Fisheye state routing in mobile ad hoc networks. In *Proceedings of ICDCS Workshop on Wireless Networks and Mobile Computing*, 2000, pp. D71–D78.
23. L. Kleinrock and K. Stevens. Fisheye: A lenslike computer display transformation. Technical Report, Department of Computer Science, UCLA, 1971.

24. G. Pei, M. Gerla, and X. Hong. LANMAR: landmark routing for large scale wireless ad hoc networks with group mobility. In: *Proceedings of the ACM MobiHoc*, 2000, pp. 11–18.
25. G. N. Aggelou and R. Tafazolli. RDMAR: A bandwidth-efficient routing protocol for mobile ad hoc networks. In *Proceedings of the IEEE WoWMoM*, 1999, pp. 26–33.
26. S.-C. Woo and S. Singh. Scalable routing protocol for ad hoc networks. *Wireless Networks*, 7(5):513–529, 2001.
27. G. Wang, Y. Ji, D. C. Marinescu, and D. Turgut. A routing protocol for power constrained networks with asymmetric links. In *Proceedings of the ACM Workshop on Performance Evaluation of Wireless Ad Hoc, Sensor, and Ubiquitous Networks (PE-WASUN)*, 2004, pp. 69–76.
28. G. Wang, D. Turgut, L. Bölöni, Y. Ji, and D. Marinescu. Improving routing performance through m-limited forwarding in power-constrained wireless networks. *Journal of Parallel and Distributed Computing (JPDC)*. To appear.
29. Y. Ko and N. Vaidya. Location-Aided Routing (LAR) in mobile ad hoc networks. In *Proceedings of the ACM MOBICOM*, 1998, pp. 66–75.
30. Y. Ko, N. Vaidya, Location-Aided Routing (LAR) in mobile ad hoc networks. *Wireless Networks*, 6(4):307–321, 2000.
31. S. Basagni, I. Chlamtac, V. Syrotiuk, and B. Woodward. A distance routing effect algorithm for mobility (DREAM). In *Proceedings of the ACM MOBICOM*, 1998, pp. 76–84.
32. B. Karp and H. Kung. GPSR: Greedy perimeter stateless routing for wireless networks. In: *Proceedings of the ACM MOBICOM*, 2000, pp. 243–254.
33. J. Li and P. Mohapatra. LAKER: Location aided knowledge extraction routing for mobile ad hoc networks. In *Proceedings of the IEEE WCNC*, 2003, pp. 1180–1184.
34. L. Blazevic, S. Giordano, and J.-Y. L. Boudec. Self organized routing in wide area mobile ad-hoc network. In *Proceedings of the IEEE GLOBECOM*, 2001, pp. 2814–2818.
35. J. Broch, D. Maltz, D. Johnson, Y.-C. Hu, and J. Jetcheva. A performance comparison of multi-hop wireless ad hoc network routing protocols. In *Proceedings of the ACM MOBICOM*, 1998, pp. 85–97.
36. G. Boato and F. Granelli. MORA: A movement-based routing algorithm for vehicle ad hoc networks. In *Proceedings of the International Conference on Distributed Multimedia Systems (DMS)*, 2004, pp. 171–174.
37. A. Valera, W. K. G. Seah, and S. V. Rao. Cooperative packet caching and shortest multipath routing in mobile ad hoc networks. In *Proceedings of the IEEE INFOCOM*, Vol. 1, 2003, pp. 260–269.
38. M. K. Marina, and S. Das. On-demand multi path distance vector routing in ad hoc networks. In *Proceedings of the IEEE ICNP*, 2001, pp. 14–23.
39. S. Lee and M. Gerla. Split multipath routing with maximally disjoint paths in ad hoc networks. In *Proceedings of the IEEE ICC*, Vol. 10, 2001, pp. 3201–3205.
40. Z. Yao, J. Jiang, P. Fan, Z. Cao, and V. Li. A neighbor-table-based multipath routing in ad hoc networks. In *Proceedings of the IEEE VTC 2003—Spring*, Vol. 3, 2003, pp. 1739–1743.

CHAPTER 6

Adaptive Backbone Multicast Routing for Mobile Ad Hoc Networks

CHAIPORN JAIKAEO

Department of Computer Engineering, Kasetsart University, Bangkok, Thailand

CHIEN-CHUNG SHEN

Department of Computer and Information Sciences, University of Delaware, Newark, DE 19716

6.1 INTRODUCTION

Mobile ad hoc networks consist of mobile nodes that autonomously establish connectivity via multihop wireless communications. Without relying on any existing, preconfigured network infrastructure or centralized control, mobile ad hoc networks can be instantly deployed and hence are useful in many situations where impromptu communication facilities are required. Typical examples include battlefield communications and disaster relief missions, in which communication infrastructures are not available or have been torn down. For many applications, nodes need collaboration to achieve common goals and are expected to communicate as a group rather than as pairs of individuals (point-to-point). For instance, soldiers roaming in the battlefield may need to receive commands from a commander (point-to-multipoint), or a group of commanders may teleconference current mission scenarios with one another (multipoint-to-multipoint). Therefore, multipoint communications serve as one critical operation to support these applications.

Many multicast routing protocols have been proposed for mobile ad hoc networks. Of them, several protocols rely on constructing a tree spanning all group members [1, 2], which may not be robust enough when the network becomes more dynamic with less reliable wireless links. In contrast, several protocols have data packets transmitted into more than one link, and they allow packets to be received over the links that are not branches of a multicast tree. These protocols fall into the category of *mesh-based* protocols because group connectivity is formed as a mesh rather than a tree to increase

---

Algorithms and Protocols for Wireless and Mobile Ad Hoc Networks, Edited by Azzedine Boukerche
Copyright © 2009 by John Wiley & Sons Inc.

robustness at the price of higher redundancy in data transmission. Flooding, where data packets are forwarded to and received from all links, is also considered a mesh protocol since the mesh is in fact the entire network topology. In highly dynamic, highly mobile ad hoc networks, flooding is a better alternative to multicast routing due to its minimal state and high reliability [3]. To the extreme, flooding provides the most robust, but the least efficient, mechanism since a multicast packet will be forwarded to every node as long as the network is not partitioned, while a tree-based approach offers efficiency but is not robust enough to accommodate highly dynamic scenarios. Furthermore, routing based on the concept of connected dominating set [4–7] can also increase the overall efficiency since the search space is reduced to only nodes in the dominating set during the route discovery and maintenance processes.

Furthermore, most proposed ad hoc multicast routing protocols may only be suitable for specific network conditions and exhibit drawbacks in other conditions. For instance, protocols that are aggressive in data forwarding and yield a high delivery ratio in highly dynamic environments may overkill when the network is relatively static. As a result, this chapter describes a mobility-adaptive multicast routing protocol, called *Adaptive Dynamic Backbone Multicast* (ADBM), which offers efficient multicast operations in static or quasi-static environments and becomes more aggressive and effective in highly dynamic environments, even within the same network. ADBM incorporates a backbone construction algorithm that autonomously extracts a subset of nodes to serve as backbone nodes and provide mobility-adaptive connectivity for multicast operations. Similar to multicast protocols based on rendezvous points [8–10], flooding of control messages can also be avoided by having each of these backbone nodes function as a *local core* node in its area to limit the amount of control traffic flooding in the process of group joining. Figure 6.1 illustrates the idea of ADBM, where a single network consists of two portions that are different in terms of mobility or dynamicity. The left portion of the network is relatively static, while nodes on the right side move with higher speed. For instance, a battlefield scenario can be envisioned where soldiers on the right side of the network are engaging the enemy, so that they are moving very quickly, while those on the left side are standing by and not moving too much. Within areas with low mobility, the multicast operation should behave using a tree-based protocol to take advantage of the efficiency of the multicast tree structure. In contrast, a multicast tree is difficult to be efficiently maintained over fast-moving nodes, and hence partial flooding would be more effective.

The remainder of the chapter is organized as follows. Section 6.2 describes the operations of ADBM, including the adaptive dynamic backbone construction algorithm, termed ADB. Performance study is briefly discussed in Section 6.3. Related works are presented in Section 6.4. Section 6.5 concludes the chapter with exercises.

6.2 ADAPTIVE DYNAMIC BACKBONE MULTICAST (ADBM)

In this section, we describe how ADBM performs multicast and how multicast connectivity adapts to various mobility conditions. The key mechanism facilitating ADBM's

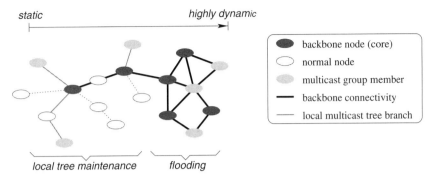

Figure 6.1. Adaptability of multicast routing inherited from the adaptive backbone construction mechanism: More backbone nodes appear in a more dynamic area than the rest of the network, causing a shift of the forwarding mechanism from tree-based forwarding to flooding in that area.

adaptability is ADB, which autonomously constructs and maintains a virtual backbone within a network.

6.2.1 Adaptive Dynamic Backbone Construction Algorithm (ADB)

Within a network, a virtual backbone infrastructure allows a smaller subset of nodes to participate in forwarding control/data packets and help reduce the cost of flooding. Typical virtual backbone construction algorithms are based on the notion of connected, one-hop dominating set (CDS), where a node is either a backbone node or an immediate neighbor of a backbone node. However, when considering a wide range of mobility conditions, even within the same network, we have the following observations:

1. Due to higher stability in more static environments, backbone nodes could be allowed to extend their coverage to more than one hop away, resulting in fewer number of nodes participating in backbone-level communications (e.g., flooding control messages over the backbone).
2. In highly dynamic environments, it may not be feasible to maintain connectivity among the backbone nodes when two backbone nodes are connected through several intermediate nodes. In addition, a larger population of backbone nodes is needed to increase reachability to other ordinary nodes.

ADB is designed to specifically address these observations. The operational framework of ADB is derived from the Virtual Dynamic Backbone Protocol (VDBP) [7], which selects a set of backbone nodes based on the one-hop dominating set property and chooses intermediate nodes to connect these backbone nodes together to form a connected backbone. However, ADB is not restricted to a one-hop

TABLE 6.1. Fields used by the Neighbor Information Table (*NIT*)

Field	Description
id	Neighbor's node ID
coreId	ID of the core to which this neighbor is associated
hops	Number of hops from this neighbor to its current core
degree	Degree of connectivity
nodeStability	Estimated stability of this neighbor's surrounding area
pathStability	Estimated stability of the path from this neighbor to its core

dominating set, but allows a node to be associated with a backbone node that is more than one hop away, depending on stability of the path between that node and its corresponding backbone node. In other words, ADB creates a *forest* of varying-depth trees, each of which is rooted at a backbone node. In high-mobility areas, small local groups of only one or two hops in radius will be formed, which reflects the fact that local topology information is changing frequently and should not be collected in a high amount, because it will be outdated soon. On the contrary, relatively static areas will result in larger local groups, since more information can be collected and remains steady. The resulting backbone will provide an infrastructure that is adaptive to mobility, which is used to support multicast service for mobile ad hoc networks. Although ADB allows more than one-hop coverage, it could behave like other CDS-based protocols, such as VDBP, by setting ADB's parameters to certain values.

ADB consists of three major components: (1) a neighbor discovery process, (2) a core selection process, and (3) a core connection process. The neighbor discovery process is responsible for keeping track of immediate neighboring nodes via HELLO packets. The core selection process then uses this information to determine a set of nodes that will become backbone nodes (or cores[1]). Since cores may not be reachable by each other in one hop, other nodes will be requested to act as intermediate nodes to connect these cores and together form a virtual backbone. This is done by the core connection process. At each node i, the three components are executed concurrently and collaboratively by maintaining and sharing the following variables and data structures:

- *parent$_i$*. This keeps the ID of the upstream node toward the core from i. If i is a core by itself, *parent$_i$* is set to i. This variable is initially set to i and is modified by the core selection process. (In other words, every node starts as a core node.)
- *NIT$_i$* (Neighboring Information Table at node i). This maintains information of all the i's immediate neighbors. Fields of each table entry are described in Table 6.1. The entry corresponding to the neighbor j is denoted by *NIT$_i$(j)*. This table is maintained by the neighbor discovery process.

---

[1]The terms *core* and *backbone node* will be used interchangeably for the rest of this chapter.

TABLE 6.2. Fields used by the Core Information Table (*CIT*)

Field	Description
coreId	Reachable core's ID
nextHop	Next hop by which the core *coreId* is reachable
hops	Distance in terms of number of hops to that core

- $nlff_i$. This stands for *normalized link failure frequency*. It reflects the dynamics of the area surrounding i in terms of the number of link failures per neighbor per second. The neighbor discovery process keeps track of link failures in every *NLFF_TIME_WINDOW* time period and update this variable as follows:

$$\text{current } nlff_i = \frac{f}{NLFF\_TIME\_WINDOW \times |NIT_i|} \quad (6.1)$$

$$nlff_i = \alpha \cdot (\text{current } nlff_i) + (1 - \alpha)nlff_i \quad (6.2)$$

where f is the number of link failures detected during the last *NLFF_TIME_WINDOW* time period, and α is the smoothing factor ranging between 0 and 1. Initially $nlff_i$ is set to zero.

- CIT_i (Core Information Table at node i). This keeps track of a list of nearby cores that are reachable by i, as well as the next hops on shortest paths to reach those cores. Each entry consists of the fields shown in Table 6.2. This table is maintained by the core connection process.

- CRT_i (Core Request Table at node i). This keeps track of a list of nearby cores which i has been requested to establish connection with. The fields used in this table are shown in Table 6.3. This table is also maintained by the core connection process.

Neighbor Discovery Process. The neighbor discovery process is responsible for keeping track of one-hop neighboring nodes. To do so, every node broadcasts a HELLO packet every *HELLO_INTERVAL* time period. In addition, HELLO packets also carry extra information to be used by other components of ADB, especially the core selection process. A HELLO packet h_i sent out by node i contains the following fields:

TABLE 6.3. Fields Used by Core Requested Table (*CRT*)

Field	Description
coreId	Requested core's ID
requester	Neighbor who has requested a connection to *coreId*

- $h_i.nodeId$. the unique ID of the sending node, that is, i.
- $h_i.coreId$. the ID of the core with which i is currently associated.

$$h_i.coreId = \begin{cases} i & \text{if } i \text{ is a core} \\ NIT_i(parent_i).coreId & \text{otherwise} \end{cases}$$

- $h_i.hops$. the number of hops away from the core.

$$h_i.hops = \begin{cases} 0 & \text{if } i \text{ is a core} \\ NIT_i(parent_i).hops + 1 & \text{otherwise} \end{cases}$$

- $h_i.degree$. the degree of connectivity (the number of neighbors), set to $|NIT_i|$.
- $h_i.nodeStability$. the estimated stability of the area surrounding i. This value is calculated from $nlff_i$ and gives an approximation of the probability that a certain wireless link between i and its neighbor will not break within the next second. The calculation of this value is explained below.
- $h_i.pathStability$. the estimated stability of the path from i to its current core in terms of the probability that this path will still exist within the next second.

$$h_i.pathStability = \begin{cases} 1 & \text{if } i \text{ is a core} \\ NIT_i(parent_i).pathStability \times nodeStability_i & \text{otherwise} \end{cases}$$

To ensure that each node and its neighbors will agree on the same information during the core selection process, all the above values are saved into temporary variables whenever a HELLO packet is sent out.

Upon receiving a HELLO packet, the receiving node updates the corresponding entry in its own *NIT*. Due to the fact that nodes may move, each entry is also associated with a timestamp recording the time at which it was last updated. When an entry is older than (*ALLOWED_LOSSES* × *HELLO_INTERVAL*), it is removed from the table, which also results in a link failure event. Every *NLFF_TIME_WINDOW* time period, the number of link failures is used to calculate *nlff* as shown in (6.1) and (6.2) above.

By maintaining *nlff*, each node i is able to estimate the stability of its surrounding area by calculating *nodeStability$_i$* which expresses the probability that a certain wireless link between i and its neighbor will not break within the next second. Although GPS could be used to obtain more accurate estimation, using GPS may not be suitable in many situations, especially for smaller devices. Without knowing nodes' geographical locations (via GPS or any other means) or mobility speeds, *nodeStability$_i$*, which is to be incorporated into each outgoing HELLO packet, is calculated as follows.

For simplicity, we model the time interval node i will wait before detecting the next link failure by the exponential distribution with the average rate of λ_i failures per second. Therefore, the probability that no link failure will be detected within the next t seconds is $e^{-\lambda_i t}$. Since *nodeStability$_i$* gives an approximation of the probability that a particular link of i will not break within the next second, if i currently has N_i

neighbors, we have

$$(nodeStability_i)^{N_i} = e^{-\lambda_i}$$

Therefore,

$$nodeStability_i = e^{-\frac{\lambda_i}{N_i}} \quad (6.3)$$

Here, λ_i can be estimated by the actual number of link failures detected per second. Since N_i is actually $|NIT_i|$, it follows from (6.1)–(6.3) that

$$nodeStability_i = e^{-nlff_i} \quad (6.4)$$

Notice that the exponential distribution used here to model link failures may not perfectly predict the actual link failure probability. However, as stated earlier, the desired behavior of ADB is to form smaller clusters in more dynamic areas and form larger clusters in more static areas. Although there exist more accurate mechanisms (e.g., reference [11]) to predict link stability, this model serves as an easy-to-calculate indicator (due to the memoryless property of the exponential distribution) effective enough[2] for ADB to achieve its objective under different mobility conditions. This has been validated by simulation, as shown in Figure 6.2 for instance.

The current implementation of the neighbor discovery process assumes that all links are bidirectional, and there is a separate mechanism in the underlying communication layer that discards all packets coming from unidirectional links. For instance, nodes might also exchange their lists of neighbors periodically. Each node then accepts only HELLO packets coming from neighbors whose neighbor lists contain its own ID.

Core Selection Process. The core selection process is responsible for extracting a subset of nodes to serve as core nodes and associating other ordinary nodes with these cores. In addition, this process ensures that each ordinary node is within the specified number of hops away from its core. Unlike most of existing cluster-head election algorithms, a hop limit is not the only parameter used to determine whether a node is allowed to associate with a certain core. The core selection process also ensures that stability of the path between each ordinary node and its own core satisfies a certain constraint.

The core selection process at each node begins after the node has started for a certain waiting period, usually long enough to allow the node to have heard HELLO packets from all of its neighbors. This process decides whether a node should still be serving as a core, or become a child of an existing core by checking for its own

---

[2] Our goal is not to have "absolute" accurate prediction, which may require more complicated computation and additional information to be collected. Instead, we would like to have a simple indicator that can differentiate node and path stability in a "relative" sense.

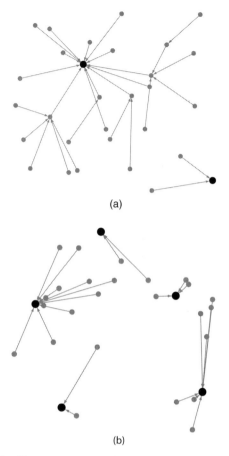

Figure 6.2. Different backbone structures under different mobility speeds (the larger black nodes represent the cores): (a) Mobility speed is 0 m/s, and (b) mobility speed is 20 m/s.

local optimality. First, each node i computes its own *height* from its current status, which is ($nodeStability_i$, $degree_i$, i). A height will also be calculated by the node i for each entry in NIT_i. As mentioned in the previous section, when calculating its own height, each node acquires its status that was saved in the temporary variables at the moment the last HELLO packet was sent out, rather than using its current status. Doing so will ensure that all surrounding nodes agree on the same height information and will avoid potential loop formation when the parent–child relationship is established among nodes. If a node has the highest height among its neighbors, it is considered a local optimal node and should continue serving as a core. If not, the node picks the local optimal node as a core and updates its *parent* variable. All subsequent HELLO packets will be changed accordingly.

While keeping the number of cores as low as possible, the core selection process has to ensure that the current backbone configuration satisfies two constraints:

the hop count limit and the path stability. These two constraints are used to ensure that the hop count and the path stability from every node to its core must not exceed the parameters *HOP_THRESHOLD* and *STABILITY_THRESHOLD*, respectively. Therefore, once the local optimality check has been made, each node must keep listening to the incoming HELLO packets. For convenience, we define a predicate *MeetConstraints$_i$(j)* to indicate whether node i would meet the backbone constraints if it chose node j as its parent. Formally, this predicate is defined as follows:

$$MeetConstraints_i(j) \iff (i = j) \lor$$
$$[NIT_i(j).hops < HOP\_THRESHOLD \land$$
$$NIT_i(j).pathStability > STABILITY\_THRESHOLD\,]$$

where $\land$ and $\lor$ denote the conventional boolean operations "**and**" and "**or**," respectively. Upon receiving a HELLO packet, h_j, from node j, node i modifies its local variables based on the following conditions:

- **Condition 1:** $(j = parent_i) \land (MeetConstraints_i(j) = FALSE)$. This condition implies that the association of node i with its current parent j has now violated the backbone constraints. Therefore, node i has to find a new parent k, where $k \neq j$ and *MeetConstraints$_i$(k)* is true, with a minimum number of hops to the core. If no satisfactory parent can be found, node i becomes a core by setting *parent$_i$* to i.
- **Condition 2:** $(j \neq parent_i) \land (i \neq parent_i) \land (h_j.hops < NIT_i(parent_i).hops) \land (MeetConstraints_i(j) = TRUE)$. Here, node i, which is currently not a core, has just heard a HELLO packet from node j, which is not its parent but has a shorter distance to the core. Node i then sets *parent$_i$* to j.
- **Condition 3:** $(i = parent_i) \land (h_j.coreId \neq i) \land (MeetConstraints_i(j) = TRUE)$. This condition tells that node i, which is currently a core, has heard a HELLO packet from node j that belongs to another core, and node i could associate itself to j without violating the constraints. If this condition holds, node i will set *parent$_i$* to j; otherwise, it remains a core.

We can see that while Condition 1 tries to ensure that all nodes that are not cores meet the constraints, it may result in more nodes becoming cores. Meanwhile, Condition 3 attempts to remove these cores by having each current core check if it can find a parent that would not violate the constraints. Therefore, the number of cores in the network will not keep increasing.

In the case of mobility, if a node detects that its parent has moved out of range (from the neighbor discovery process), it will do the same thing as if its parent violates the constraints (i.e., Condition 1). With different mobility conditions, the core selection process will result in different numbers of cores and depths of trees due to the path stability constraint. Figure 6.2a depicts a snapshot of the core selection process on a network of 30 stationary nodes, where there are only two cores and a larger tree is formed. In contrast, node mobility causes formation of smaller trees due to the path stability constraint, which results in more cores, as shown in Figure 6.2b.

Core Connection Process. Nodes that are selected to become cores by the core selection process cannot form a connected backbone by themselves, since they may not be reachable by one another within a single hop. The core connection process is responsible for connecting these cores together by designating some nodes to take the role of intermediate nodes. Consequently, the cores and intermediate nodes jointly comprise a virtual backbone, as illustrated in Figure 6.3. To do so, each core relies on nodes that are located along the border of its coverage, or *border nodes*, to collect information about surrounding nearby cores. A node is said to be a border node if and only if it is able to hear HELLO packets from nodes that are associated with different cores. Each border node then builds a list of cores reachable by itself, including the next hop and the number of hops to each core, and reports this information to its own core so that a core connection request can be performed. To reduce overhead, a border node only needs to collect and report information regarding nearby cores with IDs lower than that of its own core. Hence, a core connection request between a pair of cores is always initiated by the core with the higher ID.

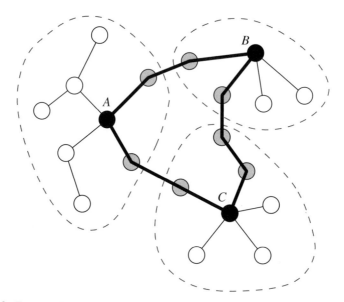

Figure 6.3. Example illustrating the core connection process: Cores *A*, *B*, and *C* (shown in black) are connected by seven intermediate nodes (shown in gray) and form a virtual backbone (represented by thick lines). Solid lines denote parent–child relationship, and dashed circles denote core coverage boundaries.

ALGORITHM 1. Node i Processing HELLO Packets

Input:
$h \leftarrow$ incoming HELLO
$src \leftarrow$ the sender of h

Begin:
$coreInfo \leftarrow$ COREINFO piggybacked by h
$coreReq \leftarrow$ COREREQUEST piggybacked by h
if $parent_i = i$ **then** /* i is a core */
   $coreId_i \leftarrow i$
   $hops_i \leftarrow 0$
else /* i is not a core */
   $coreId_i \leftarrow NIT_i(parent_i).coreId$
   $hops_i \leftarrow NIT_i(parent_i).hops + 1$
end if

remove all entries $\langle cid, nextHop, hops \rangle$ from CIT_i **where** $nextHop = src$
remove all entries $\langle cid, requester \rangle$ from CRT_i **where** $requester = src$

if $h.coreId < coreId_i$ **then**
   insert $\langle h.coreId, src, h.hops + 1 \rangle$ into CIT_i
else if $h.coreId = coreId_i$ & $h.hops > hops_i$ **then**
   for each $\langle cid, hops \rangle$ in $coreInfo$, **where** $cid < coreId_i$ **do**
     insert $\langle cid, src, hops + 1 \rangle$ into CIT_i
   end for
end if

for each $\langle cid, nextHop \rangle$ in $coreReq$ **where** $nextHop = i$ **do**
   insert $\langle cid, src \rangle$ into CRT_i
end for

In ADB, the core connection process relies on HELLO packets to carry two extra pieces of information: COREINFO and COREREQUEST tables. Algorithm 1 (presented in pseudo code) illustrates how this information is processed at receiving nodes and Algorithm 2 describes how this information is piggybacked onto outgoing HELLO packets. The following is the detailed explanation. Let $coreId_i$ be $NIT_i(parent_i).coreId$ if i is not a core, or i otherwise. When node i hears a HELLO packet, h_j, from another node j and $h_j.coreId_j < coreId_i$, node i will insert into its CIT_i (Core Information Table) an entry $\langle h_j.coreId_j, j, h_j.hops+1 \rangle$, which represent $coreId$, $nextHop$, and $hops$ fields, respectively, as described in Table 6.2. This step corresponds to lines 16–17 of Algorithm 1. To propagate this information to the core, node i piggybacks onto its outgoing HELLO packet a COREINFO table containing a list of entries $\langle coreId, hops \rangle$. Each entry indicates the shortest distance in terms of the number of hops from node i to each unique core in CIT_i, as shown in lines 21–24 of Algorithm 2. Note that CIT_i may contain multiple entries for the same core, but node i will choose only one with the least number of hops to that core. When any node j hears the HELLO packet containing the COREINFO from node i, node j looks for entries with

ALGORITHM 2. Node i Piggybacking CoreInfo and CoreRequest

```
 1: Input:
 2: h ← outgoing Hello
 3: Output:
 4: outgoing Hello with piggybacked info

 5: Begin:
 6: coreInfo ← {}
 7: coreReq ← {}
 8: if parent_i = i then /* i is a core */
 9:     for each ⟨cid, nextHop, hops⟩ in CIT_i do
10:         d ← distance to the core cid from i
11:         best ← best next hop to the core cid from i
12:         if d > 1 then
13:             coreReq ← coreReq ∪ {⟨cid, best⟩}
14:         end if
15:     end for
16: else /* i is not a core */
17:     for each ⟨cid, requester⟩ in CRT_i do
18:         best ← best next hop to the core cid from i
19:         coreReq ← coreReq ∪ {⟨cid, best⟩}
20:     end for
21:     for each ⟨cid, nextHop, hops⟩ in CIT_i do
22:         d ← distance to the core cid from i
23:         coreInfo ← coreInfo ∪ {⟨cid, d⟩}
24:     end for
25: end if
26: piggyback coreInfo and coreReq onto h
27: return h
```

lower core IDs than its own and inserts/updates the corresponding entries in CIT_j using i as the *nextHop* field as long as the following two conditions hold: (1) Nodes i and j belong to the same core, and (2) the distance from node j to the core is less than that of node i. Lines 18–21 of Algorithm 1 denote these steps. Nodes that are not cores that and have nonempty *CIT* tables must always generate and piggyback CoreInfo tables onto outgoing Hello packets so that each core can collect all information it needs and establish connectivity to nearby cores (again, in lines 21–24 of Algorithm 2). Due to the dynamics of the core selection process and the network topology itself, the *CIT* table is kept updated by having each node remove all the entries with *nextHop* field set to k when it receives a link failure notification regarding to the node k. In addition, it also removes all the entries in *CIT* that match the pattern $\langle c, k, \cdot \rangle$ when it hears a CoreInfo from k without any entry regarding the core c.

Once a core has had information about other surrounding cores stored in its *CIT* table, it is able to initiate the connection request to those cores by piggybacking a

COREREQUEST table onto each of its outgoing HELLO packets. A COREREQUEST table contains a list of entries ⟨coreId,nextHop⟩, each of which indicates which node is being requested to connect to each unique core from the *CIT* table, as shown in lines 8–15 of Algorithm 2. Similar to the construction of a COREINFO table, only one next hop with the shortest distance in terms of the number of hops is chosen for each core. Entries from *CIT* with one hop count are excluded from the COREREQUEST table since they represent other cores with direct contact, which do not require any intermediate nodes with which to establish connectivity. When node i receives a HELLO packet containing a COREREQUEST table from node j, node i checks for entries where its ID appears on the *nextHop* field. For each entry ⟨c, i⟩ found in the COREREQUEST, node i inserts into CRT_i (Core Request Table) an entry with *coreId* and *requester* set to c and j, respectively (Algorithm 1, lines 23–25). Each node i that is not a core and has at least one entry in its CRT_i is required to construct a COREREQUEST table with entries corresponding to all the cores in CRT_i (Algorithm 2, lines 17–20). For each entry ⟨c, n_c⟩ in the COREREQUEST table that node i is to generate, if node c is the current core of node i, node i will use $parent_i$ as the value of n_c; otherwise, node i will consult its CIT_i for the best next hop to node c.

Similar to the maintenance of *CIT*, when node i detects that the link to node k has failed, node i removes from CRT_i all the entries whose *requester* fields are equal to k. If CRT_i contains an entry ⟨c, k⟩ but node i hears a COREREQUEST table from node k without an entry ⟨c, i⟩, the entry ⟨c, k⟩ is removed from CRT_i.

Node i determines if it is currently on the backbone by checking if it is either a core ($parent_i = i$) or an intermediate node ($|CRT_i| > 0$). These nodes on the backbone play an important role for facilitating multicast operations, which will be described in the next section.

6.2.2 Group Joining

The group joining mechanism in ADBM is based on the concept of rendezvous points, where control messages are directed toward the core of the group instead of being broadcast to other irrelevant nodes or the entire network. Here, the rendezvous point for every node is not a single node, but it consists of the cores that have already formed a connected backbone as a result from ADB. A node that is willing to join a multicast group sends a request toward its core via its parents. Intermediate nodes that receive and forward these requests then become forwarding nodes for the group, which, together with the connected backbone, eventually form multicast connectivity among all group members. We explain the group joining operation in more details as follows.

In addition to the data structures for ADB described in Section 6.2.1, each node maintains a *join table* that has the same structure as a core request table (Table 6.3) with a group ID field instead of a core ID, as shown in Table 6.4. For node i, its join table is denoted by JT_i. Node i, which is willing to join a multicast group, informs its parent by piggybacking a JOINREQUEST, along with its parent ID, $parent_i$, onto its outgoing HELLO packets. The JOINREQUEST contains a list of all multicast group

TABLE 6.4. Fields Used by the Join Table (*JT*)

Field	Description
groupId	Multicast group ID
requester	Node that has requested to join this group

IDs that node *i* has joined and is willing to join. The JOINREQUEST also includes group IDs, of which *i* is currently a forwarding node. When a node *j* receives a HELLO with a JOINREQUEST, it checks if its ID is specified as the intended receiver of the JOINREQUEST. If so, node *j* updates its JT_j with all the entries found in the JOINREQUEST received from node *i*, using *i* as the *requester* field. It is required that nodes that are members of some groups, and any nodes whose join tables are not empty, with the exception for cores, periodically broadcast JOINREQUEST with their HELLO messages. A node that has an entry for the group ID *g* in its join table then marks itself as a forwarding node for group *g*, and nodes on the backbone are considered to be forwarding nodes for any group. As a result, these forwarding nodes form a multicast tree within each core's coverage, where multiple trees are connected by the backbone, hence establishing connectivity among the group members, as illustrated in Figure 6.4.

To deal with network dynamics, the join table is maintained in the same way as ADB does to *CIT* and *CRT* in that entries whose *requester* fields are *i* are removed from the table whenever a link to node *i* has broken. The entry with *groupId* and *requester* set to *g* and *i*, respectively, is also removed when a JOINREQUEST received from node *i* contains no entry for the group *g*.

6.2.3 Multicast Packet Forwarding

Although multicast connectivity establishment within a core's coverage resembles a tree formation, ADBM protocol allows forwarding nodes and members to accept data packets that arrive from any node. Hence it is considered a mesh-based protocol.

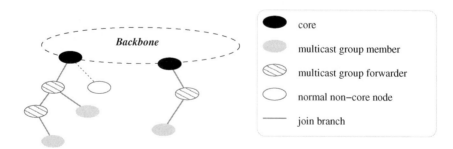

Figure 6.4. An example showing nodes forming of multicast trees, which are connected by the backbone, during group joining process.

Since there can exist more than one path between a sender and a receiver, each data packet is assigned a unique sequence number when it is leaving its source. The sequence numbers are checked by each forwarding node and member node to ensure that no duplicate data packets are rebroadcast or delivered to the application. When a node i receives a nonduplicate data packet for the group g, it checks whether it is currently a forwarding node of the group—that is, whether one or more of the following conditions hold:

- Node i is a core; that is, $parent_i = i$.
- Node i is an intermediate node on the backbone; that is, $|CRT_i| > 0$.
- Node i has an entry for g in its join table.

If so, node i rebroadcasts the packet. Otherwise, the packet is silently discarded. If node i is a member of the group g and none of the above conditions hold, node i will deliver the packet to the application without rebroadcasting.

6.3 PERFORMANCE DISCUSSION

When studying performance of most routing protocols, either unicast or multicast, we are usually interested in two different aspects: *effectiveness* and *efficiency*. The former captures how effective the protocol is in delivering packets to the destination(s). The latter typically focuses on how much more overhead is spent to get the job done. Base on these two aspects, we have measured the performance of ADBM using the following metrics.

- *Packet Delivery Ratio.* This is the ratio of the number of non-duplicate data packets successfully delivered to the receivers versus the number of packets supposed to be received. This metric reflects the effectiveness of a protocol.
- *Number of Total Packets Transmitted per Data Packet Received.* This is the ratio of the number of data and control packets transmitted versus the number of data packets successfully delivered to the application. HELLO packets are also considered as packets transmitted. This measure shows efficiency of a protocol in terms of channel access. The lower the number, the more efficient the protocol.
- *Number of total bytes transmitted per data byte received.* This metric is similar to the second metric except that the number of bytes is considered instead. Here, bytes transmitted include everything that is sent to the MAC layer (i.e., IP and UDP headers, as well as HELLO packets), where data bytes received involve only the data payloads. This metric presents efficiency of a protocol in terms of bandwidth utilization. Similar to the second metric, the lower the number, the more efficient the protocol.

Now let us briefly discuss ADBM's performance. As the protocol was designed to be adaptive with wide range of node mobility, we have studied its behavior under various mobility speeds. Overall, the study showed that ADBM performs efficiently (i.e., most packets are successfully delivered) under low mobility and still effectively under extremely high mobility with various network and application settings. However, choosing appropriate values for *HOP_THRESHOLD* and *STABILITY_THRESHOLD* is the key to achieve good tradeoff between effectiveness and efficiency of the protocol. For *HOP_THRESHOLD*, too low values will cause many nodes to become cores even when the nodes are stationary, which is not good in terms of efficiency. On the contrary, higher hop thresholds should give better results in general because fewer nodes will be on the backbone under low mobility; and with high mobility, more nodes will be forced to join the backbone, which sustains the protocol's effectiveness, given that *STABILITY_THRESHOLD* is set properly.

STABILITY_THRESHOLD is another important parameter we need to adjust. Setting this threshold too low makes the nodes very sensitive to packet losses (which are caused not only by mobility, but packet collisions as well), and the network could end up with many more forwarding nodes than necessary. In contrast, high *STABILITY_THRESHOLD* values will potentially reduce packet delivery ratio as mobility increases since large coverage of cores prevent core connectivity (as well as multicast connectivity) from reacting to mobility quickly enough.

6.4 RELATED WORK

This section surveys related multicast as well as virtual backbone protocols for ad hoc networks. There have been a number of techniques proposed for multicast support over ad hoc networks, ranging from a simple flooding scheme to tree-based or mesh-based approaches. AMRoute (Ad Hoc Multicast Routing Protocol) [1] follows the tree-based approach by relying on an underlying ad hoc unicast routing protocol to provide unicast tunnels for connecting the multicast group members together. AMRIS (Ad Hoc Multicast Routing with Increasing Sequence Numbers) [2] and MAODV (Multicast Ad Hoc On-Demand Distance Vector Protocol) [12] are also tree-based, but they are independent from unicast routing protocols while maintaining multicast trees. ODMRP (On-Demand Multicast Routing Protocol) [13] and CAMP (Core-Assisted Multicast Protocol) [10] allow data packets to be forwarded to more than one path, resulting in mesh structures to cope with link failures and increase robustness in dynamic environments. However, one major disadvantage of ODMRP and other on-demand approaches is their frequent flooding of control messages. For example, in ODMRP, senders periodically flood the entire network with query messages to refresh group membership. Dynamic Core-Based Multicast Routing Protocol (DCMP) [14] mitigates this problem by having only a small subset of senders act as core for the other senders to limit the amount of flooded messages. Although CAMP also suggests the use of cores to direct the flow of join request messages, a mapping service from multicast groups to core addresses is assumed, and a core may not be available at all times. Furthermore, CAMP still relies on an underlying unicast routing protocol that provides correct distances to known destinations within a finite time. Instead of specifying an explicit core address and relying on a unicast protocol, ADBM employs the parent–child relationship to direct a join request from a node toward its core, which prevents the message from being globally flooded.

For scalability, a hierarchical approach or a hybrid approach is often employed. MZR (Multicast Routing Protocol Based on Zone Routing) [15] adopts a hybrid approach using the same mechanism provided by the Zone Routing Protocol (ZRP) [16]. A *zone* is defined for each node with the radius in terms of the number of hops. A proactive protocol is used inside each zone, while reactive route queries are done at the zone border nodes on demand, resulting in a much smaller number of nodes participating in the global flooding search. CGM (Clustered Group Multicast) [17] and MCEDAR (Multicast Core-Extraction Distributed Ad Hoc Routing) [6] employ a similar concept by having only a subset of network involved in multicast data forwarding. This is done by extracting an approximated one-hop dominating set from the network to be used for control and/or data messages. Also based on a dominating set concept, a number of virtual backbone protocols, such as B-protocol [18] and VDBP [7], provide a generic support for creating and maintaining a virtual backbone over an ad hoc network. They can therefore serve as a mechanism to reduce the communication overhead. Amis et al. [19] proposed a generalized clustering algorithm based on a d-hop dominating set, which allows a node to be at most d hops away, rather than only one hop, from its backbone node. However, these protocols have a similar restriction in terms of local group sizes. Clustering algorithms based on a one-hop dominating set always require each ordinary node to be exactly one hop away from the nearest backbone node. In relatively static environments, a node should be allowed to connect to a backbone node that is more than one hop away to lower the number of backbone nodes and reduce overhead of multicast data forwarding. On the other hand, a d-hop clustering algorithm has a fixed value for the parameter d, which always attempts to associate nodes to backbone nodes as long as they are within d hops away, regardless of their surroundings' mobility condition. Since each pair of nearby backbone nodes must be connected through several intermediate nodes (up to $2d + 1$ hops), it may be infeasible to maintain backbone connectivity in highly dynamic environments. Thus, these existing protocols are not adaptive in a situation where a network has the level of mobility changing from time to time, or where there are dynamic and static portions coexisting in the same network. In contrast, ADBM uses two parameters, *HOP_THRESHOLD* and *STABILITY_THRESHOLD*, to determine whether a node should be attached to an existing backbone node. In static environments, where nodes do not experience many link failures, the stability criterion is always satisfied. The *HOP_THRESHOLD* parameter is, therefore, the only parameter controlling the backbone formation, making ADBM's backbone construction algorithm (i.e., ADB) similar to a typical d-hop clustering algorithm. On the contrary, *HOP_THRESHOLD* rarely affects the backbone formation in highly dynamic environments due to decreasing of node and path stability that makes nodes likely to violate the stability criterion before the hop-limit criterion is violated. Hence, nodes in high-mobility areas are likely to be controlled by the *STABILITY_THRESHOLD* parameter when deciding which backbone nodes they should attach to, or whether they should become backbone nodes themselves. As a result, ADBM achieves adaptability by allowing nodes in static areas to be up to *HOP_THRESHOLD* away from their backbone nodes (to reduce overhead) and by implicitly lowering the hop limit for nodes in dynamic areas (to increase robustness of virtual backbones).

McDonald and Znati introduced an adaptive clustering framework, called (α, t) *cluster* [20], to support a mobility-adaptive hybrid routing architecture. This framework partitions the network into clusters of nodes and ensures that every node in each cluster has a path to every other node in the same cluster for the next time period t with probability α. Hence, with different mobility conditions, cluster sizes are adjusted accordingly. Although ADB and (α, t) cluster share similar objectives in providing mobility-adaptive infrastructures for hybrid routing, there are several differences between the two protocols. First, the design of the (α, t) cluster framework mainly aims to support unicast routing, while ADB is intended for multicast routing. Second, in the (α, t) cluster framework, a node has to ensure that its associativity with every other node in the same cluster meets certain criteria. ADB, on the other hand, requires a node to maintain stability constraints with only one node in the cluster, that is, the backbone node. And finally, to maintain cluster membership, the (α, t) cluster framework relies on routing information provided by an underlying proactive routing protocol, while ADB is independent of any routing protocol to operate.

6.5 CONCLUSION

We have introduced a mobility-adaptive multicast protocol for mobile ad hoc networks, called Adaptive Dynamic Backbone Multicast (ADBM). The protocol, based on a two-tier architecture, incorporates the Adaptive Dynamic Backbone (ADB) algorithm to construct a virtual backbone infrastructure that facilitates multicast operations. ADB extracts a subset of nodes, called backbone nodes or cores, out of the network. These cores then establish connectivity among one another to form a virtual infrastructure. Each core is allowed to have larger coverage in relatively static areas, while the coverage is reduced in more dynamic areas so that connectivity among cores can be efficiently established and quickly reestablished under high mobility. ADBM employs the provided backbone infrastructure to achieve adaptive multicast routing that integrates a tree-based scheme and a flooding scheme together to provide both efficiency in static areas and robustness in dynamic areas, even within the same network. Since every node is either a core or has an association to a core via its parent node, cores are also used to avoid global flooding of control messages triggered by a node's joining a group or maintaining group membership, as found in other on-demand multicast protocols.

6.6 EXERCISES

1. Consider a network of six stationary nodes running ADB, as shown below. Gray lines indicate that nodes are neighbors. Assume that all nodes have already heard

of their neighbors' *HELLO* packets at least once and never encountered a lost *HELLO*, and also assume that the core selection process has *not yet* been triggered. What is the content of a *HELLO* packet sent out by node 6?

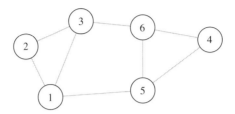

2. The diagram below illustrates a network of 10 nodes with a *stable* backbone formation. Gray lines indicate that nodes are in each other's vicinity, black lines indicate the current backbone formation, and gray nodes are current serving as cores. Given that every node has the normalized link failure frequency (*nlff*) of 0.2 link failure per neighbor per second, show the Neighbor Information Tables (NIT) of nodes with IDs 1 and 7.

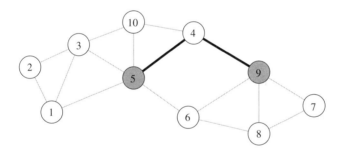

3. The following table summarizes current node information right before the core selection process is triggered. Which of the five nodes will continue serving as cores *immediately after the core selection process was triggered*?

Node	Neighbors	nlff
1	{2}	0
2	{1,3,5}	0
3	{2,4}	0.1
4	{3,5}	0
5	{2,4}	0.1

4. Suppose ADB is running on a network represented by the graph below:

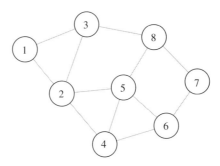

Let every node have *nlff* of 0, and let ADB's parameters in the following table be used:

Parameter	Value
HOP_THRESHOLD	3
STABILITY_THRESHOLD	0.9

Compute the final backbone formation. (In case there are more than one possible formations, show only one.)

5. Repeat the previous exercise, but this time suppose that every node now has *nlff* of 0.1 link failure per neighbor per second instead of 0.

6. Consider a network of 9 nodes with partial ADB-related information shown in the table below. If nodes 1 and 9 join the same multicast group, which of the nine nodes will be considered forwarding nodes by ADBM?

Node	Neighbors	Parent Computed by ADB
1	{4}	4
2	{3}	3
3	{2,4,5}	3
4	{1,3,5}	3
5	{3,4,6}	3
6	{5,7}	7
7	{6,8}	7
8	{7,9}	7
9	{8}	8

7. Given a network of C cores running ADBM with a multicast group of M members, suppose (on average) the coverage of each core is H hops and is connected to D other cores, and the average hop count from an individual node to its core is $\overline{H}$. How many times is a multicast packet transmitted when it is sent from a source node to all other members, assuming that there is no packet loss in the delivery?

REFERENCES

1. Mingyan Liu, Rajesh R. Talpade, and Anthony McAuley. AMRoute: Adhoc multicast routing protocol. Technical Report 99, The Institute for Systems Research, University of Maryland, 1999.
2. C. W. Wu and Y. C. Tay. AMRIS: A multicast protocol for ad hoc wireless networks. In *IEEE Military Communications Conference (MILCOM)*, Atlantic City, NJ, November 1999, pp. 25–29.
3. C. Ho, K. Obraczka, G. Tsudik, and K. Viswanath. Flooding for reliable multicast in multihop ad hoc Networks. In *The 3rd International Workshop on Discrete Algorithms and Methods for Mobile Computing and Communications (DIAL-M'99)*, August 1999.
4. R. Sivakumar, B. Das, and V. Bharghavan. Spine routing in ad hoc networks. *ACM/Baltzer Publications Cluster Computing Journal*, vol. Special Issue on Mobile Computing, 1998.
5. P. Sinha, R. Sivakumar, and V. Bharghavan. CEDAR: A core-extraction distributed ad hoc routing algorithm. In *IEEE INFOCOM '99*, March 1999.
6. P. Sinha, R. Sivakumar, and V. Bharghavan. MCEDAR: Multicast core-extraction distributed ad hoc routing In *IEEE Wireless Communications and Networking Conference (WCNC) '99*, New Orleans, LA, September 1999, pp. 1313–1317.
7. U. C. Kozat, G. Kondylis, B. Ryu, and M. K. Marina. Virtual dynamic backbone for mobile ad hoc networks. In *IEEE International Conference on Communications (ICC)*, Helsinki, Finland, June 2001.
8. T. Ballardie, P. Francis, and J. Crowcroft. Core-based trees (CBT): An architecture for scalable inter-domain multicast routing. In *Communications, Architectures, Protocols, and Applications*, San Francisco, CA, September 13–17, 1993, pp. 42–48.
9. R. Royer and C. Perkins. Multicast using ad-hoc on demand distance vector routing. In *Mobicom'99*, Seattle, WA, August 1999, pp. 207–218.
10. J. Garcia-Luna-Aceves and E. Madruga. The core-assisted mesh protocol. *IEEE Journal on Selected Areas in Communications*, **17**(8):00–00, 1999.
11. M. Gerharz, C. de Waal, M. Frank, and P. Martini. Link stability in mobile wireless ad hoc networks. In *Proceedings of the IEEE Conference on Local Computer Networks (LCN) 2002*, November 2002.
12. E. M. Royer and C. E. Perkins, Multicast operation of the ad-hoc on-demand distance vector routing protocol. In *Mobile Computing and Networking*, 1999, pp. 207–218.
13. S. Bae, S. Lee, W. Su, and M. Gerla. The design, implementation, and performance evaluation of the on-demand multicast routing protocol in multihop wireless networks. *IEEE Network, Special Issue on Multicasting Empowering the Next Generation Internet*, **14**(1):70–77, 2000.
14. S. K. Das, B. S. Manoj, and C. Siva Ram Murthy. A dynamic core based multicast routing protocol for ad hoc wireless networks. In *The ACM Symposium on Mobile Adhoc Networking and Computing (MOBIHOC 2002)*, Lausanne, Switzerland, June 9–11, 2002.
15. V. Devarapalli and D. Sidhu. MZR: A multicast protocol for mobile ad hoc networks. In *IEEE International Conference on Communications (ICC)*, Helsinki, Finland, June 2001.
16. Z. J. Haas and M. R. Pearlman. The zone routing protocol (ZRP) for ad hoc networks. 1997, IETF Internet Draft. http://www.ietf.org/internet-drafts/draft-ietf-manet-zone-zrp00.txt.

17. C. R. Lin and S.-W. Chao. A multicast routing protocol for multihop wireless networks. In *IEEE Global Telecommunications Conference (GLOBECOM)*, Rio de Janeiro, Brazil, December 1999, pp. 235–239.
18. S. Basagni, D. Turgut, and S. K. Das. Mobility-adaptive protocols for managing large ad hoc networks. In *IEEE International Conference on Communications (ICC)*, Helsinki, Finland, June 2001, pp. 63–68.
19. A. D. Amis, R. Prakash, T. H.P. Vuong, and D. T. Huynh. Max–Min D-cluster formation in wireless ad hoc networks. In *IEEE INFOCOM 2000*, Tel-Aviv, Israel, March 2000, pp. 32–41.
20. A. B. McDonald and T. F. Znati, A mobility-based framework for adaptive clustering in wireless ad hoc networks. *IEEE Journal on Selected Areas in Communications*, **17**(8):1466–1487, 1999.

CHAPTER 7

Effect of Interference on Routing in Multihop Wireless Networks[1]

VINAY KOLAR and NAEL B. ABU-GHAZALEH

Computer Science Department, Binghamton University, Binghamton, NY 13902

7.1 INTRODUCTION

Multihop wireless networks (MHWNs) are emerging as a critical technology that will play an important role at the edge of the Internet. Mesh networks [1, 2] provide an extremely cost-effective last-mile technology for broadband access (for example, a mesh network providing broadband access to the city of Philadelphia, covering 135 square miles, at less than half the cost of traditional broadband access is currently being built [3]); ad hoc networks have many applications in the military, industry, and everyday life [4]; and sensor networks hold the promise of revolutionizing sensing across a broad range of applications and scientific disciplines—they are forecast to play a critical role as the bridge between the physical and digital worlds [5]. This range of applications results in MHWNs with widely different properties in terms of scales, traffic patterns, radio capabilities, and node capabilities. Thus, effective networking of MHWNs has attracted significant research interest.

One of the primary challenges in networking MHWNs is the routing problem; how to construct efficient routes for a network that is self-configuring and potentially mobile. The initial set of proposed routing protocols focused on deriving efficient routes, using conventional metrics such as hop count, at low overhead. Many flavors of protocols including proactive protocols [6, 7], reactive protocols [8, 9], hybrids [10], as well as geometric protocols [11] emerged. However, the wireless channel in MHWNs introduces challenges that make routing different from conventional wired networks. Thus, due to signal propagation effects, the quality of the wireless channel can vary significantly from link to link and within the lifetime of a single link. The second

---

[1]This work was partially supported by AFRL grant FA8750-05-1-0130 and NSF grant CNS-0454298.

Algorithms and Protocols for Wireless and Mobile Ad Hoc Networks, Edited by Azzedine Boukerche
Copyright © 2009 by John Wiley & Sons Inc.

effect is the shared nature of the wireless channel which causes interference among nearby links; link capacity is shared, among nearby sources in a manner unlike wired links. Accordingly, this initial group of protocols generally do not achieve effective routing because they ignore link quality. Section 7.3 briefly overviews hop-count-based protocols and discusses their shortcomings in more detail.

The emergence of more realistic simulation models [12] and real test-bed experiments [13–17] show that the link quality plays a critical role in determining route quality: Hop count is not a sufficient metric for determining route quality. The available bandwidth between a pair of communicating nodes is influenced not only by the nominal communication bandwidth, but also by ongoing communication in nearby regions of the network because of the shared nature of the medium. More specifically, other ongoing transmissions contribute interference power that can make it impossible to exchange packets between a given pair of nodes, reducing the capacity of the link. As a result, a second generation of routing protocols, called *link-quality aware routing*, was proposed. In link-quality-aware routing the link quality is estimated and exposed to the routing protocol, which then can combine the link qualities to estimate route quality; this estimate is used in place of hop count to guide route selection decisions. Several metrics for measuring link quality and for combining them to estimate route quality have been proposed; these works are reviewed in Section 7.4.

The routing configurations achieved by link-quality-aware routing can be suboptimal for two primary reasons: (1) *Limitations of link-quality and route-quality estimation approaches*. Existing metrics relying on source-based measurements cannot capture scheduling effects accurately. (2) *Per-connection greedy approach*: The protocols remain greedy in nature, taking local decisions without coordination. An optimal routing configuration cannot be achieved by making greedy decisions on a connection-by-connection basis without coordination to best map the connections to the available space and achieve maximal spatial reuse. While some-link-quality-aware routing protocols attempt to adaptively switch away from paths where the quality of the links is low, it is not clear that this approach will converge to a globally effective configuration. In Section 7.5 we motivate modeling and optimization efforts for capturing the behavior of MHWNs and using this model to derive globally effective routing configurations.

Finding a globally effective routing configuration is a problem similar to traffic engineering in conventional networks—a problem that is well understood and that has effective solutions (e.g., reference 18). However, the nature of the wireless channel introduces interference and significantly complicates the problem. Section 7.6 overviews efforts in modeling MHWNs to enable globally coordinated routing. It also explores models for estimating link quality based on interference taking into account the scheduling interactions introduced by the MAC protocol (we focus on CSMA MAC protocols). Having access to such models is invaluable in analyzing and optimizing MHWNs. In addition, it is a first step toward future distributed routing protocols that can coordinate to achieve effective traffic engineering in MHWNs. In addition to overviewing the modeling work, this section presents initial results in using one of them to quantify the available improvement from coordinated routing. Section 7.7 discuss presents some final thoughts and overviews future work.

7.2 BACKGROUND

This section discusses how interference occurs in a wireless medium and their implications on link quality. It also discusses the role played by the MAC protocol in regulating interference and the overall effect on routing protocols.

In wireless transmission, as the signal from a sender propagates over the channel, it attenuates with distance; it also suffers from physical propagation due to interactions with the physical environment (e.g., passing through obstacles). A receiver receives the signal after attenuation and other propagation effects, and it attempts to decode the signal. If the received signal strength is sufficiently higher than the sum of the noise and signal from interfering signals, the signal can be decoded successfully (with low error rate); otherwise, the transmission cannot be received. Thus, interference from concurrently transmitting nodes plays an important effect in determining whether correct reception or a collision occurs.

In order to reduce the occurrence of collisions, while maintaining distributed access, Carrier Sense is often used. More precisely, senders are able to sense that the channel is busy if the sum of the noise and the signal from interfering transmissions are above a given power threshold, called the *Receiver Sensitivity Threshold (RxThreshold)*. If the channel is busy, the MAC protocol can defer transmission. The *Channel State* at a given node for a given point of time represents if the channel sensed at the node is busy or idle. A single strong signal or multiple weak signals over the same channel may together add up to create a busy channel state.

In the presence of competition for accessing the medium from interfering senders, the Medium Access Control (MAC) protocol plays the role of moderating access to reduce collisions while maintaining concurrency and fairness. Under an ideal scheduler, the amount of throughput between a sender–receiver pair depends upon the number of active nodes (senders or receivers) that are interfering with the data transmission. However, practical schedulers are rarely able to archive this ideal behavior. The IEEE 802.11 MAC protocol [19] is the de facto standard for Wireless LAN (WLAN) and MHWNs, and much of the existing research is based on this protocol. IEEE 802.11 uses *Carrier Sense Multiple Access with Collision Avoidance (CSMA/CA)*. In contention-based protocols such as IEEE 802.11, the channel state at the receiver is not known at the sender, which contributes to the well-known hidden and exposed terminal problems [20]. To counter these effects, IEEE 802.11 uses an aggressive CSMA (low receiver sensitivity) to attempt to prevent far-away interfering sources from transmitting together (but potentially preventing noninterfering sources from transmitting; hidden terminal is reduced, but exposed terminal increased). Optionally, the standard allows the use of Collision Avoidance (CA), which consists of Request-to-Send (RTS)/Clear-to-Send (CTS) control packets, to attempt to reserve the medium. After a successful RTS-CTS handshake, the actual DATA packet is sent. The successful reception of DATA packet is acknowledged by an ACK packet.

As a result of these mechanisms, many, but not all, collisions are prevented. Depending on the relative location of contending sources to each other, either they are able to effectively handshake using the MAC protocol or they continue to collide.

The problem is complicated because multiple concurrent transmissions may occur and collectively result in a collision, when individually they do not. Similarly, the aggressive collision avoidance mechanisms can prevent some possible concurrent transmissions from proceeding even though they do not cause a collision. In summary, the MAC protocol and the relative locations of contending sources play an important role in how a set of nearby sources interact and the resulting quality of the link observed by each of them.

7.3 FIRST GENERATION: GREEDY HOP-COUNT-BASED ROUTING

Routing protocols for MHWNs can be broadly divided into two categories based on *when the routes are constructed*: (1) proactive routing protocols, and (2) reactive routing protocols. Proactive routing protocols construct the "routing tables" a priori to packet transmission for all possible destinations; examples of proactive routing protocols in MHWNs include DSDV [7] and WRP [21].

Reactive routing protocols compute routes on-demand for destinations toward which they have packets by conducting a route discovery process. Examples of reactive routing protocols include AODV [9] and DSR [8]. Reactive routing protocols can be more efficient than proactive ones because they only search for routes on-demand. However, because route construction is only started when a packet transmission is needed, there is a delay before the route is available. In a dynamic topology where the nodes are mobile, proactively computing routes may be counterproductive as many routes become invalid before they are used.

Initial routing protocols such as DSDV [7], DSR [8] and AODV [9] select routes strictly favoring the shortest hop ones. The routing protocols makes local route selection decisions greedily, attempting to find the shortest hop routes for its active connections. The advantage of this approach is the distributed operation that allows simple local decisions to be taken. The intuition behind choosing the minimum hop-count path is that a shorter path translates to the following: (1) *Higher Capacity*: The number of transmissions necessary to move a packet from source to a destination is a function of the number of hops. A shorter path means fewer retransmissions, reducing competition for the channel and increasing end to end capacity. (2) *Lower energy consumption*: Fewer retransmissions also lead to lower energy expenditure in delivering the packet to the destination. (3) *Shorter end-to-end delay*: With a smaller number of hops, each packet experiences fewer retransmissions, as well as fewer queueing delays at each intermediate hop. This results in a reduction in the end-to-end delay. This general observation is true for all packet-switched networks; for this reason, hop count is used as a path-quality discriminator by some wired routing protocols.

The effectiveness of hop count as a measure of route quality has recently been brought into question [22]. More specifically, there is a large variation in link qualities, and a metric-like hop count that does not account for this variation will not accurately predict a route quality. In fact, hopcount tends to pick poorer-quality links: Since it uses fewer hops, these hops tend to be *longer*, where length is the physical distance between the sender and receiver forming a hop. As the distance between the sender and receiver increases, the signal strength received at the receiver decreases, which induces higher packet error rates: Longer hops tend to be poorer in quality than shorter ones.

The prevalence of hop count as a routing metric was assisted by simulation results that relied on simplistic propagation models. Most simulators initially had a simple *Protocol model* [23] which specified that all the nodes within a reception range receive a transmission correctly. This model favors hop count as an effective parameter since it does not model variations in link quality; it was shown by Li et al. [24] that the number of hops does determine the effectiveness of the route if all the other parameters (like channel capacity of links in different routes) are the same.

Several early works recognized that that not all the routes of the same hop count are equal. Goff et al. [25] analyzed mobile scenarios and proposed mechanisms to stop using the route when the signal strength of a link goes below a fixed threshold. New routes were discovered before a route breaks down, thus switching to a better-quality route. As more sophisticated simulation models evolved [12, 26], these effects were also clear in simulations. The added evidence from experimental testbeds finally demonstrated that link quality cannot be ignored when carrying out routing decisions. Yarvis et al. [27] show that the hop count provides an unrealistic estimation as a routing metric in a real-world sensor network and motivates the need for a routing metric that quantifies the route based on the quality of the links present in the route. De Couto et al. [22] discovered in the experimental mesh network testbed that there is substantial difference in throughput among the routes that had the same number of hops and proposed to use link quality to estimate the route effectiveness rather than just the hop count. Other studies have reached similar conclusions [13, 28, 29]. These observations motivated a second generation of routing protocols which take into account link quality. These are discussed in the next section.

7.4 LINK-QUALITY-BASED ROUTING

The advantages of link-quality-based routing was explained in the earlier section. In this section we discuss the ideas behind link-quality-based routing, popular protocols, and their limitations.

The core ideas of a link-quality-aware routing protocol are (1) to estimate the quality of each link and (2) to measure the quality of the overall route as a weighted sum of the quality of the intermediate hops. In the below sections, we discuss these ideas with respect to certain key link-quality-aware routing protocols.

7.4.1 Link-Quality-Aware Routing

In this section, we overview efforts in developing effective link-quality-aware routing protocols. The general philosophy of these protocols are to replace hop count in conventional protocols with an estimate of route quality, which itself is based on a function that combines the qualities of the links making up the route. Thus, the primary challenges in these protocols are (1) how to estimate link quality and (2) how to combine these estimates into a route quality metric. In this section, we focus on these issues and also discuss the general properties of this class of routing protocols.

7.4.2 Factors Affecting Link Quality

The problem of estimating link quality in the presence of a fading wireless channel and interference is difficult. Channel parameters such as the nominal bandwidth play a role in determining the link quality and are often or slowly changing. Other parameters like the amount of contention due to interference from competing links and the effectiveness the MAC handshake are topology- and traffic-specific, which makes them difficult to estimate. In the remainder of this section, we assume that wireless channel nominal quality, which accounts for issues such as fading, is known. We focus on the interference-related factors that influence link quality.

Assume an ideal channel where the channel state is predictable, and assume an ideal MAC protocol that is able to schedule the packets collisions. The throughput for each link is affected by the contention for the common channel by neighboring transmissions. Hence, the quality of the link depends only upon the *available capacity*. CSMA-based schedulers (like IEEE 802.11) are not ideal and are therefore unable to eliminate collisions completely under realistic carrier sense levels, due to the inherent difference in the channel state observed at the sender and the receiver. The limitations of CSMA-based schedulers were identified by Xu et al. [30]. Thus, under such schedulers, the *percentage of packet losses* are an additional factor determining the link throughput.

7.4.3 Impact of CSMA Scheduling on the Link Throughput

To illustrate the effect of packet losses in a practical CSMA-based scheduler, Kolar and Abu-Ghazaleh [31] simulated different sets of 144 uniformly distributed nodes with 25 arbitrarily chosen *one-hop* connections. The simulation was carried out in a QualNet simulator [32]. The aim of this experiment is to show the effect of CSMA-based scheduling at different interference levels. One-hop connections were chosen to eliminate multihop artifacts such as self-interference and pipelining and isolate link-level interactions. Each source sends CBR data at a rate high enough to ensure that it always has packets to send. Each node measures the amount of time the channel is busy (busy time). The ideal available capacity at a node can be approximated by the idle time observed at either source or destination.

Figure 7.1a plots the busy time at the source of a link[2] against the observed throughput of the link. In general, as interference increases, the achievable capacity decreases. However, it can be seen that at higher busy times (enlarged for clarity in Figure 7.1b), large variations in throughput arise for the same observed busy time. The scheduling effectiveness starts to play a determining effect on the throughput of the link at high interference levels.

---

[2]Other metrics such as busy time at destination were also explored with very similar results.

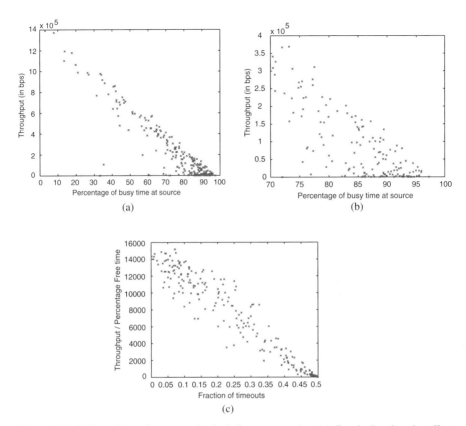

Figure 7.1. Effect of interference and scheduling on capacity. (a) Graph showing the effect of source busy time on throughput. (b) Graph enlarging the high-interference region of (a). (c) Graph showing that normalized throughput is a function of scheduling/MAC level timeouts. (The results were presented in the study [31].)

Let t_i be the throughput achieved by an ideal scheduler; intuitively t_i is proportional to the amount of *available capacity*, which is in turn proportional to the busy time observed at the link. Let t_o be the observed throughput in the CSMA-based scheduler, IEEE 802.11. The ratio of t_o/t_i, called *normalized throughput* provides a measure of the scheduling efficiency relative to an ideal scheduler independently of the available transmission time. Figure 7.1c plots the normalized throughput as a function of observed percentage of IEEE 802.11 MAC level transmissions that experience RTS or ACK timeouts (which represent the packet losses) for *all* the links. It can be seen that as the fraction of packet timeouts increases, the normalized throughput decreases almost linearly. Thus *the reason for variations in observed capacity from the nominal capacity predicted by the interference metric is the scheduling as observed in MAC level timeouts*. This shows that aggregate interference metrics quantifying the available capacity (such as busy time) cannot predict scheduling effects and correlate poorly with scheduling efficiency.

Measuring the link quality is complicated by other MAC layer effects. We discuss two key factors under such secondary effects:

- *Data Rate.* Most of the advanced MAC protocols, like IEEE 802.11, are enhanced to use different data rates based on the estimated channel noise [33–35]. Since the channel noise variance over time is significant, due to either dynamic interference from the neighboring nodes or environmental variations, the data rate between the sender and the receiver is not constant. Many protocols that use broadcast packets to notify the neighbors about the link quality may be vulnerable to the data rate variations. All the broadcast packets are sent at the lowest possible data rate. Receiving such a packet does not always ensure that the neighbor will be able to receive the actual data packets (which may be sent at higher data rates). Thus, an added requirement for the metric is to estimate the link quality given that the actual packets will be transmitted at a higher data rate.
- *Packet Size.* The efficiency of the packet transfer on a link is also a factor of the size of the packet being transmitted. Larger packets require a longer transmission time and suffer a correspondingly larger chance of errors. The measurement of the link quality with smaller control packets may be inaccurate for larger data packets. Hence, the link-quality metric should also account for the packet size that is being transmitted over the link.

The explicit measurement of the available capacity and the percentage of packet losses in a distributed routing protocol is a hard problem because the measurement at routing layer will be distorted by secondary effects (such as the amount of time the packet resides in the queue) and inability of accessing the information present at the MAC layer (since, typically, MAC is implemented on the hardware). To summarize, the important properties that the link-quality metric has to capture under the assumptions of a predictable channel are:

- *Property 1*: The available channel capacity of a link
- *Property 2*: The packet loss observed on a link

While the first property estimates the amount of contention for a given link, the second property measures the efficiency of the MAC protocol in using the available capacity.

7.4.4 Environment Effects on Link Quality

In the above discussion, we assumed a predictable channel and focused on the challenges in capturing interference-related effects that affect link quality. However, link quality in a realistic environment is subject to unpredictable channel state changes. Experiments in the real-world testbeds [13, 14, 27, 36, 37] show that the quality of the link between a pair of nodes regularly fluctuates, thus making it harder to statically assign a single-quality metric for the link[3].

---

[3]While most of the testbeds show this variation, the results from other testbeds ([45, 46]) show that the error rate and the quality of a link is relatively constant. We conjecture the effect of topology and traffic account for differences in conclusions.

One of the first approaches attempted to estimate link quality was to measure parameters of the channel such as the signal strength [25, 38, 39]. De Couto et al. [13, 22] provided experimental evidence from testbeds that signal strength is a poor predictor of the loss rate. Aguayo et al. [14] also found that the use of simple metrics like *signal-to-noise ratio* (SNR) is not a good estimator of the quality of the link. The experiment measured the average SNR at the nodes, and the delivery probability of the packets were found not to conform with the observed SNR results. Various routing metrics that accounted for the link quality were proposed to solve the routing issues in testbeds [13, 28, 29, 40–42]. Other research studies [43, 44] evaluate various link-quality-aware routing protocols and summarize certain key properties of the routing metrics. The routing metrics that are proposed for networks using multiple channels and multiple radios (e.g. references 28 and 29) are different from a single-channel wireless network, even though they share the key ideas. Since single-channel and single-radio networks are predominantly used in static wireless multihop networks (like mesh networks) and share certain fundamental ideas with the multiple-channel and multiple-radio counterpart, we focus on the routing metrics that are suitable for such networks.

7.4.5 Review of Existing Link-Quality Metrics

In this section, various link-quality metrics are studied and their effectiveness and deficiencies are summarized.

***Expected Transmission Count (ETX)* [13].** The intuition behind ETX is that the cost of a link is proportional to the expected number of attempts required to successfully transmit a packet. The ETX of a link is defined as $1/d_f d_r$, where d_f is the probability that the packet is successfully received over a link in forward direction (for data transmission) and d_r is the success probability of packet transfer in the reverse direction (for ACK transmission). Each node periodically broadcasts a probe packet at a constant rate of W packets per second. The nodes record the number of probe packets received from each neighbor. This count is included in the periodic probe packet sent by the node in order to return the link information to the sources. The packet loss rate is calculated from the data recorded by the node and the data from the probe packet of neighboring nodes as described in Algorithm 1.

The metric for the route is the sum of the ETXs of its links. While ETX measures packet loss (Property 2), it fails to account for the available capacity (Property 1). A link with a higher handshake effectiveness of the MAC protocol is always favored, without considering the contention present over the link. Also, since the ETX uses "probe" broadcast packets to inform the neighbors about their reception, the metric does not measure the link quality of the actual packets which are usually larger and transmitted at a higher data rate. While transmitting control packets of different lengths will overcome the problem, this leads to an added control overhead. Also, the estimation of the route quality based on summing the ETX of the links is a simple metric which does not account for the bottleneck links. However, the

ALGORITHM 1. Algorithm to Compute ETX of a Link

Require: Node X receives an ETX probe packet from Node Y
{// W = Number of probe packets sent by a node in 1 second}
{// n_{xy} = number of probe packets heard by Y that was sent by X in the last w seconds}
{// n_{yx} = Number of probe packets heard by X that was sent by Y in the last w seconds}
Extract n_{xy} from the received packet
$r_{xy} \leftarrow \frac{n_{xy}}{wW}$ {// Rating of XY}
Update n_y
$r_{yx} \leftarrow \frac{n_y}{wW}$ {//Rating of YX}
etx$_{xy} \leftarrow \frac{1}{r_{xy}r_{yx}}$ {///ETX for the link XY}

---

performance studies by Draves et al. [44] has shown that ETX outperforms other metrics.[4]

Round-Trip Time (RTT) **[29].** RTT is measured by sending a *probe packet* by a sender (say X). The time of packet generation (t_x) is included in the packet. The receiver (say Y) will acknowledge the probe packet by transmitting a *probe-ACK* packet back to the sender. It includes the t_x inside the probe-ACK packet. The sender estimates the time required for the packet to traverse the link (rtt$_{xy}$); this measurement incorporates the combined effect of the available capacity (Property 1), the packet losses (Packet 2), and the queueing delay of the packet. In order to avoid oscillations and temporary link-quality variations, *Smoothed RTT(SRTT)* is calculated by associating a weight of α to the most recently measured RTT. Algorithm 2 demonstrates the steps involved in calculating the RTT of a link. Study of RTT under different traffic

ALGORITHM 2. Algorithm to Compute RTT of a Link

Require: Node X has sent the probe packet and receives a probe-ACK packet from Node Y
{// α = Weight given for accounting for the most recent round-trip time}
{// t_x = Time when the probe packet was sent by X. This is marked in the probe packet}
Extract t_x from the probe-ACK packet
rtt$_{xy} \leftarrow$ Current time, t_x
{// Update the Smoothed RTT (SRTT)}
srtt$_{xy} \leftarrow \alpha$ rtt$_{xy} + (1 - \alpha)$srtt$_{xy}$

---

[4]Under a multiple-channel scenario, each with different bandwidths and data rate, an extended version of ETX called *Weighted Cumulative Estimated Transmission Time* (WCETT) was proposed by Draves et al. [28]. However, under a single-channel model, which is the focus of the chapter, WCETT measurement will be identical to ETX measurement

pattern in [44] indicates that RTT is useful for measuring links with very high loss rates, but not for other links. Since the RTT uses fixed-size packets for measurement it is susceptible for packet size. Also, the control packets are unicast to each of the neighbor and flows twice between a pair of nodes. This leads to higher measurement overhead, making it unscalable for dense networks.

Packet Pair. This is a standard approach used in estimating the bandwidth in traditional networks [47]. The sender transmits two packets back-to-back, one with smaller packet size and another with a larger packet size. The measurement of the delay between the packets at the receiver will indicate the one-way contention across the link, and this delay is conveyed to the sender. Since the receiver measures the time-interval between the packet pair, it is possible to avoid the effect of queueing delays in measuring the contention at the link. Packet pair is a close relative of RTT that eliminates the queue delay factor from the metric. Draves et al. [44] compared packet pair metric with the above metrics. While the packet pair metric eliminates the queueing delays, the other drawbacks of the RTT are still present. An additional problem of delays associated with possible packet retransmission of the second control packet can cause this approach to underestimate the link quality.

Estimated Data Rate (EDR) [40]. EDR proposes to address the deficiency of the ETX to measure the available capacity (Property 1). EDR assumes that the effectiveness of the route depends upon the available capacity and the packet loss of the bottleneck link. The bottleneck link is determined by considering the amount of queue buildup at the nodes of a route. The ETX of the bottleneck link and the MAC layer effects like backoff is approximated and the EDR metric is proposed for measuring the quality of the route. While EDR accounts for the available capacity, it bases the metric on the bottleneck link, thus not optimizing the other hops of the route. EDR is not sensitive to adaptive data rate and varying packet sizes of the actual data.

Required Number of Packets (RNP) [41]. Cerpa et al. [41] studied temporal properties of links and observed that there is a significant variation of the link properties over larger time frames. The above-mentioned metrics are insensitive to temporal variations and thus fail to depict the link quality. They proposed a metric called Required Number of Packets (RNP), which is sensitive to distribution of the packet losses and thus will avoid links with temporal instability. RNP is estimated by periodically broadcasting control packets for a small duration of time and measuring the temporal variation of the reception rate at all the neighbors. While this metric captures the temporal properties, the measurement overhead of such broadcasts over a period of time is high. Observing the effect of interfering links using RNP is difficult since it requires all the interfering links to transmit in a *time-synchronized* manner. Such measurements fail to capture the real traffic on the contending links, thus estimating the available capacity (Property 1) and packet loss (Property 2). Moreover, since broadcast is being used, such measurement is prone to the effects of adaptive data rate changes and packet size variations. However, temporal changes in link quality will place a higher emphasis on deciding the measurement time frames for all the above metrics and thus play a significant role in deciding the throughput of the link.

***Expected MAC Latency (ELR)* [42].** ETX and EDR not only require additional beacon-based broadcast packets, but also fail to account for the changes of the data rate (by the MAC protocol) and packet sizes (by the application). The unicast link properties, which are inherently different from the broadcast link properties, cannot be accurately estimated using broadcast packets. Also, the temporal changes in the traffic, which occurs frequently, are not captured by the above metrics. Expected MAC latency (ELR) proposes to overcome this problem by learning the link quality as the data packets are forwarded. The key idea of ELR is to eliminate the broadcast control packets by combining the properties of the geography-unaware routing protocols with the geography-aware routing protocols [11].

ELR measures *expected MAC latency per unit distance (LD)* to the destination by a one-time initialization of the node locations and periodically calculates the link effectiveness by switching to different probable next hops. Continuous measurement of the link effectiveness is used to calculate the probability with which the neighbor may be the best forwarder. Upon a successful packet transfer to a next hop, the LD is updated to account for the newly observed latency. If the link observes a packet loss, the previously observed latency is incremented based on the *delivery rate* on the link. Since the LD is log-normally distributed, the logarithm of LD ($\log(LD)$) is estimated (described in Algorithm 3).

ALGORITHM 3. Algorithm to Compute MAC Latency per Unit Distance (LD) of a Link

Require: Node X has sent the unicast packet to send and the set of possible forwarders is nonempty
{// α is the weight associated for the older samples to avoid random fluctuations}
{// p_{xy} = Unicast delivery rate on the link XY}
{// LD_{xy} = Estimated MAC latency between the link XY}
Probabilistically choose the next hop forwarder Y among the set of possible forwarders and transmit the packet to Y.
if Packet is successfully transmitted to Y, **then**
  $p'_{xy} \leftarrow$ Update the unicast delivery rate on the link due to packet success
  $l_{xy} \leftarrow$ Compute the latency of this packet delivery X to Y
  $\log(LD_{xy}) \leftarrow \alpha \log(LD_{xy}) + (1 - \alpha) \log(LD_{xy})'$
else {Packet transmission is not successful}
  $p'_{xy} \leftarrow$ Update the unicast delivery rate on the link due to packet failure
  {//Expected number of retries for packet success $= \frac{1}{p_{xy}}$. Increase the latency accordingly}
  $l_{xy} \leftarrow (1 + \frac{1}{p_{xy}}) l_{xy}$
  $\log(LD_{xy}) \leftarrow \alpha \log(LD_{xy}) + (1 - \alpha) \log(l_{xy})'$
end if

TABLE 7.1. Comparing Different Link Estimation Metrics

Metric	Property			
	Contention	Packet Loss	Varying Data Rates	Varying Packet Sizes
ETX	No	Yes	No	No
RTT	Yes	No	No	No
Packet pair	Yes	No	No	No
EDR	Yes	Yes	No	No
RNP	No	Yes	No	No
ELR	Yes	No	Yes	Yes

ELR accounts for the temporal variations of the link quality by sampling different neighbors adaptively. The drawback of ELR is the inability to calculate the the packet loss information since unicasts are associated with packet retries that cannot be obtained from the MAC hardware in realistic deployment.

Table 7.1 summarizes the properties of the different link quality estimation metrics. It is to be noted that some metrics may reflect some components affecting link quality, but in a very coarse form. For example, RTT may capture the packet drops because the packet drops will lead to a higher RTT. ELR cannot capture the required number of retransmissions (even though theoretically it is possible) because it uses unicast packets and the packet loss due to retransmission cannot be captured. We do not explicitly consider such secondary effects in the table.

7.4.6 Route-Quality Estimation

Once the individual link qualities are estimated, the next problem is how to combine them to reach a measure of an overall route quality. The quality of the route depends upon the quality of the individual links present in the route. The multihop route can be imagined as pipe, with each link representing a part of the pipe with constant diameter which represents the link quality. The overall throughput of the route is the amount of data that can be pushed in the pipe. It can be seen that the throughput is affected by the *bottleneck link* which has the least diameter. Multiple routes may exist between a given source and destination. The routing protocol is responsible for the choice of the route which can support maximum throughput for the connection. Yang, Wang, and Kravets [43] suggest that "isotonicity" is an important property for combining the link quality into a single route quality. Isotonocity refers to the property that the cost of a route strictly increases when it is prefixed with additional hops. For example, consider two routes between node A and node B, say R_1 and R_2 as shown in Figure 7.2. Let the combined link metrics of R_1 and R_2 be represented as r_1 and r_2. Let there be another path that is prefixed to route $A \ldots B$, say from node X to A denoted by $X \ldots A \ldots B$. Consider the two routes between X and B in $X \ldots A \ldots B$, one passing through R_1 and the other through R_2. Let us denote them by $X \ldots R_1$ and $X \ldots R_2$, respectively, and denote their metrics by xr_1 and xr_2, respectively. The isotonicity property says that if

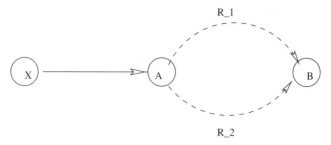

Figure 7.2. Isotonicity.

we have the metric $r_1 \leq r_2$, then we obtain $xr_1 \leq xr_2$. In essence, this property says that each part of the route should have an equal value when included in a larger route.

7.4.7 Limitations of Link-Quality-Aware Routing

In this section, we summarize the disadvantages of the link-quality estimation techniques. The majority of the link-quality estimation mechanisms require probe packets (broadcast or unicast) to sample the link quality. The accuracy of the estimates obtained via probes is limited because the actual data packets can be transmitted in different data rates and have different packet sizes; both these factors affect the perceived quality of the channel. The use of broadcast control packets fails to capture the link quality at different data rates. Capturing the link quality for all the data packet sizes makes the measurement technique infeasible.

The second disadvantage of such measurement techniques is the inability to predict the future load of the networks. Let us assume that the link quality reflects the data packet transfer and that the variance of link quality is captured at a fine level. Based on the metrics, a route with a set of links is selected and the data packets flow through this route. The transmission of these data packets will change the interference patterns and packet loss rates at other links. Thus, the past estimation of the link qualities are invalid and the measurements have to be reinitiated at the nodes that are affected.

Zhang et al. [42] discuss the disadvantages of measurement of packet losses using a broadcast medium (other than the known issues of variable data rate and packet sizes). The number of packets lost can be measured in a broadcast environment. However, the MAC layer uses retransmissions if the packet transmission is unsuccessful in unicast packets. The discrepancy arises due to the inability of the routing layer to know the number of retransmissions that have occurred at the MAC layer. The MAC protocol, which is implemented in the hardware, does not expose such information to the above layers. Hence, measurements such as ETX, which use broadcast packets for measuring packet loss, cannot be easily transformed to measure the packet loss rate of unicast traffic.

In summary, the link-quality-aware routing protocols discussed above are based on a methodology for estimating the quality of links. The links quality of the links forming a route are then combined to produce an estimate of the quality of the route.

This estimate is used, in place of hop count, in what are otherwise traditional routing protocols.

7.5 GLOBALLY COORDINATED ROUTING

While the objective of link-quality-aware routing is to maximize the efficiency of a route, these protocols operate greedily and do not coordinate routing across routes to achieve overall network performance optimization. In an MHWN, multiple connections exist that are each made of multiple links. These links compete with other links in their neighborhood for access to the common channel. The aim of the globally aware routing is to find the routing configuration that would provide optimal (or near-optimal) overall network performance across all connections. The optimality criteria can be defined in multiple ways that combine the performance of the individual connections. For example, the objective may be defined as maximizing the overall network throughput or, alternatively, in terms of the expected end-to-end delay or even in terms of fairness.

Globally coordinated routing requires global (or at least, nonlocal) state information. Thus, developing such solutions for dynamic networks is challenging. However, modeling the problem and evaluating the global solutions relative to the local solutions are important for the following reasons: (1) They are directly applicable for traffic engineering in static, or slowly dynamic, networks. Static MHWNs are an important subset of MHWNs that includes mesh networks, some sensor networks, and potentially portions of ad hoc networks; (2) this provides realistic tight limits on the achievable performance for different networks, which can be used to guide provisioning decisions and provide an upper limit on protocol performance; and (3) experience with the nature of optimal configurations and understanding the type of coordination required across connections can provide insight into designing a next generation of distributed routing protocols. Thus, this provides a flexible and effective tool for evaluating and optimizing MHWNs.

Globally coordinated routing is a component in the "optimal transmission schedule" assumed by Gupta and Kumar in deriving their asymptotic limit on MHWN performance [23]. The remaining component is optimal scheduling. However, in a contention-based protocol, guaranteeing optimal scheduling is impossible. Thus, it is important for the routing decisions to be aware of the scheduling implications. In this section, we motivate globally coordinated routing, discuss the basic model for expressing the globally coordinated routing problem, and express the desired objective function. In the next section, we discuss the problem of modeling MHWNs for purposes of globally coordinated routing.

Configurations obtained via globally coordinated routing will generally achieve a substantial improvement over greedy protocols. For example, in the small scenario depicted in Figure 7.3, coordinated routing yields a 33% more throughput *without* a decrease in any of the individual connection's throughput. The reduction in interference bottlenecks will have other beneficial advantages such as reducing the end-to-end delays and packet drops, which benefit upper-layer protocols such as TCP. It is likely

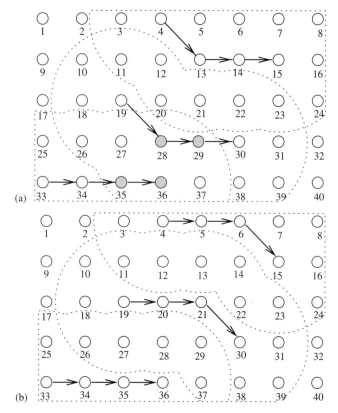

Figure 7.3. Example illustrating coordinated routing. (a) Suboptimal routes. (b) Optimal routes.

that the advantages of this approach will be amplified by larger, more complex scenarios where greedy solutions are likely to be far off from optimality. Furthermore, the objective function can be manipulated to incorporate QoS, fairness, or other desired considerations.

As a motivating example, consider the network shown in Figure 7.3. There are three connections between nodes *4* and *15*, *19* and *30*, and *33* and *36*. Consider the routing configuration in Figure 7.3a. The dotted lines denote the interference range of the active nodes. An ideal unit disk interference range is used for illustrative purpose. Let us assume that all the links used by the connection are of the same quality and connections get started in the sequence *4–15, 19–30*, and *33–36*. The connection *19–30* is able to choose an interference separated path from the connection *4–15*. However, when the connection *33–36* starts, the nodes *28, 29, 35,* and *36* experience interference from the other connection. The routing configuration shown in Figure 7.3b shows the optimal routes that do not interfere with each other. In the suboptimal route configuration (Figure 7.3a), there is no incentive for the connection

28–29 to change the route, since any route will be interfered with by either connection *4–15* or *33–36*. The connection *4–15* will also not change route because none of the nodes in the connection face interference from the neighboring connection. Hence, in the suboptimal scenario, ideally both the connections *19–30* and *33–36* will have half the throughput of the ones in the optimal scenario. This example illustrates that a greedy local view of the network scenario does not always lead to a globally optimal routing configuration.

7.6 MODELING MHWNs

In this section, we discuss the different efforts to model MHWNs to enable characterization of network performance as well as derivation of effective routing configurations. Given that the problem is similar to classical traffic engineering, with the exception of the effect of the wireless channel (susceptible to fading and the effect of interference), a critical component of these schemes is how they model the channel and capture the interference effects.

7.6.1 Asymptotic Capacity Estimation of MHWNs

Some of the earliest efforts in modeling MHWNs targeted formulation of asymptotic limits of capacity. Gupta and Kumar [23] considered a general wireless network and derived bounds on the capacity of such networks. They studied two kinds of network: (1) arbitrary networks where each node can choose an arbitrary destination, packet sending rate, and transmission range, but nodes transmit at a constant rate of W bits per second over the channel; and (2) random networks where the transmission range for each node is fixed, but each node chooses a random destination and transmission rate. In their derivation, n nodes are located in a unit disk area of a plane in the case of arbitrary networks. Under random network, n nodes are placed either on a surface of a three-dimensional surface of sphere with surface area 1 m^2 or on a plane disk of area 1 m^2. While the analysis under arbitrary networks provide the capacity of the network allowing flexibility for the nodes to choose the network parameters, random networks reflect a more realistic network setting by fixing the transmission ranges (or transmission power) and variable data sizes that need to be sent to different destinations. They proposed two models to depict the reception success (and the effect of interference): (1) *Protocol model*: Under this model, all the nodes that are present within a factor of reception range of an active sender observe interference from the sender. (2) *Physical model*: A node can successfully receive a packet if the *signal to interference and noise ratio* (SINR) is above a certain threshold.

Using the above assumptions, the bounds on the channel capacity were derived under different networks and models. Of interest to the study of the effect of interference on routing in realistic setting of networks, we focus the results in random networks. With fixed reception range, the Protocol model gives the interference region as a factor of reception range. The interference from all the currently active sources are added up in the Physical model. The upper bound on the random network is shown to be $\theta(W/\sqrt{n \log n})$ under the Protocol model and is $\theta(W/\sqrt{n})$ in the Physical model, where W is the channel capacity in $bits/second$. This quantifies the reduction in the capacity as the number of active nodes per unit area increase. The above study was based on all single-hop connections between two nodes which are within the reception range of each other. Explicit capacity estimation for a multihop connection is treated by Gastpar and Vetterli [48], who indicate that under a relay-based traffic the achievable capacity is $O(\log n)$, a more encouraging result than the study by Gupta and Kumar [23]. However, both studies consider a random network and propose the asymptotic bounds; the throughput bound for a given network under a given set of connections is desirable in most practical cases.

7.6.2 Modeling for Interference-Aware Routing

Jain et al. [49] revisit the problem of finding the bounds on the throughput with the goal of quantifying the capacity for a given topology and traffic pattern, rather than asymptotic bounds for general networks. The main idea is to identify the set of links that cannot be active together considering the effect of interference. Each link in the network is represented as a vertex in a transformed graph called *Conflict Graph*. If two links interfere with each other, then there is an edge present between the two vertices representing the links.

Consider a scenario with four active links as shown in Figure 7.4a. Figure 7.4b shows the conflict graph for the given scenario. A vertex $v_{(i,j)}$ in Figure 7.4b represents the link (i, j) in the scenario, and the edge between two vertices in the conflict graph represents the interference between the edges. An *independent set* is a set of vertices that are not connected to any of the vertices in the set (e.g., $\{v_{(1,2)}, v_{(5,6)}\}$). An independent set of vertices in a conflict graph represents the set of links that do not interfere and thus can be scheduled simultaneously. A *maximal independent set* is an independent set such that no other vertex can be added to the set that results in another independent set. Thus, the maximum number of links that can be scheduled together can be represented as a maximal independent set of a conflict graph. Let $I = \{I_1, \ldots, I_k\}$ be the set of all maximal independent sets I_i, where $1 \leq i \leq k$. In the above example, $I = \{\{v_{(1,2)}, v_{(5,6)}\}, \{v_{(1,2)}, v_{(11,12)}\}, \{v_{(9,10)}\}\}$.

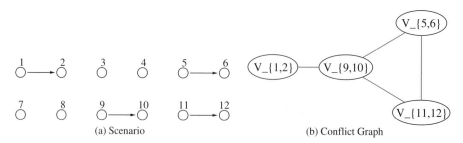

Figure 7.4. Conflict graphs.

Let $d_{i,j}$ denote the distance between the two nodes i and j and let R'_j represent the interference range of node j. Conflicting links are found by the Protocol model [23] of interference. Two edges $l_{i,j}$ and $l_{p,q}$ interference with each other if $d_{iq} \leq R'_i$ or $d_{pj} \leq R'_p$.

Basic Max-Flow Formulation. The problem of finding the throughput is modeled as a max-flow problem that maximizes the amount of flow out of the source node (maximum sending rate). The basic model for finding the maximum flow from a source to the destination considers maximizing the sum of outgoing flows from the source (or incoming flows at destination) by constraining that (1) the source generates the traffic, destination is the flow sink, and no other nodes generate other traffic and (2) the flow at each link is non-negative and does not exceed the capacity.

Optimal Throughput. The maximum throughput depends upon the number of links that can be active concurrently. At any instant of time, only one maximal independent set $I_i \in I$ can be active. Let λ_i denote the amount of time allocated to a I_i. The utilization of a maximal independent set is the amount of time each maximal independent set is active. If we normalize the overall utilization of all the maximal independent sets, then $0 \leq \lambda_i \leq 1$. A schedule to activate the maximal independent sets restricts the λ values. The constraint in Eq. (7.1) restricts the scheduling policy by not allowing multiple maximal independent sets to be active concurrently.

$$\sum_{i=1}^{k} \lambda_i \leq 1 \qquad (7.1)$$

While Eq. (7.1) restricts the schedule of the maximal independent sets, it does not limit the usage of each link. The maximum flow at the link should not exceed the utilization periods of all the maximum independent sets to which it belongs. If C_{ij} denotes the capacity of the link l_{ij}, then Eq. (7.2) expresses this constraint.

$$f_{ij} \leq \sum_{l_{ij} \in I_i} \lambda_i C_{ij} \qquad (7.2)$$

The optimal throughput is given by adding the constraints in Eq. (7.1) and (7.2) to the basic max-flow formulation.

Lower and Upper-Bound Constraints. While the above formulation yields the optimal throughput, calculation of *all* the maximal independent sets takes an exponential time. Hence, maximal independent sets are heuristically calculated. The resulting routing and scheduling formulation gives the *lower bound* of the throughput.

In a conflict graph, a clique (set of vertices that are all connected to each other) represents the links that mutually interfere. A *maximal clique*, similar to maximal independent sets, is a clique for which an additional vertex cannot be added such that the resulting set forms another clique. Due to mutual interference, only one vertex of a clique can be active at a given time. Such *Clique constraints* are used to restrict the interfering links. A clique constraint states that the sum of normalized utilization of all the vertices in the maximal clique is less than 1, which restricts the upper bound of throughput.

Kodialam and Nandagopal [50, 51] consider the same problem but use a different approach. They model the problem as a linear programming problem identifying the constraints that specify the interference between the links. They also propose algorithms to obtain the set of feasible schedules of channels (in a multichannel environment) based on the result obtained from the linear program. However, unlike the model in Jain et al. [49], the model proposed by Kodialam and Nandagopal [51] can identify the sets of links that interfere with each other in $O(|E|)$, where E is the number of edges in the conflict graph.

Accounting for Interaction Among Connections. While studies by Jain et al. [49] and Kodialam and Nandagopal [50, 51] consider the problem of estimating the throughput of a single multihop connection, they propose a simple extension to the model to handle multiple connection scenarios. The main component of a globally aware routing protocol is the interaction between the routes of different connections to enhance the network performance. The proposed frameworks in the above studies were constructed and are best used to study the throughput of a single connection. The linear programming model has to considerably altered to account for the various interactions between the connections. The resulting effects lead to a *nonlinear programming* model that is very difficult to solve computationally. Also, the above studies only consider aggregate throughput as a metric, focusing on feasible throughput bounds, rather than providing usable routing configurations.

Kolar and Abu-Ghazaleh [52] formulate a model for finding the optimal routing configurations in multiple connection MHWNs with a single channel. The model is based on a *Multi-Commodity Flow (MCF)* formulation [53], a network flow modeling approach that has been applied to traffic engineering in conventional networks.

Basic MCF Formulation. In this paragraph, the classical MCF formulation is described. Let $G(N, E)$ represent the network graph where N is the set of nodes and E is the set of links that can communicate. Let (s_n, d_n, r_n) denote source, destination, and the rate of the nth connection. The rate of connection, r_n, is the number of bits to be sent per unit time. Let C be the set of connections. The demand for a given node is the difference between the total outflow from the node and total amount of inflow to the node. The demand at a node for nth connection is represented by b_i^n and is given by Eq. (7.3).

$$b_i^n = \begin{cases} r_n, & \text{if } i = s_n \\ -r_n, & \text{if } i = d_n \\ 0, & \text{otherwise} \end{cases} \quad (7.3)$$

Let x_{ij}^n denote the flow at edge (i, j) for the nth connection. Let $u_{i,j}$ denote the maximum capacity of an edge (i, j). The feasibility constraints for the classical MCF formulation is given in Eq. (7.4)–(7.6). Equation (7.4) describe the *limiting bound* of each flow to be the maximum rate of the connection. For a given connection, each edge can carry a maximum load corresponding to the rate of the given connection. The *bundle constraint* for the given graph is given by Eq. (7.5) which limits the total flow at an edge not to exceed its capacity. The *flow constraint* in Eq. (7.6) specifies the demand requirement to be met at each node as the difference between the outflow and inflow [Eq. (7.3)].

$$0 \leq x_{ij}^n \leq r_n \forall n \in C, \forall (i, j) \in E \tag{7.4}$$

$$0 \leq \sum_{n \in C} x_{ij}^n \leq u_{ij} \forall (i, j) \in E \tag{7.5}$$

$$b_i^n = \left(\sum_{(i,j) \in E} x_{ij}^n \right) - \left(\sum_{(j,i) \in E} x_{ji}^n \right)$$
$$\forall n \in C, \forall i \in N \tag{7.6}$$

Since a single route per connection is the desirable in the majority of the networks, the authors add *integer flow constraints* to facilitate such a routing configuration.

Model Components. In contrast to the conflict graph models used in Jain et al. [49] and the edge-based approach in Kodialam and Nandagopal [51] the formulation in Kolar and Abu-Ghazaleh [52] adopts *node-based interference model* where interference at a given node, rather than at a link, is calculated. Such a formulation aids in approximation of the distributed protocol rules based on the observed behavior at the nodes. The busy time of each node i is calculated accounting for all the traffic on the interfering links which is denoted by I_i. Since interference at edges that are not active is immaterial, the notion of *normalized interference* (denoted by $\hat{I}_i$) is introduced. Normalized interference ($\hat{I}_i$) at a node i is set to the interference observed at the node (I_i) if the node is active, otherwise, it is set to zero. The total amount of time to accommodate the local traffic carried by the node (S_i) in addition to *normalized* interference observed is termed the *commitment period* of the node and is represented by A_i. Commitment period represents the amount of time the node is busy to accommodate local traffic or interfering traffic.

Optimality Formulation. Kolar and Abu-Ghazaleh [52] also explore objective function issues in their formulation. For an interference aware routing formulation, reasonable objectives include: (1) minimizing the interference experienced by each route and (2) reducing the interference interactions among different routes. The interference across each individual flow has to be minimized, thus resulting in an objective function for each flow. However, solving linear programming formulations against multiple objectives is known to be computationally difficult. Thus, there is a need for a single objective function that captures the overall network performance without sacrificing individual connection performance.

Equation (7.7) states a simple objective function to minimize the commitment period of all the nodes, and Eq. (7.8) is an objective function to minimize the hot-spot node that has the highest commitment period.

$$\text{Minimize} \quad \sum_{i \in N} A_i \qquad (7.7)$$

$$\text{Minimize} \quad \max\{A_i | \forall i \in N\} \qquad (7.8)$$

In the following paragraphs, we discuss the deficiencies of such simple objective functions and propose linear objective functions to solve them.

Conjoint Node Effect. Objective functions represented in Eq. (7.7) would lead to assigning the minimum number of nodes so that the commitment periods are minimized. In order to reduce overall network interference, multiple connections will be routed through a single node. This would lead to a scenario where a single node becomes the bottleneck and can be drained out of energy. This is referred to as the *conjoint node effect*.

Connection Coupling. A subset of connections might get an unfair amount of weight due to the very high interference and thus not optimizing the other connections. This is termed *connection coupling*. Equation (7.8) depicts such undesirable property. Connection coupling can be eliminated by considering the commitment period of each connection.

Path Inflation. The objective function is insensitive to the number of nodes in the path, thereby resulting in an inefficient route with *path inflation*. In Eq. (7.7) it can be observed that the objective reduces the hot-spot at the node, while it fails to control the number of active nodes. Path inflation can be controlled by a weighted average objective function that minimizes the commitment period and enforces a weight for the sum of the traffic carried all the nodes.

The formulation of the objective function for accounting for optimization of multiple routes is a challenging problem and may yield a nonlinear programming formulation. Kolar and Abu-Ghazaleh [52] studied the various aspects of optimizing such multiple objectives into a single linear objective function, while at the same time measuring the effectiveness of the objective function under different topologies. It concludes that an objective function that minimizes the maximum hot-spot across all connections works best; it provides a single metric for overall network throughput while not suffering from connection coupling or the conjoint node effect. It is likely that specializing objective functions for other metrics (such as QoS or fairness) would require equally careful consideration.

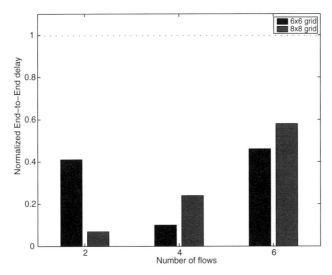

Figure 7.5. Comparison of the end-to-end delay in a globally aware routing. (This figure is taken from the results obtained in Kolar and Abu-Ghazaleh [54].)

In the following, we show sample results from the model developed by Kolar and Abu-Ghazaleh, to illustrate the performance difference between globally coordinated routing, and traditional greedy routing approaches. The results of the routes using the model were simulated in a Qualnet [32] simulator using an IEEE 802.11b MAC protocol. Figure 7.5 shows 6×6 and 8×8 grid topology results. The end-to-end delay of the routes from formulation were compared with the routes obtained by the most stable routes from DSR. The routing effects were eliminated by loading such routes as static routes. The y-axis shows the ratio of the end-to-end delay using the formulation routes to the ones used by the DSR, and the x-axis depicts the number of connections. A phenomenal improvement in the delay can be found while using the globally aware routing. Similar results were found for random scenarios also. However, the throughput improvement at higher contention was marginal, the reason being the greater packet drops while using the formulation routes. At higher contention, the substantial effect of packet drops on throughput is not captured by the above formulation. This signifies the importance of modeling the scheduling interactions.

7.6.3 MAC Modeling—Accounting for the Effect of Scheduling

The formulations discussed thus far ignore the effect of the scheduler; they assume that aggregate interference metrics such as busy time are the sole factor influencing link capacity (as would be the case, for example, under an ideal scheduler). Such formulations measure the *available capacity* of the links (Property 1 discussed in Section 7.4.3), but do not account for the effect of imperfect CSMA-based schedulers

(Property 2 in the same section). The effect of packet drops is considerable in realistic environments that operate under a contention-based MAC scheduler like IEEE 802.11 (as explained in Section 7.4.3). Modeling the effect of routing by considering such MAC protocol would require not only the formulation discussed above, but also a model of characterizing the MAC protocol that can then be combined with the nominal capacity to estimate link quality more accurately.

IEEE 802.11, a popular CSMA/CA-based MAC protocol, uses a two-way handshake between the source and the receiver to let the neighbors of both nodes know about the communication going on between the source and the receiver. The nodes that are able to *receive* the packet will defer their transmission. Under the Protocol model, all the nodes within the receiving range of source and receiver will listen to the handshake and defer their transmissions. The only nodes that interfere with the ongoing transmission are the nodes that are within the interference range, but not within the receiving range. A crude model to estimate the destructive effect of these interferences that are not prevented by the MAC protocol is to estimate the channel busy time contributed by such interferences as a measure of their scheduling impact on each link with which they interfere.

Modeling the contention-based MAC protocol is an active research area. Bianchi [54] modeled the IEEE 802.11 protocol and a throughput estimation using the proposed model was analyzed. Various other research papers such as references 55 and 56 propose new frameworks for modeling the CSMA/CA-based MAC protocols. Garetto, Salonidis, and Knightly [55] model the fairness and throughput. They formulate the starvation of the link in a multihop wireless network. Such estimation of MAC factors can be used in the formulation of the routing.

Studies in references 54–56 approximate the busy time of the node at a higher layer and do not model the MAC-specific interactions in great detail. This has two key disadvantages: (1) The throughput of the link correlates well with busy time in low-interference regions. However, as seen in Figure 7.1 in Section 7.4.3, at moderate to high interference regions the busy time and the throughput of the link fail to correlate. Thus the estimation of throughput by using the above models at high-interference regions may be incorrect. (2) The variation of the quality of the links on a small-scale time frame in a real testbed was studied in reference 2. Capturing such interactions will need a precise formulation of the MAC protocol.

Kolar and Abu-Ghazaleh [31] propose a model to overcome the above problems and to capture a detailed link behavior. The idea of capturing the MAC interactions is by identifying the patterns of interactions between the links. The model is based on the Conflict graphs [50, 51] where each link is represented as a vertex and the links that interfere has an edge between them. The *maximal independent set (MIS)* (as defined earlier in Section 7.6.2) of such a graph results in a set of links that can be initiated concurrently; hence, theoretically, any snapshot of the ongoing transmissions in a network should be the links that form a subset of an MIS. Since each MIS denotes the links that can contend with each other, it is termed the *maximal independent contention set (MICS)*. The study gives a framework to capture the low-level interactions of any CSMA/CA-based protocol and validate the model by using IEEE 802.11.

ALGORITHM 4. Calculate the Set of Maximal Independent Sets M

Require: Θ is a $n \times n$ matrix where $n =$ number of nodes. $\Theta[i][j] = 1$ if i can hear node j. Else $\Theta[i][j] = 0$.
Require: N is an array of all the nodes.
{// L denotes the number of iterations we attempt to compute unique independent sets.}
{// M is the set if all MICS}
Initialize M to empty set
for $i = 1$ to L **do**
   Randomize the ordering of N
   $S = \{N[1]\}$ {// S denotes a maximal independent set being computed}
   for $i = 2$ to $N.size()$ **do**
     $src \leftarrow N[i]$
     for $j = 1$ to $S.size()$ **do**
        if $\Theta[src][S[j]] == 1$ **then**
           Break the *for* loop i
        end if
     end for
     Add *src* to to independent set S
   end for
   {// At this point, S will be one of the maximal independent sets}
   if S is not present in M **then**
     Add the maximal independent set S to M
   end if
end for
return M

The computation of *maximal independent sets* plays an important role in the analysis of CSMA protocols [31, 57–59], as well as routing protocols [49]. The problem is NP-hard, and approximation techniques are often used for practical purposes. A simple way of approximating the maximal independent sets of nodes that can concurrently transmit under the Protocol Model of interference is given in Algorithm 4.

The model is validated by observing the interactions in a simulated environment. The results show that the model captures a vast majority of the interactions. Two or more links initiating at the same instant of time without the knowledge of each other accounts for a miniature fraction of the packet losses, which is not captured by the model in reference 31.

In the IEEE 802.11 [19] MAC protocol, a packet loss is observed when the sender does not receive a CTS for the RTS (which is called as RTS timeout), or if the ACK packet is not received for the DATA (an ACK timeout). In Figure 7.6a, the formulation is validated against simulation results to see if the model captures such losses. It was seen that less than 3% of the packet losses occurred between different MICS and thus were not captured by the model. Accordingly, truly concurrent transmissions are rare, and most collisions occur due to the failure of the MAC protocol to prevent interfering sources from concurrent transmissions. This result supports the use of the

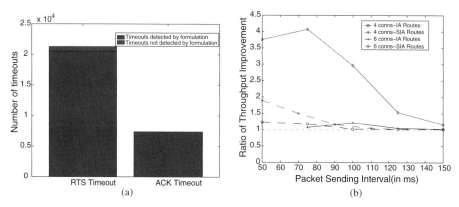

Figure 7.6. Accuracy of the model to capture interactions. This graph shows the observed number of packet drops and the accuracy of the model in predicting these drops. (b) Effectiveness of scheduling formulation. This graph compares the throughput observed in simulation by (i) most stable routes predicted by DSR (ii) routes obtained from interference-aware formulation, and (iii) routes obtained from an interference- and scheduling-aware formulation. (Figure 7.6 represents the results presented in reference 31.)

MICS-based formulation to estimate the effect of the low-level interactions of the MAC protocol.

The study also estimates the fraction of packet drops that occurs at each link due to ineffective handshake of the MAC protocol. Such a metric is similar to the ETX metric which captures the amount of packet losses. The integration of such MAC modeling with routing formulations provides a more accurate globally aware routing model and is a challenging and important research goal. Kolar and Abu-Ghazaleh [31] also performed a global scheduling and interference-aware (SIA) routing and measured the performance gains. SIA was implemented by first getting the best route from the Interference-aware formulation (IA Routes) from the basic MCF linear program model [52]. Then, the packet losses at different links were estimated using the scheduling model discussed above. The links that had the highest conflicts were mutually excluded by adding further constraints into the LP formulation; the extended model is then re-executed. These steps were repeated until the solution converged or the number of runs exceeded a predetermined threshold, at which time the best routes found are used.

Figure 7.6b compares general IA routes proposed in reference 52 with SIA routes. The IA routes work on the Protocol model of interference which assumes a constant interference range. The performance is reported normalized to the performance of the traditional routes obtained using the best DSR routes. It can be seen that the SIA routes perform better than the IA routes. The effectiveness of the scheduling formulations can be seen in high contention areas; as described above in Figure 7.1 in Section 7.4.1, it is in that area that scheduling plays a higher role in determining link capacity. It can be seen in Figure 7.6b that SIA advantages are more pronounced under higher sending rates, or for more connections. The above results justify the higher effect of scheduling at greater contention and for the need to consider scheduling formulation for an effective routing at interference hot-spots.

7.7 CONCLUSIONS AND FUTURE WORK

In this chapter, we discussed the effect of interference on the design of routing protocols. We showed the progression of routing protocol development with respect to this problem. The first-generation routing protocols were based on hop count. In realistic settings, it became evident that hop count does not provide high-quality routes. This led to the development of link-quality-aware routing—the second generation of routing protocols. These protocols estimate link quality in a number of ways and expose this estimate to the routing layer to enable more informed routing decisions. We described the problems of estimating link quality and combining link qualities into route quality. We also overviewed and classified existing approaches to this problem. While link-quality-aware routing provide significantly improves on the performance of hop-count-based routing, it still fails to provide an optimal routing configuration that optimizes the network usage. We believe that a third generation of routing protocols is needed—one that coordinates routing decisions to provide better overall network performance.

We discussed several models to represent MHWNs to enable analysis and optimization of these networks, starting from estimating the throughput of a single-hop link to estimation of an interference-aware routing configurations for multiple multihop connections. Globally aware routing formulations were studied for effective route configurations under a simplistic physical model. The effectiveness of such formulations when translated directly into a realistic environment were studied. While significant improvements were observed, it was also seen that formulation results from the simplistic formulations can lead to performance limitations under high traffic. The assumptions of the underlying physical model and the ideal scheduling policy in the formulations were found to be inaccurate. The need to capture the lower-level MAC interactions under an advanced physical model was motivated and the existing models to capture the scheduling effects were discussed. Superior performance of the globally aware routing formulations with such advanced model were demonstrated. The tight coupling of the link quality with the scheduling effects indicates that an integrated routing and MAC formulation is necessary to capture the optimal routing configurations. Analyzing the scheduling interactions in a realistic wireless testbed experiments and comparing them to the advanced physical models used in the simulators is an essential step in verifying the accuracy of the models in realistic testbeds. The statistical modeling of physical layer from the testbeds will enable capturing the intricate details like multipath and fading, thus predicting the scheduling interactions with greater accuracy and is an area of future work.

While globally aware routing formulations cannot be directly transformed into distributed protocols, it gives the fundamental insight into the parameters required in the design space of the next-generation routing protocols for MHWNs. Capturing the link quality and the temporal variations due to intricate scheduling interactions is one aspect of the third-generation protocols. Such research studies in a distributed environment have a direct implication on the globally aware routing protocol. The challenging part and the primary contribution of the third-generation routing protocols is predicting and adapting to the link-quality variations due to the route changes. While link-quality measurements quantify the present state of the link, an additional need for globally aware routing protocols is to predict the effect of the new route on the links, and thus on the route quality. The effect of changing traffic on a single link can affect the interference maps and the scheduling interactions of the network. Hence, the problem requires communication of the routing decisions and link-quality predictions between multiple connections. The scalability issues of such a distributed protocol, which needs nonlocal information, is an interesting research domain.

The third-generation distributed routing protocols will aid in the development of the dynamic QoS needs in MHWNs. The prediction of the route quality and the interaction between different routes can benefit the network utilization having dynamic connections with varying rates and QoS needs. An active area of future research is to develop the heuristics of provisioning such connections.

7.8 EXERCISES

1. Compare and contrast shortest-hop-based routing and link-quality-based routing for (i) static networks and (ii) mobile networks. Discuss issues specific to (a) routing protocol overheads and (b) measurement accuracy.

2. *Link-Quality Source Routing (LQSR)* is a modification of DSR routing protocol that uses ETX for measuring link-quality and routes according to the ETX metric instead of choosing the shortest number of hops. Explain and show with an example how (i) LQSR can outperform DSR and (ii) globally coordinated routing can outperform LQSR.

3. Consider a delay-sensitive multimedia application in a MHWN that are characterized by smaller packet size and require a lower end-to-end delay. Chart out the characteristics of an effective link-quality metric for such a network. Discuss the route quality estimation characteristics for such applications. Assume that an ideal scheduler with no packet losses and all the traffic present on MHWN are of the above nature.

4. For link-quality-aware routing, most of the protocols include mechanisms for updating the quality for links being used. However, there is no mechanism to update the cost of links that are not being used, which can vary in cost depending on the quality of the channel, and the variability in interference that is experienced. (a) Explain implications of this behavior. (b) Assuming that the cost of unused links can somehow be tracked as well, how would the behavior change? Can this result in instability? Show using an example.

5. The implicit assumption in most of the link-quality routing protocols is that the link capacity is constant. Some wireless cards adapt their link speed up or down to match to observed channel quality, such that the achieved quality in terms of loss rate remains relatively constant. Explain how, if possible, you would change the multi-commodity flow formulation to account for this effect.

6. Commitment period is used to measure of amount of interference at a given node in the multi-commodity formulation [52] under the assumption of an ideal scheduler. Derive the equation for the commitment period when such an assumption is relaxed. Assume that the probability of link error is a constant (p) that is known.

7. Networks with energy-constrained devices have to consider the issue of devices running out of energy. When an active node runs out of energy, the network topology is effectively changed, and routes must adapt to the new topology. Discuss the modifications needed to the multi-commodity formulation [52] for maximizing the lifetime of such networks. Assume that send, receive, and idle energy costs are known per bit (or per unit time) and that the probability of collisions increases with the degree of interference. The cost of a collision is equal to the cost of a reception.

8. Discuss briefly about how the multi-commodity flow model would be changed to:

    Capture multiple routing per study by Nasipuri et al. [60]

    Fading models

    Directional antennas

    Support for multiple channels

REFERENCES

1. I. F. Akyildiz, X. Wang, and W. Wang. Wireless mesh networks: A survey. *In Computer Networks,* **47**(4):445–487, 2005.
2. Meshdynamics Inc. http://www.meshdynamics.com/.
3. John Cox. Philadelphia wireless win launches earthlink new strategy, *Network World.* October 2005. http://www.networkworld.com/news/2005/100605-earthlink-wireless.html.
4. Mesh Networking Forum. Building the business case for implementation of wireless mesh networks, *Mesh Networking Forum* 2004, San Francisco, CA, October 2004.
5. J. Hill, M. Horton, R. Kling, and L. Krishnamurthy. The platforms enabling wireless sensor networks. In *Communications of the ACM*, 2004.
6. V. Park and S. Corson. Temporally-ordered routing algorithm (TORA) version 1 functional specification. Internet Draft, Internet Engineering Task Force, November 2000. http://www.ietf.org/internet-drafts/draft-ietf-manet-tora-spec-03.txt.
7. C. E. Perkins and P. Bhagwat. Highly dynamic destination-sequenced distance-vector routing (DSDV) for mobile computers. *ACM Computer Communications Review,* **24**(4): 234–244, 1994. SIGCOMM '94 Symposium.
8. A. Boukerche. *Handbook of algorithms and Protocols for Wireless Networking and Mobile Computing.* CRC Chapman Hall, 2005.

9. C. E. Perkins, E. M. Royer, and S. R. Das. Ad hoc on demand distance vector routing (AODV), http://www.faqs.org/rfcs/rfc3561.html. Internet RFC 3561, 2003.
10. Z. Haas, M. Pearlman, and P. Samar. Zone routing protocol (zrp) for ad hoc networks. Internet Draft, Internet Engineering Task Force, December 2002. http://www.ietf.org/internet-drafts/draft-ietf-manet-zone-zrp-04.txt.
11. B. Karp and H. Kung. Greedy perimeter stateless routing for wireless networks. In *Proceedings of the Sixth Annual ACM/IEEE International Conference on Mobile Computing and Networking (MobiCom 2000)*, August 2000, pp. 243–254.
12. M. Takai, J. Martin, and R. Bagrodia. Effects of wireless physical layer modeling in mobile ad hoc networks. In *MobiHoc*, 2001.
13. D. S. J. De Couto, D. Aguayo, J. Bicket, and R. Morris. A high-throughput path metric for multi-hop wireless routing. In *MobiCom '03: Proceedings of the 9th Annual International Conference on Mobile Computing and Networking*, ACM Press, New York, 2003, pp. 134–146.
14. D. Aguayo, J. Bicket, S. Biswas, G. Judd, and R. Morris. Link-level measurements from an 802.11b mesh network. In *SIGCOMM '04: Proceedings of the 2004 Conference on Applications, Technologies, Architectures, and Protocols for Computer Communications*, New York, ACM Press, New York, 2004, pp. 121–132.
15. M. Yarvis, D. Papagiannaki, and S. Conner. Characterization of 802.11 wireless networks in the home. In *Proceedings of Wireless Network Measurements (WiNMee)*, 2005.
16. J. Bicket, D. Aguayo, S. Biswas, and R. Morris. Architecture and evaluation of an unplanned 802.11b mesh network. In *MobiCom '05: Proceedings of the 11th Annual International Conference on Mobile Computing and Networking*, ACM Press, New York, 2005, pp. 31–42.
17. H. Lundgren, K. Ramachandran, E. Belding-Royer, K. Almeroth, M. Benny, A. Hewatt, A. Touma, and A. Jardosh. Experiences from the design, deployment, and usage of the ucsb meshnet testbed. In *Wireless Communications, IEEE*, IEEE Computer Society, New York, 2006, pp. 18–29.
18. A. D. J. Malcolm, J. Agogbua, M. O'Dell, and J. McManus. Requirements for traffic engineering over MPLS, 1999. RFC 2702.
19. The IEEE Working Group for WLAN Standards. IEEE 802.11 Wireless Local Area Networks. http://grouper.ieee.org/groups/802/11/, 2002.
20. V. Bharghavan, A. Demers, S. Shenker, and L. Zhang. MACAW: A media access protocol for wireless LANs. In *SIGCOMM*, 1994, pp. 18–27.
21. S. Murthy and J. J. Garcia-Luna-Aceves. An efficient routing protocol for wireless networks. *Mobile Network Applications* **1**(2):183–197, 1996.
22. D. S. J. De Couto, D. Aguayo, B. A. Chambers, and R. Morris. Performance of multihop wireless networks: Shortest path is not enough. *SIGCOMM Computer Communications Review*, **33**(1):83–88, 2003.
23. P. Gupta and P. Kumar. The capacity of wireless networks. In *IEEE Transactions on Information Theory*, 2000.
24. J. Li, C. Blake, D. S. J. D Couto, H. I. Lee, and R. Morris. Capacity of ad hoc wireless networks. In *MobiCom*, 2001.
25. T. Goff, N. Abu-Ghazaleh, D. Phatak, and R. Kahvecioglu. Preemptive routing in ad hoc networks. In *Proceedings, ACM Mobicom 2001*, 2001.

26. D. Cavin, Y. Sasson, and A. Schiper. On the accuracy of MANET simulators. In *POMC '02: Proceedings of the Second ACM International Workshop on Principles of Mobile Computing*, ACM Press, New York, pp. 38–43.
27. M. D. Yarvis, W. S. Conner, L. Krishnamurthy, J. Chhabra, B. Elliott, and A. Mainwaring. Real-world experiences with an interactive ad hoc sensor network. ICPP Workshop, 2002, pp. 143–147.
28. R. Draves, J. Padhye, and B. Zill. Routing in multi-radio, multi-hop wireless mesh networks. In *MobiCom '04: Proceedings of the 10th Annual International Conference on Mobile Computing and Networking*, ACM Press, New York, 2004, pp. 114–128.
29. A. Adya, P. Bahl, J. Padhye, A. Wolman, and L. Zhou. A multi-radio unification protocol for ieee 802.11 wireless networks. In *BROADNETS '04: Proceedings of the First International Conference on Broadband Networks (BROADNETS '04)* (Washington, DC, 2004), IEEE Computer Society, New York, pp. 344–354.
30. K. Xu, M. Gerla, and S. Bae. How Effective is the IEEE 802.11 RTS/CTS Handshake in Ad Hoc Networks? In *IEEE Globecom*, 2002.
31. V. Kolar and N. Abu-Ghazaleh. The effect of scheduling on link capacity in multi-hop wireless networks. *Technical report under arxiv.org: cs.NI/0608077*, 2006.
32. Qualnet network simulator, version 3.6. http://www.scalable-networks.com/.
33. A. Kamerman and L. Monteban. WaveLAN -II: a high-performance wireless LAN for the unlicensed band. In *Bell Labs Technical Journal*, 2:118–133, 1997.
34. G. Holland, N. Vaidya, and P. Bahl. A rate-adaptive mac protocol for multi-hop wireless networks. In *MobiCom '01: Proceedings of the 7th Annual International Conference on Mobile Computing and Networking*, ACM Press, New York, 2001, pp. 236–251.
35. B. Sadeghi, V. Kanodia, A. Sabharwal, and E. Knightly. Opportunistic media access for multirate ad hoc networks. In *MobiCom '02: Proceedings of the 8th Annual International Conference on Mobile Computing and Networking*, ACM Press, New York 2002, pp. 24–35.
36. H. Lundgren, E. Nordströ, and C. Tschudin. Coping with communication gray zones in ieee 802.11b based ad hoc networks. In *WOWMOM '02: Proceedings of the 5th ACM International Workshop on Wireless Mobile Multimedia*, ACM Press, New York, 2002, pp. 49–55.
37. K.-W. Chin, J. Judge, A. Williams, and R. Kermode. Implementation experience with manet routing protocols. *SIGCOMM Computer Communications Review* **32**(5):49–59, 2002.
38. R. Dube, C. D. Rais, K.-Y. Wang, and S. K. Tripathi, Signal stability based adaptive routing (ssa) for ad-hoc mobile networks. *IEEE Personal Communications*, **3**(2):35–40, 1997.
39. J. Jubin and J. D. Tornow. The darpa packet radio network protocols. In *Proceedings of the IEEE 75*, 1987.
40. J. C. Park and S. K. Kasera. Expected data rate: An accurate high-throughput path metric for multi-hop wireless routing. In *SECON: Second Annual IEEE Communications Society Conference on Sensor and Ad Hoc Communications and Networks*, 2005.
41. A. Cerpa, J. L. Wong, M. Potkonjak, and D. Estrin. Temporal properties of low power wireless links: Modeling and implications on multi-hop routing. In *MobiHoc '05: Proceedings of the 6th ACM International Symposium on Mobile Ad Hoc Networking and Computing*, ACM Press, New York, 2005, pp. 414–425.
42. H. Zhang, A. Arora, and P. Sinha. Learn on the fly: Data-driven link estimation and routing in sensor network backbones. In *25th IEEE International Conference on Computer Communications (INFOCOM)*, 2006.

43. Y. Yang, J. Wang, and R. Kravets. Designing routing metrics for mesh networks. In *WiMesh: First IEEE Workshop on Wireless Mesh Networks*, 2005.
44. R. Draves, J. Padhye, and B. Zill. Comparison of routing metrics for static multi-hop wireless networks. In *SIGCOMM*, 2004.
45. D. Eckhardt and P. Steenkiste. Measurement and analysis of the error characteristics of an in-building wireless network. In *SIGCOMM '96: Conference Proceedings on Applications, Technologies, Architectures, and Protocols for Computer Communications*, ACM Press, New York, 1996, pp. 243–254.
46. D. Kotz, C. Newport, and C. Elliott. The mistaken axioms of wireless-network research. Technical Report TR2003-467, Department of Computer Science, Dartmouth College, July 2003.
47. S. Keshav. A control-theoretic approach to flow control. In *SIGCOMM '91: Proceedings of the Conference on Communications Architecture & Protocols*, ACM Press, New York, 1991, pp. 3–15.
48. M. Gastpar and M. Vetterli. On the capacity of wireless networks: The relay case. In *Twenty-First Annual Joint Conference of the IEEE Computer and Communications Societies. Proceedings, IEEE INFOCOM 2002*, 2002.
49. K. Jain, J. Padhye, V. N. Padmanabhan, and L. Qiu. Impact of interference on multihop wireless network performance. In *MobiCom*, 2003.
50. M. Kodialam and T. Nandagopal. Characterizing achievable rates in multihop wireless networks: the joint routing and scheduling problem. In *MobiCom*, 2003.
51. M. Kodialam and T. Nandagopal. The Effect of Interference on the Capacity of Multihop Wireless Networks. In *Bell Labs Technical Report, Lucent Technologies*, 2003.
52. V. Kolar and N. B. Abu-Ghazaleh. A multi-commodity flow approach to globally aware routing in multi-hop wireless networks. In *IEEE Pervasive Computing and Communications (PerCom)*, 2006.
53. R. K. Ahuja, T. L. Magnanti, and J. B. Orlin. *Network Flows: Theory, Algorithms, and Applications*, Prentice-Hall, Englewood Cliffs, NJ, 1993.
54. G. Bianchi. Performance analysis of the ieee 802.11 distributed coordination function. *IEEE Journal on Selected Areas in Communications*, 18:535–547, 2000.
55. M. Garetto, T. Salonidis, and E. W. Knightly. Modeling per-flow throughput and capturing starvation in csma multi-hop wireless networks. In *IEEE INFOCOMM*, 2006.
56. M. M. Carvalho and J. J. Garcia-Luna-Aceves. A scalable model for channel access protocols in multihop ad hoc networks. In *MobiCom '04: Proceedings of the 10th Annual International Conference on Mobile Computing and Networking*, ACM Press, New York, pp. 330–344.
57. R. R. Boorstyn, A. Kershenbaum, B. Maglaris, and V. Sahin. Throughput analysis in multihop csma packet radio networks. *IEEE Transactions on Communication*, 1987.
58. F. A. Tobagi and J. M. Brazio. Throughput analysis of multihop packet radio network under various channel access schemes. In *IEEE INFOCOM*, 1983.
59. X. Wang and K. Kar. Throughput modelling and fairness issues in csma/ca based ad-hoc networks. In *INFOCOM*, 2005.
60. A. Nasipuri, R. Castaneda and S. R. Das. Performance of multipath routing for on-demand protocols in mobile ad hoc networks. *Mobile Network Applications*, 2001.

CHAPTER 8

Routing Protocols in Intermittently Connected Mobile Ad Hoc Networks and Delay-Tolerant Networks

ZHENSHENG ZHANG

San Diego Research Center, San Diego, CA 92121

8.1 INTRODUCTION

In the last few years, there has been much research activity in mobile (wireless) ad hoc networks (MANETs). MANETs are infrastructureless, and nodes in the networks are constantly moving. In MANETs, nodes can directly communicate with each other if they enter each other's communication range. A node can terminate packets or forward packets (serve as a relay). Thus, a message traverses an ad hoc network by being relayed from one node to another, until it reaches its destination. As nodes are moving, this becomes a challenging task, since the topology of the network is in constant change. How to find a destination, how to route to that destination, and how to ensure robust communication in the face of constant topology change are major challenges in mobile ad hoc networks. Routing in mobile ad hoc networks is a well-studied topic. To accommodate the dynamic topology of mobile ad hoc networks, an abundance of routing protocols have recently been proposed, such as OLSR [1], AODV [2], DSR [3], LAR [4], EASE [5, 6], ODMRP [7], and many others [8, 9]. For all these routing protocols, it is implicitly assumed that the network is connected and there is a contemporaneous end-to-end path between any source and destination pair. However, in a physical ad hoc network, the assumption that there is a contemporaneous end-to-end path between any source and destination pair may not be true as illustrated below. In MANETs, when nodes are in motion, links can be obstructed by intervening objects. When nodes must conserve power, links are shut down periodically. These events result in intermittent connectivity. At any given time, when no path exists between source and destination, network partition is said

---

Algorithms and Protocols for Wireless and Mobile Ad Hoc Networks, Edited by Azzedine Boukerche
Copyright © 2009 by John Wiley & Sons Inc.

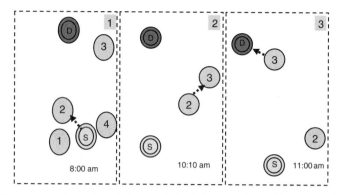

Figure 8.1. Illustration of time-evolving behavior of the ad hoc networks.

to occur. Thus, it is perfectly possible that two nodes may never be part of the same connected portion of the network. Figure 8.1 illustrates the time evolving behavior in intermittently connected networks (ICNs). In Figure 8.1, there is no direct path from node S to node D at any given time. Packets from node S can be delivered to node D if intermediate nodes can hold/carry the packets (at 8:00 am, node S sends the packets to node 2, at 10:10 am, node 2 forwards the packets to node 3, and at 11:00 am, node 3 forwards the packets to node D). Examples of an intermittently connected network are as follows:

(a) An interplanet satellite communication network where satellites and ground nodes may only communicate with each other several times a day.
(b) A sensor network where sensors are not powerful enough to send data to a collecting server all the time or scheduled to be wake/sleep periodically.
(c) A military ad hoc network where nodes (e.g., tanks, airplanes, soldiers) may move randomly and are subject to being destroyed.

Applications in ICNs must tolerate delays beyond conventional IP forwarding delays, and these networks are referred to as delay/disruption tolerant networks (DTNs). *Routing protocols such AODV and OLSR do not work properly in DTNs, since under these protocols, when packets arrive and no contemporaneous end-to-end paths for their destinations can be found, these packets are simply dropped. New routing protocols and system architectures should be developed for DTNs.*

There are many potential applications in DTNs, such as interplanetary network (IPN), Zebranet, DataMule and village networks. IPN [10] consists of both terrestrial and interplanetary links, which suffers from long delays and episodic connectivity. In Zebranet [11], wild-life researchers drive through a forest collecting information about the dispersed zebra population. In the DataMule project [12], DataMules randomly move and collect data from low power sensors. For village networks, for example, a recent project in developing nations uses rural buses to provide Internet connectivity to otherwise isolated and remote villages that do not have any communication infrastructure [13]. Another example of village networks is presented in reference 14, in which the Wizzy digital courier service provides disconnected Internet access to people/students in remote villages of South Africa. A courier on a motorbike, equipped with a USB storage device, travels from a village to a large city which has high-speed Internet connectivity. Typically, it takes a few hours for the courier to travel from the village to the city.

There are many different *terminologies* used for DTNs in the literature, such as eventual connectivity, space-time routing, partially connected, transient connection, opportunistic networking, extreme networks, and end-to-end communication. The data unit in DTNs can be a message, a packet, or bundle, which is defined as a number of messages to be delivered together. For simplicity, throughout this chapter, we use bundles, messages, and packets interchangeably.

The characteristics of DTNs are very different from the traditional Internet in that the latter implicitly has some well-known assumptions: (1) continuous connectivity, (2) very low packet loss rate, and (3) reasonably low propagation delay. DTNs do not satisfy all of these assumptions, and sometimes none. The challenges in designing efficient protocols in the DTNs are extremely long delay (up to days), frequent disconnection, and opportunistic or predicable connections. Consequently, the existing protocols developed for the wired Internet are not able to handle the data transmission efficiently in DTNs. In DTNs, end-to-end communication using TCP/IP protocol may not work, because packets whose destinations cannot be found are usually dropped. If packet dropping is too severe, TCP eventually ends the session. UDP provides no reliable service and cannot *hold and forward. New protocols and algorithms need to be developed.* There are several different types of DTN due to their different characteristics. For instance, the satellite trajectories in example (a) are predictable, while the movement of a solider or tank in example (c) may be random. Therefore, for different types of DTN, different solutions may need to be proposed. Recently, the DTN research group under the Internet Research Task Force (IRTF) has proposed several research documents including a DTN architecture [10, 15–17]. This architecture addresses communication issues in extreme networks or networks encompassing a wide range of architectures. The architecture is a network of regional networks, with an overlay on top of a transport layer or on top of these regional networks. It provides key services, such as in-network data storage and retransmission, interoperable naming, and authenticated forwarding. The DTN solutions in reference 15 are concerned with message transport between infrastructures of disparate architectures by using gateways that handle bundles of messages between these infrastructures. The DTN architecture addresses the issues of eventual connectivity and partitioned networks by the use of a store and forward mechanism, and it handles the diverse addressing needs of the overlay architecture by using an addressing scheme that exploits the late binding of addresses. Local addresses are not bound to nodes until the message is in the local area of the destination. This creates a hierarchical routing structure that makes routing across networks easier to implement. There are many activities in the DTN working group; and due to space limitation, readers are referred to reference 15 (and the references therein) for more information.

Routing in DTNs is one of the key components in the architecture document. Based on different types of DTN, deterministic or stochastic, different routing protocols are required. Due to intermittent connectivity, it is likely that paths to some of the destinations may not exist from time to time. When a packet arrives and its destination cannot be found in the routing table, the packet is simply dropped under the routing protocols mentioned above (developed with the assumption that the network is connected). Therefore, these routing protocols will not work efficiently in DTNs. In this chapter, we provide an overview of the state of the art in DTN routing protocols.

To cope with intermittent connectivity, one natural approach is to extend the store-and-forward routing to a store-carry-forward (SCF) routing. In store–carry–forward routing, a next hop may not be immediately available for the current node to forward the data. The node will need to buffer the data until the node gets an opportunity to forward the data, and it must be capable of buffering the data for a considerable duration. The difficulty in designing a protocol for efficiently and successfully delivering messages to their destinations is to determine, for each message, the best nodes and time to forward. If a message cannot be delivered immediately due to network partition, the best carriers for a message are those that have the highest chance of successful delivery—that is, the highest delivery probabilities. Because ad hoc networks could be very sparse, SCF routing could mean that the node may have to buffer data for a long period of time. This condition can get worse if the next hop is not selected properly. A bad forwarding decision may cause the packets to be delayed indefinitely. If messages must be stored somewhere, a buffer management scheme should be proposed.

If all the future topology of the network (as a time-evolving graph) is deterministic and known, or at least predictable, the transmission (when and where to forward packets) can be scheduled ahead of time so that some optimal objective can be achieved. If the time-evolving topology is stochastic, SCF routing performs routing by moving the message closer to the destination one hop at a time. If the nodes know nothing about the network states, then all that the nodes can do is to randomly forward packets to their neighbors, protocols in this category are referred to as epidemic. If one can estimate the forwarding probability of its neighbors, a better decision could be made. Protocols in this category are referred to as history- or estimation-based forwarding. Furthermore, if the mobility patterns can be used in the forwarding probability estimation, an even better decision may be made. Protocols in this category are referred to as model-based forwarding. In some cases, network efficiency can be achieved if the movements of certain nodes are controlled, and these protocols are in the category of controlling node movements. Recently, coding-based routing protocols are also proposed for DTNs. These protocols can be categorized as follows. In this chapter, we will review some of the routing protocols.

8.2 DETERMINISTIC ROUTING

In this section, we review a few routing protocols assuming that future movement and connections are completely known (that is the entire network topology is known ahead of time).

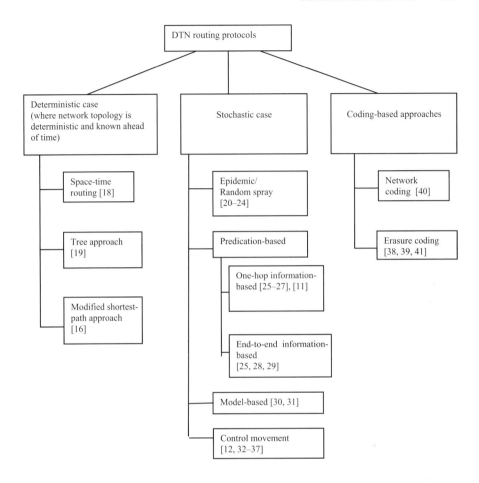

In a tutorial article [42], Ferreira describes a simple combinatorial reference model that captures most characteristics of the time-varying networks. A notion of evolving graphs, which consists of formalizing time domain graphs, is introduced. Modeling time in mobile ad hoc networks gives rise to several different matrices that may serve as objective functions in routing strategies, such as *earliest time to reach one or all the destinations or minimum hop paths (with or without the condition that packets arrive before a predefined time period)*. The readers are referred to reference 42 for more details and references therein. In Figure 8.2, originally shown in reference 42, the min-hop path from S to D takes 4 hops at time interval 1 while it takes one hop at time interval 4, where the number next to each link denotes the time interval during which the link is active.

In reference 19, algorithms selecting the path of message delivering are presented, depending on the available knowledge about the motion of hosts. Three cases are considered. In the first case, it assumes that global knowledge of the characteristic profiles with respect to space and time (that is, the characteristic profiles of the motion and availability of the hosts *as functions of time*) are completely known by all the

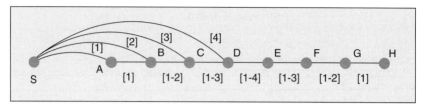

Figure 8.2. Illustration of different metrics.

hosts. Paths are selected by building a tree first, such an approach is referred to as the tree approach. Under the tree approach, a tree is built from the source host by adding children nodes and the time associated with nodes. Each node records all the previous nodes the message has to travel and records the earliest time to reach it. A final path can be selected from the tree by choosing the earliest time (or minimum hop) to reach the desired destination. The results for the time-evolving graph in Figure 8.3 using the tree approach are given in Figure 8.3a. From the figure, one can easily notice that a path from A to D can be set up at time period 1 with 3 hops or at time period 4 with one hop.

In the second case, it assumes that characteristic profiles are initially unknown to hosts, hosts gain this information through learning the future by letting neighbor hosts exchange the characteristic profiles available between them. Paths are selected based on this partial knowledge. In the third case, to enhance the algorithm in the second case, it also requires hosts to record the past, that is, it stores the sequence of hosts a message has transited within the message itself.

For DTNs, several routing algorithms are proposed in reference 16, depending on the amount of knowledge about the network topology characteristics and traffic demand. They define four *knowledge oracles*; each oracle represents certain knowledge of the network. *Contacts Summary Oracle* contains information about aggregate statistics of the contacts (resulting in time-invariant information). A contact is defined as an opportunity to send data. *Contacts Oracle* contains information about contacts between two nodes at any point in time. This is equivalent to knowing the time-varying networks. *Queuing Oracle* gives information about instantaneous buffer occupancies

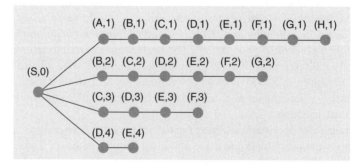

Figure 8.3. Illustration of the tree approach.

(queuing) at any node at any time. *Traffic Demand Oracle* contains information about the present or future traffic demand. Based on the assumption of which oracles are available, the authors present corresponding routing algorithms. For example, if all the *oracles* are known, a linear programming is formulated to find the best route. If only the Contacts Summary Oracle is available, Dijkstra with time-invariant edge costs based on average waiting time is used to find the best route. If only the Contact Oracle is available, modified Dijkstra with time-varying cost function based on waiting time is used to find the route. All of the algorithms developed (except the zero-knowledge case) in reference 16 are for the deterministic case.

Assuming that the characteristic profile is known over an infinite time horizon may not be realistic in ad hoc networks. In reference 18, it is assumed that the characteristic profile can be accurately predicted over the time interval of T. They model the dynamic of the networks as *space–time graph*. Routing algorithms in the constructed space–time graph are developed using dynamic programming and shortest-path algorithm. The routing algorithm finds the best route for messages by looking ahead. The idea of the time layers in the space–time graph comes from time-expanded graphs [43]. The time-expanded graphs approach translates a problem of network flow over time to a classical static network flow, and standard tools of graph theory such as the Floyd–Warshall algorithm can be applied to compute the shortest path for a source destination pair. Figure 8.4 gives a link active schedule and for link AC, a space-time graph is illustrated in Figure 8.5.

In all these approaches under the deterministic case, an end-to-end path (possibly time-dependent) is determined before messages are actually transmitted. However, in certain cases, the topology of the network may not be known ahead of time. In the following section, we review some of the protocols designed for stochastic or random networks.

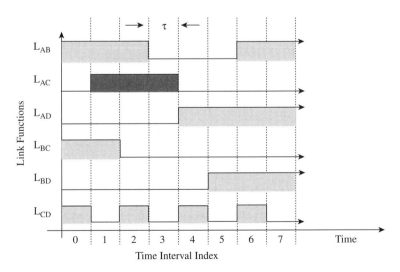

Figure 8.4. link active schedule.

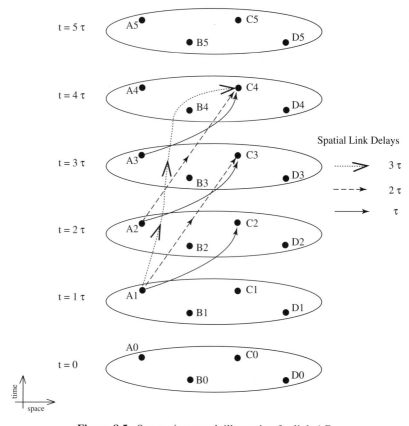

Figure 8.5. Space–time graph illustration for link AC.

8.3 STOCHASTIC OR DYNAMIC NETWORKS

In this section, we review some of the routing protocols when the network behavior is random and not known. These protocols depend on decisions regarding where and when to forward messages. The simplest decision is to forward to any contacts within range, while other decisions are based on history data, mobility patterns, or other information.

8.3.1 Epidemic (Or partial) Routing-Based Approach

In the epidemic routing category, packets received at intermediate nodes are forwarded to all or part of the nodes neighbors (except the one who sends the packet) without using any predication of the link or path forwarding probability. Epidemic routing is a natural approach when no information can be determined about the movement patterns of nodes in the system.

Vahdat and Becker [20] propose an epidemic routing protocol for intermittently connected networks. When a message arrives at an intermediate node, the node floods the message to all its neighbors. Hence, messages are quickly distributed through the connected portions of the network. Epidemic routing relies on carriers of messages coming into contact with another node through node mobility. When two nodes are within the communication range, they exchange pairwise messages that the other node has not seen yet. The epidemic algorithm works as follows. Assume that host A comes into contact with host B and initiates an anti-entropy session.

Step 1. A transmits it summary vector, SV_a, to B. SV_a is a compact representation of all the messages being buffered at A.

Step 2. B performs a logical AND operation between the negation of its summary vector, SV_b, (the negation of B's summary vector, representing the messages that it needs) and A. That is, B determines the set difference between the messages buffered at A and the messages buffered locally at B. It then transmits a vector requesting these messages from A.

Step 3. A transmits the requested messages to B.

This process is repeated transitively when B comes into contact with a new neighbor. Given sufficient buffer space and time, these anti-entropy sessions guarantee eventual message delivery through such pairwise message exchange.

Their simulation results show that, in the special scenarios considered, epidemic routing is able to deliver nearly all transmitted messages, while existing ad hoc routing protocols fail to deliver any messages because of the limited node connectivity when the buffer capacity is sufficiently large.

Another extreme is to let the source hold the message and deliver to the destination only when they are within the communication range. This approach obviously has the minimal overhead, but the delay could be very long. Grossglauser and Tse [21], propose a two-hop forwarding approach and have explored a theoretical framework where nodes with infinite buffer move independently around the network and every node gets close to any other node for some short time period per time slot. Within this framework, a node s gives a message addressed to node t to another *randomly* chosen node one hop away in the network called a *receiver*. When the receiver happens to be within the range of the destination node t, the receiver sends the message to the destination. Hence a message will only make *two hops*, and no message will be transmitted more than twice. They prove that a message is guaranteed to be delivered, even if its delivery time is averaged over many timeslots. This result sets a theoretical bound, since it assumes a complete mixing of the trajectories so that every node can get close to another one.

In the Infostation model [44], users can connect to the network in the vicinity of Infostations which are geographically distributed throughout the area of network coverage. Infostations provide strong radio signal quality to small disjoint geographical areas and, as a result, offer very high rates to users in these areas. However, due to the lack of continuous coverage, this high data rate comes at the expense of providing intermittent connectivity only. Since a node that wishes to transmit data may be located outside the Infostations' coverage areas for an extended period of time and must always transmit to an Infostation directly, large delays may result. Upon arrival in a coverage zone, the node can transmit at very high bit-rates. Thus, *Infostations trade connectivity for capacity, by exploiting the mobility of the nodes.* It is assumed that the *Infostations* are connected. Small and Haas [22] propose a Shared Wireless Infostation Model (SWIM), where SWIM is a marriage of the *Infostation*s concept with the (epidemic) ad hoc networking model (propagation of information packets within SWIM is identical to epidemic routing protocol [20]). The only difference is that any one of the many Infostations could serve as a destination node, while in reference 20, there is only one destination node for a given packet. A real-world application based on the Infostation model is presented in reference 22. One of the benefits SWIM has, by allowing the packet to spread throughout the mobile nodes, is that the delay for the replicas to reach an Infostation can be significantly reduced. However, this comes at a price: Spreading the packets to other nodes consumes network capacity. Again, there is a capacity–delay tradeoff. This tradeoff can be controlled by limiting the parameters of the spread—for example, by controlling the probability of packet transmission between two adjacent nodes, transmission range of each node, or the number and distribution of the Infostations. However, how to choose these controlling parameters is not discussed in reference 22. Let T be the time from the packet generation until the first time the packet reaches one of the infostations. For random waypoint and group mobility patterns, a three-state Markov Chain model can be presented as in Figure 8.6, where S denotes a node with no packets, I denotes a node having packets, R represents nodes offload their packets to infostations, β is the nodes contact rate, and γ is the contact rate per infostation. Using the three-state Markov chain model, the cumulative distribution of T can be easily obtained.

A relay-based approach to be used in conjunction with traditional ad hoc routing protocols is proposed in reference 23. This approach takes advantage of node mobility to disseminate messages to mobile nodes. The result is the Mobile Relay Protocol (MRP), which *integrates message routing and storage in the network.* The basic idea is that if a route to a destination is unavailable, a node performs a *controlled local broadcast* (a relay) to its immediate neighbors (that is the only time that broadcast is used in the protocol). All nodes that receive this packet store it and enter the relaying mode. In the relaying mode, the MRP first checks with the (traditional) routing protocols to see if a route of less than d hops exists to forward the packet. If so, it forwards the packet and the packet is delivered. If no valid route exists for the packet, it enters the storage phase, which consists of the following *steps*:

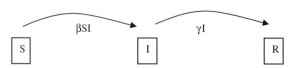

Figure 8.6. Markov chain model of an infectious disease with susceptible, infected, and recovered states.

Step 1. If the packet is already stored in the node's buffer, then the older version of the packet is discarded.

Step 2. Otherwise, the node buffer is checked. If it is not full, then the packet is stored and the time-to-live parameter h in the MRP header of the packet is decremented by 1.

Step 3. If the buffer is full, then the least recent packet is removed from the buffer and it is relayed to a single *random* neighbor if $h > 0$.

In a network with sufficient mobility, it is quite likely that one of the relay nodes to which the packet has been relayed will encounter *a node* that has a valid, short (conventional) route to the eventual destination, thereby increasing the likelihood that the message will be successfully delivered. It is not clear why the authors choose to broadcast only once and then select one node *randomly* to forward in case the buffer is full (to keep the packet in the network). Why not broadcast twice or three times?

To limit the amount of broadcasting to all its neighbors as in references 20 and 22, the Spraying protocol [24] restricts the forwarding to a ray in the vicinity of the destination's last known location. In reference 24, it is assumed that the destination's last location is known and there is a separate location manager in the system. To deal with high mobility, Spray routing [24] multicasts traffic within the vicinity of the last known location of a session's destination. The idea is that even though a highly mobile node may not be in the location last reported by the location tracking mechanism, it is likely to be in one of the surrounding locations. By *spraying* to the vicinity of the last-known location of the destination, the algorithm attempts to deliver packets to the destination even if it moves to a nearby location during the location tracking convergence time. A sprayed packet is first unicast to a node close to the destination, and then it is multicast to multiple nodes around the destination. The magnitude of the spraying depends on the mobility: The higher the mobility, the larger the vicinity. Upon a change in affiliation, a node sends a location update to its location manager. In order to communicate with a destination node, D, a source node sends a location subscribe to the location manager. The current location and changes (in location) thereafter are sent by the location manager to the source using a location information message. Note that it is possible that the destination gets duplicate packets, and it is assumed that there is an end-to-end duplicate detection mechanism that will discard such packets. Spray routing is an integrated location tracking and forwarding scheme. Both location managers and switches/routers participate in spray routing. How to choose the first node at the beginning to forward to is not clearly explained in reference 24. An illustration of the spraying routing algorithm is given in Figure 8.7.

8.3.2 Estimation (of the Link Forwarding Probability)-Based Approach

Instead of blindly forwarding packets to all or some neighbors, intermediate nodes estimate the chance, for each outgoing link, of eventually reaching the destination.

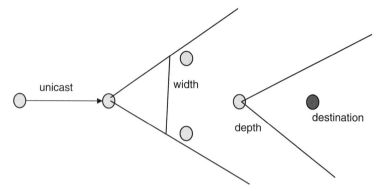

Figure 8.7. Illustration of the spraying algorithm.

Based on this estimation, the intermediate nodes either (a) decide whether to store the packet and wait for a better chance or (b) decide to which nodes (and the time) to forward. A first theoretic work on link estimation is given in reference 45. Some protocols make the per contact forwarding decisions based on only the next-hop information, such as next hop forwarding probability, while some other protocols make the per contact forwarding decisions based on average end-to-end metrics, such as expected shortest path or average end-to-end delay.

Per-Contact Routing Based on Next-Hop Information only. A follow-up work to reference 20 is presented in reference 25, which extends Vahdat's work to situations with limited resources. In their work, though still using flooding-like propagation, they enhance the drop strategy in epidemic routing when caches or buffers are filled. Their algorithm works as follows. When a node A meets another node B, they perform a bundle exchange through a number of steps. First, node A gives to node B a list of the bundles that node A carries with their destinations. Each bundle also contains a likelihood of delivery by node A. Node A receives the same list from B and calculates the likelihood of delivering B's bundles. Node A now sorts the combined lists by the likelihood of delivery, removes node A's own bundles, and also deletes bundles that B has a higher likelihood of delivering. Node A then selects the top *n* bundles remaining, and it *requests from B all the bundles (up to n) that node A does not already have*. They proposed four types of drop strategies for deciding which bundles to exchange when two nodes meet and simulation results show that Drop-Oldest (DOA) and Drop-Least-Encountered (DLE) yielded the best performance. The encounter value is an estimation of meeting likelihood of two given nodes. The DLE algorithm has peers keep track of the other peers they meet regularly over time. Peers initialize their likelihood of delivery a bundle to a moving peer as 0. When peer A meets another peer B, the former sets the likelihood of delivering bundles to B as 1. These values degrade over time, such that they are reinforced only if A and B meet periodically. More specifically, the DLE algorithm calculates the likelihood of delivery as follows:

$$M_{t+1}(A, C) = \begin{cases} \lambda M_t(A, C) & \text{if none are co-located} \\ \lambda M_t(A, C) + 1 & \text{if B = C} \\ \lambda M_t(A, C) + \alpha M_t(B, C) & \text{for all C} \neq \text{B} \end{cases}$$

where $\alpha = 0.1$ is the B's value that A should add, $\lambda = 0.95$ is the decay rate, and M0(A,C) = 0. DLE orders packets according to the relative ability of two agents to pass a packet to the destination. When the buffer is full, packets with less likelihood of delivery will be dropped first.

Similar to the work in reference 25, a probabilistic routing protocol, PROPHET (Probabilistic Routing Protocol using History of Encounters and Transitivity), is proposed in reference 26. PROPHET first estimates a probabilistic metric called delivery predictability, $P(a, b)$, at every node a, for each known destination b. This indicates how likely this node will be able to deliver a message to that destination. The operation of PROPHET is similar to that in reference 25. When two nodes meet, they exchange summary vectors and also exchange a delivery predictability vector containing the delivery predictability information for destinations known by the nodes. The summary vectors are obtained in the same way as in reference 25 (where the vector is called a list). This additional information is used to update the internal delivery predictability vector as follows:

$$P_{(a,b)} = \begin{cases} P_{(a,b)old} + (1 - P_{(a,b)old})P_{init} & \text{if a and b meet} \\ P_{(a,b)old} \bullet \gamma^k & \text{if a and b have not met for k unit times} \\ P_{(a,b)old} + (1 - P_{(a,b)old}) \bullet P_{(c,b)} \bullet \beta & \text{Transitive property} \end{cases}$$

where P_{init} in (0,1) is an initialization constant (with all $P_{(a,b)}$ being set at P_{init}) and γ in (0,1) is an aging constant. The information in the summary vector is used to decide which messages to request from the other node. Simulation results show that for the network considered, the improvement of packet delivery ratio under PROPHET over the epidemic routing can be up to 40%.

In reference 27, the authors study a relay-based routing scheme for ad hoc *Satellite* networks where nodes are required to buffer data for a certain period of time until the node gets an opportunity to forward it. They propose Interrogation-Based Relay Routing (IBRR), where the nodes interrogate each other to learn more about network topology and nodal capacity to make intelligent routing decisions. The main issue in interrogation-based relay routing is the next hop selection process. Given the dynamic topology and heterogeneity of an ad hoc satellite constellation, it may be difficult to decide whether or not to forward the data to a given node. To make effective routing decisions, satellites are expected to track, to their best extent, the positions of neighboring nodes and even some distant nodes. Moreover, a satellite encountered at one moment by a node may not be the best candidate for node A to forward the data to, but it may be the only choice in the foreseeable future. In such a case, node A has to decide whether to forward the data or bet its luck and wait for the next opportunity. Optimistic forwarding proposed in reference 31 (discussed later) is used in IBRR. To select the next hop, a node needs to know not only the present and future connectivity relation with its current time, but also the same information of its current neighbors. This one-hop *look-ahead* is necessary for making routing decisions in the relay based routing framework. Look-ahead beyond one hop can prove to be time-consuming and counterproductive. To select the next hop, a node needs to evaluate the potential candidates and select the most promising forwarding node and forward the data to that node. The selection is based on the following metrics:

- Spatial location and orbital information of the candidate nodes
- Bandwidth of the intersatellite link to the candidate nodes
- Relative velocity/mobility between two nodes
- Vicinity of this candidate to other satellites and ground stations
- Capability of the candidate satellites
- Data transmission time

The authors propose to use interrogation where the nodes do not transmit any hello messages except to initialize a session between two neighboring nodes. After the initialization, the IBRR protocol proceeds to exchange orbital and routing information between nodes in the form of *queries and responses* and may or may not continue with actual data transmission. Instead of discovering the entire path to the destination, a node puts more effort into acquiring information about the immediate neighbors and that of the neighbors' neighbors (one-hop look-ahead). *The best next-hop candidates for node A* are nodes that have sent out the most number of replies to node A's beacons. In the simulations, the authors choose the number of best next hops to forward for each node to be 3 without giving any reason.

In ZebraNet [11], wireless sensor nodes, namely collars (attached to Zebras), collect location data and opportunistically report their histories when they come in radio range of base stations, or the researchers or data collection objects, which periodically drive through (or fly-over) with receivers to collect data. Collars operate on batteries with/without solar recharge. The goal is to study the animal behaviors through designing a collar and communication protocols that works on Zebras (high data collection rate). They study two routing protocols: flood-based routing protocols and history-based protocols. In the flood-based routing protocol, data are flooded to their neighbors whenever they meet. It is expected that as nodes move extensively and meet a number of neighbors, given enough time, data will eventually reach to the base station. In the history-based routing protocol, each node is assigned a likelihood of transferring data to the base station based on its past success. A higher value corresponds to a higher probability of eventually being within the range of the base station. Data are forwarded to its neighbor with the highest transferring probability. Experimental results indicate that the flood-based protocol yields higher system throughput if the buffer capacity at each node is large enough, but the energy consumed by the flood-based protocol can be eight times that of the history-based protocol. There is a tradeoff between throughput and energy consumption. Their conclusion is that while flooding makes sense at low-radio-range and low-connectivity points in the design space, it is not a good choice in a high-connectivity regime.

Per Contact Routing Based on Average End-to-End Performance Metrics. In the protocols reviewed in the previous section, decisions about forwarding packets (when and where) are based on the likelihood of the delivery of each neighbor. No end-to-end performance is considered. In this section, we review three protocols in which decisions are based on end-to-end performance: one on the probability to deliver to the destination, one on the expected shortest path to the destination, and one on the average end-to-end delay.

Extending the previous work in reference 25, the *meets and visits* (MV) protocol [33] uses the same exchange scheme as in reference 25, but presents a new way to estimate the likelihood of forwarding. MV learns the frequency of meetings between nodes and visits to certain regions. The past frequencies are used to rank each bundle according to the likelihood of delivering a bundle through a specified path. MV determines a probability, $P_n^k(i)$, that the current node, k, can successfully deliver a bundle to a region i within n transfers. The probability is estimated by the following formula, assuming an infinite buffer at each node and N being the number of nodes in the network,

$$P_n^k(i) = 1 - \prod_{j=1}^{N} [1 - m_{j,k} P_{n-1}^j(i)]$$

where $P_o^k(i) = t_i^{(k)}/t$, and $t_i^{(k)}$ is the number of rounds node k visited region i during the previous t rounds and the meeting probability based on the *meetings* in the last t rounds is and $t_{j,k}$ is the number of meets between nodes j and k in the same region. MV does not blindly forward to neighbors, but only forwards these bundles upon request from next-hop neighbors. Simulation results show that the packet delivery rate can be 50% higher under MV than an FIFO scheme.

The *shortest expected path routing* (SEPR) protocol is proposed in reference 28. SEPR first estimates the link forwarding probability based on history data using the following formula:

$$P_{i,j} = \frac{Time_{connection}}{Time_{window}},$$

where $Time_{connection}$ is the time period the nodes i and j are connected and $Time_{window}$ is the sampling time window length. The shortest expected path is calculated based on the estimated link probability. When two nodes meet, the following routing protocol is performed. Each message stored in the cache is assigned an *effective path length*, EPL:

$$EPL_m = min(EPL_m, E_{path}(B, D))$$

The value is set at infinity when the message is first inserted in the cache. When the message is propagated to another node B, EPL is updated if the expected path length from B to the destination is smaller. A smaller value of EPL indicates a higher probability of delivery. Therefore, during the cache replacement process, those messages with smaller EPLs are removed first. EPL is also used in deciding which nodes to forward the messages. For each neighbor, B, of node A, if

$$E_{path}(B, D) <= minimum(EPL_m(A), E_{path}(A, D))$$

then a message is forwarded from A to B. Their protocols were evaluated through simulation assuming certain types of mobility models. The probability estimation does not rely on any location information. Under the algorithm, the same message could be forwarded to multiple nodes to increase reliability and reduce delay. Numerical results indicate that under SEPR, a 35% improvement of deliver rate and 50% reduction in resource cost can be achieved compared with epidemic [20] and DLE routing [25]. The gain comes from the fact that SEPR considers end-to-end performance, while DLE only considers one hop performance.

Similar to SEPR, minimal estimated expected delay (MEED) routing is proposed in reference 29. It is an extension of the work by Jain et al. [16] under the *contact summary oracle* assumption discussed in Section 8.2. The later model is a time-independent model (minimal expected delay, MED). MEED computes the expected delay using the observed contact history, in which a node records the connection and disconnection time of each contact over a sliding history window. When local link state information changes, updates must be propagated to all nodes in the network. Epidemic link state protocol is used for link state exchange. The routing table is recomputed each time a contact arrives and before a message is to be forwarded, resulting in per contact routing. The difference between MEED and MED is that under MEED a decision is made with the most recent information possible, while under MED a decision is made offline using average information and will not change over time. MEED may have higher overhead and may result in loops. Care must be given to prevent loops. How to choose the window size is not discussed in reference 29, and the window size is treated as a control parameter. Numerical results show that in buffer constrained networks, MEED performs much better than Epidemic [29].

8.3.3 Model-Based Approach

In previous work, in estimating the forwarding probability, it is assumed that mobile devices move *randomly* without any specific knowledge about the trajectories. In the real world, however, devices do move following *certain known patterns* such as

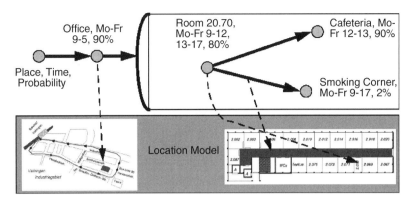

Figure 8.8. Local model and user profile example.

walking along a street or driving down the highway. Once users describe their motion pattern, the intermediate nodes have a more accurate estimation of which nodes move toward the destination with higher probability.

Model-Based Routing (MBR) [30] uses world models of the mobile nodes for a better selection of relaying nodes and the determination of a receiver location without flooding the network. World models contain location information (e.g., road maps or building charts) and user profiles indicating the motion pattern of users, Figure 8.8 (originally shown in reference 30). The key idea of the approach is to take into account that mobile devices typically do not follow the random walk motion pattern but are carried by human beings. Once humans describe their motion pattern or some sort of monitoring deduces it, MBR can rely on this information in the form of user profiles to choose a relay that moves toward the target with higher probability. With the information of the receiver location, each intermediate node can determine the next relaying node based on the user profile. Each node offers an interface that emits the probability that the user will move toward a given location. Hence the routing algorithm can choose fewer relays if a small number of relays have been found that will move near or to the location with high probability. However, only a sketch of the algorithm is described in reference 30 (no detail is given). Obtaining the user profile is an open research question. Their work relies on the known receiver location that is provided by a central location service, an unrealistic assumption.

A model of nodes moving along on a highway is described in reference 31. With ad hoc networks deployed on moving vehicles, network partitions due to limited radio range become inevitable when traffic density is low, such as at night, or when few vehicles carry a wireless device. A key question to ask is whether it is possible to deliver messages in spite of partitions, by taking the advantage that predictable node movement creates opportunities to relay messages in a store-and-forward fashion. Chen et al. [31] test the hypothesis that the motion of vehicles on a highway can contribute to successful message delivery, provided that messages can be relayed and stored temporarily at moving nodes while waiting for opportunities to be forwarded further. Messages are propagated greedily each time step, by hopping to the neighbor closest to the destination. Two kinds of transmission scheme are used. One is *pessimistic forwarding* and one is *optimistic forwarding*, which are distinguished by how long the messages are permitted to stay in intermediate nodes. In *pessimistic forwarding*, a message is dropped whenever no next hop exists for its destination. This is how forwarding works in most ad hoc network implementations. In *optimistic forwarding*, messages without next hops may remain on intermediate nodes for some time, hoping that physical movement of network nodes eventually creates a forwarding opportunity. Using vehicle movement *traces* from a traffic micro-simulator, the authors measure average message delivery time and find that it is shorter than when the messages are not relayed.

8.3.4 Node Movement Control-Based Approaches

The approaches discussed in the previous sections let the mobile host wait *passively* for the network to reconnect. This may lead to unacceptable transmission delays for some application. Some works, therefore, have proposed approaches that try to limit these delays by exploiting and controlling node mobility. As illustrated in Figure 8.9, the trajectories of some nodes (or special nodes) can be controlled so that overall system performance metrics, such as delay, can be improved.

Li and Rus [32] explore the possibility of changing the host trajectories in order to facilitate communication in ad hoc networks. In contrast to letting the mobile host wait passively for reconnection, the mobile hosts actively modify their trajectories to minimize transmission delay of messages. Given an ad hoc network of mobile computers where the trajectory of each node is known, they develop an algorithm for computing a trajectory for sending a message from host A to host B by asking intermediate hosts to change their trajectories in order to complete a routing path between hosts A and B. The communication protocol proposed is an application-layer protocol (rather than a network-layer protocol). When the network cannot route

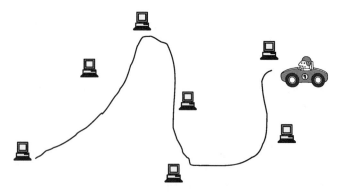

Figure 8.9. Control of node movement.

ALGORITHM 1. Sketch: The Optimal Relay Path to All Hosts in the System

Input:
1. Initial time when host h_0 begins to send a message.
2. The moving function of host h_i, which gives the position of h_i at time t.

Output: The optimal moving path from host h_0 to all other hosts $h_1, h_2, \sum\sum\sum, h_n$.

Steps:
1. Compute the optimal trajectory for host h_0 to reach all the other hosts directly, and record the earliest time point $t[k]$ for h_k.
2. Choose the unmarked host h_i with the least $t[i]$, mark h_i, Ready$[h_i] = 1$.
3. Compute the optimal trajectory (use the Optimal Trajectory algorithm) for host h_0 to reach all the unmarked hosts, such as h_j by way of h_i. If the time point computed for the optimal path from h_0 to h_j by way of h_i is less than the original t[j], update t[j] with the newly computed time point.
4. Go to 2 until all the hosts have been marked.

---

a message to the destination due to a network partition, it will try to do an *up-call* for the scheme presented. Algorithms that minimize the trajectory modifications are developed under two different assumptions: (a) The movements of all the nodes in the system are known and (b) the movements of the hosts in the system are not known. In the first case, the problem is, given a mobile ad hoc network (which may be disconnected) and the motion descriptions of the hosts (which is assumed to be known for all the movement of hosts), finding the shortest time strategy to send a message from one host to another. An optimal relay path algorithm is proposed and presented in Algorithm 1, which computes a sequence of intermediate hosts that can relay the message to the destination. Intermediate nodes modify their trajectories in the smallest possible way. In the second case, they propose a method in which hosts *inform the other hosts* of their current positions. The key issues that need to be considered to make this approach work are (1) when a host should send out information about its location update; (2) to whom the host should send out this information; and (3) how the host should send out this information. They model the communication problem in unknown mobile network environments by constructing a minimum spanning tree (MST), which contains the shortest edges in the graph that provide *full connectivity* in the graph. Each host has the responsibility of updating its location by informing all the hosts connected to it in the MST. However, their work assumes that the network is almost fully connected; it is not quite clear what happens if no such MST is found.

In wireless networks, there usually are mismatches between available capacity and demand. When such a mismatch occurs in such networks, one way to add capacity is to increase the number of participants carrying bundles in the network. To achieve this, the work in reference 33 suggests the addition of a limited number of autonomous agents to the network area and studies the problem of augmenting the capacity of a DTN through autonomous agents which move in the network with the purpose of increasing network performance. The addition of these agents requires a control algorithm that can coordinate agent movements in order to optimize the performance of the network according to quality of service metrics desired by the network administrator. The authors present a control-based approach and develop multiobjective controllers to control the mobility of autonomous agents. The design of control strategies *assumes the use of autonomous agents that can move to arbitrary locations in the physical environment*. Four controllers—*latency, bundle latency, unique bandwidth and bandwidth*—are defined in reference 33. Two approaches to multiobjective control—*subsumption* and *nullspace*—have been implemented and explored. Both techniques are from robotic research; nullspace controllers use linear algebra to coordinate controllers. *Nullspace* is defined as the set of inputs to a function where the value of the function does not change. *Nullspace composition* is used to coordinate collections of controllers. The controllers are ordered in a way such that subordinate controller is forced to operate in the nullspace of controllers above it according to the order. The thresholded nullspace approach extends the *nullspace* approach to handle the networking situation that needed thresholded control. The *subsumption* approach differs from the *nullspace* approach in how the controllers dominate one another. Experimental results show that the *thresholded nullspace* approach outperforms the *subsumption* approach when resources are limited.

Zhao et al. [34] describe a Message Ferrying (MF) approach for data delivery in sparse networks. MF is a proactive mobility-assisted approach that utilizes a set of special mobile nodes called *message ferries* to provide communication services for nodes in the network. Similar to their real-life analog, message ferries move around the deployment area and take responsibility for carrying data between nodes. The main idea behind the Message Ferrying approach is to introduce nonrandomness in the movement of nodes and exploit such nonrandomness to help deliver data. Two variations of the MF schemes were developed, depending on whether ferries or nodes initiate nonrandom proactive movement. In the Node-Initiated MF (NIMF) scheme, ferries move around the deployed area according to known specific routes and communicate with other nodes they meet. With knowledge of ferry routes, nodes periodically move close to a ferry and communicate with that ferry. In NIMF, the ferry route is known by nodes—for example, periodically broadcast by the ferry or conveyed by other out-of-band means. Nodes take proactive movement periodically to meet up with the ferry. As the sending node approaches the ferry, it forwards its messages to the ferry, which will be responsible for delivery. The trajectory control mechanism of the node determines when it should proactively move to meet the ferry for sending or receiving messages. The difference between NIMF and VMN is that the nodes' movements are not controlled in VMN, whereas they are controlled in NIMF. In the Ferry-Initiated MF (FIMF) scheme, ferries move proactively to meet nodes. When a node wants to send packets to other nodes or receive packets, it generates a service *request* and transmits it to a chosen ferry using a *long-range radio*. Upon reception of a service request, the ferry will adjust its trajectory to meet up with the node and exchange packets using short-range radios. In both schemes, nodes can communicate with distant nodes that are out of range by using ferries as relays. It is assumed that the ferry moves faster than nodes. In addition, it is assumed that nodes are equipped with a long-range radio that is used for transmitting control messages. Note that while the ferry is able to broadcast data to all nodes in the area, the transmission range of nodes' long-range radios may not necessarily cover the whole deployment area due to power constraints.

Zhao et al. [35] study the problem of using multiple ferries to deliver data in networks with stationary nodes and designing ferry routes so that average message delay can be minimized. Multiple ferries offer the advantages of increasing system throughput (reducing message delay) and robustness to ferry failures. On the other hand, the route design problem with multiple ferries is more complicated than the single ferry case considering the possibility of interaction between ferries. The authors present ferry route algorithms for single-ferry and multiple-ferrys cases, respectively.

In the single-ferry case, solutions for the well-studied traveling salesman problem (TSP) are adopted. The algorithm (sketch) for computer the best rout in the single ferry case is presented below.

```
Compute an initial route using TSP heuristic algorithm;
 do
   Apply 2-opt swaps;
   Apply 2H-opt swaps;
while (weighted delay is reduced);
Extend ferry route to meet bandwidth requirements;
```

Basically, the single route algorithm (SIRA) first generates an initial route using some TSP heuristic algorithm (e.g., the nearest-neighbor heuristic) and then refines the initial route using local optimization techniques. The following *2-opt* and *2H-opt swap* operations to improve the route are involved.

- **2-opt swap.** Consider the route as a cycle with edges that connect consecutive nodes in the route. A 2-opt swap removes two edges *AB* and *CD* from the route and replaces them with edges AC and BD while maintaining the route as a single cycle.
- **2H-opt swap.** A 2H-opt swap moves a node in the route from one position to another.

The algorithm tries to reduce the weighted delay of the route by applying 2-opt swaps and 2H-opt swaps until no further improvement can be found.

In the multi-route algorithm (MURA), ferries may follow multiple routes to carry data between nodes. But ferries do not relay data between themselves. So data are carried by at most one ferry. The MURA is presented below.

```
EWD(op): EWD of node assignment after operation op
Set the number of ferries to n;
Assign each node to a ferry;
```

```
while number of ferries >  m or EWD is reduced do
  Identify the best overlap or merge operation ops;
  Identify the best merge- or reduce operation opl;
  if EWD(ops) <  EWD(opl) and
      EWD(opl)Y < Ycurrent EWD then
        Perform ops;
else
  Perform opl;
Refine node assignment to maintain feasibility;
Compute each ferry route;
```

EWD refers to estimated weighted delay.

The MURA uses a greedy heuristic for assigning nodes to ferries. MURA starts with n ferries, and each node is assigned to a ferry. That is, each ferry route consists of one node. MURA refines the node assignment and reduces the number of ferries to **m** by using four types of operations. In each step, MURA estimates the weighted delay of the resulting node assignment for each operation and chooses to perform the best one until the number of ferries is m and no further improvement can be found. The details on how to estimate the weighted delay for a node assignment are given in reference 34. Then MURA modifies the node assignment, if necessary, to ensure feasibility; that is, there is a path between each sender/receiver pair and the total traffic load on each route is lower than its capacity. Given the node assignment, we can apply the algorithms in SIRA to compute each ferry route. Simulation results are obtained to evaluate the performance of route assignment algorithms, especially on the effect of the number of ferries on the average message delay. Numerical results indicate that when the traffic load is low, the improvement in delay due to the increased number of ferries is modest. This is because the delay is dominated by the distance between nodes. However, when the traffic load is high, an increase in the number of ferries can significantly reduce the delay.

While the MV and FIMF approaches control the individual virtual nodes' trajectories, Chatzigiannakis et al. [36] present a snake protocol, where a snake like sequence of carriers (called *supports* in reference 36; see Figure 8.10a) or virtual nodes always remain pairwise adjacent and move in a way determined by the snake's head.

The main idea of the snake protocol is as follows.

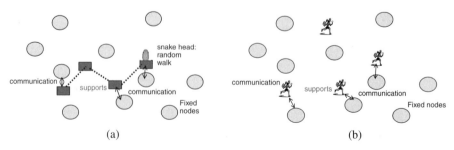

Figure 8.10. Illustration of the (a) snake protocol and (b) runner's protocol.

There is a setup phase of the ad hoc network, during which a predetermined number, k, of hosts become the nodes of the support. The members of the support perform a leader selection, which is run once and imposes only an initial communication cost. The elected leader, denoted by MS_0, is used to coordinate the support topology and movement. Additionally, the leader assigns local names to the rest of the members $MS_1, MS_2, \ldots, MS_{k-1}$. The head moves by executing a random walk over the area covered by the network. The nodes of supports move fast enough in a coordinated way so that they sweep (in sufficiently short time) the entire motion graph. Their motion and communication are accomplished in a distributed way via a support management subprotocol. The supports play a moving backbone subnetwork through which all communication is routed. The protocol consists of three components: *support motion subprotocol P_1, sensor subprotocol P_2, and synchronization subprotocol P_3. Subprotocol P_1* controls the motion of the supports in a distributed way. *Sensor subprotocol P_2* notifies a sender that it may send its messages. *Synchronization subprotocol P_3* synchronizes all the nodes in the support. Essentially, the motion subprotocol P_1 enforces the support to move as a snake, with the head (the elected leader MS_0) doing a random walk on the motion graph and each of the other nodes MS_i executes the simple protocol "move where MS_{i-1} was before". When some node of the support is within the communication range of the sender, the sensor subprotocol P_2 notifies the sender that it may send its message. The messages are stored in every node of the support using the synchronization protocol P_3. When a receiver comes within the communication range of a node of the support, the receiver is notified that a message is waiting for him and the message is then forwarded to the receiver. Duplicate copies of the message are then removed from the other members of the support.

The average delay or communication time of the protocol is given by

$$\frac{2}{l_2(G)} Q(n/k) + Q(k)$$

where G is the graph, λ is the second eigenvalue, k is the number in the support, and n is the number of nodes. Results derived from an implementation show that only a small number of carriers is required for efficient communication.

Chatzigiannakis et al. [37] extend the work in reference 36 by presenting a new protocol, called the *runners* (see Figure 8.10b), where *each carrier* performs a random walk sweeping the whole area covered by the network, which is the only difference between the two protocols. The authors perform an experimental evaluation and comparison between the snake protocol and the runner's protocol. It turns out that the runner's protocol is more efficient (smaller message delays and memory requirements) and robust than the snake protocol. The authors also note that while the snake protocol is resilient only to one carrier failure, the runner protocol is resilient to up to M failures, where M is the number of carriers.

In DataMules [12], a three-tier architecture is proposed which connects spare sensors at the cost of high latency. At the top tier, there are access points or repositories that can be set at convenient places. The middle tier consists of DataMules, which are mobile nodes (whose mobility pattern is not known) that can communicate with sensors and access points. DataMules have large storage capacity and renewable power. As Data Mules move, they collect data from sensors and forward these data to the access points. The bottom tier consists of sensors which are randomly distributed across a region. To save energy, work performed by the sensors is minimal. DataMules are assumed to be capable of short-range wireless communication and can exchange data from a nearby sensor access point they encounter as a result of their motion (the movements of the DataMules are not predictable). Thus DataMules can pick up data from sensors when in close range, buffer it, and drop off the data to wired access points when in proximity. The main advantage of the three-tier approach is the potential of large power savings that sensors can have because communication now takes place over a short range. Simple analytical models are presented to study the scaling of system characteristics as the system parameters, number of sensors, or number of DataMules change. Numerical results provide some relationship between the buffer requirements at the sensors (and at the DataMules) and the number of sensors (and the number of DataMules), respectively. It is observed that the change in the buffer capacity on each sensor should be greater than the number DataMules so that the same success rate can be maintained.

8.3.5 Coding-Based Approaches

To cope with wireless channel loss, erasure coding and network coding techniques have recently been proposed for wireless ad hoc networks and DTNs. The basic idea of erasure coding is to encode an original message into a large number of coding blocks. Suppose the original message contains k blocks; using erasure coding, the message is encoded into n ($n > k$) blocks such that if k or more of the n blocks are received, the original message can be successfully decoded. Here, $r = n/k$ is called the replication factor and determines the level of redundancy. Network coding comes from information theory and can be applied in routing to further improve system throughput. More on network coding will be described later.

To extend the work in reference 16, Jain et al. [38] assume the probability, P_i, that the transmission over link i is successful (independent of other transmissions) is known. Given that the replication factor is r, they study the following allocation problem: to determine an optimal fraction, x_i, of the erasure code blocks that should be sent over path i, such that the probability of successful reception is maximized. The formal definition of the problem is given below.

Definition. Consider a node s sending a message of size m to node d, and let there be n feasible paths from s to d. For each path i, let V_i be the volume of the path, and let S_i be a random variable that represents the fraction of data successfully transmitted on path i. Assume that an erasure coding algorithm can be used (with a replication factor r) to generate $b = (mr)/l$ code blocks of size l such that any m/l code blocks can be used to decode the message.

The Optimal Allocation problem is to determine what fraction (x_i) of the b code blocks should be sent on the *i*th path, subject to the path volume constraint, to maximize the overall probability of message delivery.

Formally, let $Y = \sum_{i=1}^{n} x_i S_i$. Find ($x_1, x_2, \ldots, x_n$) that maximize Prob($Y \geq r - y1$), where $\sum_{i=1}^{n} x_i = 1$ and $.i \in 1\ldots n,\ 0 \leq x_i \leq V_i/mr$.

By fixing the replication factor and treating delay as a constraint, they formulate the problem as an optimal allocation problem and consider two cases of path failure scenarios: Bernoulli (0–1) path failure and partial path failures. In the Bernoulli path failure case, when a path fails, all the messages sent over the path are lost. In the partial path failure case, some messages can be recovered with certain probabilities. They proved that the optimal allocation problem under the Bernoulli path failure case, which is formulated as a mixed integer programming, is NP hard. For the partial path failure case, they first show that maximizing the successful probability is equivalent to maximizing the Sharpe ratio, which plays an important role in the theory of allocation assets in investment portfolios (see references in reference 38 for more details). They propose to use approximation approaches from economic theory to maximize the ratio. The solution of the optimal allocation problem is static and does not change over time because it is assumed that the underlying path failure probability does not change over time.

Instead of optimally allocating a fixed portion of the coded blocks on each path from source to destination, it is proposed in reference 39 that coded blocks with replication factor *r* are equally split among the *first mr relays* (or contacts), for some constant *m*. And those relays must deliver the coded blocks to the destination *directly*. The original message can be decoded as soon as *m* contacts deliver their data (that is, as soon as *1/r* of the coded blocks have been received). The difference between this approach and the one presented in reference 38 is that this approach sends data dynamically to the first *m* contacts the node meets (in other words, the allocation of the coded blocks is not fixed). It also differs from the estimation-based approaches discussed in Section 8.3.2 in that it does not attempt to find which contacts have better chances to deliver the data. Instead, it simply forwards to the first *m* contacts the node meets (all contacts are equally good relays). Both analytic and simulation results show that the erasure coding-based forwarding in DTNs significantly improves the worst-case delay (compared with several other simple forwarding schemes).

To further improve the performance of the forwarding protocols-based erasure coding, Liao et al. [41] propose to combine erasure coding and estimation based forwarding, which is referred to as estimation-based erasure coding (EBEC). The original messages are first encoded (using an erasure coding scheme). The encoded messages are forwarded to different relays that have a higher chance of delivering the messages. Numerical results show that EBEC outperforms the scheme studied in reference 39.

A probabilistic forwarding approach based on network coding is proposed for DTNs in reference 40. Recently, network coding has drawn a lot of attention in the networking research community. Instead of simply forwarding packets received, intermediate nodes can combine some of the packets received so far and send out as a new packet. For example, suppose that there are three nodes, A, B, and C. Nodes A and C want to exchange information through the middle node B. Node A first transmits packet x to node B, and node C transmits packet y to node B. Node B broadcasts x *or* y (not x and y in sequence). Since node A has packet x and node C has packet y, node A can decode y and node C can decode packet x. For this example, it is easy to see that the number of transmissions is reduced when network coding is used. The basic idea in reference 40 is to use network coding to generate new packets. A coding vector is attached to each new packet. When a packet is received at a node, d new packets are generated and broadcast to the neighbors of the node, where d is referred to as a forwarding factor. When enough packets are received at the receiver, the original packet can be decoded. The value of d depends on the node density. Simulation results show that, for the given network setting in reference 40, for example, the packet deliver ratio using network coding is much higher than that under probabilistic forwarding and most of the packets are delivered with lower forwarding factor, see Figure 8.11 (originally shown in [40]).

8.4 DISCUSSIONS AND FUTURE RESEARCH ISSUES

In the previous sections, we classify routing protocols into different categories: deterministic routing, random forwarding, history-based forwarding, and so on. Based on the existing work and the unique characteristics of the DTN, it is apparent that many research issues remain to be solved in the area of DTNs. In this section, we list some of research issues that should be addressed and we hope that this will stimulate activity in the research community.

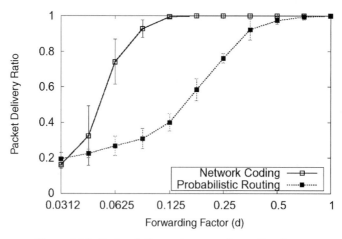

Figure 8.11. Packet deliver ratio versus forwarding factor.

- A critical question in designing a protocol in DTNs is, What is the proper objective function, short delay, or high throughput or others? Another related question is, How can we define the system capacity in such an intermittently connected network?
- Methods to determine how many nodes to forward should be developed. There is a tradeoff. The larger the number of nodes forwarded, the better the chance for packets to reach their destination, but, the more the network resource (bandwidth and buffer space) is needed. Analytical models should be developed if possible, and simulation results should be obtained to quantify the tradeoff.
- When multiple copies of the packets are in the network, duplication of packets occurs and such duplication requires a way of eliminating unnecessary copies to reduce the buffer occupancy. Where should the duplication reduction be done, at the destination or intermediate nodes and how? When original packets are received successfully at the receiver, how do we inform intermediate nodes to discard these packets? Informing intermediate nodes requires extra resources. Again, there is a tradeoff between efficiency and additional overhead.
- Scheduling becomes much more complex in DTNs than in IP-centric networks, because connections in DTNs are intermittent while those in normal IP networks are not. Appropriate buffer management schemes (which packets to discard when full) and scheduling should be developed. One possible approach is to have separate queues for different outgoing links. Those packets whose destination will be disconnected soon (if known) should be scheduled to transmit first.
- Whenever possible, information about node location and future movement should be utilized in designing the protocols. The forwarding protocols (Section 8.3) should leverage simple and accurate link availability estimation methods to make intelligent decisions if feasible. There are some papers dealing with estimation of link availability for ad hoc networks [45, 46]. How to define user profiles and how to use them to estimate the deliver probability is also an open issue.
- New security mechanisms must be developed because techniques that rely on access to a centralized service cannot be used or the assumption that all intermediate nodes are trusted is not valid.
- Self-learning and automation algorithms should be developed so that the underlying network is cognitive so that intelligent decisions on scheduling and forwarding can be made automatically.
- Open spectrum [47] allows secondary users to opportunistically explore unused licensed band on a noninterfering basis. New algorithms of how to utilize those unused channels (resulting in intermittent connectivity) dynamically and efficiently should be developed.
- Transmissions in networks with directional antennas are often prescheduled and may result in intermittent connectivity [48]. Power management in energy-aware network (range and/or wake/sleep periods control) may also result in intermittent

connectivity. Therefore, scheduling transmissions with directional antennas and power management should take the DTN requirements or characteristics into account.

- As the mobility of nodes in a mobile ad hoc network might lead to network partitions, and directional antennas can transmit over longer distances, Saha and Johnson [49] propose to the use of directional antennas to bridge such partitions when needed. The basic idea behind this method is to use the capability of a directional antenna to transmit over longer distance, but to adaptively use this capability only when necessary for selected packets. Methods to close broken links in ad hoc networks should be developed to cope with partial connectivity.

8.5 CONCLUSIONS

In this chapter, we provide an overview of the state of the art on routing protocols in DTNs. Many excellent approaches to addressing the unique problems in DTNs have been reported in the literature. Each approach has its own advantages and disadvantages in terms of network efficiency and resources required. Even though the network behavior may not be known, the performance of these protocols developed under deterministic case may serve as bounds on the network performance and as guidelines in designing protocols for stochastic case. It seems that dynamic forwarding protocols based on the latest contact information combining with history information may perform better in the stochastic case. Two other related research areas (not covered here) are estimation methodologies for link availability and partition prediction (see references 45, 46, and 50) and how to provide efficient data dissemination in partially connected networks, see [51].

DTN research, in general, is still in its early stage, and there are still many open issues that need to be resolved before the benefits of the DTNs can be fully utilized. Wireless networks with high mobility and short radio range will become a common phenomenon; therefore, it is imperative that those issues be fully understood and studied. Because it is still in its early stage of research, the purpose of this chapter is to summarize the current status of an evolving research community. It is the author's desire that this summary can provide a jump-start for other new researchers in this area and motivate the research community in developing new efficient and better protocols.

8.6 EXERCISES

1. Describe what a DTN is? Why are there intermittently connected networks? (Please list at least three reasons.)
2. Explain why TCP and UDP will not work effectively in DTNs.
3. In Figure 8.e-1, find all the paths from node A to F in time period 1 and in time period 4, respectively. How many different paths are there from A to C in time periods 1, 2, and 4, respectively?

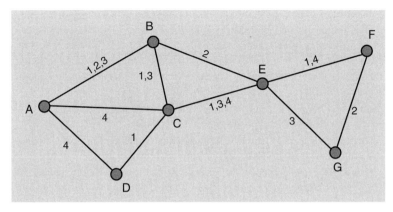

Figure 8.e-1. Time-evolving graph. The numbers on the line denote the time periods during which the link is active.

4. Suppose for a given network and for some metric used, the two shortest paths from node A to Z are A-C-G-Z and A-D-F-H-Z, and their expected lengths are 20 and 45, respectively. Furthermore, the expected lengths (again for the metric selected) of the links AC and AD are 3 and 35, respectively. Using this information as an example, develop a *dynamic* forwarding scheme that gives overall performance. (*Hint*: One simple forwarding scheme is to always select the expected shortest path no matter when two nodes meet.)

REFERENCES

1. T. Clausen and P. Jacquet. Optimized Link State Routing Protocol (OLSR), RFC 3626, IETF Network Working Group, October 2003.
2. C. Perkins, E. Belding-Royer, and S. Das. Ad Hoc On-Demand Distance Vector (AODV) Routing, RFC 3561, IETF Network Working Group July 2003.
3. D. B. Johnson and D. A. Maltz. *Dynamic Source Routing in Ad Hoc Wireless Networks*, X. Imielinski Y. Korth, editors, *Mobile Computing*, Vol. 353. Kluwer Academic Publishers, Norwell, MA, Chapter 5, 153–181, 1996.
4. Y.-B. Ko and N. Vaidy. Location-aided routing in mobile ad hoc networks, *ACM Wireless Networks Journal*, 6:307–321, 2000.
5. M. Grossglauser and M. Vetterli. Locating nodes with EASE: Mobility diffusion of last encounters in ad hoc networks. In *INFOCOM*, 2003.
6. H. Dubois-Ferriere, M. Grossglauser, and M. Vetterli. Age matters: Efficient route discovery in mobile ad hoc networks using encounter age. In *MobiHoc*, 2003.
7. S.-J. Lee, W. Su, and M. Gerla. Wireless ad hoc multicast routing with mobility prediction. *Mobile Networks and Applications Journal*, 6:351–360, 2001.
8. Y. Ge, T. Kunz, and L. Lamont. Quality of service routing in ad hoc networks using OLSR. In *HICSS*, 2003.

9. J. J. Garcia-Luna-Aceves, M. Mosko, and C. Perkins. A new approach to on-demand loop-free routing in ad hoc networks. In *Proceedings Twenty-Second ACM Symposium on Principles of Distributed Computing (PODC 2003)*, Boston, MA, July 13–16, 2003.
10. S. Burleigh et al. Delay-tolerant networking: An approach to interplanetary internet. *IEEE Communications Magazine*, **June**:128–136, 2003.
11. P. Juang et al. Energy-efficient computing for wildlife tracking: Design tradeoffs and early experiences with ZebraNet. In *Proceedings ASPLOS*, October 2002.
12. R. Shah, S. Roy, S. Jain, W. Brunette. Data MULEs: Modeling a three-tier architecture for sparse sensor networks. In *IEEE SNPA Workshop*, May 2003.
13. A. Pentland, R. Fletcher, and A. A. Hasson. A road to universal broadband connectivity. In *Second International Conference on Open Collaborative Design for Sustainable Innovation; Development by Design*, December 2002.
14. Wizzy Project. http://www.wizzy.org.za/.
15. DTN Research Group. http://www.dtnrg.org/.
16. S. Jain et al. Routing in delay tolerant network. In *ACM SIGCOM 04*, Portland, OR, 2004.
17. K. Fall. A delay-tolerant network architecture for challenged internets. In *Proceedings of SIGCOMM'03*, August 2003.
18. S. Merugu et al. Routing in space and time in networks with predicable mobility. Georgia Institute of Technology, Technical Report, GIT-CC-04-7, 2004.
19. R. Handorean et al. Accommodating transient connectivity in ad hoc and mobile settings. In *Pervasive 2004*, April 21–23, 2004, Vienna, Austria, pp. 305–322.
20. A. Vahdat and D. Becker. Epidemic routing for partially connected ad hoc networks. Technical Report CS-200006, Department of Computer Science, Duke University, Durham, NC 2000.
21. M. Grossglauser and D. Tse. Mobility increases the capacity of ad hoc wireless networks. In *INFOCOM*, 2001.
22. T. Small and Z. J. Haas. The shared wireless Infostation model—A new ad hoc networking paradigm (or where there is a whale, there is a way). In Mobihoc 2003, June 1–3, 2003.
23. D. Nain et al. Integrated routing and storage for messaging applications in mobile ad hoc networks. In *Proceedings of WiOpt*, Autiplis, France, March 2003.
24. F. Tchakountio and R. Ramanathan. Tracking highly mobile endpoints. *ACM Workshop on Wireless Mobile Multimedia (WoWMoM)*, Rome, Italy, July 2001.
25. A. Davids, A. H. Fagg, and B. N. Levine. Wearable computers as packet transport mechanisms in highly-partitioned ad-hoc networks. In *Proceedings of the International Symposium on Wearable Computing*, Zurich, October 2001.
26. A. Lindgren et al. Probabilistic routing in intermittently connected networks. In *Proceedings of the First International Workshop on Service Assurance with Partial and Intermittent Resources (SAPIR 2004)*, August 2004, Fortaleza, Brazil.
27. C. Shen et al. Interrogation-based relay routing for ad hoc satellite networks. *IEEE Globecom'02*, 2002.
28. K. Tan, Q. Zhang, and W. Zhu. Shortest path routing in partially connected ad hoc networks. In *IEEE Globecom*, 2003.

29. E. Jones, L. Li, and P. Ward. Practical routing for delay tolerant networks. In *SIGCOMM05-DTN Workshop 05*, 2005.
30. C. Becker and G. Schiele. New mechanisms for routing in ad hoc networks. In 4th Plenary Cabernet Workshop, Pisa, Italy, October 2001.
31. Z. Chen et al. Ad hoc relay wireless networks over moving vehicles on highways. In *ACM Mobihoc*, 2001.
32. Q. Li and D. Rus. Communication in disconnected ad hoc networks using message relay. *Journal of Parallel Distributed Computing*, 63:75–86, 2003.
33. B. Burns et al. MV Routing and capacity building in disruption tolerant networks. In *IEEE INFOCOM 2005*, Miami, FL, March 2005.
34. W. Zhao et al. A message ferrying approach for data delivery in sparse mobile ad hoc networks. In *Proceedings of the 5th ACM International Symposium on Mobile Ad Hoc Networking and Computing*, ACM Press, New York, 2004, pp. 187–198.
35. W. Zhao, M. Ammar, and E. Zegura. Controlling the mobility of multiple data transport ferries in a delay-tolerant network. In *INFOCOM*, 2005.
36. I. Chatzigiannakis et al. Analysis and experimental evaluation of an innovative and efficient routing protocol for ad-hoc mobile networks. In *Lecture Notes in Computer Science*, **1982**: 99–111, 2001.
37. I. Chatzigiannakis et al. An experimental study of basic communication protocols in ad-hoc mobile networks. *Lecture Notes in Computer Science*, **2141**:159–169, 2001.
38. S. Jain, M. Demmer, R. Patra, and K. Fall. Using redundancy to cope with failures in a delay tolerant network. In *SIGCOMM05*, 2005.
39. Y. Wang, S. Jain, M. Martonosi, and K. Fall. Erasure-coding based routing for opportunistic networks. In *SIGCOMM DTN Workshop*, 2005.
40. J. Widmer and J. Le Boudec. Network coding for efficient communication in extreme networks. In *SIGCOMM DTN Workshop*, 2005.
41. Y. Liao, K. Tan, Z. Zhang and L. Gao. Combining erasure-coding and relay node evaluation in delay tolerant network routing. In *Proceedings of the International Wireless Communications and Mobile Computing (IWCMC)*, July 2006.
42. A. Ferreira. Building a reference combinatorial model for MANETs. In *IEEE Networks*, September/October 2004.
43. L. Ford and D. Fulkerson. *Flows in Networks*, Princeton, University Press, Princeton, NJ, 1962.
44. A. Iacono and C. Rose. Infostations: New Perspectives on Wireless Data Networks. WINLAB, technical document, Rutgers University, 2000.
45. A. B. McDonald and T. Znati. A mobility-based framework for adaptive clustering in wireless ad hoc network. *IEEE Journal of Selected Areas in Communication*, **August**:1466–1487, 1999.
46. S. Jiang et al. A prediction-based link availability estimation for mobile ad hoc networks. In *IEEE INFOCOM*, 2001.
47. R. Berger. Open Spectrum: A path to ubiquitous connectivity. In *ACM Queue*, May 2003.
48. Z. Zhang. DTRA: Directional transmission and reception algorithms in WLANs with directional antennas for QOS support. In *IEEE Networks*, May/June 2005.

49. A. Saha and D. Johnson. Routing improvement using directional antennas in mobile ad hoc networks. In *IEEE Globecom*, 2004.
50. P. Samar and S. Wicker. On the behavior of communication links of a node in a multi-hop mobile environment. In *ACM MobiHoc*, 2004.
51. G. Karumanchi et al. Information dissemination in partitionable mobile ad hoc networks. In *Proceedings of IEEE Symposium on Reliable Distributed Systems*, Lausanne, Switzerland, October 1999.

CHAPTER 9

Transport Layer Protocols for Mobile Ad Hoc Networks

LAP KONG LAW

Trapeze Networks, Pleasanton, CA, 94588-4084

SRIKANTH V. KRISHNAMURTHY and MICHALIS FALOUTSOS

Computer Science Department, University of California, Riverside, California 92521, USA

9.1 INTRODUCTION

The Transmission Control Protocol (TCP) is an efficient transport layer protocol designed for wired networks (such as the Internet). However, it has been known to perform badly in the wireless and mobile ad hoc network (MANETs) environments [1–4]. This is not surprising since the network characteristics of MANETs are substantially different from that of the wired networks. Wired networks are characterized by high-bandwidth links, infrequent changes in network topology, and extremely low link bit error rate (BER). In contrast, MANETs are characterized by limited bandwidth, dynamic network topology, and an error-prone, shared wireless channel. These characteristics render MANETs to be substantially different from the wired networks. Since TCP was originally developed and optimized under the wired network assumptions, one may anticipate that TCP will not perform well in MANETs.

The fundamental problem of using TCP in MANETs is its misinterpretation of packet losses due to mobility or wireless errors as a sign of network congestion. As mentioned previously, TCP was designed for relatively robust wired networks, and thus network congestion is almost assumed to be the only reason to cause packet loss and delays.[1] Therefore, when TCP detects a packet loss (either by the reception of three duplicate acknowledgments called DUPACKs or the occurrence of a timeout), it will execute the congestion control algorithms (discussed in Section 9.2) in an attempt

---

[1]For TCP, excessive delay is equivalent to packet loss: The protocol assumes that a packet is lost if a transmitted packet is not acknowledged within a timeout.

Algorithms and Protocols for Wireless and Mobile Ad Hoc Networks, Edited by Azzedine Boukerche
Copyright © 2009 by John Wiley & Sons Inc.

to alleviate the congestion on the network. While this mechanism works well for wired networks, it does not work well for MANETs since route failures and wireless channel errors are indeed the dominating factors that cause packet loss. TCP, without any proper mechanisms to detect and handle the cause of the packet loss, erroneously executes the congestion control algorithms to reduce the sending rate unnecessarily. As a result, the available bandwidth of the network is heavily underutilized.

Despite its poor performance, TCP has been used extensively in MANETs because many network applications such as FTP, HTTP, Telnet, and SMTP are developed on top of TCP. The use of TCP over MANETs not only allows mobile nodes to seamlessly interconnect to hosts in the wired infrastructure but also makes applications portable to MANETs. Therefore, most of the research efforts have been focused on investigating techniques to improve the TCP performance instead of proposing a new transport layer protocol for MANETs.

Even though numerous TCP enhancements have been proposed for MANETs, there does not exist (a) an extensive study that compares these proposed schemes and (b) a transport layer protocol that is emerging as the ultimate solution. Most of the existing schemes address a limited set of problems that arise due to the use of TCP, and therefore they are usually not sufficient to provide a complete transport layer solution for MANETs. It may be worthwhile to combine several of these schemes together, but this requires an extensive understanding of their assumptions and their relative strengths and weaknesses. Furthermore, these schemes are not even compatible with each other due to their distinctive mechanisms of operation.

In this chapter, an extensive study of the major transport layer enhancements for MANETs is conducted. The enhancements are classified and compared in terms of the way in which they attempt to improve the TCP performance. This is in contrast to the previous studies on TCP [5, 6] which classify enhancements into *end-to-end* proposals (only the sender and the receiver attempt to handle losses and thus maintain the end-to-end semantics of TCP), *link-layer* proposals (the link layer tries to hide the characteristics of lossy wireless links from the transport layer), and *split-connection* proposals (splitting the TCP connection between the mobile host and the fixed host into two separate connections and handling them separately).

The goal of this chapter is to provide a thorough understanding of the problems of TCP in MANETs and present the principles of the proposed TCP enhancements. Most of the current TCP versions are not sufficient (in isolation) in providing a complete transport layer solution for MANETs. In contrast, the design of new transport protocols specially designed for MANETs could effectively lead to substantial performance gains.

The rest of the chapter are organized as follows. In Section 9.2, the background on conventional TCP and specifically its congestion control mechanisms are described. In Section 9.3, the problems that arise while using TCP in MANETs are discussed. In Section 9.4, some of the representative TCP enhancements that have been proposed to tackle the problems are presented. Strengths and weaknesses of each enhancement are also discussed in this section. In Section 9.5, we discuss a new transport layer protocol for MANETs. In Section 9.6, a discussion on the appropriateness of the TCP enhancements is provided. A conclusion is presented in Section 9.7.

9.2 BACKGROUND

A brief overview of the TCP congestion control mechanisms is provided in this section. A detailed description of the TCP congestion control algorithms can be found in reference 7.

The TCP congestion control mechanism is a loss-based mechanism in the sense that it uses either packet loss or excessively delayed packets to trigger corresponding congestion-alleviating actions. Basically, the size of the congestion window at the sender (and the receiver advertised window) regulates the maximum number of outstanding packets (transmitted but not acknowledged packets) that can be transmitted. TCP detects the congestion level of the connection and manipulates the size of the congestion window appropriately so as to maintain the sending rate at an appropriate level. The TCP congestion control mechanism consists of three different mechanisms, and they are briefly described below:

Slow-Start. When a TCP connection is first initiated (or after a prolonged disconnection), the size of the congestion window ($cwnd$) is set to "one" sender maximum segment size ($SMSS$). By this, only one packet or segment as it is called is allowed to be sent. From now on, for each arrival of a nonduplicate (or unique) acknowledgment (ACK), $cwnd$ is increased by one. Slow-start is essentially a mechanism that estimates the available bandwidth by progressively probing for more bandwidth. During this phase, $cwnd$ increases exponentially over time until a packet loss event occurs or $cwnd$ reaches the slow-start threshold ($ssthresh$). When a packet loss occurs, TCP will respond by reducing the size of $cwnd$, which in turn slows down the sending rate. After the first packet loss, $ssthresh$, which is used to determine the start of the congestion avoidance phase, is set to half of the current $cwnd$. Then, $cwnd$ will be reset to one and will grow according to the aforementioned procedure until $ssthresh$ is reached. Simply put, the slow-start procedure can be regarded as a *coarse-grained* means of estimating the available bandwidth of networks.

Congestion Avoidance. Congestion avoidance, as its name implies, tries to avoid network congestion by restricting the growth of $cwnd$. During this phase, $cwnd$ increases approximately by one $SMSS$ for each round-trip time (RTT) until a packet loss event occurs. This conservative increase of the congestion window provides a *fine-grained* estimation of the available bandwidth.

Fast Retransmit and Fast Recovery. TCP uses two indications to identify possible packet loss. One of the indications is the occurrence of a timeout event, which corresponds to a failure of receiving an ACK within a predefined time interval.

However, waiting for a timeout event to trigger retransmission is considered inefficient when the retransmission timeout (RTO) value is large. In order to provide a timely detection of a lost packet, fast retransmit was proposed and is triggered when the sender receives N DUPACKs[2] from the receiver. TCP, upon the reception of these DUPACKs, retransmits the (possibly) missing packet as indicated in the DUPACKs. In TCP-Reno and later versions, TCP incorporates the fast recovery algorithm into the congestion control mechanism. TCP executes fast recovery after fast retransmit: It first sets *ssthresh* to half of the current *cwnd* and then reduces *cwnd* to *ssthresh* + 3 (the constant 3 reflects the segments that have possibly left the network). Further reception of DUPACKs increases *cwnd* by one $SMSS$; if possible, it transmits a new data segment. When a nonduplicate acknowledgment is received, TCP exits the fast recovery,[3] sets *cwnd* equal to *ssthresh*, and enters the congestion avoidance phase. Fast recovery eliminates the need to restart from the slow-start phase and thus enables a quicker catchup to the optimum value of *cwnd*.

9.3 PROBLEMS OF USING TCP OVER MANETs

In this section, the problems that arise with TCP, when used in MANETs, is discussed. Before going into the details of the problems, a list of the major factors that particularly affect TCP performance in MANETs is provided. These factors are listed below:

1. *Mobility*. The mobility of nodes causes routes to change and disconnect frequently which leads to low route stability and availability.
2. *High Bit Error Rate (BER)*. The use of the wireless channel is vulnerable to errors due to weather conditions, obstacles and interference.
3. *Unpredictability and Variability*. The time-varying nature of wireless channel quality creates uncertainty, which causes substantial difficulty in measuring the RTT and estimating a proper timeout value.
4. *Contention*. The use of the shared wireless channel limits the ability of a node to send packets. Nodes within a local neighborhood have to compete for wireless channel access. Therefore, the bandwidth obtained by a node depends on the sending need of its neighbors.

Due to the above factors, the performance of TCP is greatly degraded. Mobility of nodes and high BER of the wireless channel cause TCP to misinterpret route failures and wireless errors as network congestion. Moreover, contention will reduce the throughput and fairness of TCP. Sudden delay spikes that are caused by the unpredictability and variability of the wireless channel, also contribute to the poor performance of the TCP. In the following, each of the above factors is described in detail.

---

[2] N is commonly set to three.

[3] TCP-NewReno requires a nonduplicate acknowledgment that acknowledges all segments that were sent prior to the reception of the third DUPACK in order to exit the fast recovery.

9.3.1 TCP misinterprets route failures as congestion

Route failures in MANETs are the norm rather than the exception. When a route failure occurs, packets that are buffered at intermediate nodes along the route will be dropped. This large amount of packet drops usually causes the TCP sender to encounter a timeout or even a series of timeouts. In this case, TCP will mistakenly interpret the loss as an indication of network congestion and trigger the congestion control mechanisms to reduce the size of *cwnd* and *ssthresh*. In addition, the retransmission timeout (RTO) value will also be doubled for any subsequent timeouts that are caused by the prolonged route disconnection. These actions have two adverse effects: (1) The small *cwnd* and *ssthresh* values reduce the initial sending rate after the route is restored. Therefore, it takes a long time for the sending rate to catchup to a high value after a new route is found. (2) The large RTO value reduces the responsiveness of TCP; even if the route is restored, TCP will take long to converge to the right level of operation.

9.3.2 TCP Misinterprets Wireless Errors as Congestion

Due to the fading and the interference effects, the wireless channel could suffer from prolonged or intermittent signal deterioration and thus cause it to be error-prone in nature. Wireless channel errors usually occur randomly and in bursts [8]. In the presence of severe wireless channel conditions, where the link-layer local recovery mechanism [9] is unable to recover the lost packets, the TCP sender could receive DUPACKs. When the number of DUPACKs observed by the sender reaches a certain threshold, TCP will erroneously take this as an indication of network congestion and trigger the fast retransmit and the fast recovery algorithms to reduce the sending rate unnecessarily. In other words, even though the network is not congested, TCP reduces the sending rate in the face of wireless errors. In case multiple losses occur within a transmission window, the use of the fast retransmit and the fast recovery algorithm can only detect and recover the first of those losses effectively. However, the TCP sender still has to encounter a timeout event before the subsequent losses are detected and the corresponding segments are recovered. In TCP-newReno, the introduction of the *fast recovery phase* can avoid the occurrence of the above timeout event. However, its fast recovery mechanism can only detect and recover one lost packet in every RTT. Therefore, it is not an efficient method to recover a large number of packet losses.

9.3.3 Intraflow and Interflow Contention Reduce Throughput and Fairness

Interflow contention refers to the contention experienced by a node due to transmissions by nearby flows. Intraflow contention (or self-contention [10]), on the other hand, refers to the contention experienced by a node due to the transmissions of the same flow (due to the forward data transmissions and the reverse ACKs transmissions). Both types of contention degrade TCP performance significantly [11]. The following discussion demonstrates the effects of both types of contention. With the use of IEEE 802.11 MAC protocol, when a node successfully obtains the channel and performs its transmission, any other node (either belonging to the same flow or other flows) that is within the node's transmission range (and the node's sensing range) should not perform any transmission and must defer their transmissions for a later time. Although this conservative transmission policy can reduce the chance of packet collisions, it introduces the *exposed node problem* [3]. An exposed node is a node that is within the transmission range of the sender but out of range of the destination. According to the above-mentioned transmission policy, the exposed node is restricted from transmitting even though its transmission will not hinder the communication between the sender and the destination. Therefore, the available bandwidth is underutilized. Moreover, contention allows an aggressive sender to capture the channel which reduces the chance of transmissions of the other senders in the vicinity.

Interflow and intraflow contention typically lead to increased packet transmission delays. Heavy contention is usually caused from using an inappropriately large congestion window or when a bad transmission scheduling policy is used. Previous studies have shown that the maximum congestion window size should be kept small and should be maintained at a level proportional to some fraction of the length (in terms of the hop count on the path) of the connection [4, 12] in order to alleviate the effect of contention. Unfortunately, as indicated in reference 4, conventional TCP does not operate its congestion window at an appropriate level and thus, its performance is affected by the contention severely.

9.3.4 Delay Spike Causes TCP to Invoke Unnecessary Retransmissions

The RTT estimation of TCP is adequate in a stable network in which the RTT fluctuations are small. However, a sudden increase in the packet transmission time will cause TCP to invoke spurious retransmissions [13]. A sudden delay spike may be caused by several factors: (1) a sudden change in the link quality that leads to a burst of transmission errors and many link-level retransmission attempts, (2) route changes or intermittent disconnections due to mobility that lead to a higher delay experienced by the transmitted packets, and (3) an increase in the contention along the route that leads to a longer waiting time before the packets in the queue are transmitted. Under these circumstances, the sender may not receive an acknowledgment within the timeout period and thus the sender will regard all the transmitted but not acknowledged packets as lost. As a consequence, the sender will unnecessarily reduce the sending rate and retransmit those packets that are deemed lost but are merely delayed.

9.3.5 Inefficiency Due to the Loss of Retransmitted Packet

Due to the instability of routes and the error-prone wireless channel, packet drops can occur more frequently in MANETs (as compared to wireline networks). In other words, a TCP sender may need to perform more retransmissions in MANETs. However, most TCP implementations do not have a mechanism to efficiently detect or recover from the loss of retransmitted packets without using the inefficient timeout mechanism. If a retransmitted packet is lost, the TCP sender can only wait until an expiration of the retransmission timer to detect the loss. Since the retransmission timeout value is usually large (due to the doubling of *RTO* for every retransmission attempt), the extended waiting period of the TCP sender will lead to poor performance.

In wireline networks, due to the low BER, one might expect that the chance of the retransmitted packet being lost again is extremely low and therefore the aforementioned problem is considered insignificant. However, this problem can severely affect the performance of the TCP in MANETs due to the high BER of the wireless channel.

9.4 VARIOUS TCP SCHEMES

This section presents some of the major TCP enhancements that have been proposed to alleviate the TCP performance issues that were discussed in the previous section.

In contrast with many of the previous efforts that classify and compare TCP enhancements by their type [5, 6] (i.e., *end-to-end* proposals, *link-layer* proposals, and *split-connection* proposals), they are classified as per the strategy used in order to improve the TCP performance.

Strategy using which they offer improvements[4]:

1. Estimating the available bandwidth.
2. Determining route failures and wireless errors.
3. Reducing contention.
4. Detecting spurious retransmissions.
5. Exploiting buffering capabilities.

For each of the above strategies, the representative schemes are presented and their operating mechanisms are described, along with discussions on their strengths and weaknesses in improving TCP performance in MANETs.

9.4.1 Estimating the Available Bandwidth

As discussed in Section 9.3, the traditional loss-based congestion control mechanism of TCP cannot accurately adjust the sending rate when it is used in MANETs. Packet loss is not always a sign of congestion; it could be due to mobility. To this end, several TCP schemes have been proposed recently to address the problem by estimating better the bandwidth of the connection. The representative schemes that take advantage of bandwidth estimation to enhance TCP performance include TCP-Vegas [14], TCP-Westwood [15], and TCP-Jersey [16].

---

[4]Some schemes can be matched to several types of strategies but are only classified according to their main strategy.

TCP-Vegas. TCP-Vegas [14] uses a rate-based congestion control mechanism. The idea is to adjust the sending rate carefully by comparing with the estimated rate. Besides, it also modifies the congestion detection and avoidance algorithms to improve the overall throughput of TCP.

Rate-Based Congestion Control Mechanism. The most significant difference between TCP-Vegas and the conventional TCP variants is the use of a rate-based technique to control the size of the congestion window. TCP-Vegas compares the *expected throughput* (calculated as the current window size divided by the minimum observed *RTT*) to the *actual throughput* measured (measured as the number of bytes transmitted between the time a distinguished segment is transmitted and acknowledged, divided by the time it takes to get the acknowledgment back) for every *RTT*. If the difference between the two values is smaller than α, TCP-Vegas increases *cwnd* linearly for the next *RTT*; and if the difference is greater than β, TCP-Vegas decreases the congestion window linearly for the next *RTT*. If the difference is between α and β, TCP-Vegas keeps *cwnd* unchanged.

Modified Slow-Start Mechanism. In traditional TCP, the exponential growth of *cwnd* in the slow-start phase usually causes packet loss. TCP-Vegas avoids this packet loss by allowing *cwnd* to grow exponentially only once in every other *RTT*. This is to allow TCP-Vegas to compare the expected and the actual throughput. If the difference is greater than the γ threshold, TCP-Vegas changes from slow-start mode to the linear increase/linear decrease mode as described above.

Even though TCP-Vegas was not intentionally designed for MANETs, it can improve the TCP throughput because its inherent rate-based congestion control algorithm could proactively avoid possible congestion and packet losses by ensuring that the number of outstanding segments in the network is small. Moreover, the modified slow-start algorithm could avoid the typical sharp increase in *cwnd* during the slow-start phase by switching TCP to the congestion avoidance phase early. In this way, TCP-Vegas could prevent packet losses due to the aggressive growth of the congestion window during the traditional slow-start phase.

However, TCP-Vegas also inherits most of the weaknesses of conventional TCP. It cannot handle the effects of route failures and wireless channel errors. Moreover, TCP-Vegas suffers from several other problems as discussed in reference [17]. First, there are inaccuracies in the calculated *expected throughput* after a route change event. A route change is likely to invalidate the *baseRTT* used in TCP-Vegas for calculating the *expected throughput*. Therefore, the *baseRTT* may not reflect the actual minimum measured round-trip time of the connection. If this inaccuracy causes the calculated throughput difference to fall below the β threshold, TCP-Vegas will reduce its rate as if congestion occurs. Another problem is the unfairness of TCP-Vegas. Unfairness occurs when TCP-Vegas competes with other TCP-Reno connections. Because of the aggressiveness of TCP-Reno and the conservativeness of TCP-Vegas, TCP-Reno can obtain a greater share of the available bandwidth. Even with TCP-Vegas operating alone, the connections that start at a later time usually obtain a smaller share of bandwidth because their observed *baseRTTs* are usually higher.

TCP-Westwood. TCP-Westwood [15] is another TCP variant that is based on bandwidth estimation. The main idea of TCP-Westwood is to keep track of the average throughput of the connection and use the information maintained appropriately when a packet loss occurs. Therefore, it is particularly effective in tackling the effects of wireless errors on the TCP performance.

Bandwidth Estimation and Faster Recovery. The bandwidth estimation is done by measuring the rate of return of ACKs. The estimation can adaptively use a higher weighting factor on the current bandwidth measurement when the interarrival times of ACKs increases. Therefore, when congestion occurs, the estimated bandwidth will capture the congestion state much faster and provide a more conservative estimate. Moreover, when there is a prolonged absence of ACKs (potentially because of congestion), the estimated bandwidth will exponentially decrease. Upon a packet loss event, instead of blindly reducing *cwnd* and *ssthresh* to heuristic values, the previously estimated bandwidth is used to compute a proper *cwnd* and *ssthresh* values. In essence, TCP-Westwood allows the sending rate to be maintained at the level just before the occurrence of a packet loss.

TCP-Westwood is especially robust to wireless errors because it effectively avoids an excessive reduction in the sending rate because of packet losses that are caused by wireless errors instead of congestion. Moreover, the use of bandwidth estimation can also potentially alleviate any excessive (heuristic) reductions of *cwnd* and *ssthresh*, when packet losses are due to congestion.

However, TCP-Westwood does not explicitly handle route failures. When a route failure occurs, the estimated bandwidth will quickly become zero as described previously. This leads to poor performance after new routes are established after failure. Similar to TCP-Vegas, TCP-Westwood uses the observed smallest round-trip time (*RTTmin*) in estimating the bandwidth. This can lead to problems, since any route change will invalidate the *RTTmin* and thus lead to incorrect bandwidth estimates.

TCP-Jersey. TCP-Jersey [16] can be considered to be an extension of TCP-Westwood. Besides the available bandwidth estimation component, TCP-Jersey utilizes explicit feedback from the intermediate nodes to distinguish wireless-induced packet losses from congestion-induced packet losses.

Available Bandwidth Estimation. TCP-Jersey adopts the bandwidth estimation strategy of TCP-Westwood (estimation is done by using the rate of the returning ACKs). The differences between the two protocols lie in how the estimated bandwidth is computed and used. Basically, TCP-Jersey uses a simpler estimator (time-sliding window estimator [18]) to estimate the available bandwidth and uses the estimated bandwidth only when a congestion warning signal is set.

Congestion Warning. Besides the available bandwidth estimation, TCP-Jersey makes use of the congestion warning (CW) signal from intermediate routers to help the TCP sender distinguish wireless-induced packet losses from congestion-induced packet losses. When the average queue size of a node exceeds a certain threshold, the node marks the so-called CW flag in all of its outgoing segments. When a sender receives a DUPACK with the CW flag set, it knows that the network is in the congestion state. Otherwise (if the CW flag of the DUPACK is not set), the DUPACK is assumed to be caused by wireless errors. TCP-Jersey uses the estimated bandwidth to adjust *cwnd* and *ssthresh* only when it receives an ACK or three DUPACKs with the CW flag set.

Explicit Retransmit. TCP-Jersey also modifies the traditional fast retransmit algorithm. Instead of halving *cwnd* upon the reception of three DUPACKs, TCP-Jersey keeps *cwnd* unchanged (as long as the DUPACKs are not due to congestion); *cwnd* is in fact adjusted separately by the rate control procedure if necessary.

TCP-Jersey has strengths that are similar to that of TCP-Westwood because both of them use the idea of bandwidth estimation. However, TCP-Jersey could perform better in some cases than TCP-Westwood [16] because TCP-Jersey uses explicit congestion warnings from intermediate nodes to notify the sender of possible congestion instances in the network. TCP-Jersey uses this extra information to distinguish wireless packet losses from congestion-related packet losses. In addition, TCP-Jersey triggers the rate control procedure upon the reception of ACKs (or DUPACKs) with the CW flag set, and thus it can proactively avoid congestion more rapidly. Besides, TCP-Jersey has a more robust bandwidth estimator since the estimation does not involve the use of the minimum observed *RTT* as in TCP-Vegas and TCP-Westwood. Therefore, its estimation is not directly affected by route changes.

However, similar to conventional TCP, TCP-Jersey does not have a mechanism to handle the effects of route failures. Moreover, TCP-Jersey requires the use of the explicit feedback information from intermediate nodes. Deploying TCP-Jersey is more difficult, since it relies on the cooperation of all (or at least many) nodes.

9.4.2 Determining Route Failure and Wireless Error

As mentioned in Section 9.3, conventional TCP does not have any mechanisms to handle route failures and wireless errors. In effect, TCP performance degrades significantly due to these factors. There is a need for having some effective means to determine these events accurately so as to allow TCP to react appropriately.

The representative schemes that determine route failures and wireless errors and distinguish these effects from congestion for improving the TCP performance include Explicit Link Failure Notification (ELFN) [1] and ADTCP [19].[5]

Explicit Link Failure Notification (ELFN). ELFN [1] is a simple scheme that provides link failure information to the TCP sender to assist the sender in

---

[5]There are schemes called TCP-Feedback [20] and ATCP [21], which employ ideas that are very similar to that of ELFN and ADTCP, respectively. Thus, the discussions of these scheme are omitted.

distinguishing packet losses that are caused by link failures from those that are caused by congestion.

Explicit Link Failure Notification. When a link failure occurs, the upstream node of the failed link will send a "host unreachable" ICMP message to the TCP sender. The sender, upon receiving this message, disables its retransmission timers and enters the "standby" mode. It then uses a periodic probe message to determine whether the route has been restored. When an acknowledgment is received (implying that the route is reestablished), the TCP sender restores its retransmission timers and invokes the previous states (prior to the failure). In this way, the TCP sender can avoid the slow-start phase after a route failure.

ELFN is a simple yet efficient scheme for improving TCP performance. ELFN prevents the TCP sender from invoking congestion control unnecessarily when it detects that a packet loss is caused by a link failure as opposed to congestion. Besides, its periodic probe mechanism allows the TCP sender to actively determine when the route is restored so that the normal transmission can be resumed quickly to avoid an underutilization of the available bandwidth.

However, ELFN requires intermediate nodes to assist in detecting and notifying the TCP sender of link failures, and this complicates its implementation and deployment. Furthermore, ELFN uses the previously stored states to resume transmission after a link failure, which may not be appropriate since the previously stored states may not reflect the characteristics of the new route. Another problem with ELFN is that only the TCP sender that triggers the ELFN message is notified of the failure; that is, all the other TCP senders that use the same link may not know that their path has failed until they trigger the ELFN message by themselves.

ADTCP. ADTCP [19] is an end-to-end scheme that uses multiple end-to-end metrics to determine the cause of packet losses and allow the TCP sender to react appropriately.

Classification of Connection States. ADTCP casts a connection into one of the following states: (1) congestion, (2) channel error, (3) route change, and (4) disconnection. ADTCP avoids an erroneous execution of congestion control mechanism when it determines that the packet loss is not caused by network congestion.

Multiple Metrics. ADTCP uses a multimetric technique to identify the reason for packet losses. The metrics under consideration are:

- *Interpacket delay difference (IDD)* is the delay difference between consecutive packet arrivals.
- *Short-term throughput (STT)* is the throughput during a short interval of observation.
- *Packet out-of-order delivery ratio (POR)* is the ratio of the number of out-of-order packets to the total number of received packets during a short interval of observation.

- *Packet loss ratio (PLR)* is the ratio of the number of missing packets within the receiving window during a short interval of observation.

These metrics can be measured end-to-end, without the need for an intervention from the intermediate nodes.

Identifying the Connection States. When a packet loss occurs, ADTCP determines the connection state as follows: If IDD is high and STT is low, the connection is in the congestion state. In contrast, if the connection is not congested but POR is high, the connection is in the route change state. Similarly, if the connection is not congested but PLR is high, the connection is in the channel error state. The disconnected state occurs if STT reaches zero.

ADTCP improves the performance of TCP by identifying the current connection state, thus allowing TCP to react to packet loss more appropriately. ADTCP does not require any feedback from the intermediate nodes. It identifies the connection state by using end-to-end measurements only, which allows for the easy deployment of the scheme.

However, the measurement of IDD requires the use of the TCP timestamp option or the sender to record the sending time of all the outstanding segments, which increases the overhead of the protocol. Furthermore, the determination of connection states requires certain metric thresholds to be carefully defined in order for ADTCP to operate efficiently.

9.4.3 Reducing Contention

In Section 9.3, the adverse effects of the interflow and intraflow contention problems on the TCP performance have been discussed. The reason for this is that conventional TCP does not operate its congestion window at an appropriate level.

The representative schemes that reduce contention and the effects thereof include Congestion Window Limit [12] and Link RED and Adaptive Pacing [4]. These schemes are described below.

Congestion Window Limit (CWL). Congestion Window Limit (CWL) [12] is a simple technique that mitigates the congestion window overshoot problem by restricting the maximum congestion window size. This scheme can reduce contention and improve the TCP performance.

Adaptive Maximum Congestion Window Size. A study [4] has shown that a maximum of $1/4$ spatial reuse can be achieved for a one-way flow along a chain of nodes. This study implies that there is a limit on the maximum sending rate of a TCP source, and in turn implies a limit on the maximum size of the congestion window. Chen et al. [12] have shown that the best throughput of TCP is actually achieved when the congestion window is set to approximately $1/5$ of the round-trip hop count (RTHC). This approximation of $1/5$ can be easily explained by considering also an increase in contention (and thus a reduced spatial reuse) contributed by the reverse ACK packets along the same path. To this end, an adaptive CWL scheme is proposed to adjust the maximum congestion window size to approximately 1/5 of the RTHC of the flow.

CWL is a simple scheme that adjusts the maximum congestion window size dynamically so that the sending rate will not exceed the maximum spatial reuse of the channel. By this, both the intraflow and interflow contention problems are reduced.

However, CWL must be used with routing protocols that are aware of the path length. Moreover, the 1/5 window limit is set by considering a single flow. For a multiplicity of flows that compete, it is unclear that this will hold always. The factor could depend on density and the number of competing connections.

Link RED (LRED) and Adaptive Pacing (AP). LRED and AP [4] are techniques that allow TCP to react preemptively link overload by adaptively delaying certain packet transmissions to reduce contention.

Link RED. LRED works by maintaining an average number of retries for recent packet transmissions. If the average retry attempt value exceeds the minimum threshold value (min_{th}), LRED will mark the outgoing packets with a probability depending on the value. TCP will then reduce its sending rate and thereby, to some extent, avoid packet loss.

Adaptive Pacing. AP works by distributing traffic among intermediate nodes in a more balanced way. The basic idea is to let some nodes wait, in addition to the normal backoff period, for an extra amount of time equal to a packet transmission time when necessary. This additional waiting period helps reduce the contention related drops caused by the exposed receivers. AP is used in coordination with LRED. When LRED starts to mark packets (the average retry value exceeds min_{th}), AP will then increase the backoff time of the pending transmission by an interval equal to the transmission time of the previous packet.

LRED provides an early sign of network overload which helps TCP improve the interflow fairness. When LRED is used in conjunction with AP, they improve the spatial reuse by reducing contention and thus improve the TCP performance.

However, LRED requires the MAC layer to maintain an average transmission retry attempt value. It also requires a RED-like algorithm to be implemented at the MAC layer, and this complicates the implementation and the deployment of the scheme.

9.4.4 Detecting Spurious Retransmissions

Spurious retransmission is a consequence of TCP's inability to tolerate and handle sudden delay spikes experienced by packet transmissions. The representative schemes that are classified into this category include TCP-Eifel [13] and Forward RTO-Recovery [22].

TCP-Eifel. TCP-Eifel [13] is a technique that is specially designed to detect and handle spurious retransmissions.

Detection of Spurious Retransmissions. TCP-Eifel uses the TCP timestamp option to solve the ambiguity of retransmissions. TCP-Eifel includes in every transmitted packet the transmission time and this timestamp is echoed back by the receiver. In addition, the sender records the time of the first retransmission. When an ACK acknowledging a recently retransmitted packet arrives, the sender can compare the timestamp on the ACK with the recorded time to determine whether the retransmission is spurious or not. If the timestamp on the ACK is smaller than the recorded time (which means that the ACK is generated due to a packet that was transmitted prior to the particular retransmission), the retransmission is likely to be spurious. If it is spurious, the sender will restore the *cwnd* and *ssthresh* values before the retransmission.

Response after Detecting Spurious Retransmission. If only one retransmission of the oldest unacknowledged segment was performed, the stored *ssthresh* and *cwnd* values will be restored. If more than one retransmission was performed, *ssthresh* is halved. If exactly two such retransmissions were performed, *cwnd* is set to the *ssthresh* (which had been halved previously). If more than two retransmissions were performed, *cwnd* is set to one. Therefore, the more retransmissions, the more conservative the sender gets.

TCP-Eifel is robust to a sudden increase in the packet delivery time. It identifies spurious retransmissions and avoids unnecessary *go-back-N* retransmissions for packets that are not lost but are just delayed.

However, TCP-Eifel requires the use of the TCP timestamp option or some modifications to the TCP header to enable the detection of spurious retransmissions. Moreover, extra memory and processing is needed to store the timestamp of each transmitted but not yet acknowledged packets.

Forward RTO-Recovery (F-RTO). F-RTO [22] can be considered as an improvement to Eifel algorithm. It is used to avoid further, unnecessary retransmissions when a spurious retransmission has been detected by the scheme.

Detecting Spurious Retransmissions after a Retransmission Timeout (RTO) Event. When the first ACK after a retransmission due to a RTO arrives, F-RTO does not immediately continue with further retransmissions but instead checks if the ACK advances the congestion window to determine whether to perform further retransmissions or to continue sending new data. There are two cases:

1. If the first ACK after a RTO-triggered retransmission advances the window, F-RTO transmits two new segments instead of continuing retransmissions. If the next ACK also advances the window, the RTO is likely to be spurious, because this second ACK must be triggered by an originally transmitted segment.

2. If either one of the two ACKs after a RTO is a duplicate ACK, the sender continues retransmissions in a manner similar to that of the operations of conventional TCP.

F-RTO facilitates the detection of spurious timeouts that are caused by a sudden increase in the packet delay. It does not require any modification to the TCP header, nor does it require the use of the TCP timestamp option.

However, F-RTO does not explicitly handle spurious retransmissions that are caused by spurious fast retransmits [13]. The spurious retransmissions that are caused by packet reordering cannot be handled by F-RTO. Besides, the determination of spurious retransmissions of F-RTO requires a default of two new segments to be sent after receiving the first ACK after a retransmission. However, it may not be possible sometimes due to the window restriction. If no new segments can be sent, the TCP sender has no choice but to follow the conventional TCP RTO recovery by entering the slow start phase.

9.4.5 Exploiting the Buffering Capability

In conventional TCP, a route failure can cause TCP to perform poorly. In fact, the poor performance of TCP, to a significant extent, is due to the inefficiency of TCP to preserve the work that was already done. For instance, when a route failure occurs, many of the transmitted but not yet acknowledged packets are dropped at the intermediate nodes (or delayed due to the broken route) and need to be retransmitted.

The representative schemes that make use of the buffering capability to avoid unnecessary retransmissions to enhance TCP performance include Split-TCP [23] and TCP-BUS [24].

Split-TCP. Split-TCP [23] is a scheme that separates the functionalities of TCP congestion control and reliability. It makes use of proxies set up along the connection to improve TCP performance and fairness.

TCP with Proxies. For a TCP connection, certain nodes along the route become proxies for the connection. The proxies become "responsible" for ensuring that packets will arrive at a subsequent proxy or at the destination. One way to see this is that a TCP connection is "split" into several small TCP connections. Each of these small connections has its own local acknowledgment mechanism that can control its own rate and ensure reliable arrival of packets at the next proxy.

With Split-TCP, the buffering of packets at the proxies allows any dropped packets to be recovered from the most recent proxy instead of the need to retransmit it all the way back from the source. Furthermore, by dividing a long connection into several short segments, Split-TCP enables better pipelining of data transmissions and alleviates the capture effect (due to the IEEE802.11 MAC) due to other coexisting short TCP connections.

However, an extra buffer is required at each proxy to store all the outstanding packets that have not been acknowledged. Furthermore, an additional state such as the local congestion window have to be maintained at the proxies and at the sender. Since proxies need to communicate with each other (acknowledge the arrival of a packet), there could be a higher overhead than with conventional TCP.

Buffering Capability and Sequence Information (TCP-BUS). TCP-BUS [24] is a scheme that remedies the effects of route failures on the TCP performance by taking advantage of the buffering capabilities of nodes.

Explicit Notifications. TCP-BUS defines two messages called Explicit Route Disconnection Notification (ERDN) and Explicit Route Successful Notification (ERSN). These two messages are sent by the intermediate nodes to notify the sender of route failure events. When a route failure occurs, the ERDN message is propagated toward the TCP sender. During this process, all the transmissions along the route of the connection are notified to stop. The TCP sender also freezes its TCP states. When an ERSN is received, the connection resumes its operations.

Extending Timeout Values. When a route failure occurs, those packets that are on-the-fly and located between the source and the node upstream of the failure are buffered. These buffered packets usually take a longer time to be delivered to the destination because they must wait for a recovery from the failed link. The transmission delay of these packets may cause the TCP sender to timeout. TCP-BUS avoids this timeout by doubling the retransmission timeout value associated with these buffered packets.

Avoiding Unnecessary Retransmissions. After the route is reestablished, the destination notifies the source of the sequence number of the last data packet that it had received successfully. Therefore, the source can advance its congestion window accordingly and selectively retransmit only the packets that are lost. Since the previously buffered packets at the intermediate nodes will arrive at the destination earlier than the selectively retransmitted packets (with smaller sequence numbers) after the route is reestablished, fast retransmit requests may be triggered at the destination. However, TCP-BUS suppresses these fast retransmit requests by performing a special procedure at the destination.

The use of the explicit route disconnection and route success messages allows the sender to react to route disconnection and the corresponding reconnection more quickly and effectively. Since the sender knows that packets will be buffered en route, it will set larger timeout values for each transmission; this could potentially reduce timeouts and unnecessary retransmissions.

However, the use of the explicit notification scheme requires assistance from the intermediate nodes, and this complicates the implementation and deployment. Furthermore, it requires modification to the routing layer protocol to cope with its operations. Moreover, it has scalability problems because it requires an intermediate node to store the buffered packets and the states of the different flows passing through it.

9.5 NEW TRANSPORT LAYER PROTOCOL FOR MANETs

A new transport layer protocol for MANETs called Ad Hoc Transport Protocol (ATP) was proposed in reference 25. It deviates from the TCP school of thought. ATP seems to outperform all the other previous TCP schemes in terms of transport layer performance.

9.5.1 Ad Hoc Transport Protocol (ATP)

ATP [25] is a transport layer protocol that is specially designed for MANETs, and therefore its operation is substantially different from that of conventional TCP.

Layer Coordination. ATP uses lower layer information and explicit feedback from intermediate nodes to assist the transport layer operations. This information includes (a) an initial rate feedback for a quick-start, (b) a regular rate-based feedback from intermediate nodes to control the sending rate, and (c) a path failure notification for the detection of a route failure.

Rate-Based Transmissions. ATP uses rate-based transmissions instead of window-based transmissions that is used with the conventional TCP. ATP not only uses the rate feedback from intermediate nodes to control the transmission rate but also uses a transmission scheduler to schedule the transmissions evenly over time to reduce the burstiness of the connection.

Decoupling of Congestion Control and Reliability. In contrast to conventional TCP, ATP decouples congestion control and reliability. ATP does not require the arrival of ACKs to clock out packet transmissions but depends on the regular feedback from the network to perform the congestion control. For reliability, ATP does not employ cumulative ACKs but solely relies on the use of selective acknowledgment (SACK) information that is periodically reported by the receiver to identify packet losses.

Assisted Congestion Control. ATP requires each intermediate node along the connection to maintain two pieces of information. The first information is the average queuing delay, Q_t, experienced by packets traversing that node. The second information is the average transmission delay, T_t, experienced by the head-of-line packet at that node. Q_t is related to the contention between packets belonging to different flows at the same node. T_t is related to the contention between packets within the vicinity of the node. The highest $Q_t + T_t$ value of the nodes in the forward path is returned by the receiver to the sender. The sender then uses this information to control the transmission rate.

ATP is well-suited for MANETs due to many reasons. ATP uses rate feedback from the intermediate nodes to identify the bandwidth bottlenecks on the forward path and thus is able to determine the maximum forward transmission rate accurately without the restriction of the constrained reverse path. This is in contrast with the bandwidth estimation of TCP-Westwood, TCP-Vegas, and TCP-Jersey, wherein the estimated bandwidth is somehow implicitly affected by the characteristics of the ACK on the reverse path. Another advantage provided by the use of the rate feedback is that ATP does not require the arrival of ACKs to clock out packet transmissions; therefore, it is more robust to ACKs losses. ATP uses a periodic SACK scheme instead of using cumulative ACKs to provide reliability; this reduces the traffic overhead on the reverse path. The use of the SACK scheme also enables ATP to recover more than one segment at a time; thus, it is more efficient and robust in high loss environments. The probing and the initial rate feedback in ATP is very important when considering that the route failure events occur frequently in MANETs. Probing allows a quicker detection of route recomputations, and the initial rate feedback allows a very fast transmission rate estimation of the connection (within only an *RTT*).

However, ATP requires the cooperation of the intermediate nodes and the lower layers to operate in conjunction with the transport layer, and these requirements could make ATP difficult to implement and deploy. Due to its incompatibility with conventional TCP, many existing applications that were originally built on top of TCP may not function without extensive modifications. The time within which ATP can detect and recover lost packets may be long and unacceptable; this is because ATP depends on the regular transmission of SACK information from the receiver (every one second by default).

A summary of the main characteristics of each of the discussed TCP enhancements is provided in Tables 9.1 to 9.4.

9.6 DISCUSSION

In this section, a discussion on the transport layer issues in MANETs is provided. In particular, the discussion addresses the certain questions that arise, given the state of the art.

Which TCP Enhancement Seems to Be the Best for MANETs? It seems that the most effective means of improving the transport layer performance in MANETs is to alleviate the effects of route failures and wireless channel errors. Therefore, those schemes that make use of explicit feedback from intermediate relay nodes to detect route failures or congestion status almost always provide significant improvements in TCP performance.

However, if we consider only TCP enhancements (with minor modifications to conventional TCP), TCP-Westwood and TCP-Jersey seem to be the preferred choice. The reason is that both use the idea of bandwidth estimation and can effectively alleviate the effects of packet losses that are not caused by congestion.

If we consider both TCP-based and non-TCP-based protocols, ATP will be the best choice from among all of the considered schemes; this is because it not only resolves the problems of route failures and wireless errors but also resolves other problems such as the spurious retransmissions and contention problems that were discussed in Section 9.3. However, the most critical problem with ATP is its incompatibility with

TABLE 9.1. Comparison of the Main Characteristics of Various TCP Enhancements

	TCP Variants		
Problems	TCP-Vegas	TCP-Westwood	TCP-Jersey
Misinterprets route failures as congestion	**No**: *ssthresh* heuristically sets to *cwnd*/2, *cwnd* reduces to 1 and enters slow-start.	**No**: *cwnd* is set to 1 but *ssthresh* is set to the estimated bandwidth instead of being adjusted heuristically.	**No**: Resets *cwnd* to 1 and enters slow-start phase.
Misinterprets wireless errors as congestion	**No**: Similar to conventional TCP, but it reduces *cwnd* only by a quarter if the loss is detected by the new faster retransmission mechanism.	**Yes**: *ssthresh* and *cwnd* is assigned based on the estimated bandwidth.	**Yes**: It uses Congestion Warning (CW) to distinguish wireless loss from congestion and keeps *cwnd* unchanged.
Intraflow and interflow contention	**Partial**: It maintains a smaller and stabler *cwnd* which reduces contention at lower layer.	**Partial**: The growth of *cwnd* is carefully controlled based on the estimated bandwidth.	**Partial**: Upon congestion indication, it adjusts the *ssthresh* and *cwnd* based on the estimated bandwidth.
Spurious retransmissions	**No**: Spurious timeouts: Go-back-N retransmissions. Spurious fast retransmits: *cwnd* is halved.	**No**: Spurious timeouts: Go-back-N retransmissions. Spurious fast retransmits: *ssthresh* and *cwnd* are adjusted based on the estimated bandwidth.	**No**: Spurious timeouts: Go-back-N retransmissions. Spurious fast retransmits: *ssthresh* and *cwnd* are adjusted based on the estimated bandwidth if CW is set
Inefficiency due to the loss of retransmitted packet	**No**: Must wait for a timeout to trigger retransmissions. Slow-start is executed thereafter.	**No**: Must wait for a timeout to trigger retransmissions. Slow-start is executed thereafter.	**No**: Must wait for a timeout to trigger retransmissions. Slow-start is executed thereafter.

TABLE 9.2. Comparison of the Main Characteristics of Various TCP Enhancements

Problems	TCP Variants		
	ELFN	ADTCP	CWL
Misinterprets route failures as congestion	**Yes**: Explicit link failure notification (ELFN) is reported to the sender. The sender disables congestion control mechanisms and freezes TCP's states. A packet is sent periodically to probe the network to see if the route has been restored.	**Yes**: Multiple metrics are used to identify the connection state. It puts the TCP sender in *disconnection state*. It keeps sending data packet to the network until a new ACK is arrived. $cwnd$ is set to $ssthresh$ after the route is reestablished.	**No**: $ssthresh$ heuristically sets to $cwnd/2$, $cwnd$ reduces to 1 and enters slow-start.
Misinterprets wireless errors as congestion	**No**: Resets $cwnd$ to 1 and enters slow-start.	**Yes**: Multiple metrics are used to identify the connection state. It puts the TCP sender in the *channel error state*. The lost packet is simply retransmitted without adjusting TCP's states.	**No**: Resets $cwnd$ to 1 and enters slow-start.
Intraflow and interflow contention	**No**: Large $cwnd$ may cause serious contention at lower layer.	**Partial**: Congestion that is caused by contention may be detected. The sender rate will be decreased accordingly.	**Yes**: $cwnd$ is adaptively adjusted to 1/5 of the $RTHC$ of the route so that the contention is reduced.
Spurious retransmissions	**No**: Spurious timeouts: Go-back-N retransmission. Spurious fast retransmits: $cwnd$ is halved.	**No**: Spurious timeouts: Go-back-N retransmission. Spurious fast retransmits: $cwnd$ is halved.	**No**: Spurious timeouts: Go-back-N retransmission. Spurious fast retransmits: $cwnd$ is halved.
Inefficiency due to the loss of retransmitted packet	**No**: Must wait for a timeout to trigger retransmissions. Slow-start is executed thereafter.	**No**: Must wait for a timeout to trigger retransmissions. Slow-start is executed thereafter.	**No**: Must wait for a timeout to trigger retransmissions. Slow-start is executed thereafter.

TABLE 9.3. Comparison of the Main Characteristics of Various TCP Enhancements

Problems	TCP Variants		
	LRED & AP	Eifel	F-RTO
Misinterprets route failures as congestion	**No**: *ssthresh* heuristically sets to *cwnd*/2, *cwnd* reduces to 1 and enters slow-start.	**No**: *ssthresh* heuristically sets to *cwnd*/2, *cwnd* reduces to 1 and enters slow-start.	**No**: *ssthresh* heuristically sets to *cwnd*/2, *cwnd* reduces to 1 and enters slow-start.
Misinterprets wireless errors as congestion	**No**: Resets *cwnd* to 1 and enters slow-start.	**No**: Resets *cwnd* to 1 and enters slow-start.	**No**: Resets *cwnd* to 1 and enters slow-start.
Intraflow and interflow contention	**Yes**: It detects early sign of congestion. The sending rate is reduced to improve spatial reuse and interflow fairness.	**No**: Large *cwnd* may cause serious contention at lower layer.	**No**: Large *cwnd* may cause serious contention at lower layer.
Spurious retransmissions	**No**: Spurious timeouts: Go-back-N retransmission. Spurious fast retransmits: *cwnd* is halved.	**Yes**: Timestamp option is used to solve the retransmission ambiguity.	**Yes**: Detect and avoid spurious retransmissions. However, spurious retransmissions due to fast retransmissions are not handled.
Inefficiency due to the loss of retransmitted packet	**No**: Must wait for a timeout to trigger retransmissions. Slow-start is executed thereafter.	**No**: Must wait for a timeout to trigger retransmissions. Slow-start is executed thereafter.	**No**: Must wait for a timeout to trigger retransmissions. Slow-start is executed thereafter.

conventional TCP and existing network applications. In other words, if ATP is used as a transport layer protocol for MANETs, many applications may have to be modified or redeveloped.

Is Any TCP Enhancement Sufficient or Do We Need to Redesign a Transport Protocol for MANETs from Scratch?

Most of the TCP enhancements can provide significant improvements to the TCP performance over MANETs but are insufficient in isolation for providing a complete transport layer solution in MANETs. The reason is that most of the schemes proposed only tackle a small set of problems. One may want to combine various schemes together to provide a much more comprehensive solution. However, different schemes may not be compatible with each

TABLE 9.4. Comparison of the Main Characteristics of Various TCP Enhancements

Problems	TCP Variants		
	Split-TCP	TCP-BuS	ATP
Misinterprets route failures as congestion	**Partially**: Only the TCP segment where the route breaks is affected. Other TCP segments can still transmit their packets. Furthermore, the missing packets at the destination can be retransmitted by proxies.	**Yes**: Explicit route disconnection notification (ERDN) is sent to the sender. The sender stops all the transmissions and doubles the timeout values of all the outstanding packets to avoid the occurrence of a timeout event.	**Yes**: Route failure is determined by either the explicit link failure notification from the intermediate nodes or the loss of three rate feedbacks. The sender periodically probes the receiver until a response is received. The rate feedback in the response enables the quick-start mechanism.
Misinterprets wireless errors as congestion	**Partially**: Only the TCP segment where the wireless errors occur is affected. Other TCP segments can still transmit their packets. The lost packet is retransmitted by the previous proxy.	**No**: Resets $cwnd$ to 1 and enters slow-start.	**Yes**: SACKs are regularly sent from the receiver to indicate possible packet losses. Lost packets are retransmitted with a higher priority. The sending rate of the sender is unaffected by the loss.
Intraflow and interflow contention	**Yes**: Long TCP flows are split into shorter flows. It ensures that long TCP flows are not starved due to the capture effect of short TCP flows.	**No**: Large $cwnd$ may cause serious contention at lower layer.	**Yes**: The level of contention is reflected from the average transmission delay (T_t) maintained in each node. This value is feedback-ed to the sender. The higher the value of T_t, the lower the sending rate resulted.
Spurious retransmissions	**No**: The impacts of spurious retransmissions do not span the entire TCP flow but only the TCP segments that are involved.	**No**: Spurious timeouts: Go-back-N retransmission. Spurious fast retransmits: $cwnd$ is halved.	**Partial**: Packets may be unnecessarily retransmitted upon the reception of SACKs but the sending rate is not affected.
Inefficiency due to the loss of retransmitted packet	**No**: Must wait for a timeout to trigger retransmissions. Slow-start is executed thereafter.	**No**: Must wait for a timeout to trigger retransmissions. Slow-start is executed thereafter.	**Yes**: The loss of retransmitted packet will be notified to the sender by SACKs from the receiver. The sending rate is not affected.

other. Due to this difficulty, one may want to consider the redesign of a new transport layer protocol for MANETs. An excellent example of a new transport protocol is ATP.

9.7 CONCLUSION

This chapter provides an extensive survey that summarizes the problems of using TCP over MANETs. The most critical problem with using TCP in MANETs is the inability of TCP to cope with the effects of route failures and wireless errors. Besides, TCP also suffers heavily from interflow and intraflow contention and the spurious retransmission problem. Several major TCP schemes that have been proposed to remedy or alleviate the problems are discussed. It is suggested that the use of explicit feedback from the intermediate nodes to assist TCP in distinguishing between various factors contributing to packet loss or determining the congestion status can significantly improve TCP performance over MANETs. In summary, existing TCP schemes are not sufficient in isolation in providing a complete transport layer solution for MANETs. In fact, the most efficient scheme that has been proposed is ATP; ATP is a transport layer protocol specifically designed for MANETs.

9.8 EXERCISES

1. What are the characteristics of mobile ad hoc networks (MANETs) that require changes in existing TCP?

2. Why does TCP perform poorly in mobile ad hoc networks (MANETs)? List all the major problems facing TCP in MANETs.

3. What are the particular advantages and disadvantages of using a bandwidth estimation approach to improve TCP performance in MANETs?

4. TCP-Eifel and F-RTO are used to avoid spurious retransmissions and improve TCP performance. What are the strengths and weaknesses with these two schemes?

5. ATP seems to be the best transport layer protocol for MANETs. What makes it particular suitable for use in MANETs?

6. List the major strategies for improving TCP performance in MANETs. Also, list two representative schemes from each of these strategies.

REFERENCES

1. G. Holland and N. Vaidya. Analysis of TCP performance over mobile ad hoc networks. In *Proceedings of ACM/IEEE MobiCom*, 1999, pp. 219–230.
2. S. Xu and T. Saadawi. Performance evaluation of TCP algorithm in multi-hop wireless packet networks. *Wireless Communications and Mobile Computing*, **2**:85–100, 2002.

3. S. Xu and T. Saadawi. Does the IEEE 802.11 MAC Protocol work well in multihop wireless ad hoc networks. *IEEE Communications Magazine*, **39**(6):130–137, 2001.
4. Z. Fu, P. Zerfos, H. Luo, L. Zhang, and M. Gerla. The impact of multihop wireless channel on TCP throughput and loss. In *Proceedings of IEEE INFOCOM*, San Francisco, CA, April 2003.
5. H. Elaarag. Improving TCP performance over mobile networks. *ACM Computing Surveys*, **34**(3):357–374, 2002.
6. H. Balakrishnan, V. N. Padmanabhan, S. Srinivasan, and R. H. Katz. A comparison of mechanisms for improving TCP performance over Wireless Links. *IEEE/ACM Transaction on Networking*, **5**(6):756–769, 1997.
7. TCP slow start, congestion avoidance, fast retransmit, and fast recovery algorithms. RFC2001, January 1997.
8. P. Bhagwat, P. Bhattacharya, A. Krishna, and S. K. Tripathi. Using channel state dependent packet scheduling to improve TCP throughput over wireless LANs. *Wireless Networks*, **3**(1):91–102, 1997.
9. A. Chockalingam, M. Zorzi, and V. Tralli. Wireless TCP performance with link layer FEC/ARQ. In *IEEE ICC*, 1999, pp. 1212–1216.
10. D. Berger, Z. Ye, P. Sinha, S. Krishnamurthy, M. Faloutsos, and S. Tripathi. TCP friendly medium access control for ad-hoc wireless networks. In *IEEE MASS*, 2004.
11. H. Zhai, X. Chen, and Y. Fang. Alleviating Intra-flow and inter-flow contentions for reliable service in mobile ad hoc networks. In *IEEE MILCOM*, 2004.
12. K. Chen, Y. Xue, and K. Nahrstedt. On setting TCP's congestion window limit in mobile ad hoc networks. In *IEEE International Conference on Communications (ICC 2003)*, Vol. 26, May 2003, pp. 1080–1084.
13. R. Ludwig and R. H. Katz. The Eifel algorithm: Making TCP robust against spurious retransmissions. In *SIGCOMM Computer Communication Review*, **30**(1):30–36, 2000.
14. L. S. Brakmo, S. W. O'Malley, and L. L. Peterson. TCP Vegas: New techniques for congestion detection and avoidance. *SIGCOMM Computer Communication Review*, **24**(4): 24–35, 1994.
15. C. Casetti, M. Gerla, S. Mascolo, M. Y. Sanadidi, and R. Wang. TCP Westwood: End-to-end congestion control for wired/wireless networks. *Wireless Networks*, **8**(5):467–479, 2002.
16. K. Xu, Y. Tian, and N. Ansari. TCP-Jersey for wireless IP communications. *IEEE Journal on Selected Areas in Communications*, **22**(4):747–756, 2004.
17. J. Mo, R. J. La, V. Anantharam, and J. Walrand. Analysis and comparison of TCP Reno and Vegas. In *IEEE INFOCOM*, 1999, pp. 1556–1563.
18. D. D. Clark and W. Fang. Explicit allocation of best-effort packet delivery service. *IEEE/ACM Transactions on Networking*, **6**(4):362–373, 1998.
19. Z. Fu, B. Greenstein, X. Meng, and S. Lu. Design and Implementation of a TCP-Friendly Transport Protocol for Ad Hoc Wireless networks. In *Proceedings of IEEE ICNP*, 2002.
20. K. Chandran, S. Raghunathan, S. Venkatesan, and R. Prakash. A feedback based scheme for improving TCP performance in ad-hoc wireless networks. In *Proceedings of ICDCS*, 1998, pp. 472–479.
21. J. Liu and S. Singh. ATCP: TCP for mobile ad hoc networks. *JSAC*, 2001.

22. P. Sarolahti, M. Kojo, and K. Raatikainen. F-RTO: An Enhanced recovery algorithm for TCP retransmission timeouts. *SIGCOMM Computer Communication Review*, **33**(2): 51–63, 2003.
23. S. Kopparty, S. V. Krishnamurthy, M. Faloutsos, and S. K. Tripathi. Split TCP for mobile ad hoc networks. In *Proceedings of IEEE GLOBECOM*, 2002, pp. 138–142.
24. D. Kim, C.-K. Toh, and Y. Choi, TCP-BUS: Improving TCP performance in wireless ad hoc networks. *Journal of Communications and Networks*, **3**(2):175–186, 2001.
25. K. Sundaresan, S. Vaidyanathan, H.-Y. Hsieh, and R. Sivakumar. ATP: A reliable transport protocol for ad-hoc networks. In *Proceedings of ACM MobiHoc*, Annapolis, MD, 2003, pp. 64–75.

CHAPTER 10

ACK-Thinning Techniques for TCP in MANETs

STYLIANOS PAPANASTASIOU, MOHAMED OULD-KHAOUA, and
LEWIS M. MacKENZIE

Department of Computing Science, University of Glasgow, Glasgow G12 8RZ, Scotland

10.1 INTRODUCTION

Recently, the demand for ubiquitous connectivity and the availability of cheap wireless transceivers has led to an interest increase in wireless networks, both in terms of commercial exploitation and with regard to research inquiry. Mobile ad hoc networks (MANETs) are a particular facet of the wireless communications mosaic; MANETs are spontaneously organized, self-healing networks consisting of mobile agents of varying processing power and bandwidth capacity which largely cooperate to provide connectivity and services. Such agents or "nodes" may both produce (as hosts) and forward (as routers) messages through wireless, possibly ephemeral, links to form spontaneous networks with varying and dynamic topologies. Existing infrastructure is not presupposed for MANETs, and centralized administration is not required. Hence, MANET applications include disaster relief operations or settings where short-term connectivity may be convenient yet infrastructure may be inadequate or missing, such as conferences and university campuses [1]. Notably, the concept of a MANET (or multihop wireless network) is somewhat independent of the wireless technology used for its implementation which may be Bluetooth [2], HiperLan [3], 802.11 [4], and so on.

The Transmission Control Protocol (TCP) is a widely used end-to-end communications protocol and has been considered in various research works as the *de facto* reliable transport protocol for use in MANETs [5–9]. In short, TCP provides a reliable, connection-oriented service over IP and includes provisions to effectively utilize the available network bandwidth while ensuring fairness and avoiding network congestion. TCP connections are initiated between two hosts and are, usually,

---

Algorithms and Protocols for Wireless and Mobile Ad Hoc Networks, Edited by Azzedine Boukerche
Copyright © 2009 by John Wiley & Sons Inc.

duplex in nature; that is, both hosts can send and receive data over the same connection. Reliability is ensured using an Automatic Repeat Request (ARQ) paradigm, cumulative acknowledgments (ACKs), and checksum verification on received segments [10].

Intuitively, the message exchange dynamic in TCP follows a send-verify pattern. Every *DATA* segment transmitted from a source to a destination triggers, upon successful reception, an *ACK* response from the destination to the source. If a segment has not been acknowledged for some time, it is considered lost and is retransmitted by the source. When considering the utilization of the available bandwidth, it is clear that both ACK and DATA segments contribute to it. However, unlike DATA segments, ACKs are considered control traffic because they are used solely to ensure reliable delivery. Therefore, it is desirable to minimize the amount of control traffic (ACKs) so as to make more bandwidth available for the actual message (DATA segments). In view of this, past research efforts have concentrated on minimizing the number of produced ACKs for TCP in wired networks without compromising reliability [11]. One such widely deployed optimization includes the piggybacking of ACKs onto regular DATA segments when communication is two-way (peer-to-peer rather than client–server).

However, in wireless multihop communications, competition for bandwidth between DATA and ACK segments is even more pronounced due to the broadcasting nature of the shared wireless medium and the limited available bandwidth [12]. Particularly, in the case of 802.11-based MANETs, the inability of the MAC mechanism to properly coordinate transmissions leads to spurious packet losses; in such a setting, reducing the number of segments in the pipe at any one time can help smooth out this link-layer inefficiency [13]. Clearly, minimizing the number of ACKs produced is one way of achieving such a reduction. These *ACK-reducing* or *ACK-thinning* proposals are the focal point of interest in this survey. Although, such proposals may differ in the means of achieving their goal, their common aim is to reduce the overhead of ACKs (control data) produced by a TCP conversation, without compromising reliability.

As defined in the relevant RFC [10], the semantics of TCP are largely end-to-end in nature. Hence, the protocol maintains and utilizes information pertinent to the end points and is mostly oblivious of the point-to-point effort necessary to deliver a bytestream between two hosts. Ideally, any modifications to the protocol which would achieve ACK reduction would maintain this paradigm. Nonetheless, there is scope for point-to-point modifications if the benefits are significant enough. This survey acknowledges the tradeoff and distinguishes proposals in *end-to-end* and *hybrid*. A further distinction made when discussing ACK-thinning schemes is between *single* and *cross-layer* approaches, with the former being ,in general, easier to deploy and the latter possibly exhibiting more significant performance merits.

Overall, this chapter provides a comprehensive survey of the main ACK-thinning techniques that have been proposed for MANET environments in recent times [5–7,14]. In the following sections we outline a useful classification of these techniques and compare their relative merits with respect to ease of implementation, backward compatibility, and performance achievement. Specifically, the rest of the survey is organized as follows. Section 10.2 briefly visits the concept of a mobile ad hoc network, its characteristics, and applications as contrasted to wired LANs.

The section also entails a description of the main TCP features in such networks with special emphasis placed on those features pertinent to ACK-thinning techniques. Section 10.3 contains a brief overview of the main challenges faced by TCP when used over a (by definition) diverse ad hoc network environment. The main points of interest in this section are (a) the effects on TCP caused by the dynamic nature of the network topology and (b) the effects of packet drops due to interference-induced spatial contention. Section 10.4 includes a description of several techniques that aim to reduce ACK production by the TCP agent while maintaining reliability. Each technique is briefly described in turn and discussed in terms of ease of deployment and scope of application. Section 10.5 provides insight into the choice of implementation level when introducing changes in the protocol stack; as mentioned before, the techniques presented here may be deployed in an end-to-end or point-to-point fashion and may be contained in a single layer of the protocol stack or span several layers. Finally, Section 10.6 summarizes this overview and offers concluding remarks and pointers to future research work.

10.2 PRELIMINARIES

10.2.1 MANETs

In contrast to the scope of the Mobile-IP specification [15], which supports nomadic host "roaming" based on a fixed infrastructure, mobile ad hoc networks (MANETs) consist of autonomous, mobile, wireless domains, where a set of nodes form the network routing infrastructure in an ad hoc fashion. This section aims to differentiate MANETs clearly from other wireless network types and contrast them with traditional wired networks.

MANET Definition. As wireless devices have become increasingly commonplace and affordable, the notion of "nomadic" computing has gained in popularity. The term refers to the use of networking in mobile devices where connectivity is not determined by location or motion; that is, devices are expected to work in a consistent and useful manner regardless of their point of connection to the Internet.

The Mobile-IP specification [15] has attempted to support freedom of motion, as associated with wireless operations, over the Internet. Its principal function is to provide protocol facilities that would enable a client to seamlessly use different points of attachment to the Internet for communication with other clients. In this scheme, the mobile node is always identified by its *home address*, which provides a fixed point of reference for communications addressed to it, no matter what actual network the mobile node is connected to at the time. The scheme takes care to redirect traffic to the mobile node while it travels between networks (i.e., points of attachment on the Internet) in a manner that is transparent to applications running on the node.

The process is perhaps best illustrated through an example by briefly considering the scenario depicted in Figure 10.1. When a roaming (mobile) node is away from its home (and thus is not attached at its home address), a home agent monitors communications. When the roaming node obtains a connection point to the Internet, it

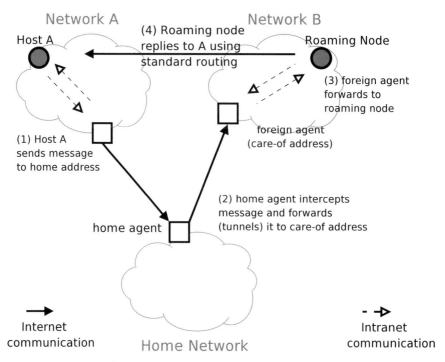

Figure 10.1. An example of Mobile-IP in effect.

registers, through a foreign agent, a care-of address with the home agent. The home agent then relays any messages for the roaming node to the foreign agent, which in turn forwards them to the roaming node. This dynamic is depicted in instances (1), (2), and (3) in Figure 10.1. Being aware that some third party is seeking communication, the roaming node may then directly communicate with the interested host, using the normal routing rules, as shown in instance (4) in Figure 10.1. A thorough overview of Mobile-IP is not within the scope of this survey, but readers requiring more information may refer to Perkins [15].

The concept of MANETs is similar to Mobile-IP with respect to connectivity of a mobile agent. However, unlike Mobile-IP, mobile ad hoc networking aims to implement connectivity without relying on existing infrastructure by allowing the mobile nodes to act as autonomous hosts/routers and form their own routing infrastructure in a spontaneous manner. This ad hoc networking paradigm leads to the creation of autonomous, mobile, wireless domains which may even merge into a greater whole or partition themselves into smaller segments [16]. Figure 10.2 depicts examples of MANETs as they are expected to be realized in various settings. In particular, instance (1) in Figure 10.2 depicts a MANET existing in isolation from other networks, while instance (2) outlines a MANET with a connection point to Network A. MANETs may also act as gateways between networks, as shown in instance (3), and may even provide extended coverage for IEEE wireless infrastructure networks [4], as illustrated in instance (4) in Figure 10.2.

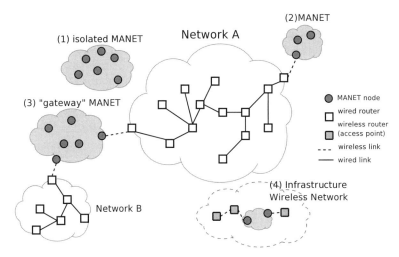

Figure 10.2. Types of MANETs.

MANET—LAN Differences. The wireless and mobile nature of MANETs results in a profile of differentiating characteristics as compared to wired networks. The following discussion outlines those differences and should aid comprehension when discussing TCP challenges in MANETs in Section 10.2.2. An even more detailed account of the unique MANET characteristics may be found in Corson and Macker [16].

Mobility. Mobility is an inherent attribute of a MANET setting. By definition, a MANET is a self-organized, self-healing wireless multihop network consisting of free roaming nodes forming randomly changing topologies with potentially ephemeral links. This characteristic is in stark contrast with the relatively stable and infrequently changing topology of a wired network. Furthermore, wireless networks may include unidirectional links; wired networks almost invariably entail bidirectional point-to-point communication between nodes.

Bandwidth Constraints. Generally, wireless links are of lower capacity than their hardwired equivalents. Furthermore, the adverse effects of interference, fading, and other throughput-degrading factors contribute to wireless links, being relatively less reliable in data transfers. With such (relatively) low throughput capabilities, it is only natural that wireless links frequently experience more congestion, yet they are increasingly expected to provide services similar to those on offer in wired networks. Worthy of note is also the bandwidth variability that is experienced among mobile nodes equipped with different transceivers and/or signal reception capabilities.

From a protocol designer's perspective, the bandwidth constraints limit the prospects of extensive control message exchanges; that is, protocol overhead must be kept to a minimum to conserve bandwidth for the actual messages the application processes need to exchange. This requirement, combined with the additional control information needed to track changes in the dynamic topology, provides a significant challenge in balancing bandwidth share between control and actual data.

Other Considerations. The freely roaming nodes that largely comprise a MANET have small energy reserves at their disposal because they are usually battery-operated or have some other limited energy source. This makes energy conservation an important criterion in MANET protocol design and, in turn, implies that protocols should be simple so as to minimize CPU use, because this, of course, translates into larger energy expenditure. Similarly, wireless transmissions are costly and should be kept to a minimum; for instance, control packet transmissions should be sporadic.

Security is another challenging aspect of the MANET mosaic. Due to the broadcast nature of wireless transmissions, the possibility of eavesdropping is increased compared to wired connections. Furthermore, the decentralised nature of MANETs and the ephemeral nature of their wireless links provides opportunities for spoofing, although these MANET characteristics also emphasize network resilience because there is no single point of failure in the network.

As an aside, it should be mentioned that when the bandwidth and energy storage capabilities of the nodes in a MANET are sufficiently small (perhaps in line with the node's physical size and computational capacity), then the wireless multihop network is classified as a *sensor network* and does not necessarily follow IP conventions. Such networks are not considered in this study but may be of interest to some readers, who are encouraged to refer to other chapters in this book or consult Marrón et al. [17] for further details.

10.2.2 TCP: End-to-End Congestion Control

The Transmission Control Protocol (TCP) has been widely and successfully deployed on the Internet and other networks as a means of end-to-end, byte-stream-oriented communication between two hosts. This section provides a brief overview of TCP mechanisms as a prelude to Section 10.3 on TCP behavior over MANETs. Readers interested in more details about TCP are advised to consider the excellent reference work in Stevens [18].

In general, TCP provides a connection-oriented service, usually on top of IP, between pairs of processes running on (usually) different hosts. TCP performs stream data transfers and achieves multiplexing by allowing several pairs of application processes on the same hosts to communicate at the same time over the same network protocol (again, usually IP). TCP further ensures reliability and in-order delivery of data even if the lower protocol does not. Moreover, TCP throttles its transmission speed to adapt to changing network conditions, in an attempt to ensure fair sharing of the communications medium with other flows. The main characteristics of TCP are now summarized in turn.

Reliability. TCP uses a send-acknowledge paradigm to ensure reliable delivery. The sender transmits its intended message to the receiver, which, upon reception, responds with an acknowledgment (ACK) segment. If such an ACK segment is not forthcoming for some time, the sender repeats the transmission. Each data byte transmitted is assigned a sequence number, but since the data are transmitted in blocks, only the sequence number of the first data byte in the segment is sent to the destination (the others can be easily deduced by the segment size). The ACKs sent by the receiver to the sender include the sequence number of the last *contiguous* byte received and help the sender decide if a retransmission should occur, but do not provide complete information on what segments have been received by the destination (for instance, which out-of-order segments have been received by the destination cannot be inferred from an ACK). TCP further includes a checksum error-detection mechanism which ensures that an arriving segment has not been corrupted in transit.

In-order Delivery. For efficient operation, TCP may transmit several discreet segments at one time, before receiving an ACK from the receiver. These may be reordered in the network or even duplicated. TCP uses the sequence numbers of the segments to rearrange them when they arrive out of order, as well as to discard duplicates. So, an application process using TCP always receives the data in order and without repetition. Note, that if a segment is not delivered in-order, the receiver responds to the sender with a *duplicate ACK*—that is, an ACK which contains the same sequence number as the ACK response to the segment delivery prior to the out-of-order one.

Flow Control. TCP implements flow control, allowing the destination to process data at its preferred pace. When responding with an ACK to the sender, the destination indicates the number of bytes it can receive/process beyond the last received segment, without overflowing its buffers. Then, the sender sends only as much data to the receiver as the receiver has indicated it is capable of processing so as not to overwhelm the destination. Such a mechanism, is called a *window* mechanism, and it is coupled with the congestion control function of TCP [19].

Congestion Control. As DATA and ACK segments are exchanged between sender and receiver, TCP records round-trip time information and calculates the time it should wait for an ACK response before it considers the sent segment to have been lost in transit. The maximum waiting time or *retransmission timeout* (RTO) is dynamically calculated, because network conditions (such as delay) can vary in an IP network. TCP also implements congestion control, meaning it tries to prevent a sender from overflowing the network with segments or utilising network resources at the expense of other TCP flows sharing the medium. Detailed discussion of the mechanisms for congestion control is included in Allman et al. [19], but for the sake of completeness we will briefly outline their principal operations here.

TCP will generally start by transmitting one or two segments and keep increasing its sending rate exponentially during this *Slowstart* phase. Then, at some appropriate point, it switches to a linear increase in the sending rate, during the *Congestion Avoidance* phase. If a *congestion indication event* occurs at any time, then the sending rate is reduced. A congestion indication event might be duplicate ACK reception or an RTO. When a certain number of duplicate ACKs are received, TCP performs a retransmission of what appears to be the missing segment, without waiting for an RTO. This is the *fast retransmit* mechanism and is followed by *fast recovery* where TCP reenters the congestion avoidance phase after reducing its sending rate by half. If the congestion event is an RTO, then TCP starts retransmission in a *Slowstart* fashion and at an initial rate of one or two segments.

As implicitly stated above, TCP injects several segments along the path at any one time (these are sometimes referred to as *in-flight*). At the same time, TCP throttles the rate of injection of new segments according to network conditions. The amount of segments that TCP maintains in-flight at any one time is controlled by TCP's *congestion window*, which is part of its congestion avoidance and control facilities.

Note that the TCP protocol has evolved significantly through its lifetime into a number of variants that differ mostly in the way they implement congestion avoidance. Some variants rely on waiting for packet loss as a congestion indication (TCP Reno, NewReno, SACK [20, 21]), while others try to avoid packet loss altogether through throttling the transmission rate by intelligently taking into account round-trip time and other available information (TCP Vegas, Westwood [22, 23]). For brevity, the outline presented here is a generic one. Some further familiarity with TCP and its operation is assumed for the rest of this survey, and readers who need additional information are urged to refer to other sources [18, 19, 24].

10.3 TCP OVER MANETs

TCP was originally designed to facilitate a reliable, connection-oriented service over IP networks. The bulk of its application scope consisted of deployment in wired networks, and a number of the design decisions made at its inception reflect that fact [19]. In a mobile, multihop, wireless environment, such as a MANET, a number of the assumptions inherent in the design of TCP are inapplicable or false, which leads to suboptimal performance in that setting. This section outlines the causes of this discrepancy and offers some insight in concert with recent research findings in the literature.

10.3.1 Congestion Avoidance and Packet Loss

As mentioned in Section 10.2.2, TCP reduces its sending rate to facilitate congestion control when a *congestion indication event* occurs (i.e., an RTO expiration is triggered or a number of duplicate ACKs are received). Essentially, congestion indication events denote a reasonable degree of certainty that segment loss has occurred in the network. In wired networks, such losses are mostly the result of overflowing buffers in the routers along the segments' path; hence, to ensure that the network is not overwhelmed, TCP reduces its sending rate. This allows the routers to process and forward the backlog of segments in their buffers so that subsequent segment arrivals may enter the router's queue rather than be rejected [10].

However, in MANETs, segment loss is not predominantly the result of buffer overflows especially over long paths [12, 25]. Furthermore, suboptimal performance can occur when TCP misinterprets segment loss as a sign of congestion, since reducing the sending rate may not be the optimal course of action in every such instance. This section aims to outline the causes of additional segment loss in MANETs which highlight the importance and goals of ACK-thinning techniques.

Mobility-Induced Segment Loss. As nodes in a MANET move freely, there may be frequent topology changes and links may be relatively short-lived compared to their hardwired counterparts. As such, a source node's transmission attempt may fail as the destination moves out of the source's transmission radius. Such failures may be accompanied by several consequent retransmission attempts at the link layer level before the node gives up altogether and declares the link unusable (as is the case with IEEE 802.11 transceivers [4]). It is then the responsibility of the routing protocol to reroute the segment from the point of link failure, in an attempt to either (a) salvage it or (6) discard it and reroute the subsequent retransmission all the way from the source.

From the end-to-end perspective of TCP, mobility-induced losses should not necessarily be met with a reduction of the transmission rate at the source when there is sufficient confidence that the new routing path is not in fact congested. However, in light of absence of feedback from the routing protocol on a new route's traffic condition, TCP should gently inject packets (i.e., start in the Slowstart phase), as it attempts to "probe" the new path's capacity. In actual implementations, when a segment encounters a break in its path, it is possible that the routing protocol will not find an alternate route in time for TCP's retransmission (after the expiration of the RTO) and so the retransmitted segment may be discarded or cached while a new route is sought out. This may lead to consecutive RTOs and long periods of inactivity which are detrimental to TCP throughput [26]. The correct course of action in this case would be for TCP to resume transmission from the Slowstart phase as soon as a new route is discovered, which, however, presupposes some sort of feedback from the routing mechanism to the TCP agent. As a consequence of node mobility, link failures causing consecutive RTOs negatively impact TCP throughput greatly. A detailed discussion of this phenomenon is included in Papanastasiou et al. [27].

Spatial Contention-Induced Segment Loss. Previous research has revealed that the way in which wireless transmissions reserve the area around the transmitter using an omnidirectional antenna leads to interference among transmitting nodes. In particular, while a node is transmitting, if another transmission occurs, the signals superimpose and *interfere* with one another, leading to an undecipherable superposition at the intended destinations. It is the responsibility of the MAC protocol, to coordinate transmissions in such a manner so as to maximize the number of concurrent transmissions that do not interfere with one another. The number of such transmissions is referred to as *spatial reuse* [12, 25] and is a measure of optimality for the shared medium.

A decentralized MAC protocol, such as the one used in IEEE 802.11 ad hoc networks [4], may fail to achieve an optimal level of concurrency (i.e., optimal spatial reuse) and may not prevent collisions for transmitted segments. Such collisions are considered to contribute to the *spatial contention* along a path [13, 25, 28, 29]. Further note that this phenomenon is rare in wired networks because there is typically little interference between physical cables. Instead, in a wired setting the main cause of packet loss would be *buffer contention* in the forwarding nodes.

When a packet is lost due to spatial contention, TCP reduces its sending rate (considering packet loss as a congestion indication event). However, this does not help alleviate spatial contention and instead can lead to repeated RTOs and unfairness in sharing the link's bandwidth [13, 28, 29]. If TCP were to be able to identify the cause of the packet loss as spatial contention, it would not need to reduce the sending rate; however, such a distinction would normally require some sort of feedback from the link layer [25].

Error-Induced Segment Loss. Wireless transmissions are, in general, more error-prone than their wired counterparts [30]. As such, the error checks at various layers in the networking stack note erroneous receptions more often in a wireless than in a wired context, which leads to the rejection of the misreceived segment. Again, such segment losses are considered by TCP as congestion indication events even though they may simply reflect variability in the transmission conditions. Intuitively, if the error conditions persist, TCP throughput suffers. In principle, it is possible for TCP to avoid reducing its sending rate if there is feedback from the link layer on the cause of the segment loss. However, the effectiveness of such feedback may be questionable under realistic wireless conditions [31].

10.3.2 Understanding the Effects of Segment Reduction

The previous section summarized causes of segment loss in multihop wireless communications, placing special emphasis on the fact that for noncongestion-related packet losses, TCP's reduction in sending rate negatively impacts throughput. Although there has been significant research interest on TCP's interaction with packet losses due to mobility and errors, as mentioned in Section 10.3.1, such losses are not a primary concern when examining ACK-thinning techniques. This holds true because employing ACK-thinning does not affect TCP's reaction to drops of segments due to errors or mobility; these may be handled separately and orthogonally to ACK-thinning [32].

Instead, the reduction in throughput as a result of spatial contention-related losses is the main motivation for the techniques presented in the next section (Section 10.4). In broad terms, spatial contention is caused by (a) the competition of DATA segments for transmission time as they travel through the multihop path and (b) the inability of the MAC mechanism to coordinate properly those transmissions. It has been shown in previous research [25] that, as the path length increases, segment loss due to spatial contention is higher than that from buffer contention. In the case of 802.11 transceivers, it has also been demonstrated that the fewer the number of segments that are present in the pipe (forwarding path), the more effectively the MAC mechanism can coordinate their transmissions [13]. It is thus desirable to minimize, if possible, the number of segments along the pipe, in order to reduce transmission competition among the segments and hence the spatial contention.

From an end-to-end perspective in general and a TCP viewpoint in particular, there are two types of segment: DATA and ACKs, injected into the network by the sender and the receiver, respectively. As stated above, to reduce spatial contention, it is desirable to reduce the number of segments in the communications pipe at any one time. In order to achieve this, a reduction in DATA segments would entail one of the following: (a) a slowdown of the sending rate [33], (b) an upper limit on the maximum outstanding number of segments at any one time [12], or (c) a reduction of the maximum segment size allowable [34]. These approaches have been examined in the literature and have been shown to be beneficial in terms of increasing TCP throughput. From the receiver's perspective, spatial contention may be reduced by introducing fewer ACK segments, say by taking advantage of their cumulative property. For instance, the destination may send a single ACK for segment s_i implying that s_{i-1}, s_{i-2}, and so on, have also been successfully received, without sending separate ACKs for each. Such techniques aiming to reduce spatial contention caused by ACKs are named *ACK-thinning* or *ACK-reducing* and are the point of focus for this survey.

10.4 ACK-THINNING TECHNIQUES

10.4.1 Introduction

As discussed in Section 10.3.2, ACK-thinning or ACK-reducing techniques aim to alleviate the effects of spatial contention caused by competition of DATA and ACK TCP segments along a multihop wireless path. When overviewing such techniques, however, each proposed implementation has to be judged on more than its performance merits; it should be practically applicable and realistically implementable, especially if the proposed changes concern the MAC layer, which may in turn imply changes in the firmware of the wireless transceiver (which makes deployment a more involving affair than, say, end-to-end alterations). Furthermore, the limited processing capabilities and power reserves of the mobile node have to be taken into account, which precludes considering mechanisms of significant time and space complexity.

Each ACK-thinning technique outlined below includes an overview of its principles of operation and explicit discussion on its implementation requirements. The coarse, initial division of the proposals is between *end-to-end* and *hybrid* approaches. The former implies that changes need only be made at the sender and/or receiver of the communications path, without altering the forwarding nodes in between. The latter indicates that alterations are required on a point-to-point basis, which is in every intermediate node in the communications path, and, possibly, at the endpoints as well. For each method a further distinction is made between methods that require feedback and interaction amongst the layers of the protocol stack (named *cross-layer* approaches) and those that do not (called *single-layer* approaches). Further insight on the categorization is included in Section 10.5.

10.4.2 End-to-End Techniques

End-to-end modifications imply changes to the TCP source and destination alone, although the communications path may span several nodes which, acting as "stepping stones," forward segments to their eventual destination (the TCP receiver).

Delayed ACKs. The delayed acknowledgments (DACKs) mechanism is an oft-enabled feature of TCP [35], as first described Clark [36]. Its principal operation is simple and relies on the cumulative nature of TCP acknowledgments; instead of immediately replying with an ACK upon receiving a DATA segment, the TCP receiver waits a short time (usually 100–500 ms [11]). If subsequent DATA segments arrive, then it is possible to inject a single ACK into the pipe, which cumulatively verifies the receipt of both DATA segments. Furthermore, since TCP connections are duplex, it is also possible to piggyback the ACK to DATA segments being sent in the other direction—that is, from the "receiver" to the "sender," thus saving bandwidth. To avoid confusing TCP estimates, such as the round trip-time (RTT), and TCP's ACK-clocking mechanism, the relevant RFC [37] dictates that ACKs should not be delayed in any case for more than a single (extra) DATA segment, or for a time period of more than 500 ms. Figure 10.3 illustrates the use of dupACKs in the case of a sender with a congestion window of two segments.

It has been shown in the literature that delayed ACKs are beneficial for TCP throughput and should be enabled by default [11]. Subsequent research in wired networks has investigated the possibility of increasing the delay response for TCP so as to alleviate the competition of ACKs for bandwidth space along with TCP DATA. However, it has been demonstrated that this might result in "burstiness" in the transmission pattern or has other shortcomings, especially with regard to wide deployment on the Internet [38]. In MANETs, the throughput enhancing property of delayed ACK has been demonstrated repeatedly in static and dynamic topologies, with reactive [34] and proactive [39] TCP agents. In special cases the improvement in TCP throughput can be up to 15–32% [40]. Recent work [41] has also analytically verified the throughput enhancement effect.

Adaptive Delayed ACK (ADA). TCP with Adaptive Delayed Acknowledgment (TCP-ADA) for MANETs is a receiver-only modification to Reno-based variants introduced by Singh and co-workers [5]. The method's goal is conceptually simple; it aims to reduce the number of produced ACKs to one per full congestion window, as opposed to one per segment received. The subsequent reduction in the number of produced ACKs alleviates some of the spatial contention caused by ACKs and DATA competing for transmission time on the same medium. The main algorithm is presented in Algorithm 1.

The authors in reference 5 first analytically demonstrate that the throughput of TCP connections may be improved by increasing the number of data packets required before TCP responds with an ACK. In short, the TCP ADA scheme maintains a running average of packet interarrival times (line 1 in Algorithm 1), which "arms" an

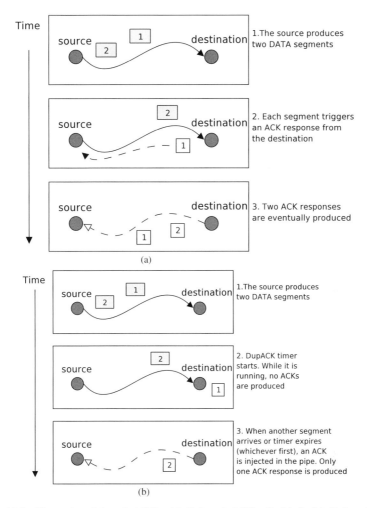

Figure 10.3. Illustrating delayed ACKs. (a) Delayed ACKs disabled. (b) Delayed ACKs enabled.

ACK expiration timer (line 1). The receiver may keep on processing incoming DATA packets without producing an ACK response for as long as a *MaxDeferTime* period, provided that additional DATA segments arrive within a $\beta * avgPktInt$ time frame. If either the MaxDeferTime or the ACK Timer expire, then a cumulative ACK is produced and sent to the sender. Note that no limit is enforced on the number of ACK packets to delay as in the case of delayed ACKs (Section 10.4.2). Experimentation on static string topologies in reference 5 has revealed a 5–25% throughput increase when this method is employed on a Reno (presumably) agent when compared to simply deploying delayed ACKs.

However, although the changes proposed in TCP-ADA are end-to-end and confined to a single layer, objections may be raised on its feasibility as a general solution for

ALGORITHM 1. TCP-ADA Principal Operation

1: $lastPktArr \leftarrow 0, avgPktInt \leftarrow 0$
2: $\alpha = 0.8, \beta = 1.2$
3: $MaxDeferTime = 500ms$
4: **if** $lastPktArr = 0$ **then**
5:    $lastPktArr \leftarrow now$
6: **else**
7:    $avgpktArr \leftarrow \alpha * avgPktInt + (1 - \alpha) * (now - lastPktArr)$
8:    $lastPktArr \leftarrow now$
9: **end if**
10: **if** MaxDeferTimer not pending **then**
11:    Schedule MaxDeferTimer for MaxDeferTime
12: **end if**
13: Reschedule ACK timer for $\beta * avgPktInt$

MANETs. Specifically, the performance advantages offered by the technique have not been confirmed in dynamic MANETs, which may generate much different results to those exhibited in static topologies [7]. There is also little discussion with respect to the choice of the α and β parameters, which define the weighing constants for the average interarrival ACK time and the wait factor for the ACK timer, respectively. Finally, the issue of overhead in terms of processing capacity of implementing a fine-grained ACK response timer per TCP connection is not addressed in the original work. Overall, the method can be considered a variation on the theme of dynamically adaptive ACKs, as described in more detail in Section 10.4.2.

Dynamic Delayed ACK (DDA). As with several of the techniques described in this section, the authors in reference 6 have also explored the possibility of utilizing the cumulative nature of TCP ACKs in order to reduce the number of acknowledgments sent to inform the sender of successful DATA exchanges. This ACK reduction aims to reduce spatial contention and mitigate packet drops as fewer packets contend for medium access time. The main idea of dynamic delayed ACKs (DDA) expands on the delayed ACK mechanism as defined in [37] by introducing the possibility of producing delayed ACKs for more than two received segments.

In particular, through experimentation in static string topologies, it is first established in reference 6 that a longer delay period for the production of ACKs is beneficial especially as the path length increases. Furthermore, it is noted that delaying several ACKs is really only beneficial if there are enough DATA segments in the pipe for the mechanism to help cumulatively acknowledge the incoming segments. For instance, delaying the ACK response until three segments are received is of limited use when the congestion window at the receiver is sized at two segments; in this case the delayed ACK would only be triggered (launched) at the expiration of the ACK timer, because the third segment would never arrive (only two segments were injected into the pipe

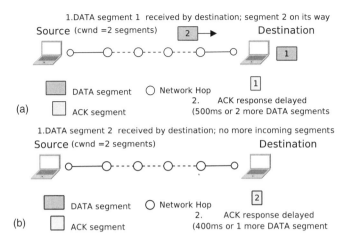

Figure 10.4. Dynamic delayed ACKs. (a) First ACK delayed. (b) Unnecessary second ACK delay for 400ms

by the sender). Such a delay in ACK transmission could be as high as approximately 500 ms (the maximum recommended time length of the expiration timer [37]). This undesirable situation is depicted in Figure 10.4.

To facilitate the delay in ACK production in a dynamic way, Altman and Jiménez [6] identify that the above two factors, namely path length and packets-in-flight, cannot be known at the receiving end without utilizing extra signaling. Hence, the technique is applied in a generally "useful" fashion, which means that it is designed to provide modest returns in the case of short paths but reasonable gains in throughput otherwise. Specifically, a delay coefficient parameter, d, is introduced to the receiving end of a TCP connection where it defines the number of ACKs the receiver should delay before responding to the reception of DATA segments. Hence, if $d = 2$ the scheme becomes equivalent to the standard delayed ACKs mechanism described in Section 10.4.2. Then, three thresholds are defined, l_1, l_2, l_3 according to the sequence number N of the received DATA segment.[1] For $N < l_1, d = 1$, for $l_1 \leq N < l_2, d = 2$, for $l_2 \leq N < l_3, d = 3$ and for $l_3 \leq N, d = 4$. The set of threshold values used in the subsequent evaluation is $\{l_1, l_2, l_3\} = \{2, 5, 9\}$. The principal notion is to increase the delay coefficient as more and more segments appear in order. It is hoped that during that time period the number of packets-in-flight will gradually increase and the increase in the delay coefficient will, thus, have a meaningful and positive impact. The algorithm is outlined in Algorithm 2.

---

[1] The initial sequence number (ISN) for each endpoint is negotiated by TCP during the three-way handshake upon connection establishment and is 32-bits wide [18]. In this context, the sequence number is assumed to be 0 initially without loss of generality (the l thresholds may be presented as full-sized segment offsets, in bytes, to the initial negotiated ISN as appropriate).

ALGORITHM 2. Estimating Number of ACKs to Delay (d) for DDA

1: $d \leftarrow 1$
2: $l_1 = 2, l_2 = 5, l_3 = 9$
3: $N \leftarrow ISN(0)$
4: **if** $N < l_1$ **then**
5:     $d \leftarrow 1$
6: **else if** $l_1 \leq N < l_2$ **then**
7:     $d \leftarrow 2$
8: **else if** $l_2 \leq N < l_3$ **then**
9:     $d \leftarrow 3$
10: **else if** $l_3 \leq N$ **then**
11:     $d \leftarrow 4$
12: **end if**

Subsequent evaluation of the technique in Altman and Jiménez [6] has revealed dramatic improvements in reducing end-to-end delay (18–22%), minimizing packet losses (up to 30%) and increasing throughput (up to 11%) when compared to a typical NewReno TCP agent utilizing delayed ACKs in long string topologies. However, the merits of the technique over short paths or in multiple flow environments remain unexplored. Furthermore, evaluation in dynamic or more diverse static topologies is also a subject for further study.

Dynamic Adaptive ACK (TCP-DAA). The Dynamic Adaptive Acknowledgment (DAA) method is a sophisticated sender/receiver modification introduced by de Oliveira and Braun in [7]. It aims to reduce the number of ACKs produced at the receiver by taking advantage of the cumulative property of ACKs, not unlike the DDA technique presented in Section 10.4.2. In contrast to DDA, however, DAA dictates changes to the sender as well as the receiver and the delay of ACK responses is performed in a dynamic manner so as to adapt to changing network conditions. There is some processing overhead associated with this method, but the tradeoff is a general increase in throughput and better utilization of the wireless channel.

The main operation of DDA is illustrated in Figure 10.5. The sender restricts its congestion window (*cwnd*) to 2...4 segments (that is, can have 2, 3, or 4 segments outstanding at any one time). The receiver maintains a dynamic delaying window (*dwin*) with size ranging from 2 to 4 full-sized segments, which determines when an ACK will be produced. Whenever a DATA segment is received, an *ack_count* variable increases by one until it reaches the current value of *dwin*. When a DATA packet is received, say, $i, i + 1, i + 2, \ldots$, for which an ACK is to be delayed, its interarrival gap with the previous DATA reception is recorded, say $\delta_i, \delta_{i+1}, \delta_{i+2}, \ldots$). Effectively, for each ACK delay *epoch*, the interarrival times of incoming DATA segments are noted. The *ack_count* variable differentiates between these *epochs*—that is, the group of DATA segments that will have their ACKs delayed. As mentioned above, whenever

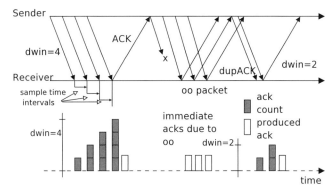

Figure 10.5. Demonstration of dynamic adaptive ACKs.

a DATA segment is received, *ack_count* increases by one. When *ack_count* = *dwin*, then an ACK response is immediately produced and *ack_count* is reset to one segment (signifying the beginning of the next *epoch*). The process is outlined in Algorithm 3.

The interarrival periods collected are used to calculate a smoothed average that signifies an "expected" interarrival time, say $\overline{\delta}_{i+1}$, for consecutive ACK segments. This average is calculated using a low-pass filter and is further used to calculate a timeout interval for the ACK response. If $\overline{\delta}_i$ is last average calculated, δ_{i+1} is the

---

ALGORITHM 3. TCP-DAA Principal Operation

1: $ack\_count \leftarrow 0, dwin \leftarrow 2$
2: **if** consecutive DATA segment received **then**
3:    **if** $ack\_count > 0$ **then**
4:       record inter-arrival time
5:    **end if**
6:    $ack\_count \leftarrow ack\_count + 1$
7:    **if** $ack\_count = dwin$ **then**
8:       produce ACK
9:       $ack\_count \leftarrow 0$
10:   **end if**
11:   **if** $dwin < 4$ **then**
12:      $dwin \leftarrow dwin + 1$
13:   **end if**
14: **else**
15:    produce ACK
16:    $ack\_count \leftarrow 0$
17:    $dwin \leftarrow 2$
18: **end if**

DATA segment interarrival time sampled, and α is an interarrival smoothing factor, with $0 < \alpha < 1$, then we obtain

$$\overline{\delta}_{i+1} = \alpha * \overline{\delta}_i + (1 - \alpha) * \delta_{i+1} \tag{10.1}$$

In order to prevent erroneous ACK responses due to delay variation, the receiver waits for a time before returning an ACK response if $dwin < ack\_count$. As per RFC 2581 [19], in the case of out-of-order segments an ACK response is immediately prompted, but otherwise the receiver waits for a time period T_i before responding with an ACK. This effective timeout interval is calculated with a timeout tolerance factor, $\kappa, \kappa > 0$ as follows, where $\overline{\delta}_i$ is calculated by Eq. (10.1):

$$T_i = (2 + \kappa) * \overline{\delta}_i \tag{10.2}$$

There is also a mechanism in the DAA method to account for variable, instead of fixed, increases in the dynamic window. The speed increase factor, μ, with $0 < \mu < 1$. If the *maxdwin* is a status indicator that turns true when the maximum possible value for the dynamic ACK window has been reached (by default set to 4 segments), then the dynamic ACK window's growth is set to

$$dwin = \begin{cases} dwin + \mu & \text{if } maxdwin\text{=false,} \\ dwin + 1 & \text{otherwise} \end{cases} \tag{10.3}$$

Equation (10.3) allows the receiver to respond immediately with ACKs in the case when the TCP sender is in the Slowstart phase and each ACK increases its congestion window by a single segment. If ACKs were delayed during this phase the sender would not receive enough ACKs to increase its sending rate effectively. Essentially, the *maxdwin* parameter signifies (at the receiver) when the Slowstart phase (at the sender) is over. Once the maxdwin is reached once, then this mechanism is not activated again for the same connection. Hence, this facility is intended for short file transfers.

The DAA method has been evaluated on chain and grid topologies of varying length and different number of flows [7]. On chain topologies of up to 8 hops and 20 flows, the method increases TCP NewReno throughput up to 50% over its regular version and 30% over the DDA method described in Section 10.4.2. In particular, it is noted that as the number of concurrent flows increases, the DAA method becomes increasingly effective. On grid topologies, taking into account 3 and 6 cross traffic sources, the performance improvement is of the same magnitude, but the difference diminishes against optimized (with respect to maximum *congestion window* size) Vegas and SACK agents. The method does not lead to a discernible performance advantage in the case of short-lived flows.

From an implementation perspective the DAA method is an end-to-end sender/receiver modification. The sender needs to be tweaked with respect to initial *cwnd* size, and the receiver needs to implement the ACK dynamic window.

However, no modifications to other layers are required, and no cross-layer feedback is assumed.

Although the performance advantages of DAA are compelling, it should be noted that the method introduces new state variables and processing at the receiver, which increases its computational requirements. Furthermore, it is not clear how the method would interact with unmodified agents—that is, if a DAA receiver would result in some penalty against a nonoptimized sender. Future enquiry might also consider the effect of each of the individual modifications proposed in the DAA scheme on throughput; that is, distinguish which of the new mechanisms is responsible for the majority of the performance improvement.

Active Delay Control (TCP-ADC). Active Delay Control (ADC) is a TCP extension introduced in Hsiao et al. [42], which imposes delays on the transmission of packets at the endpoints, with a view to avoiding timeouts (RTOs). As such, it is primarily useful for applications that cannot tolerate long periods of transmission inactivity, such as video streaming. The proposed solution is aimed at wired networks, but its general principles of operation still apply to MANET environments, albeit with certain caveats, mentioned at the end of this section.

Hsiao et al. [23] have investigated implementing delay at either the sender or the receiver. In particular, two receiver-based methods are proposed: the "Exact Receiver-Based Delay Control (RDC)" and the "1-bit RDC." Both aim to help TCP avoid timeouts by simulating the effects of growth in the router's buffer (i.e., increasing packet delay). Essentially, the receiver witholds generating an ACK after a successful segment reception so as to "slow down" the sending rate increase at the sender and avoid synchronised losses at the routers.

The difference between the two schemes lies in how the delay is computed. Under the Exact RDC scheme, as depicted in Figure 10.6a, routers are responsible for computing the delays based on their congestion levels, which are then appended to the segment as delay notification. It is this notification which finally determines the degree of delay, a receiver will impose on a segment. In 1-bit RDC as shown in Figure 10.6b, the receiver approximates the delay according to the Explicit Congestion Notification (ECN) congestion signal [43] received by the routers, which presupposes the existence of an active queue management mechanism (AQM) such as RED [44]. The receiver then estimates the amount of time to delay each segment. This delay may increase if a congestion signal is received, or it may decrease otherwise.

In Kung et al. [9] a One-Bit Sender-Based Delay Control (SDC) is also proposed which works similarly to the 1-bit RDC method, differing only that the delay is implemented at the sender rather than the receiver. However, since the sender is privy to more information than the receiver—namely, round-trip time (RTT) and congestion window (*cwnd*) size—it can more accurately adjust the delay to approximate traditional TCP.

The performance evaluation of SDC over a bottleneck topology with dense, long-lived traffic (50 TCP flows) is used in Hsiao et al. [42] as a means of demonstrating its performance merits. In particular, it has been shown that for TCP NewReno agents,

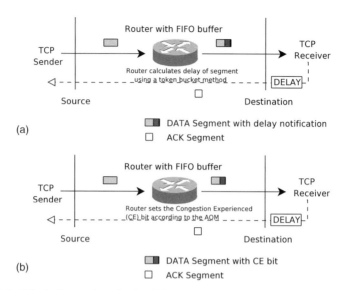

Figure 10.6. Principal operation of active delay control. (a) Exact receiver-based delay control. (b) One-bit sender-based delay control.

timeouts are reduced significantly (~80%) when the queue size at the routers is small. Furthermore, the SDC method is also deemed beneficial for short-lived TCP flows which, on average, have shorter transfer times for small queue sizes, compared with traditional TCP flows.

Although all the ADC methods are based on the same principle, the techniques differ in their implementation requirements. Specifically, Exact RDC requires point-to-point alterations, while 1-bit RDC and SDC are end-to-end in nature. Intuitively, since the Exact RDC technique benefits from feedback from the routers along the path, it has the most up-to-date information of all the variants and does not need to make coarse-granularity guesses as to the amount of delay for each segment. To achieve this effect, intra(single)-layer communication between TCP and the routing layer is required for the Exact and 1-bit RDC variants.

Notably, it is not clear if the ADC techniques would function as intended in a MANET setting. As previous research has suggested [25], it is spatial contention rather than overfull buffers that are the leading cause of packet loss in even static MANETs. However, spatial contention is somewhat alleviated by the ACK-thinning properties of the ADC techniques, because ACKs are delayed at the receiver. An interesting point of future research would be to reveal whether the ACK reducing philosophy of ADC is also useful in a MANET environment.

10.4.3 Hybrid Techniques

Hybrid modifications include proposals that involve changes at entities other than the endpoints of the connection. Such modifications may be network-wide, or confined to a few nodes in the network, and can span several layers in the networking stack. When the changes involve every node in the communications path, they are termed *point-to-point*. This section surveys such approaches.

ACK Bundling with DATA. For a given TCP connection, combining the acknowledgment with a segment carrying data can reduce the number of segments in the communications pipe, and indeed this is a goal of the original TCP RFC [10]. In particular, it is stipulated that an ACK segment (which is really an empty DATA segment with the acknowledgment number field filled-in) may be piggybacked on a DATA segment if one is forthcoming. So for an $A \leftrightarrow B$ communicating pair, an $A \to B$ DATA segment may elicit a $B \to A$ ACK response piggybacked (embedded) on a $B \to A$ DATA segment if B happens to have a DATA segment to transmit to A at an appropriate time. Evidently, in such an occasion, a single *combined* (ACK and DATA) transmission is used instead of two *discrete* ones, which results in savings in transmission time (bandwidth). Notably, this technique is usable per individual TCP connection; that is, an ACK for connection $A_1 \leftrightarrow B_1$ (connection 1 between hosts A and B) cannot be piggybacked on a DATA segment for connection $A_2 \leftrightarrow B_2$. Yuki et al. [14] have extended this scheme and have proposed an IP-level mechanism that achieves the merging of DATA and ACK TCP segment transmissions regardless of whether they are part of the same connection. In doing so, the number of *discrete* ACK transmissions may be reduced to an even greater degree.

The ad hoc network configuration considered in Yuki et al. [14] is the Flexible Radio Network (FRN), presented in detail in Sugano et al. [45]. Although the details of the setup of such a network are not within the scope of this survey, a brief overview follows; interested readers may refer to Sugano et al. [45] for more information. In short, an FRN configuration involves a spontaneous network setup with an emphasis on the automatic initialization of nodes joining in, rather than with the effects of node mobility; hence, the scheme is targeted at primarily *static* ad hoc networks. Routes are discovered and maintained using a table-driven approach similar to DSDV [46] or OLSR [47], featuring periodic communication between nodes to update their routing information. For each potential destination in the network, multiple routes are recorded. If a route has the minimum hop count to a destination, say n hops, it is recorded as a *forward* route. If its hop count is $n + 1$, it is labeled as a *sideward* route, while if the hop count is larger or equal to $n + 2$, the route is considered a *backward* route. An example of this classification is illustrated in Figure 10.7. Intuitively, the idea is to try transmitting along the forward and sideward routes, in that order, before resorting to transmitting to a backward node.

As mentioned previously, TCP ACK and DATA segments are combined, wherever possible, so as to be sent at the same transmission slot. Note that for the scheme described bellow to function, nodes are assumed to be equipped with omnidirectional antennas of equal transmission range. Every node maintains an ACK and DATA queue, as shown in Figure 10.8a,b for node C, which forwards segments for TCP connections between nodes A and E. If a segment is contained only in either of the queues, then transmission proceeds as normal. However, if there are segments in both the ACK and DATA queues, then a combined transmission occurs, whereas the DATA and ACK

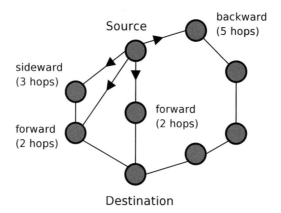

Figure 10.7. Classification of routes in an FRN setup.

segments are sent as part of the same transmission. If the node receiving this combined segment is not part of the route for one of the subsegments, it discards the subsegment not addressed to it (recall that nodes in FRN maintain full knowledge of all possible routes between nodes and, as such, know if they are part of a segment's route toward any destination). In the example considered in Figure 10.8, node C transmits the combined DATA and ACK segment to node D, as depicted in Figure 10.8c. Node D then discards the ACK segment because it is not part of the segment's route toward its destination (node E). Node B overhears the transmission of node C and realizes that it is part the ACK's route toward its destination (node A). Therefore, node B discards the DATA part of the combined segment and forwards the ACK toward node A, as shown in Figure 10.8d. Node D also continues its transmission of the DATA segment toward its destination (node E) as illustrated in Figure 10.8(e).

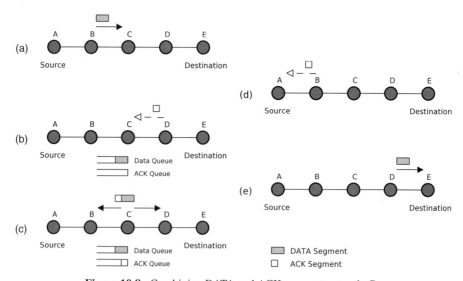

Figure 10.8. Combining DATA and ACK segments at node C.

As a result of the above modifications, the authors note a 20–60% throughput improvement with the combined segment optimization in a 5-node chain topology using TCP Reno over an FRN. The improvement in throughput is 15–20% in a 10-node cross-chain topology and about 10% for 9 connections in a 19-node mesh topology.

The modifications required to implement the combined segment approach are network-wide: Since the routing protocol is required to make forwarding decisions for each combined segment and separate the segments if required, it is necessary to add this functionality in a point-to-point fashion (i.e., for each node). Furthermore, incremental deployment of the changes is not possible because each forwarding node must know how to interpret the combined segment in order to distinguish the ACK and DATA segments in it. Although the proposed technique is not end-to-end, it is worth noting that the transport agent itself need not be modified in any way, because the routing agent can handle the dissection itself and merge and forward the combined segment.

A potential problem that may hamper the deployment of such a method is the requirement for omnidirectional antennas of equal transmission radius. This latter constraint is imperative as the technique is based on what is essentially a single broadcast of the combined segment in order for it to be forwarded by the appropriate node. Avenues of future research may include implementation in other ad hoc settings where mobility is taken into consideration as is the case with the AODV [48] and OLSR [47] protocols over IEEE 802.11 transceivers or otherwise.

Different Routes for ACK and TCP Data. The Contention-Based Path Selection (COPAS) [49] is a routing algorithm mechanism that does not aim to reduce the number of ACKs produced at the receiver and as such is not, by a strict definition, an ACK-thinning technique. However, the technique is aimed to alleviate the effects of DATA and ACK segment-induced spatial contention, by reducing the competition between DATA and ACK segments along a TCP connection. Although it does not decrease the number of acknowledgments produced, the aims of the technique are the same as those of the other ACK-reducing paradigms surveyed in this study, which justifies COPAS' inclusion here.

To achieve its goal, COPAS collects information on how "busy" the medium is around each node. It then dictates different paths for forward and reverse traffic. As suggested in Cordeiro et al. [49], the technique can be applied to any protocol making use of Route Requests (RREQs) and Route Replies (RREPs), so that multiple routes to a given destination may be probed.

The method operates as follows: Unlike other proactive routing protocols (such as AODV) that are designed to obtain only one path during route discovery, COPAS dictates the collection of multiple paths at the destination by collecting more than one RREPs. Then, two paths are chosen (at the destination node) on the criteria of disjointness and minimal route congestion, as will be explained later on. Eventually, the destination replies with two Route Reply messages to the source, which then obtains enough information to route ACK and DATA segments along separate paths.

Route disjointness can be measured because each Route Request is required to carry a complete description of the route traveled in its header until the destination

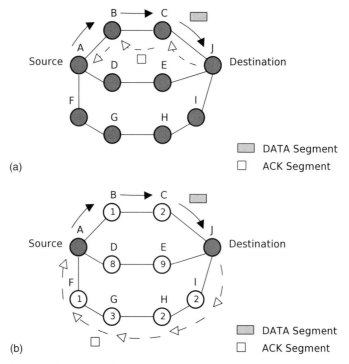

Figure 10.9. Example of COPAS routing path selection. (a) ACK and DATA segments use the same path in DSR. (b) ACK and DATA segments use disjoint paths in COPAS while also considering network contention.

is reached; this requirement makes COPAS a natural fit to the DSR protocol. To measure congestion, each node monitors its MAC layer behaviour and embeds this in the Route Request packet as it travels to the destination. The metric used here is a weighted average of MAC backoffs due to activity in the medium, where a higher average implies a busier medium.

Figure 10.9 succinctly illustrates the process. For nodes A, J to communicate there are three disjoint routes to consider. The DSR protocol chooses one route (one of the shortest ones), namely $A \to B \to C \to J$ to use for both ACK and DATA segments, as shown in Figure 10.9(a). COPAS, on the other hand, chooses two disjoint routes, namely $A \to B \to C \to J$ and $A \to F \to G \to H \to I \to J$ for DATA and ACK segments, respectively. Note that route $A \to D \to E \to J$ is dropped from consideration as its congestion metric $(8 + 9)$ is higher than the alternatives.

Consequently, using separate paths for ACK and DATA segments helps alleviate spatial contention, because competition for transmission is avoided over the same path. Because congestion may be transient, COPAS keep track of it by piggybacking congestion information on packets along the forward and reverse paths. If contention is noted to rise above a certain threshold on some route, a new, less congested route is sought out. From an implementation perspective, COPAS requires alterations almost solely at the IP layer but also requires cross-layer feedback (from the link layer) to maintain the congestion information at the nodes.

In Cordeiro et al. [49] the method is evaluated over static topologies using DSR. A throughput improvement of 90% using TCP Reno is observed compared to vanilla DSR, but a slight delay penalty is noted for end-to-end delivery as COPAS tends to circumvent the shortest route. As a prospect for future work, COPAS may need be evaluated under conditions of some mobility and be examined in concert with other TCP throughput enhancing techniques, orthogonal to it.

Split TCP. Split TCP [28] is a modification applicable to TCP connections spanning several hops, that is, connections utilising long paths. The main concept is to divide a long end-to-end TCP conversation into multiple, smaller subconversations through the use of proxies. The advantages of such an approach are twofold. Firstly, since it is more likely for long paths to experience link failures due to mobility, shorter connection spans are at an advantage over long ones. This technique aims to alleviate the disadvantage of long end-to-end conversations. Secondly, flows on a long path suffer from the effects of hidden terminal interference and experience unfairness when mixed with shorter flows. The placement of the proxies in the Split TCP scheme is fixed by the inter-proxy distance parameter and determines which nodes will be acting as proxies.

The mechanism of Split TCP is a simple insertion of store-and-forward TCP proxies along the initial end-to-end TCP conversation. Each proxy intercepts TCP packets, buffers them, and acknowledges their receipt to the source (or previous proxy) by using local acknowledgments (LACKs). In turn, when a LACK is received, the transmitting proxy can expunge the packet as its delivery to the next proxy has been confirmed. The regular end-to-end ACK response of TCP is still maintained, however, as means of maintaining TCP reliability. Because the distance between proxy nodes may be significant, proxy nodes maintain flow and congestion control for the exchange of segments between them. Notably, the source node obeys both a local and an end-to-end congestion window. The evolution of the local congestion window depends on the LACKs received by the next proxy segment, while the evolution of the end-to-end congestion window depends, in on the rate of arrival of end-to-end ACKs from the destination. Intuitively, the congestion window is always smaller than the end-to-end window; in particular, Kopparty et al. [28] suggest limiting end-to-end ACKs to approximately one in every 100 packets because the possibility of non-delivery from proxies is low (because there is little distance between them and the end-to-end mechanism of TCP already regulates the traffic rate). Figure 10.10 illustrates the principal operation of Split TCP. Note that the proposed modifications do not require cross-layer information but have to be implemented in a point-to-point fashion—that is, at every node along the communications path.

Simulation results in Kopparty et al. [28] show that Split TCP with an inter-proxy distance between 3 and 5 hops can increase throughput in a range of 5–30% and may also improve fairness. On the downside, in order for the proxies to be able to function, large buffer space may be required. Also note that Split TCP overrides the end-to-end semantics of TCP and requires transport protocol level manipulation of segments at

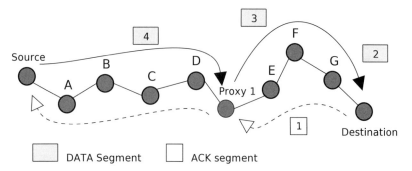

Figure 10.10. The functionality of Split TCP over an 11-hop string topology featuring an inter-proxy distance of 4.

the proxies, which places extra computational demands on these forwarding nodes. This latter requirement may place a strain on the (already limited) power reserves of mobile devices.

Disabling RTS/CTS for Short Segments. In the 802.11 protocol, the exchange of unicast messages is conducted at the link layer, first with an RTS/CTS handshake and then with a DATA-ACK exchange [4]. The Request to Send/Clear to Send (RTS/CTS) exchange is a bilateral point-to-point transmission between successive nodes along a path to "reserve" the space around the nodes, avoiding the "hidden" and "exposed" terminal effects [4]. The DATA/ACK exchange at the link layer is an attempt to ensure reliable delivery of DATA on a point-to-point basis, although there is a limit to such efforts before the link layer gives up and drops the frame [2]; specifically up to four retries for DATA and seven times for RTS frames are attempted.[3]

Figure 10.11 illustrates the principles of the RTS/CTS exchange. In Figure 10.11a the source and destination nodes communicate by a simple DATA/ACK exchange, as denoted by instances 1 and 2. The source node senses the channel to determine if there are ongoing transmissions, in which case it defers its own until such time as there are no more transmissions occurring. Assuming similar transceivers and signal propagation conditions, the source can only be certain that there are no transmissions within its sensing range (the "reserved" area denoted by the shaded section in Figure 10.11a). However, nodes transmitting within the nonreserved area may interfere with the source's signal because they have no means of detecting the source's transmission and thus will not defer in turn.

The RTS/CTS exchange solves this issue as shown in Figure 10.11b. The usual DATA/ACK transmissions in instances 3 and 4, respectively, are preceded by an RTS

---

[2]The term "frame" is employed here when discussing issues related to Link/MAC layer messages as per IEEE practice [4].

[3]The number of retries for segment retransmission are defined by the *dot11ShortRetryLimit* and *dot11LongRetryLimit* variables in the IEEE 802.11 specification.

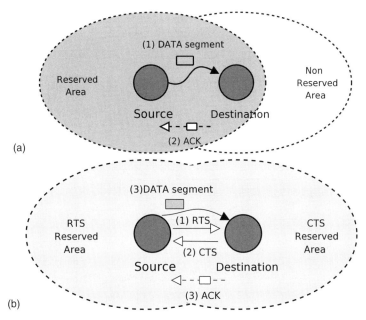

Figure 10.11. Illustrating communications with the plain and RTS/CTS transmissions scheme. (a) Plain DATA/ACK exchange does not "reserve" the whole space around the transmitter and receiver. (b) RTS/CTS followed by a DATA/ACK exchange "reserves" more space around the communicating nodes.

probe from the source which is accompanied by a CTS response from the destination. The RTS frame informs nodes within the *source's* transmission range that a DATA frame will follow as a transmission, while the CTS response informs nodes within the *destination's* transmission radius that they should withhold transmission so as not to interfere with the reception of the DATA frame from the source. The RTS and CTS frames are very small in size so as to reduce the overhead of this preamble to the "real" transmission. If a CTS response is not forthcoming from the destination, the source can then surmise that the destination did not receive its RTS because another transmission near the destination has interfered with its signal. Hence, the source defers its transmission to a later time. Details of this exchange in the context of the 802.11 protocol may be found in reference 4.

However, previous research has revealed that enabling the RTS/CTS exchange may in fact reduce throughput and lead to unfairness if the source and destination nodes are sufficiently far apart [50]. The identifying cause for this effect is the discrepancy between a transceiver's transmission and interference ranges. Readers interested in more details should consult references 50 and 51. It should also be observed that eliminating the RTS/CTS exchange altogether is not an optimal solution because it allows for "hidden" and "exposed" terminal effects to occur, thus leading to frequent frame collisions [4].

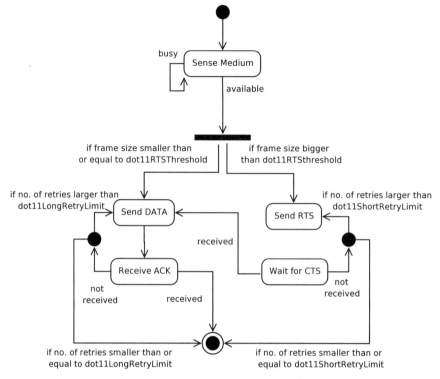

Figure 10.12. State diagram of selective activation of RTS/CTS in 802.11 transceivers.

For clarification purposes, note that the DATA/ACK exchange in the above context referred to MAC layer units and not the TCP DATA and acknowledgment units. Since the following paragraph will need to refer to both concepts, each will have as a subscript the layer it refers to; so, for instance, $DATA_{TCP}$ will refer to a TCP DATA segment and $DATA_{MAC}$ will refer to a MAC layer frame.

Because both $DATA_{TCP}$ and ACK_{TCP} segments make use of the RTS/CTS handshake, there is substantial area reserved for transmission around the source and destination nodes impairing effective spatial reuse (i.e., increasing the number of simultaneous transmissions). Moreover, it is not efficient to disable the RTS/CTS handshake completely (as mentioned above). However, it is possible to disable the handshake when transported frames ($DATA_{MAC}$) are of small size. The rationale is that for small-sized frames, it is not worth incurring the overhead of reserving the whole transmission area around the sender and receiving nodes, but instead these may be transmitted, admittedly with less certainty, by using a simple $DATA_{MAC}/ACK_{MAC}$ exchange. Because the frames are small, it is less likely that they will interfere with or be interfered with by other transmissions.

Provisions for such a mechanism have been included in the 802.11 standard [4] where the *dot11RTSThreshold* variable determines up to which frame size the RTS mechanism will not be in effect. A state diagram of the mechanism guiding the decision on whether to use RTS/CTS is depicted in Figure 10.12. However, most of the ACK-thinning research in MANETs has ignored this mechanism, instead opting to use the RTS/CTS exchange regardless of frame size. In applying the concept to TCP, the threshold variable might be set so as to use RTS/CTS for $DATA_{TCP}$ segments, but disable it for the much shorter ACK_{TCP} segments. Such changes would have to be point-to-point because they need be propagated to every node in the path, but they are confined in the link layer. Further note that this technique is orthogonal to the other ACK-thinning techniques outlined in this section, and studying the way in which it might interact with the latter would be a prospect for future work.

10.5 COMPARISON OF TECHNIQUES

This section discusses the tradeoffs involved in implementing the aforementioned proposed ACK-thinning techniques along the end points of a TCP communicating pair against implementing such alterations at intermediate nodes as well. Furthermore, the option of utilizing cross-layer communication between various layers in the protocol stack is also explored for each ACK-thinning method.

10.5.1 End-to-End Versus Point-to-Point

In most configurations, TCP operates on top of the IP protocol. However, there is no coupling implied or required [37]; TCP as a transport-oriented protocol requires only that the protocol encapsulating it routes segments to the intended destination. Such a clear separation between TCP and IP aids in providing other transport services such as DCCP [52] and UDP [53] over IP, without overburdening IP semantics with TCP-specific requirements.

As such, TCP semantics are end-to-end; that is, the transport agent is relatively oblivious of the circumstances encountered by segments while on route. The information available to the transport agent includes the round-trip time of segments (reliably estimated if the timestamp option is used [54]) and the sending/receiving rate, from the sender and receiver's perspective respectively. With such limited information, the TCP agent has to apply "rules of thumb" to received input and deal with substantial uncertainty. Examples of this may be found in TCP's heuristic for segment loss. Segment loss is not explicitly indicated, but there is a Retransmission Timeout (RTO) timer based on observed round-trip times; when an ACK has not arrived within a reasonable time frame at the sender, some segment launched within the current transmission window is considered lost. Since the ACK mechanism is simply a cumulative counter of consecutive bytes successfully received, the sender cannot be certain (if selective ACKs are not used [20]) *which* of the segments within the batch has been lost.

From the end-to-end techniques outlined previously, the Delayed ACKs, ADA, DDA, and DAA, techniques in Section 10.4.2 involve some static (in the case of Delayed ACKs) or dynamic heuristic (in the other cases) which requires that a receiver wait for a number of DATA segments to appear before responding with an ACK. However, with so much uncertainty in traffic conditions between the end points, the effectiveness of the proposed techniques may be very situation-specific. Note, however, that end-to-end mechanisms are easily deployed because they only require alterations at two nodes on a path, namely the source (root node) and the destination (sink node).

It should be emphasized that in the context of this survey the term end-to-end does not preclude gathering feedback from other layers in the networking stack. However, because other layers may operate on a point-to-point manner, the available information is of limited use. For instance, obtaining link-layer feedback, say with regard to congestion conditions, at the source node is of limited use because information would only be forthcoming concerning the node itself and not the routing path (which would typically be of more interest). The Active Delay Control technique (described in Section 10.4.2) is indicative of end-to-end versus point-to-point tradeoffs; the idea of imposing an artificial delay on packet transmission to avoid RTOs is implemented on the one hand by altering end-to-end agents only (in the cases of 1-bit RDC and SDC) and on the other by point-to-point changes (in the case of Exact RDC).

The point-to-point approaches outlined in this chapter primarily involve changes-coupling with some network layer (IP) mechanism. For instance, the DATA/ACK-bundling approach suggested in Section 10.4.3 and the COPAS method in Section 10.4.3 require processing of TCP segments at intermediate nodes, which implies functionality not normally implemented on the Internet; this fact precludes the use of these techniques in a mixed setting where "legacy" nodes might be present. The Split TCP approach in Section 10.4.3 involves the creation of TCP proxies along the communications path which breaks the end-to-end paradigm of TCP, but requires only a simple modification at each node in the path. Although incremental adoption of this technique is not taken into account in the original proposal, it is likely that the scheme could be adapted to set the inter-proxy distance in such a way as to establish proxies only at nodes (if any) that implement Split TCP. Finally, the last scheme discussed, which defines an optimization in the default 802.11 link layer mechanism, does not involve TCP or IP level changes. Instead, the modifications are MAC layer-oriented, which in turn means they can be gradually implemented, can maintain the end-to-end TCP paradigm, and do not interfere with segment-forwarding if segments enter a more conventional networking environment (say, a gateway to the Internet).

10.5.2 Cross Layer Versus Single Layer

To allow the proposed solutions to be ubiquitous and flexible enough to implement in a variety of situations, it is desirable to confine them to a single layer in the protocol stack. This would contribute to the ease of deployment and allow for a wider scope of application (that is, it would allow the technique to be combined with others) by isolating any changes to the particular layer involved.

The proposed solutions presented in Section 10.4 may thus be categorized as *single* or *cross*-layer. Note that cross-layer solutions also include proposals that require feedback from a nonadjacent layer, as, for instance, when the transport protocol obtains information from the link layer. Furthermore, it should be emphasized that end-to-end approaches, in general, cannot be cross-layer since in order for changes to be end-to-end, they must be confined at the transport layer. An apparent exception to this rule may be feedback obtained from the routing protocol, if that feedback does not require special collaboration from intermediate nodes (i.e., is standard in the routing protocol—say, for example, hop count information).

The end-to-end techniques examined in Section 10.4.2 fall into the category of single layer modifications, because transport layer changes are necessitated. The Active Delay Control (ADC) method, contained in the same section (10.4.2), consists of three variants: the Exact Receiver-Based Delay Control (RDC), the 1-bit RDC, and the Sender-based Delay Control (SDC). Of these, the Exact RDC technique requires explicit changes in the intermediate nodes and the TCP agent which makes it cross-layer. The 1-bit RDC technique makes use of the ECN bit, but since this is an often-used feature of an Active Queue Management (AQM) mechanism, there are no special cross-layer requirements.

Of the hybrid techniques, the ACK-bundling facility described in Section 10.4.3 requires changes that are confined to the routing layer; therefore, it is single layer in nature. Similarly, COPAS routing, as described in Section 10.4.3, is a network-layer-only enhancement. Split TCP requires both transport and network layer alterations because the proxies require special signaling (with their particular acknowledgment scheme) and the transport layer needs to be aware of this. The discussion on the RTS/CTS scheme in Section 10.4.3 is only pertinent to the link layer and is thus single-layered.

As may be apparent from the previous observations, the vast majority of the mechanisms proposed in the literature apply to a single layer in the networking stack. The rationale for setting such a scope is based on pragmatism. Because the main goal of the authors of the above works is to propose solutions that may be *applicable* in real-world MANETs, it is prudent to propose changes that can be implemented realistically within the bounds of the system's configuration. As such, while changes may be extensive, if they are confined to a single layer, then a single, interchangeable module may be constructed and deployed to implement the intended system functionality. Furthermore, single-layer approaches provide a scope for testing orthogonal techniques in adjacent layers, without affecting the functionality of the new module. Finally, proposed changes within a single layer ensure some degree of longevity for the method because it can continue to be deployed as protocols in other layers evolve and change over time.

10.6 CONCLUSIONS

This survey has outlined the latest research activity on reducing the effect of frequent acknowledgment feedback on TCP connections over mobile Ad hoc networks (MANETs). Previous work reported in the literature has shown clearly that TCP's assumption that segment losses are congestion indication events leads to suboptimal performance in MANET environments. It has further been shown that this negative effect may be somewhat avoided if a TCP agent reduces the number of segments injected into the pipe at any one time, thereby reducing the competition for transmission time over the shared medium. In the end-to-end dynamic of TCP, both DATA and

ACK segments propagate along the path length in each direction. Previous research has been concerned with optimizing the transmission of DATA segments in the pipe, so as to achieve optimal throughput for a given path length. The work on which this survey has focused has tried to develop techniques introduced to reduce the impact of ACK segment transmission on TCP throughput. Reducing the number of ACKs injected into the network is commonly referred to as ACK-thinning.

Methods aiming to reduce competition between DATA and ACK segments may be categorized into two sets: solutions deployable in an *end-to-end* fashion, which affect only the TCP sender and receiver, and those deployable *point-to-point*, involving some alterations at intermediate nodes along the communications path. Depending on whether a solution requires changes to layers other than the transport (TCP layer), modifications may further be distinguished into *single layer* or *cross-layer*. From a deployment perspective, end-to-end modifications are easier to implement because they only involve changes at two points, the sender and the receiver; nonetheless, because of the limited amount of information available at the endpoints, point-to-point mechanisms may be more effective. Further note that the term ACK-thinning may be somewhat misleading because not all of the proposed modifications involve a reduction in the number of ACKs entering the pipe; however, all of them aim to reduce the competition of ACK and DATA segments.

ACK-thinning techniques in MANETs have recently received more attention as the properties of MANETs, and their effects on TCP behavior have become better understood. Although substantial progress has been made, there is a large scope for future work. As the capabilities of wireless transceivers increase, it becomes important to ask if the applicability of these techniques retain their performance benefits on more advanced hardware. Furthermore, since most of the proposed ACK-thinning mechanisms are orthogonal to techniques restricting the injection of DATA segments, it would be interesting to examine if cumulative performance improvement is possible.

Even though researchers have simulated and analyzed the effects of their proposed techniques in supposedly typical MANET scenarios, there is little data from real testbed measurements. The analysis of the algorithms in a real-life setting would, of course, increase confidence in the accuracy of simulation results, validating previous work. While the question as to whether MANETs of any size will, in fact, be deployed widely in the real world remains open, benefits of the proposed techniques, even on a more modest scale, would help instill confidence that the conclusions drawn from these efforts would be applicable in a larger scale.

10.7 EXERCISES

1. **Evaluating the Effects of ACKs on Spatial Contention**: As mentioned throughout this survey, spatial contention is an important concept in MANETs and a prevalent concern with respect to TCP throughput performance. Previous research work has considered the amount of ACK-produced traffic to contribute significantly to spatial contention, but there have been few attempts to assess this contribution.

Using a network simulator (such as NS-2 available at http://nsnam.isi.edu/nsnam or otherwise), try to *quantify* the amount of spatial contention caused by ACK replies. This enquiry might require the introduction of new metrics that should have their use justified. A possible approach would involve measuring the amount of frame drops caused by DATA and ACK segment transmissions to roughly gauge the contribution of each segment type to spatial contention.

2. **Exploring the Effect of SACK Responses on Spatial Contention**: As a follow-up to the previous exercise, consider assessing the effect of selective ACKs (SACKs) [20] on spatial contention and contrast it to that of plain ACKs. SACK blocks add substantial overhead (a typical SACK response is 60 bytes long) to the length of a plain ACK segment (20 bytes) that is offset by the smarter segment recovery mechanism that becomes possible with SACK use. Investigate with the use of a simulator whether the tradeoff is still a favorable one in the case of MANETs by measuring throughput performance when using both techniques in a variety of mobility scenarios.

3. **Analyzing Spatial Contention Under Peer-to-Peer Two-Way Communication Traffic Patterns**: The vast majority of studies on TCP behavior under spatial contention have been conducted using heavily one-sided DATA traffic patterns (such as an FTP source) that follow a distinct client–server paradigm. Other traffic patterns do exist, however, such as peer-to-peer data exchanges that involve the dispatch of DATA from both TCP communicating parties in relatively equal measures. Using a simulator or otherwise, evaluate the contribution of ACK segments to spatial contention under such conditions (of denser DATA traffic) and appraise whether ACK-thinning techniques can still improve TCP throughput substantially. An introduction of a metric such as the one mentioned in exercise 1 may be useful.

4. **Choices in Solution Deployment**: When a problem becomes quantifiable (so that its severity may be appraised) and reproducible (so that the conditions under it occurs are understood), a solution may then become evident. In the case of spatial contention, assume that it is possible to deal with the problem of degrading TCP performance by deploying new wireless transceivers that contain a link-layer mechanism and MAC protocol that deal with the issue. Discuss the pros and cons of deploying such a solution at the following settings:

   - A proposed MANET network where all hardware and software will have to be newly purchased
   - An existing MANET network consisting of a few dozen nodes which is already in use and should have minimal downtime.
   - An existing MANET network consisting of a few dozen nodes which is already in use but can incur substantial downtime.
   - An existing MANET network consisting of about a hundred nodes.

   To aid discussion, consider advantages/disadvantages in terms of ease of deployment and financial costs associated with hardware replacement and software development (if needed).

5. **Effects of Spatial Contention at Different Layers**: Skim through the survey and briefly discuss the effects of spatial contention on the following layers in the protocol stack:

- Application layer (e.g., FTP)
- Transport layer (e.g., TCP)
- Network layer (e.g., Routing protocol)
- Link layer (e.g., wireless transceiver/MAC mechanism)

Suppose you were to design a new solution for this problem; choose a level of implementation and then justify your choice.

REFERENCES

1. P. Johansson, T. Larsson, N. Hedman, B. Mielczarek, and M. Degermark. Scenario-based performance analysis of routing protocols for mobile ad-hoc networks. In *Proceedings of the Fifth Annual ACM/IEEE International Conference on Mobile Computing and Networking*, ACM Press, New York, 1999, pp. 195–206.
2. The Bluetooth Special Interest Group. http://www.bluetooth.com.
3. BRAN Technical Committee. HiperLAN2 specification. http://portal.etsi.org/radio/HiperLAN/HiperLAN.asp.
4. IEEE Standards Association. IEEE P802.11, The Working Group for Wireless LANs. http://grouper.ieee.org/groups/802/11/index.html.
5. A. K. Singh and K. Kankipati. TCP-ADA: TCP with adaptive delayed acknowledgement for mobile ad hoc networks. In *Wireless Communications and Networking Conference (WCNC)*, Vol. 3, IEEE Computer Society Press, New York, 2004, pp. 1685–1690.
6. E. Altman and T. Jiménez. Novel delayed ACK techniques for improving TCP performance in multihop wireless networks. In *Personal Wireless Communications*, Vol. 2775, Springer-Verlag, Heidelberg, 2003, pp. 237–250.
7. R. de Oliveira and T. Braun. A dynamic adaptive acknowledgment strategy for TCP over multihop wireless networks. In *Proceedings of Twenty-Fourth Annual Joint Conference of the IEEE Computer and Communications Societies (INFOCOM 2005)*, Vol. 3, March 2005, pp. 1863–1874.
8. C. Peng Fu and S. C. Liew. ATCP: TCP for mobile ad hoc networks. *IEEE Journal on Selected Areas in Communications*, 21(2):216–228, 2003.
9. H. T Kung, K.-S. Tan, and P.-H. Hsiao. TCP with sender-based delay control. *Computer Communications*, 26(14):1614–1621, 2003.
10. University of South California Information Sciences Institute. *Transmission Control Protocol*. Internet Draft, http://www.ietf.org/rfc/rfc793.txt, September 1981. Request for Comments.
11. M. Allman. On the generation and use of TCP acknowledgments. *SIGCOMM Computer Communication Review*, 28(5):4–21, 1998.
12. K. Chen, Y. Xue, S. H. Shah, and K. Nahrstedt. Understanding bandwidth-delay product in mobile ad hoc networks. *Computer Communications*, 27(10):923–934, 2004.

13. S. Xu and T. Saadawi. Revealing the problems with 802.11 medium access control protocol in multi-hop wireless ad hoc networks. *Computer Networks*, **38**(4):531–548, 2002.
14. T. Yuki, T. Yamamoto, M. Sugano, M. Murata, H. Miyahara, and T. Hatauchi. Performance improvement of TCP over an ad hoc network by combining of data and ACK packets. In *The 5th Asia-Pacific Symposium on Information and Telecommunication Technologies (APSITT 2003)*, IEEE Computer Society, New York, 2003, pp. 339–344.
15. C. Perkins. *IP Mobility Support*. Internet Draft, http://www.ietf.org/rfc/rfc2002.txt, October 1996. Standards Track.
16. S. Corson and J. Macker. *Mobile Ad hoc Networking (MANET): Routing Protocol Performance Issues and Evaluation Considerations*. Internet Draft, http://www.ietf.org/rfc/rfc2501.txt, January 1999. Informational RFC.
17. P. J. Marrón, A. Lachenmann, D. Minder, M. Gauger, O. Saukh, and K. Rothermel. Management and configuration issues for sensor networks. *International Journal Network Management*, **15**(4):235–253, 2005.
18. W. R. Stevens. *TCP/IP Illustrated*, Vol. 1, Addison-Wesley, Reading, MA, 1994.
19. M. Allman, V. Paxson, and W. Stevens. *TCP Congestion Control*. Internet Draft, http://www.ietf.org/rfc/rfc2581.txt, April 1999. Request for Comments.
20. E. Blanton, M. Allman, K. Fall, and L. Wang. *A Conservative Selective Acknowledgment (SACK)-Based Loss Recovery Algorithm for TCP*. Internet Draft, http://www.ietf.org/rfc/rfc3517.txt, April 2003. Experimental RFC.
21. S. Floyd, T. Henderson, and A. Gurtov. *The NewReno Modification to TCP's Fast Recovery Algorithm*. Internet Draft, http://www.ietf.org/rfc/rfc3782.txt, April 2004. Standards Track.
22. L. S. Brakmo and L. L. Peterson. TCP Vegas: End to end congestion avoidance on a global Internet. *IEEE Journal on Selected Areas in Communications*, **13**(8):1465–1480, 1995.
23. C. Casetti, M. Gerla, S. Mascolo, M. Y. Sanadidi, and R. Wang. TCP Westwood: End-to-end congestion control for wired/wireless networks. *Wireless Networks*, **8**(5):467–479, 2002.
24. W. Stevens. *TCP Slow Start, Congestion Avoidance, Fast Retransmit, and Fast Recovery Algorithms*. Internet Draft, http://www.ietf.org/rfc/rfc2001.txt, January 1997.
25. Z. Fu, P. Zerfos, H. Luo, S. Lu, L. Zhang, and M. Gerla. The impact of multihop wireless channel on TCP throughput and loss. In *Proceedings of Twenty-Second Annual Joint Conference of the IEEE Computer and Communications Societies (INFOCOM 2003)*, Vol. 3, March 2003, pp. 1744–1753.
26. T. D. Dyer and R. V. Boppana. A comparison of TCP performance over three routing protocols for mobile ad hoc networks. In *Proceedings of the 2001 ACM International Symposium on Mobile Ad Hoc Networking & Computing*, ACM Press, New York, 2001, pp. 56–66.
27. S. Papanastasiou, L. M. Mackenzie, M. Ould-Khaoua, and V. Charissis. On the interaction of TCP and routing protocols in MANETs. In *International Conference on Internet and Web Applications and Services/Advanced International Conference on Telecommunications (AICT-ICIW '06)*, Guadeloupe, French Caribbean, IEEE Computer Society Press, New York, 2006, pp. 62–69.
28. S. Kopparty, S. V. Krishmniurthy, M. Faloutsos, and S. K. Tripathi. Split TCP for mobile ad hoc networks. In *Global Telecommunications Conference, 2002. GLOBECOM '02*, Vol. 1, IEEE Computer Society Press, New York, 2002, pp. 138–142.

29. K. Xu, M. Gerla, L. Qi, and Y. Shu. Enhancing TCP fairness in ad hoc wireless networks using neighborhood RED. In *Proceedings of the 9th Annual International conference on Mobile Computing and Networking*, ACM Press, New York, 2003, pp. 16–28.
30. H. Liu, H. Ma, M. El Zarki, and S. Gupta. Error control schemes for networks: An overview. *Mobile Networks Applications*, **2**(2):167–182, 1997.
31. K. G. Seah and A. L. Ananda. TCP HACK: A mechanism to improve performance over lossy links. *Computer Networks*, **39**(4):347–361, 2002.
32. T. Plesse, J. Lecomte, C. Adjih, M. Badel, and P. Jacquet. Olsr performance measurement in a military mobile ad-hoc network. In *Proceedings of the 24th International Conference on Distributed Computing Systems Workshops (ICDCSW '04)*, Vol. 6, IEEE Computer Society, New York, 2004, pp. 704–709.
33. S. Papanastasiou, M. Ould-Khaoua, and L. M. Mackenzie. Exploring the effect of inter-flow interference on TCP performance in MANETs. In *Second International Working Conference of Performance Modelling and Evaluation of Heterogeneous Networks (HET-NETs '04)*, Bradford UK, 2004, p. 41.
34. S. Xu and T. Saadawi. Performance evaluation of TCP algorithms in multi-hop wireless packet networks. *Wireless Communications and Mobile Computing*, **2**(1):85–100, 2002.
35. M. Allman, S. Floyd, and A. Medina. Measuring the evolution of transport protocols in the Internet. http://www.icir.org/tbit/TCPevolution-Dec2004.pdf.
36. D. D. Clark. *Window and Acknowledgment strategy in TCP*. Internet Draft, http://www.ietf.org/rfc/rfc813.txt, July 1982.
37. R. Braden. *Requirements for Internet Hosts—Communication Layers*. Internet Draft, http://www.ietf.org/rfc/rfc1122.txt, October 1989.
38. V. Paxson, M. Allman, S. Dawson, W. Fenner, J. Griner, I. Heavens, K. Lahey, J. Semke, and B. Volz. *Known TCP Implementation Problems*. Inter Draft, http//www.ietf.org/rfc/rfc2525.txt, March 1999. Request for Comments.
39. S. Papanastasiou and M. Ould-Khaoua. Exploring the performance of TCP Vegas in mobile ad hoc networks. *International Journal of Communication Systems*, **17**(2):163–177, 2004.
40. S. Xu and T. Saadawi. Evaluation for TCP with delayed ACK option in wireless multi-hop networks. In *Proceedings of IEEE Vehicular Technology Conference (VTC 2001)*, Vol. 1, 2001, pp. 267–271.
41. A. A. Kherani and R. Shorey. Performance Improvement of TCP with Delayed ACKs in IEEE 802.11 Wireless LANs. In *Wireless Communications and Networking Conference (WCNC)*, Vol. 3, IEEE Computer Society Press, New York, 2004, pp. 1703–1708.
42. P.-H. Hsiao, H. T. Kung, and K.-S. Tan. Active delay control for TCP. In *Global Telecommunications Conference, 2001. GLOBECOM '01*, Vol. 3, IEEE Computer Society Press, New York, 2001, pp. 25–29.
43. K. Ramakrishnan, S. Floyd, and D. Black. *The Addition of Explicit Congestion Notification (ECN) to IP*. Internet Draft, http://www.ietf.org/rfc/rfc3168.txt, September 2001.
44. Sally Floyd and Van Jacobson. Random early detection gateways for congestion avoidance. IEEE/ACM Transactions on Networking (TON), **1**(4):397–413, 1993.
45. M. Sugano, T. Araki, M. Murata, T. Hatauchi, and Y. Hosooka. Performance evaluation of a wireless ad hoc network: Flexible radio network (FRN). In *Proceedings of IEEE International Conference on Personal Wireless Communications*, IEEE Computer Society, New York, 2000, pp. 350—354.

46. C. E. Perkins and P. Bhagwat. Highly dynamic destination-sequenced distance-vector routing (DSDV) for mobile computers. In *Proceedings of the Conference on Communications Architectures, Protocols and Applications*, ACM Press, New York, 1994, pp. 234--244.
47. T. Clausen and P. Jacquet. *Optimized Link State Routing Protocol (OLSR)*. http://www.ietf.org/rfc/rfc3626.txt, October 2003. Experimental RFC.
48. C. E. Perkins, E. M. Belding-Royer, and S. R. Das. *Ad hoc On-Demand Distance Vector (AODV) Routing*. Request for Comments, http://www.ietf.org/rfc/rfc3561.txt, July 2003. Experimental RFC.
49. C. Cordeiro, S. R. Das, and D. P. Agrawal. COPAS: Dynamic contention-balancing to enhance the performance of TCP over multi-hop wireless networks. In *Proceedings of the 10th International Conference on Computer Communication and Networks (IC3N)*, 2002, pp. 382–387.
50. K. Xu, M. Gerla, and S. Bae. How effective is the IEEE 802.11 RTS/CTS handshake in ad hoc networks? In *Global Telecommunications Conference, 2002. GLOBECOM '02*, Vol. 1, November 2002, pp. 72–76.
51. E.-S. Jung and N. H. Vaidya. A power control MAC protocol for ad hoc networks. In *Proceedings of the eighth annual international conference on Mobile computing and networking*, ACM Press, 2002, pp. 36–47.
52. E. Kohler, M. Handley, and S. Floyd. *Datagram Congestion Control Protocol*. Internet Draft, http://www.ietf.org/rfc/rfc4340.txt, March 2006. Standards Track.
53. J. Postel. *User Datagram Protocol*. Internet Draft, http://www.ietf.org/rfc/rfc768.txt, August 1980. Standards Track.
54. V. Jacobson, R. Braden, and D. Borman. *TCP Extensions for High Performance*. Internet Draft, http://www.ietf.org/rfc/rfc1323.txt, May 1992. Request for Comments.

CHAPTER 11

Power Control Protocols for Wireless Ad Hoc Networks

JUNHUA ZHU and BRAHIM BENSAOU

Department of Computer Science and Engineering, The Hong Kong University of Science and Technology, Hong Kong

FARID NAÏT-ABDESSELAM

LIFL-UMR CNRS USTL 8022, University of Sciences and Technologies of Lille, France

11.1 INTRODUCTION

With the proliferation of portable wireless terminals, wireless ad hoc networks have attracted significant attentions in recent years as a method to provide data communications among these terminals without support from infrastructure. Nodes in a wireless ad hoc network operate not only as hosts but also as routers, forwarding packets on behalf of other nodes whose destinations are not in their direct transmission range. Such network is self-configured; that is, nodes in the network automatically establish and maintain connectivity with other nodes. In this way, flexible and untethered multihop communications are allowed, which is useful in many applications, such as battlefield communication and disaster recovery.

Transmission power control problem in wireless ad hoc networks is the selection of transmit power level for packet transmission at each node in a distributed manner [1]. The transmit power determines the signal strength at the receiver, the transmission range, and the interference it creates for other receivers in the network. Therefore, transmission power control affects the sharing of wireless medium, thereby affecting many aspects of the operation of wireless ad hoc networks. For example, transmission power control can determine the contention region at the MAC layer, determine the set of candidate nodes for next-hop selection in routing protocols, and affect the operation of congestion control in the transport layer by affecting the congestion level of wireless medium.

Algorithms and Protocols for Wireless and Mobile Ad Hoc Networks, Edited by Azzedine Boukerche
Copyright © 2009 by John Wiley & Sons Inc.

Transmission power control is a key factor to several performance measures such as throughput delay, and energy efficiency. Reducing transmit power level can reduce the energy consumption for communication and increase the spatial reuse of wireless medium, which improves the throughput of wireless network. On the other hand, increasing transmit power level can increase the transmission range of nodes, which reduces the average number of hops each route needed in the network. In this situation, the total transmission delay along each route decreases. Moreover, higher transmit power level improves the quality of the signal received at the receiver, which reduces the bit error probability (BER) of the packets and hence avoids the additional delay induced by the link layer retransmissions.

Based on the above observations, transmission power control problem can be regarded as a typical example of cross-layer design problems in wireless ad hoc networks. Intensive research has been conducted to improve the network performance in terms of network throughput, delay, and energy efficiency with transmission power control. The first wave of such approaches is power-controlled MAC protocols—for example, BASIC schemes [2–5], PCM [6, 7], PCMA [8], and so on. Then other approaches with transmission power control are exploited to improve the network performance. These early works only consider the interplay between power control and single layer; hence they may not necessarily lead to performance improvement when combining power-aware algorithms across multiple layers. Simulation results in reference 9 show that this phenomenon can be realized under certain conditions. Therefore, power control problem should be tackled in a systematic approach [10–15].

Energy is a precious resource in wireless ad hoc networks because most nodes in the network are driven by battery and cannot be recharged in most cases. In order to keep the network functional as long as possible, energy-efficient algorithms should be devised. By reducing transmit power level, energy consumption for packet transmission at each node can be reduced. Thus, transmission power control is an important technique to design energy-efficient algorithms for conserving the battery energy of nodes and prolonging the functional lifetime of the network. Techniques such as energy-efficient routing [16–18], topology control [19–23], and power management [24–27] also help to improve energy efficiency of wireless ad hoc networks.

The remaining part of this survey is organized as follows. Section 11.2 describes the impact of power control on the wireless ad hoc networks, and it derives some basic guidelines for transmission power control problem. In Section 11.3, we review the transmission power control problem following a single-layer approach; that is, we only consider interactions between power control and single layer. Transmission power control in MAC, network, and transport layer is investigated respectively. In Section 11.4 we consider the transmission power control problem from the holistic perspective, and in Section 11.5 we review energy efficient algorithms and protocols joint with power control to improve energy efficiency. Finally, we conclude the survey in Section 11.6. Open issues are discussed at the end of each section.

11.2 DESIGN PRINCIPLES FOR POWER CONTROL

In this section we first describe the fundamental network architecture of wireless ad hoc networks. Then we introduce the problem of power control in wireless ad hoc

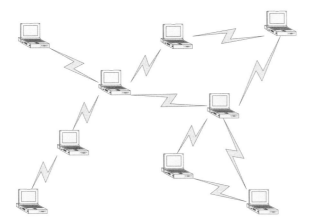

Figure 11.1. Wireless ad hoc network topology.

networks. Based on the works conducted by Kawadia and Kumar [1], we describe the impacts of power control on the layers in the protocol stack and then derive some guidelines to design efficient power control algorithms in wireless ad hoc networks.

11.2.1 Wireless Network Architecture

Wireless ad hoc networks are multihop wireless networks in which a set of mobile or stationary terminals cooperatively maintain the network connectivity [28]. This on-demand architecture is completely from the wired and wireless local networks. An example of a wireless ad hoc network topology is shown in Figure 11.1. Ad hoc networks are characterized by dynamic, unpredictable, random, multihop topologies with no infrastructure support. Since wireless ad hoc networks are self-organized, they are helpful in situations where communications are needed but no infrastructure is available or the cost to build such infrastructure is not affordable. For example, they can be used in military environments, emergence operations, and so on. Wireless ad hoc networks have attracted considerable attentions in recent years as evidenced by the IETF working group MANETs (Mobile Ad Hoc Networks).

Due to the success of the layered architecture of current Internet, wireless ad hoc networks are also developed based on the same layered structure with slight modifications. The typical protocol stack for a generic wireless ad hoc network is shown in Figure 11.2. The hierarchy of layers provides natural abstractions to deal with the natural hierarchy present in networks.

- *Physical Layer.* The physical layer is responsible for converting bits into radio signals, and it disseminates them in the wireless medium. It consists of-radio frequency (RF) circuits, modulation, and coding systems.
- *Data Link Layer.* The data link layer is responsible for establishing a reliable and secure logical link over the unreliable wireless medium. The data link layer is thus responsible for sharing the wireless medium, link error control, and security. The data link layer have two sublayers: Media Access Control (MAC) sublayer

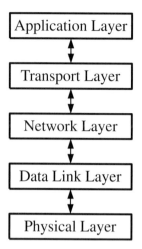

Figure 11.2. Protocol stack for a generic wireless network.

and Logical Link Control (LLC) sublayers. The MAC sublayer is responsible for allocating the time, frequency, or code space among terminals to share the same wireless channel in a region. Transmission power control can affect the wireless medium sharing among a set of terminals and thus has a significant impact on the MAC layer.

- *Network Layer.* The network layer is responsible for routing packets, thus providing host-to-host communication between a pair of terminals. Incorporating the transmission power control function into the routing protocols, energy can be saved for terminals, which is an important metric to assess the performance of wireless ad hoc networks.
- *Transport Layer.* The transport layer is responsible for providing communications for applications residing in different terminals in the network. One of the main functions at the transport layer is congestion control, which mitigates the congestion level in wireless ad hoc networks and thus improves the end-to-end throughput of the networks. Transmission power control has a significant influence on this layer because it determines the interference level in the wireless networks, which creates congestion region in wireless networks.
- *Application Layer.* The main function of application layer is to support network applications. In addition, it may provide some network management services to monitor or manage the performance of wireless ad hoc networks.

11.2.2 Impact of Power Control on Protocol Stack

The transmit power level determines the quality of the signal received at the receiver, the range of transmission, and the magnitude of the interference it creates for other receivers. These factors have considerable influences on several layers in the protocol stack—for example, MAC, network, and transport layer. Therefore, transmission power control problem can affect many aspects of the operation of the wireless ad hoc networks.

At the logical link level, transmit powers of a node and its neighbors affect the signal-to-noise ratio (SINR) at the receiver, so transmission power control affects the capacity of a link. Also, transmission power control may lead to asymmetric links if pairs of nodes use different transmit power levels. This happens when the transmit power of node i is high enough to reach a node j, but not vice versa.

At the MAC layer, transmit power determines the transmission and carrier sensing range of the sending node. Any node in the transmission range can decode the frame successfully, and any node in the carrier sensing range can detect this transmission. Thus, for contention-based random access MAC protocols, transmission power control can affect the contention region of a node, where any neighboring nodes should be silenced when it is receiving packets from another node. For interference-limited MAC protocols, transmission power control affects the interference level that nodes create for other nodes; hence, it influences the degree of simultaneously transmissions allowed. Moreover, most MAC proposed for wireless ad hoc network so far assumes the bidirectionality of links (e.g., 802.11). Since transmission power control can create unidirectional links, it can influence the performance of MAC protocols.

Transmit power determines the transmission range of a node, so transmission power control affects the topology of wireless network, which determines the ability of packet delivery. It also affects the routing protocol because it determines the set of candidate nodes for next-hop selection in route discovery procedure. Moreover, most existing routing protocols for wireless ad hoc networks (e.g., AODV [29], DSR [30]) assume a bidirectional link between nodes. Because transmission power control can create asymmetric links, less links are available for route choice.

At the transport layer, congestion control mechanisms, such as those in TCP, regulate allowed source rates so that the aggregate traffic load on any link does not exceed the available capacity. Transmission power control affects the interference level in the wireless network and determines the available link capacities, so it affect the operation of congestion control in transport layer.

11.2.3 Design Principles for Power Control

Since transmission power control can affect several layers of the protocol stack, it is a prototypical cross-layer design problem in wireless ad hoc networks. Cross-layer design, in general, should be approached holistically with some caution, keeping in mind the longer-term architectural issues [31]. Kawadia and Kumar [1] enumerate several principles that guides the design of transmission power control algorithms.

- *Effect of Power Control on Capacity and Throughput.* As transmit power causes interference in the surrounding region due to the sharing nature of wireless medium, reducing the transmit power can reduce the area of interference. However, a lower power level results in shorter links, which means that more hops are required per route. This puts more relaying burden on nodes. For a transmission range of r, the area of the interference is proportional to r^2, whereas the relaying

TABLE 11.1. Components of Power Consumption in a Wireless Card

$P_{Rx_{elec}}$	The power consumed in receiver electronics for processing.
$P_{Tx_{elec}}$	The power consumed in transmitter electronics for processing.
$P_{Tx_{Rad}}$	The power consumed by the power amplifier in the transmitter for transmitting the packet.
P_{Idle}	The power consumed when the radio is on but no signal is being received.
P_{Sleep}	The power consumed when the radio is turned off.

burden (i.e., the number of hops) is inversely proportional to r, implying that reducing the transmission power level can increase network capacity [32, 33].

- *Effect of Hardware on Power Control.* Table 11.1 shows the various components of power consumption in a wireless card. If $P_{Tx_{rad}}$ dominates the power consumption, then an efficient way to conserve energy is to implement energy-efficient routing algorithms with low power levels for commonly used inverse αth ($\alpha \geq 2$) law path loss model. The energy-efficient routing algorithms ensure that the traffic is routed through a path that minimizes the total energy consumed for this data delivery. When P_{Sleep} is much less than P_{Idle}, then a possible way to save energy is to turn off the radio whenever nodes are not scheduled to transmit or receive [24–26].

- *Effect of Load on Power Control* [34]. The end-to-end delay is the sum of propagation delay, processing delay, transmission delay, and queueing delay. The propagation delay is proportional to the distance between the source and destination and generally is negligible. The processing and transmission delay is the time spent to receive, decode, and retransmit a packet and is proportional to the number of hops between source and destination. Queueing delay is the time spent by a packet waiting in a queue for transmission when the channel is busy. It depends on the congestion level of wireless channels. When load in the network is low, then queueing delay is quite small since the number of nodes contending for channel access at same time is small. In this situation, processing and transmission delay dominate the end-to-end delay, and the performance of the wireless network can be improved by increasing the transmit power level, which reduces the average number of hops in a route. On the other hand, if the network is highly loaded, then queueing delay is the dominant part of the end-to-end delay. In this situation, reducing transmit power level can alleviate the congestion of wireless medium. End-to-end delay decreases by decreasing queueing delay at each hop. Therefore, to improve the network performance, transmit power level at each node should be adapted according to the load in the network.

11.3 POWER CONTROL: THE SINGLE-LAYER APPROACH

In this section, we discuss the transmission power control problem in the MAC, network, and transport layer, respectively. We study the interactions between transmission power control and these layers, and we describe several typical algorithms.

11.3.1 Power Control: The MAC Layer Problem

The Inefficiency of 802.11 Standard. Carrier sensing MAC protocols (CSMA) such as IEEE 802.11 [35] protocol families are the dominating methods for sharing the wireless medium in wireless ad hoc networks. This protocol uses a four-way handshake based on the CSMA/CA mechanism to resolve the channel contention. If node A wants to send data to another node B, it first sends a request-to-send (RTS) packet to B, which replies with a clear-to-send (CTS) packet. The data transmission $A \rightarrow B$ can now proceed; and once completed, node B sends back an acknowledgment (ACK) packet to node A. The RTS and CTS packets includes the duration of DATA and ACK transmission and are used to reserve a transmission floor for the subsequent DATA and ACK packets. Any other node that hears the RTS and/or CTS message defers its transmission until the ongoing transmission is over. The CTS message prevents collisions with the data packet at the destination node B, while the RTS message prevents collisions with the ACK message at the source node A. Nodes transmit their control and data packets at a *fixed (maximum) power level*.

The carrier sensing MAC protocols are simple, but overly conservative. This leads to low spatial reuse, low energy efficiency, and high cochannel interference. For example, consider the situation in Figure 11.3, where node A uses its maximum transmit power (TP) to send packets to node B. If omnidirectional antennas are used, the region reversed for the communication between node A and B is the union of the regions circled by the RTS transmission range, the CTS transmission range, and the physical carrier sensing range. According to CSMA/CA, since nodes D and E fall into the reserved region, they have to refrain from transmission (either data or control packet) to avoid interfering with the ongoing transmission between A and B. However, it is easy to verify that the three data transmission $A \rightarrow B$, $D \rightarrow C$, and $F \rightarrow E$ can be active simultaneously if nodes are able to synchronize locally and select appropriate transmission powers. Furthermore, all the necessary transmit power will be less than the maximum transmit power level defined in 802.11, which means that much energy can be saved.

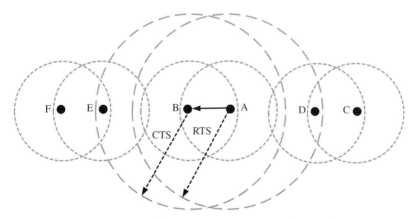

Figure 11.3. Inefficiency of 802.11 RTS/CTS approach.

Due to the benefits of increasing spatial reuse and energy conservation, power-controlled MAC protocols have been intensively studied. The basic idea of power-controlled MAC protocols in the literature is as follows. Nodes exchange their control packets (e.g., RTS/CTS packets) at the maximum allowable power level in order to reduce the collision probability of DATA and ACK packets, but send their DATA and ACK packets at the minimum power level necessary for reliable communication. They can be classified into three categories: contention-based, interference-based, and scheduling-based schemes, which will be discussed in detail in the following subsections.

Contention-Based MAC Schemes

BASIC Schemes. In BASIC power control schemes [2–5], RTS and CTS packets are sent at the highest(fixed) power level (P_{max}), and DATA and ACK packets are sent at the minimum power level necessary to communicate with each other. The BASIC schemes are designed to improve energy efficiency. The RTS/CTS handshake is used to decide the transmit power for subsequent DATA and ACK packets. This can be done in two different ways as described below. Let P_{max} be the maximum possible transmit power level.

- Suppose node A wants to send a packet to node B. Node A transmits the RTS at power level P_{max}. When B receives RTS from A with received power level P_r, B can calculate the minimum necessary transmit power level $P_{desired}$, for the DATA packet based on P_r, P_{max} and the noise power level. Node B then specifies $P_{desired}$ in its CTS packet to A, which will be used as the transmit power level for the following data transmission.

- In the second alternative, when a destination node receives an RTS, it responds by sending a CTS as usual. When the source node receives the CTS, it calculates $P_{desired}$ based on received power level P_r and transmit power level P_{max}. For example, computing $P_{desired}$ as follows

$$P_{desired} = \frac{P_{max}}{P_r} \times Rx_{thresh} \times c$$

where Rx_{thresh} is the minimum required received signal strength and c is a constant.

Though BASIC schemes can reduce the energy consumption, it can lead to network throughput degradation as shown in references 6 and 7. The reason is that reducing power for DATA packet transmission also reduces the carrier sensing range so that ACK (as well as DATA) packets are more likely to be collided. This leads to retransmission of this packet and reduces throughput. Moreover, more energy is consumed for this packet delivery.

PCM. A Power Control MAC (PCM) [6, 7] works similar to the BASIC schemes in the way that it also uses maximum allowable power level P_{max} for the RTS/CTS packet transmissions and the minimum necessary power level P_{min} for DATA/ACK packet transmissions. In order to avoid a potential collision caused by the reduced carrier sensing range during the DATA packet transmission, PCM periodically increases the transmit power to P_{max}.

In this way, PCM effectively reduces the amount of possible collisions induced by BASIC schemes, thereby reducing the number of retransmissions as much as possible. Thus, PCM can achieve the goal of energy saving. Results show that PCM can achieve a throughput comparable to the IEEE 802.11 but with less energy consumption. However, PCM requires a frequent increase and decrease in transmit power levels, which is not easy in the implementation.

δ-PCS. The δ-PCS [36] is a power control scheme that shows significant improvement over IEEE 802.11 in network throughput, energy efficiency, and fairness simultaneously while still following the same single-channel and single-transceiver design. The main idea is that the sending node computes an appropriate transmit power based on its traffic distance d and an estimate of the interference level it experiences, instead of collision avoidance information (CAI) advertised by neighboring nodes used in references 37 and 38. The power function, δ-PCS, is defined as $P^{(t)} = P_{max}(d/d_{max})^\delta$, where d_{max} is the transmission range of P_{max}, δ is a constant between 0 and α, and α is the power attenuation factor. Each δ value corresponds to a different power control scheme, and in one scheme the value δ is uniform to all the nodes in the network. By choosing an appropriate value of δ, this scheme can improve energy efficiency and channel spatial reuse; furthermore, it yields a fair throughput distribution among different destination ranges. Simulation results show that δ-PCS can achieve improvement of over 40% in throughput, energy efficiency by a factor of 3, and better fairness with respect to channel utilization.

Interference-Based MAC Schemes. The power control algorithms discussed above can achieve good reduction in energy consumption. However, it cannot improve the throughput. This is because the exchange of RTS and CTS messages, which are used to silence neighboring nodes, prohibits potential concurrent transmissions in the neighborhood during the DATA/ACK packet transmission.

To increase spatial reuse, and thus increase the throughput, references 8 and 37–39 have proposed various interference-limited media access control algorithms. In these algorithms, new transmission is admitted as long as it does not corrupt ongoing neighboring transmissions. The basic idea is that by exchanging collision avoidance information (CAI) in the control packet, nodes compute the maximum allowable power levels that will not corrupt any ongoing transmissions in the neighborhood, and then they use this power level as the upper bound of the transit power level for subsequent packet transmissions including the control packets. This is completely different from the idea of collision avoidance in current carrier sensing medium access control algorithms, where nodes in the carrier sensing range of an ongoing transmission should defer their potential transmissions.

PCMA. The Power Controlled Multiple Access Protocol (PCMA) [8] proposes a flexible "variable bounded-power" collision suppression model with two separate channels (data channel and busy tone channel), which is generalized from the transmit-

or-defer "on/off" collision avoidance model of CSMA/CA. The power control component in PCMA has two main mechanisms:

- A request-power-to-send (RPTS)/acceptable-power-to-send (APTS) handshake between sender and receiver is used to determine the minimum transmit power level required for successful packet transmission. The RPTS/APTS handshake occurs in the data channel and precedes the data transmission.
- The noise tolerance advertisement is used by each active receiver to advertise the maximum addtional noise it can tolerate, given its current received signal and noise power level. The noise tolerance advertisement is periodically pulsed in the busy tone channel, where the signal strength of the pulse indicates the tolerance to additional noise.

In PCMA, a potential transmitter first "senses the carrier" by listening to the busy tone channel for a minimum time period to determine the upper bound of its transmit power for all control (RPTS, APTS, ACK) and DATA packets. By reducing the transmission power, more concurrent transmission are allowed; hence the spatial reuse of wireless medium increases.

PCMA works effectively in terms of energy conservation while increasing the network throughput. Results show that PCMA can improve the throughput performance by more than a factor of 2 compared to the IEEE 802.11 for highly dense networks.

Similar to PCMA, reference 39 also combines the mechanism of power control, RTS/CTS handshake and busy tones to increase the channel utilization while saving energy. The main difference is that reference 8 introduces *interference margin* (noise tolerance), which can help to admit more concurrent transmissions. This is a tradeoff between throughput and energy consumption. The larger the interference margin, the larger the number of simultaneous transmissions allowed, but also the larger the transmission power. In PCMA, the interference margin is advertised by the receiver over the separate busy tone channel.

PCDC. To avoid using busy tone channel to locally broadcast the interference margin used in reference 8, The Power-Controlled Dual Channel (PCDC) Medium Access Protocol [37] advertises this value by RTS and CTS control packets, which are transmitted on a separate control channel. In addition, to further increase the spatial reuse and provide better protection of ACK packets, reference 37 proposes the use of subcontrol channel for transmitting ACK messages. Simulation results show that PCDC achieves improvements of up to 240% in channel utilization and over 60% in throughput, along with a reduction of over 50% in energy consumption.

Moreover, PCDC is the first one that utilize the interplay between MAC layer and network layer to provide an efficient and comprehensive solution to power control problems. By controlling the transmit power of RREQ packets, the MAC layer can effectively control the set of candidate next-hop nodes in route discovery. PCDC maintains a connectivity set (CS) for each node. Any node in the set cannot find an indirect route that consumes less energy than direct communication. Thus, if the connectivity sets of all nodes can maintain the network connectivity, power-efficient routes can be found with nodes only in the connectivity sets. PCDC proposes a distributed algorithm to maintain such sets. However, the algorithm requires that a node continuously cache the estimated channel gain and angle of arrival (AOA) of every signal it receives over the control channel regardless of the intended destination of this signal, which is hard to achieve and can impose a lot of computing workload in real implementation.

POWMAC. The POWMAC protocol is a comprehensive, throughput-oriented, and single-channel MAC protocol for wireless ad hoc networks [38]. Instead of exchanging only one pair of control packets (RTS/CTS) during one four-way handshake, POWMAC uses an access window (AW) to allow a series of RTS/CTS exchanges to occur before the multiple, concurrent data packet transmissions. The length of an AW is dynamically adjusted based on local traffic load information such that the maximum number of concurrent transmissions in the vicinity can be achieved. Collision avoidance information is inserted into the CTS packet and is used to bound the transmit powers of potential interfering nodes, rather than to silence these nodes. The basic operation of POWMAC is described in Figure 11.4.

The collision avoidance information in the CTS packet sent by node i is the value of the maximum tolerable interference $P_{\text{MTI}}^{(i)}$ and is defined as follows:

$$P_{\text{MTI}}^{(i)} = \frac{P_{\text{MAI-add}}^{(i)}}{(1+\alpha)N_{\text{AW}}^{(j)}}$$

where $P_{\text{MTI-add}}^{(i)}$ is the maximum total interference power that node i can endure from future transmitters, $N_{\text{AW}}^{(j)}$ is the size of access window of the intended transmitter j,

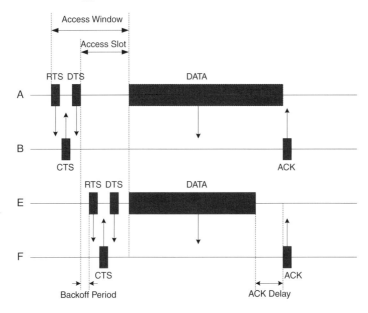

Figure 11.4. Basic operation of POWMAC.

and α is a constant that takes into account interference from future transmitters outside of the carrier sensing range of node i and j. This guarantees that the SINR requirement will be satisfied in receiver i in the subsequent data transmissions. Simulation results demonstrate the achievable, significant throughput and energy gains.

Scheduling-Based MAC Schemes. Because collisions in the MAC layer constitute one of the main sources for low performance (e.g., throughput, delay) and energy waste in wireless multihop networks, power-controlled and scheduling-based MAC protocols are proposed in wireless multihop networks. The main motivation is in two ways. First, scheduling can avoid collisions in the MAC layer. scheduling in frequency (FDMA), time (TDMA) or code domain (CDMA) is essential to coordinate the transmissions of independent nodes in the network. This mechanism assigns each user a predetermined and fixed portion of the wireless physical layer resource and can eliminate the interference among different users. With scheduled medium access, each node transmits packets only with the preassigned frequency band or code, or at the preassigned timeslots. Thus no collison in the MAC layer can occur. Therefore, scheduling can alleviate energy waste and improve network throughput by collision avoidance. Moreover, the scheduling algorithm is based on graph theory, and it is easier to analyze its performance and interplay with other layers in the protocol stack than to analyze random-access MAC protocols.

ElBatt and Ephremides [40] propose a joint scheduling and power control algorithm to maximize the MAC layer throughput while conserving energy. The joint design is formulated as a constrained optimization problem, and the objective function is to minimize the total power of the scheduled links, and the performance metric concerned is the throughput of neighboring transmissions in the network. The concept of admissible transmission scenario with m links is introduced, which satisfies the condition that there exists a set of powers, $P_{ij} > 0$, such that $SINR_{ij} > \beta, \forall ij$ links. The algorithm runs in two alternating phases—the scheduling phase and power control phase—and terminates only when an admissible set of users along with their powers is reached. In the scheduling phase, a simple scheduling algorithm coordinates independent users' transmissions to eliminate strong levels of interference that cannot overcome by power control. In the second phase, a distributed power control algorithm determines the set of powers that could be used by the scheduled users to satisfy their transmissions, if one exists. One distributed power control algorithm used in the cellular system is used as the heuristic solution to the constrained optimization problem. The form of the algorithm is as follows:

$$P_i(N+1) = \min\left[P_{\max}, \frac{\beta}{SINR_i(N)} P_i(N)\right]$$

where P_i is the transmission power of node i, and β is the SINR threshold for reliable communications.

The benefit of this algorithm is that it reduces the computational overhead significantly and simplifies the structure of power control problem, while still maintaining high performance. However, it requires a centralized controller to do the scheduling task, which is not realistic in wireless ad hoc networks.

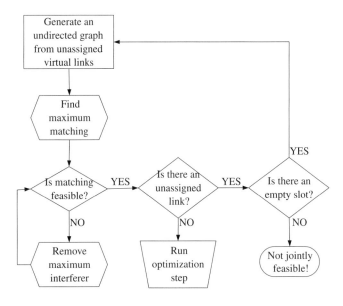

Figure 11.5. Flowchart of the joint scheduling and power control algorithm in Kozat et al. [41].

Kozat et al. [41] formulate the joint problem of power control and scheduling similar to that of ElBatt and Ephremides [40]. The authors also try to minimize the total transmit power in each timeslot while guaranteeing the end-to-end quality of service for sessions in terms of bandwidth and bit error rate probabilities. They prove that the problem of finding feasible power assignment and schedule under given source data rate constraint and SINR requirements is NP-completeness, hence focus should be put on finding the near-optimal and efficient algorithms to perform the joint scheduling and power allocation. Two heuristics are proposed: One is top-down design and the other is bottom-up design. The first approach tries to find the feasible maximal matching of links and then tries to further reduce the energy consumption by moving links from congested timeslots to the empty or less congested one. The procedure of this algorithm is shown in Figure 11.5. The second approach is a bottom-up method by adding one link at each iteration. The adding of links follows the water-filling method, where links are added to timeslots where the links are still feasible and the total power level increased is minimized.

Behzad and Co-workers [42, 43] propose a cross-layer design of power control and scheduling based on a novel interference graph (the Power-Based Interference Graph). The set of vertices in this graph is the set of links in the network, and two vertices (i_1, j_1) and (i_2, j_2) in the graph are connected if and only if one of the following conditions hold:

- Nodes i_1, j_1, i_2, and j_2 are not distinct.
- Transmissions $i_1 \to j_1$ and $i_2 \to j_2$ cannot begin simultaneously.

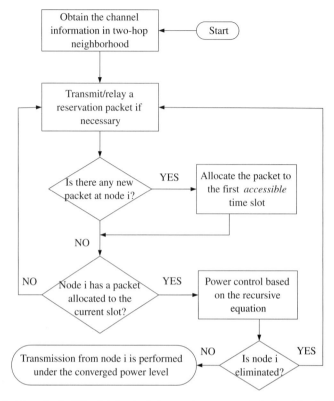

Figure 11.6. Flowchart of the joint scheduling and power control algorithm in Behzad and Co-workers [42, 43].

Thus, any connected pair of nodes in the graph means that they cannot start transmissions simultaneously. With this interference graph, a distributed power control and scheduling algorithm is introduced. The scheduling policy decides whether two links can activate at the same time according to the graph, and the distributed power control algorithm finds the transmit power level for the scheduled links in the way similar to the one used in reference 40. The procedure of this cross-layer design is given in Figure 11.6.

Fang and Rao [44] formulate the resource allocation (power) problem of maximizing the sum of transmitter utilities subject to a minimum and maximum data rate constraint per link and peak power level constraint per node in a wireless multihop network. The Additive Gaussian White Noise (AWGN) channel is assumed to be the channel model, so the system utilities defined as functions of link rates enjoy the same properties as functions of transmission powers. Because the system utilities are functions of the transmit power levels of nodes, they propose a distributed power control algorithm based on the gradient method. The update of transmit power is as follows:

$$P_i(t+1) = \left[P_i(t) - \gamma \frac{\partial L(P(t), q_k)}{\partial P_i} \right]_0^{P_{\max}(i)}$$

where $\partial L(P(t), q_k)/\partial P_i$ is the gradient of the objective function in each iteration, and γ is the step size. The scheduling policy is a greedy degree-based scheduling algorithm, trying to limit the total interference at the receivers by scheduling a small number of transmitters around them. Simulation results shows that this algorithm achieves significant throughput improvement.

11.3.2 Power Control: The Network Layer Problem

Transmission power control can be considered as a network layer problem. Transmission power control in MAC layer approach tries to send packets at the minimum necessary power such that the SINR at receiver is just above predefined threshold for successful transmission, which leads to improvement of spatial reuse in wireless medium and energy saving for nodes. However, the intended receivers of the transmission are determined by the network layer through routing protocols. The job of lower layers (MAC and physical layers) is simply to send the packet to destination specified by the network layer. Therefore, placing power control functionality at the MAC layer does not give routing protocols the opportunity of choosing optimal next-hop node. This means that MAC approach to transmission power control problem only leads to local optimization of network performance, and hence it is desired to placed transmission power control functionality at the network layer (or higher) to obtain the global optimization.

Routing is one of the main functionalities in the network layer. Transmit power determines the neighboring nodes that a node can communicate with, and thus it affects the next hop selection in the routing protocols. The initial routing protocols proposed for MANET (e.g., AODV [29], DSR [30]) only deal with the problem of discovering available routes between source and destination; they didn't consider the energy consumption of nodes in the network. Thus, they may consume too much energy, which is usually a precious resource for battery-driven nodes in the network. This motivates the design of energy-aware routing protocols in wireless ad hoc networks. So far, most proposed schemes use the minimum necessary transmit power for reliable communication at each hop, and they use this power level as the link metric in next hop selection. Various link metric functions based on transmit power level are proposed to achieve different design goals of energy-efficient routing protocols.

Neely et al. [45] consider dynamic routing and power allocation for a wireless network with time-varying channels. The time-varying channels are modeled as stochastic processes $\mathbf{S}(t)$—taking values on a finite-state space—and ergodic with time average probabilities $\pi_\mathbf{S}$ for each state $\mathbf{S}$. Time is slotted with slots normalized to integral units. At the beginning of each timeslot, nodes determine transmission rate on each link by allocating a power matrix $\mathbf{P}(t)$ subject to the power constraints. Link rates are determined by a corresponding rate–power curve $\mu(\mathbf{P}(t), \mathbf{S}(t))$. The authors first characterize the notion of network layer capacity region, and then they propose dynamic routing and power control (DRPC) algorithm to achieve the capacity region and offer delay guarantees with consideration of the full effects of queueing. In the network, packets randomly enter into the system at each node and wait in output queues to be transmitted through the network to their destinations. The DRPC algorithm is based on the maximum differential backlog algorithms, and the details are described below:

1. For all links (a, b), find commodity $c_{ab}^*(t)$ such that

$$c_{ab}^*(t) = \arg \max_{c \in \{1,\ldots,N\}} \{U_a^{(c)}(t) - U_b^{(c)}(t)\}$$

and define

$$W_{ab}^*(t) = \max[U_a^{(c_{ab}^*(t))} - U_b^{(c_{ab}^*(t))}(t), 0]$$

2. *Power Allocation*: Choose a matrx $\mathbf{P}(t)$ such that

$$\mathbf{P}(t) = \arg \max_{\mathbf{P} \in \Pi} \sum_{a,b} \mu_{ab}(\mathbf{P}, \mathbf{S}(t)) W_{ab}^*(t)$$

3. *Routing*: Define transmission rates as follows:

$$\mu_{ab}^{(c)} = \begin{cases} \mu_{ab}(\mathbf{P}(t), \mathbf{S}(t)), & \text{if } c = c_{ab}^* \text{ and } W_{ab}^* > 0 \\ 0, & \text{otherwise} \end{cases}$$

Here $U_a^{(c)}(t)$ is the backlog of commodity c in node a at the beginning of timeslot t, and W_{ab}^* represents the maximum differential backlog between nodes a and b. In DRPC policy, routing is implicit based on the information of backlogs at different nodes. The authors prove that they DRPC algorithm can stabilize any arriving rate vectors in the capacity region, and they offer delay guarantee, though the delay guarantee is very loose.

Savas et al. [46] consider the problem of joint routing and power control as a constrained optimization problem. They construct an auxiliary graph $G(V, E)$, where $V = R \cup T$. R is the set of nodes in the network, and $T = \cup_{i \in R} T(i)$ where $T(i) = \{t(i, l) | i \in R, l = 1, 2, \ldots, L\}$. l is the possible transmit power levels for nodes i. Edges in this graph satisfy either of the following two conditions:

(C1) $i \in R$ and $j \in T(i)$
(C2) $i \in T$ and $j \in R$

Condition (C1) provides flow paths for relayed data, and condition (C2) indicates that a given transmission can be received by all receivers. Then they formulate the studied problem as a mathematical optimization problem; and numerical results show that by joint routing and power control, this can lead to better channel utilization and increase the network throughput.

Besides considering the transmit power level as link metrics for routing protocols, there are other perspectives considering the interaction between power control and network layer. For example, the authors in references 33, 47, and 1 and propose COMPOW, CLUSTERPOW, and LOADPOW schemes, which tackle this problem

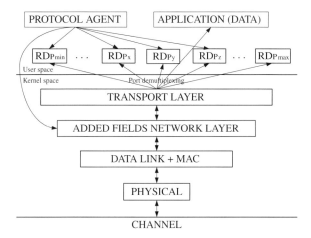

Figure 11.7. Software architecture of COMPOW, CLUSTERPOW, MINPOW protocols.

from a totally different perspective. They give a clear implementation of these protocols at the network layer. Figure 11.7 describes the software architecture design of these protocols.

- **COMPOW**: The COMPOW power control scheme is the simplest power control at the network layer. In COMPOW, each node chooses a common power level. The power level is chosen such that the network remains connected. This can be achieved by running multiple independent routing protocols, one for each feasible power level, and a COMPOW agent figures out the lowest possible power level that ensures the connectivity of the network. The protocol has nice features such as bidirectional links, which allow all OSI-based protocols to operate normally over wireless networks. This protocol provides good performance when nodes are uniformly distributed in the networks. However, for constantly moving nodes, the protocol incurs significant overheads, and convergence to a common power level may not be possible. Moreover, the performance of the COMPOW scheme would be highly degraded in the presence of few isolated nodes that require large transmit power for ensuring connectivity. Gomez and Campbell [48] provide an analysis on the pros and cons of common range power control and variable range power control. Their results show that a variable-range transmission approach can outperform a common-range transmission approach in terms of energy savings and capacity improvement, and this motivates the design of the CLUSTERPOW scheme.
- **CLUSTERPOW**: In the CLUSTERPOW scheme, each node runs a routing protocol at each feasible power level p_i, and it maintains a routing table RT_{p_i} for each power level. In order to route a packet to destination d, a node looks up the routing table for d in increasing order of the corresponding transmit power level. So the network layer makes sure that a packet is transmitted with the lowest possible

power at each hop. Clustering in CLUSTERPOW is implicit because it is based on the transmit power levels rather than on addresses or geographical locations.
- **LOADPOW**: The LOADPOW scheme is similar to the COMPOW and CLUSTERPOW schemes, and it tries to adapt the transmit power according to the network load. It opportunistically uses a higher transmit power level whenever the network load is low, and vice versa. The LOADPOW algorithm attempts to avoid interference with ongoing communications in the neighborhood. It keeps track of the busy (sending or receiving) nodes in the neighborhood. When deciding to transmit a packet, it looks up the routing table and determines the safe power level that can be used without corrupt any ongoing transmission. In this way, LOADPOW can achieve the design goal as the contention of wireless medium that indicates the traffic load in the network. However, LOADPOW may create temporary routing loops, and it is hard to determine the safe power level.

11.3.3 Power Control: The Transport Layer Problem

As mentioned in Section 11.2.2, transmission power control can affect congestion control in the transport layer. Thus, in order to achieve in a power-efficient manner high end-to-end throughput in wireless ad hoc networks with multihop communications, both congestion control and power control need to be optimally designed and distributedly implemented.

It has been shown in Kelly et al. [49] that congestion control mechanisms can be viewed as distributed algorithms solving the following network utility maximization problem [50]:

$$\begin{aligned}
\text{maximize} \quad & \sum_{s} U_s(x_s) \\
\text{subject to} \quad & \sum_{s \in L(s)} x_s \leq c_l, \quad \forall l \\
& \mathbf{x} \succeq 0
\end{aligned} \quad (11.1)$$

where source rates $\mathbf{x} = \{x_s\}$ are the optimization variables, link capacities $\{c_l\}$ are the constant parameters, $L(s)$ denotes the set of links l in the path originated from source s, and the utility U_s for each source can be any increasing, concave function.

However, in this standard form of network utility maximization framework, the link capacities $\{c_l\}$ are assumed to be constant, which is not true in wireless ad hoc networks where the transmission power control can change the attainable data rates on the links. Intuitively, a proper power-level algorithm would allocate the right amount of power at the right nodes to alleviate the bandwidth bottlenecks by increasing capacity on the appropriate links, which will then induce an increase in end-to-end TCP throughput. What complicates this approach is that changing the transmit power on one link would affect the data rates available on other links, due to interference in wireless networks. With this intuition, the network utility maximization framework can be modified to fit into the wireless ad hoc network with elastic link capacities:

POWER CONTROL: THE SINGLE-LAYER APPROACH

$$\text{maximize} \quad \sum_s U_s(x_s)$$

$$\text{subject to} \quad \sum_{l \in L(s)} x_s \leq c_l(\mathbf{P}) \quad \forall l, \quad (11.2)$$

$$\mathbf{x} \succeq 0$$

where the optimization variables are both source rates $\mathbf{x} = \{x_s\}$ and transmission powers $\mathbf{P} = \{P_l\}$. Here the link capacities $\{c_l\}$ are nonlinear and global functions of the transmission powers $\mathbf{P}$.

Chiang [10, 51] proposes a jointly optimal congestion control and power control (JOCP) algorithm in wireless multihop networks based on the above observations. CDMA-based medium access and fixed single-path routing are assumed in the algorithm, thus avoiding the problem of contention resolution in the MAC layer and routing in the network layer. The congestion control algorithm is based on TCP Vegas [52], where the source adjusts the window size according to the delay between the source and destination. The weighted queueing delay is used as the implicit price. During each timeslot t, the following four updates are carried out simultaneously until convergence.

1. At each intermediate node, a weighted queueing delay λ_l is implicitly updated, where $\gamma > 0$ is a constant.

$$\lambda_l(t+1) = \left[\lambda_l(t) + \frac{\gamma}{c_l(t)} \left(\sum_{s:l \in L(s)} x_s(t) - c_l(t) \right) \right]^+$$

2. At each source, total delay D_s is measured and used to update the TCP window size w_s. Consequently, the source rate x_s is updated.

$$w_s(t+1) = \begin{cases} w_s(t) + \dfrac{1}{D_s(t)} & \text{if } \dfrac{w_s(t)}{d_s} - \dfrac{w_s(t)}{D_s(t)} < \alpha_s \\ w_s(t) - \dfrac{1}{D_s(t)} & \text{if } \dfrac{w_s(t)}{d_s} - \dfrac{w_s(t)}{D_s(t)} > \alpha_s \\ w_s(t) & \text{else} \end{cases}$$

$$x_s(t+1) = \frac{w_s(t+1)}{D_s(t)}$$

3. Each transmitter j calculates a message $m_j(t) \in R^+$ based on locally measured quantities, and it passes the message to all other transmitters by a flooding protocol.

$$m_j(t) = \frac{\lambda_j(t) SIR_j(t)}{P_j(t) G_{jj}}$$

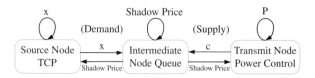

Figure 11.8. Nonlinearly coupled dynamics of joint congestion control and power control.

4. Each transmitter updates its power level based on locally measured quantities and the received messages, where $\kappa > 0$ is a constant.

$$P_l(t+1) = P_l(t) + \frac{\kappa \lambda_l(t)}{P_l(t)} - \kappa \sum_{j \neq l} G_{lj} m_j(t)$$

The power control algorithm updates the power level of a link in two ways: First, it increases power directly proportional to the current link price and inversely proportional to the current power level, and then it decreases the power level by a weighted sum of the messages from all other transmitters. The intuition is that if local queueing delay is high, transmit power should increase, with more moderate increase when the transmit power is already high. On the other hand, if the queueing delay on the other links are high, transmit power should decrease so that the interference on those links can be reduced. The interaction between power control and congestion control is shown in Figure 11.8. The author has proved that this joint algorithm can converge to one optimal solution.

11.3.4 Open Issues

The power control problem is mainly considered as a MAC layer problem because of their tight interactions. Due to the emergence of new physical layer technologies, many critical issues remain unresolved. Several of them are listed below.

- *The efficiency of transmission power control.* Many works on joint scheduling and power control design claim that the optimal power allocation policy is $0 - P_{\max}$ policy; that is, power control is not needed. However, no analysis is given so far on the pros and cons of power control in the contention-based design. Thus, whether power control is needed or not in wireless multihop networks is still an open problem, and a rigorous analysis on this problem is strongly desired.
- *Power control with advanced physical layer technologies,* with the continuous emergence of new physical layer technologies (e.g., MIMO, directional antennas). It is worth taking advantage of these techniques combined with power control to further improve performance or reduce energy consumption of the stations in the networks. The relationship between these techniques and transmit power should be characterized and should be used as a guide to the design of transmission power control algorithms.

- *Power control in interference-limited medium access.* Since current CSMA algorithms such as the 802.11 protocol families have been demonstrated to be inefficient in terms of link layer throughput, the collision avoidance may need to be shifted from the carrier-sensing approach to the interference-limited approach. In the latter one, multiple nodes sharing the same wireless channel are allowed to send data the corresponding receiving nodes concurrently as long as the SINR at these receiving nodes is satisfied. In this approach, the power control algorithm needs to be carefully designed because it will directly affect the SINR at receivers. There is a tradeoff between the throughput and energy consumption in this type of MAC protocol. Increasing power can improve the throughput but consumes more energy, and vice versa. With the design of efficient power control algorithms, a balance between these two goals can be achieved.
- *Practical Implementations.* So far, most power-controlled MAC schemes that intend to exploit spatial reuse or reduce energy consumption still suffer from the limitations that the assumptions on the physical layer are somewhat unrealistic or introduce significant complexity into the design of MAC protocols. Therefore, works are still needed to find practical, simple and efficient designs of MAC protocols with transmission power control, and this may be achieved at the expense of slight performance degradation.

11.4 POWER CONTROL: THE SYSTEMATIC APPROACH

Kawadia and Kumar [31] suggest that cross-layer design in wireless networks should be considered holistically because one cross-layer interaction can affect not only the layers concerned, but also other layers of protocol stack. Two loosely correlated cross-layer designs may lead to disastrous unintended consequences on overall performance. Simulation results in reference 53 support this argument. The authors compared the performance of a power-aware routing protocol PARO [9] with the performance of minimum hop routing protocol (MHRP), using basic-like power control in the MAC layer, and showed that the latter one saved more energy. Moreover, since basic-like power control does not increase the spatial reuse of medium, and PARO would introduce more hops in the routes, it can cause a degradation of network throughput. Thus, in order to improve the network performance or save energy, we need to tackle transmission power control problem in a systematic approach, taking into account the MAC, network, and transport layers together.

11.4.1 Power Control as Constrained Optimization Problems

Recently, there has been a trend that models the network problems as mathematical optimization problems. This approach can help to clearly find the interactions between parameters in the network problems, and it can serve as a guide to design the optimal or near-optimal algorithms. Because power control can affect most layers in the protocol stack, it is useful to model the power control problem as a constrained optimization problem. In this way, the power control problem can be tackled holistically, and thus the negative impact of careless cross-layer design with power control can be avoided.

Generic Formulation. The transmission power control problem in wireless networks can be modeled as a constrained optimization problem with utility maximization framework. One form of the generalized formulations [10, 11] is as follows:

$$\text{maximize} \quad \sum_s U_s(x_s, P_{e,s}) + \sum_j V_j(w_j)$$
$$\text{subject to} \quad \mathbf{Rx} \preceq \mathbf{c}(\mathbf{w}, \mathbf{P}_e),$$
$$\mathbf{c} \in \mathcal{C}_1(\mathbf{P}_e) \cap \mathcal{C}_2(\mathbf{F}),$$
$$\mathbf{R} \in \mathcal{R}, \quad \mathbf{F} \in \mathcal{F}, \quad \mathbf{w} \in \mathcal{W}.$$

Here, x_s denotes the rate for source s, and w_j denotes the physical layer resource at network element j. The utility functions U_s and V_j may be any nonlinear, monotonic functions. $\mathbf{R}$ is the routing matrix and $\mathbf{c}$ are the logical link capacities as functions of the physical layer resources $\mathbf{w}$ and the desired decoding error probabilities $\mathbf{P}_e$. The issue of signal interference and power control can be captured in this functional dependency. The rates must also be constrained by the interplay between physical-layer decoding reliability and upper-layer error control mechanisms like ARQ in the link layer. This constraint set is denoted as $\mathcal{C}_1(\mathbf{P}_e)$, and it captures the issue of rate-reliability tradeoff and coding. Constraints on the rates by the medium access success probabilities is represented by the constraint set $\mathcal{C}_2(\mathbf{F})$, where $\mathbf{F}$ is the contention matrix [54]. The issue of packet collision and MAC is captured in this constraint set. The set of possible physical-layer resource allocation schemes is represented by $\mathcal{W}$, that of possible scheduling- or contention-based medium access schemes by $\mathcal{F}$, and that of single-path or multipath routing schemes by $\mathcal{R}$. The optimization variables are $\mathbf{x}, \mathbf{w}, \mathbf{P}_e, \mathbf{R}, \mathbf{F}$.

Five layers in the current standard protocol stack are modeled in problem (11.3).

- *Application Layer.* The utility functions U_i and V_j represent the resource requirements of applications.
- *Transport Layer.* The source rate x_s of each user s represents the end-to-end throughput.
- *Network Layer.* The routing algorithm can be expressed by varying $\mathbf{R}$ within the constraint set $\mathcal{R}$.
- *Link Layer.* Through scheduling, antenna beamforming, or spreading code assignment, the contention matrix F can be expressed within the constraint set $\mathcal{F}$. The rates are then constrained by contention-free or contention-based medium access schemes, which are described by the constraint set $\mathcal{C}_2$.

- *Physical Layer.* Adaptive resource allocation can be achieved by power control, adaptive modulation, and coding. This will lead to different logical link capacities **c**, which are functions of the decoding error probabilities $\mathbf{P}_e$ and the physical layer resource **w**.

Related Schemes. Radunovic and Le Boudec [12] study the problem of optimal scheduling, routing, and power control that achieves max–min fair rate allocation in a multihop wireless network. They restrict their study to symmetric, one-dimensional networks with line and ring topologies, and they numerically solve the problem for a large number of nodes. The point-to-point link is modeled as single-user Gaussian channels where nodes cannot send and receive at the same time, and both direct and minimum-energy routing techniques are investigated in the simulation. Numerical results reveal several interesting design guidelines for optimal scheduling, routing, and power control design:

- For small power constraints, it is better to relay, for both direct and minimum-energy routing techniques.
- For high powers, it is optimal to send data directly to the destination.
- For symmetric networks, it is optimal to schedule only one link in each slot and rotate these timeslots. Moreover, the optimal power allocation policy is 0 or full power policy.

Radunovic and Le Boudec [13] give a mathematical analysis on the problem of optimal power allocation, rate adaption, and scheduling in a multihop wireless network. Assumptions are made that the rate-SINR function is linear, and fine-grained rate adaption is allowed (e.g., TH-UWB, low-gain CDMA wireless networks). The authors prove that the optimal power allocation is 0 or full power policy, and hence power control is not required. However, if the number of possible link rate is small, it is desirable to use power control and scheduling to increase the network throughput.

Cruz and Santhanam [14] propose a joint routing, scheduling, and power control design to support high data rates for broadband wireless multihop networks. They divide it into two subproblems: (a) joint power control and scheduling subproblem and (b) routing subproblem. The aim of the problem of joint scheduling and power control is to minimize the total average transmit power in the wireless multihop networks, subject to given constraints regarding the minimum average data rate per link, as well as the peak transmit power constraints per node. The problem is solved in the duality approach, and thus a byproduct is the link sensitivities that tell the change in the minimal total average power with respect to a perturbation of the required data rate on a link. The routing policy is shortest path routing with link sensitivities as the link metrics. Since the minimal total average power is a convex function of the required minimum average data rates, shortest path routing can lead to the search for globally optimal routes. They find the optimal power control policy is such that each node is either transmitting at peak power to a single receiver or not transmitting at all, and the optimal link schedule time-shares a small number of optimal subsets of links ($\leq L + 1$, where L is the number of links) in order to achieve the required data rates.

Radunovic and Le Boudec [55] studied the problem of optimal power control, scheduling, and routing in UWB wireless networks. In UWB wireless network, the link capacity is a linear function of the SINR at the receiver. For this particular physical-layer characteristic, the authors prove that the optimal power control policy is sending at the full power whenever data are sent over a link; otherwise, the link should remain silent. The reason is that the rate on a UWB link can always increase when the SINR at the receiver is increased. Even though an increase of a transmit power will increase interference at other receivers, this will always be compensated by the increase of the rate on the link itself. The optimal MAC design should be a combination of rate adaptation and mutual exclusion. The rates of senders should be adapted to the amount of noise and interference at the receiver; and when a node is receiving, it should maintain an exclusion region where any nodes in this region should remain silent during the reception. The size of this exclusion region depends only on the power constraints of the source of the transmission, and not on the length of this link or the position of other nodes. The optimal routing algorithm is minimum energy routing (MER) not only from the energy but also from throughput performance viewpoint for static networks. Another interesting result is that MAC layer is insensitivity to the choice of routes; that is, the optimal MAC protocol does not depend on the choice of routes.

Bhatia and Kodialam [15] consider the joint scheduling, routing, and power control design for power-efficient communication in wireless multihop networks. The overall problem is formulated as an optimization problem with nonlinear objective function and nonlinear constraints, and the derived solution is a performance-guaranteed polynomial time approximation algorithm for jointly solving these three problems. The authors first consider the joint scheduling and power control problem under given flow rates and fixed routes. They assume the Additive White Gaussian Noise (AWGN) channel in the physical layer, and rate-power function is $R = W \log_2(1 + \frac{P}{N_0 W})$. Thus, the joint scheduling and power control problem is also the problem of joint scheduling and rate adaptation. First they get the minimum power levels and data rates required to support the flow rates. However, this does not guarantee that we will find a link schedule to satisfy the obtained solution. Thus, a scale factor is computed to scale-up the data rates, in order to find a schedule that support the given flow rate. Therefore, the algorithm is a near-optimal approximation algorithm that consumes more energy, and the authors prove that they can approximate the optimal solution by a factor of 3. By appropriate transformation of the original optimization problem, the original problem can be transformed into a Quadratically Constrained Programming problem [56] with only flow rates as the optimization variables. The transformed problem can be solved in polynomial time—for example, using the interior point method. The authors also propose an iterative algorithm based on the Frank–Wolfe method [57], which is efficient in practice and has the nice property that it can be computed in distributed way. In each iteration, the optimal routing policy is shortest path routing.

11.4.2 Power Control: Other Approaches

As mentioned in Section 11.3.1, PCDC considers the interplay between power control, MAC, and routing. The scheme can efficiently save energy and increase the throughput at the same time. Li and Ephermides [58] propose a heuristic solution to the joint scheduling, power control, and routing algorithm for wireless ad hoc networks. They assume a TDMA-based wireless ad hoc network where each node has one transceiver. They consider the joint scheduling, routing, and power control problem as two subproblems: (a) joint scheduling and power control subproblem and (b) joint scheduling and routing subproblem.

The joint scheduling and power control algorithm is similar to the one in reference 40, and the main idea is to iteratively update the power level of links while making sure that all the scheduled links satisfy their SIR requirement. The link metrics are based on queue size of links, and the link with lower link metric has higher priority in the scheduling to occupy the timeslot and update the power level as follows:

$$P_i^{(n+1)} = \beta \left(\sigma^2 + \sum_{k \neq i} P_k^{(n)} G_{kj} \right) \bigg/ G_{ij}$$

where $P_k^{(n)}$ is the power level of node k at timeslot n, G_{ij} is the path loss factor of link (i, j), σ is the ambient noise power level, and β is the SIR requirement. It means that when adding one link into the scheduled link set, all selected links will compute their minimum necessary power level again. A New link would be admitted if and only if all the power levels of scheduled links don't exceed the maximum power level P_{max}.

The joint routing and scheduling algorithm is based on the Bellman–Ford algorithm, and the link metric is defined as

$$D(i, j) = d \left(\frac{Q_{ij}}{Q_{max}} \right) + e \left(\frac{R_{ij}}{R_{max}} \right)^\gamma$$

where Q_{ij} is the queue size of link (i, j), and R_{ij} is the distance between node i and j. d and e are constants satisfying $d + e = 1$. The first term is the queue size, encouraging the usage of less congested links, hence alleviating congestion. The second term is related to energy consumption, encouraging transmission over short distance. Simulation results show that this joint design can improve the network performance compared with only joint scheduling and power control algorithm, and they also show that there is a tradeoff between the network performance and energy consumption. However, the joint algorithms are centralized, and they are hard to be implemented in wireless ad hoc networks.

11.4.3 Open Issues

Transmission power control can affect the MAC layer, network layer, and transport layer in several ways; hence, it is a prototypical cross-layer design in wireless multihop networks. Since the number of possible operations of a wireless network is large, the formulation and solutions of this problem is large, and critical challenges still exist. Here we list several major challenges.

- *The Goodness of Power Control.* Many works on joint optimal scheduling, routing, and power control show that if fine-grained rate adaptation is adopted in the wireless network, the optimal power allocation policy is $0 - P_{\max}$ policy. Therefore, it is not necessary to incorporate power control into layers of protocol stack. Studies are needed to be conducted to find in what situation power control is helpful and in what situation power control is helpless. Analysis on this problem is still a critical issue that disturbs the research of transmission power control problem in wireless networks.
- *The Model of Cross-Layer Power Control Problem.* So far, all the works on the cross-layer power control problem involving MAC layer and network layer assume scheduling as the MAC scheme and fixed single-path routing as the routing protocol, which is not sufficient or realistic in wireless networks. Because contention-based MAC schemes (e.g., 802.11) are the dominating MAC schemes in current wireless multihop networks, it is desired to formulate power control, contention-based MAC, and routing as a constrained optimization problem and study their interactions and performance.
- *Practical Algorithms for Cross-Layer Power Control Problem.* In recent years, works on cross-layer power control are mainly focused on formulating the problem as a constrained optimization problem and deriving an algorithm from the formulated problem, then analyzing its performance. The proposed algorithms are either centralized or induce significant computational overhead. Both should be avoided in wireless ad hoc networks. Thus, this problem is still open, and a lot of works can be done in this direction.
- *The Overhead of Cross-Layer Power Control.* Transmission power control will induce additional computational and communication overhead to the operation of wireless ad hoc networks. Usually the relationship between benefits and costs of power control is not linear. Thus, in some situations, power control makes no difference to the performance of network; even worse, it may degrade the performance. Therefore, clear understanding of the tradeoff between the benefit obtained and the overhead required can help to tune the operation of wireless networks into its optimal performance.

11.5 POWER CONTROL: THE ENERGY-ORIENTED PERSPECTIVE

Most terminals in wireless ad hoc networks are driven by battery, thus energy is a precious resource in such networks. Transmission power control can reduce the power level of transmitters and hence can save energy. Since power control can affect many layers in ad hoc networks, there are also many approaches to design energy-aware protocols at different layers combined with power control in order to conserve energy–

for example, power-controlled MAC protocol (Section 11.3.1). Energy conservation gained in this approach is due to the reduction of transmit power to the minimum necessary power level for reliable transmission.

Nodes in wireless ad hoc networks serve as relaying nodes for other nodes requiring communication. Therefore, the routing protocol will affect the energy consumption of nodes by next-hop node selection in the network. This motivates the design of energy-aware routing, which aims to prolong the network lifetime or conserve energy for nodes. Generally, there are two categories, minimum energy routing and maximizing network lifetime routing. The former one intends to minimize the energy consumed for a single source-destination communication, and the latter one intends to extend the network lifetime by proper traffic distribution. Combined with transmission power control, energy-aware routing can further reduce the energy consumption of terminals.

Power-efficient topology control is another approach to reduce energy consumption. By choosing the transmit power carefully, the ad hoc network maintains a moderate connectivity, and it conserves energy with a lower power level.

Inspired by the non-negligible energy consumption of nodes in idle mode in ad hoc networks, power management schemes are proposed to save energy wasted in idle mode. Nodes in network turn off their radio when there is no data to send or receive.

In this section, we discuss the energy-oriented power control problem in terms of these issues. The main objective is to reduce energy consumption of nodes and prolong the lifetime of networks. Throughput and delay are second objectives in such approaches.

11.5.1 Energy-Aware Routing

The first generation of routing protocols in ad hoc networks are essentially minimum hop routing protocols (MHRP) that do not consider energy efficiency as the main goal. While energy conservation becomes a major concern for the ad hoc network, many energy-aware routing algorithms have been proposed in recent years. Singh et al. [59] propose several metrics for energy-aware routing:

- *Minimize Energy Consumed/Packet.* In this way, the total energy consumption of this network is minimized. However, it may cause some nodes to drain energy out faster since it tends to route packet around areas of congestion in the network.
- *Maximize Time to Network Partition.* Given a network topology, there exists a minimal set of nodes, the removal of which will cause the network to partition. The routes between these two partitions must go through one of these critical nodes. A routing procedure therefore must divide traffic among these nodes to maximize the lifetime of the network.
- *Minimize Variance in Node Power Levels.* The intuition behind this metric is that all nodes in the ad hoc network are of equal importance, and no node must be penalized more than any other nodes. This metric ensures that all the nodes in the network remain up and running together.

- *Minimize Cost/Packet*. In order to maximize the lifetime of all nodes in the network, metrics other than energy consumed/packet need to be used. The paths selected when using these metrics should be such that nodes with depleted energy reserves do not lie on many paths.
- *Minimize Maximum Node Cost*. This metric ensures that node failure is delayed. Unfortunately, there is no way to implement this metric directly in a routing protocol. However, minimizing the cost/node does significantly reduce the maximum node cost in the network.

The early energy-aware routing algorithms are based on shortest path algorithms. Instead of using delay or hop counts as the link metrics, energy-oriented link metrics such as signal strength [60], battery energy level at each node [61, 62], power level [63–65], and energy consumption per transmission [9, 66] are used in these algorithms. The link condition and power level of nodes are exchanged periodically in order to keep these metrics up-to-date. In this way, energy consumption for routes can be minimized. However, certain nodes in the network may drain energy out faster than other nodes since these nodes may lie on many paths. Therefore, energy consumption must be balanced among nodes to increase network lifetime.

Chang and Tassiulas [16, 17] formulate the problem of maximizing the lifetime of network as a linear programming problem. By solving the optimization problem, the network lifetime can be maximized. However, the optimal solutions either are not distributed or induce significant computational and communication overhead. Chang and Tassiulas [17] also propose a heuristic based on shortest path routing. The link cost is proportional to the energy consumed per packet and is inversely proportional to the normalized resident battery energy at sender and receiver. The battery status of nodes should be broadcast periodically in order to keep the link cost up-to-date. Simulation results show that the heuristic can approximate the maximum network lifetime if the frequency of information update is large enough. Zussman and Segall [18] formulate the energy-efficient anycast problem that follows the method in references 16 and 17, and they propose an iterative algorithm to obtain the optimal solution.

The energy-aware routing algorithms discussed above do not directly interact with power control, although power control is always used in the MAC layer to further reduce the energy consumption by decreasing the transmission power. An alternative is to incorporate power control directly into routing algorithm [64, 65]. This method requires that each source can put the transmit power level P_{TX} at a suitable format field in the transmitted packet. It also requires that the radio transceiver can measure the received signal strength P_{RX}. With these two values, the node that received the packet can estimate the link attenuation. Upon the received signal power, the node can adjust its transmit power to the remote node by

$$\tilde{P}_{TX} = P_{TX} - P_{RX} + S_R + Sec_{th}$$

Here S_R is the minimum power level required for correct packet reception, and Sec_{th} is a power margin introduced to take into account channel and interference power-level fluctuations. When a packet from this node is sent and received by the next-hop node, the next-hop node can do the same adjustment.

11.5.2 Power-Efficient Topology Control

Power-efficient topology control [19–23, 67, 68] in wireless multihop networks maintains a moderate connectivity by adapting transmission range of nodes. Reducing the transmission range reduces the energy consumed for per-packet transmission, and it also promotes the spatial reuse. The objective of power-efficient topology control is to minimize the transmit power while maintaining certain connectivity. Ramanathan and Rosales-Hain [23] formulate this problem as a constrained optimization problem under the constraints that the result topology should be connected and bi-connected. Their goal is to minimize the maximum transmit power of all the nodes in the network, and two centralized algorithms are proposed based on a new analytical representation of wireless multihop networks. Haiiaghayi et al. [68] studied the problem of minimizing power while keeping k-fault tolerance; that is, each node has at least k neighboring nodes within its transmission range. Two centralized approximation algorithms are proposed, and the upper bound of the approximation factor is proved. A distributed heuristic for 2- and 3-connectivity is also given based on the Minimum Spanning Tree (MST) algorithm. Li and Sinna [20] designed a framework for evaluating (a) the performance of topology control algorithms using the overall network throughput and (b) total energy consumption per packet delivered as the metrics. Based on this framework, scenarios in which topology control improves the network performance can be identified.

Besides the efforts on theoretical analysis of power-efficient topology problem, distributed heuristics are also proposed to improve the network performance in terms of energy and throughput. ElBett and Krishnamurthy [67] proposed a simple distributed algorithm to find the minimum transmit power for maintaining k-connectivity of each node. The algorithm needs a global signaling channel to facilitate the computation of optimal transmission range, and this range will be dynamically changed in a distributed manner in order to adapt to channel variations. Rodoplu and Meng [19] proposed a distributed position-based topology control algorithm that consists of two phases. Phase one is used for link setup and configuration. Each node broadcasts its position to its neighbors and uses this position information of its neighbors to build a spare and strongly connected graph called *enclosure graph*. The key point is that all the globally optimal links (for the minimum power consumption for communication to the destination) are included in the enclosure graph. Therefore, in phase two, nodes find these optimal links on the enclosure graph by applying the distributed Bellman–Ford shortest path algorithm with power consumption as the link cost. Each node broadcasts its cost to its neighbors, where the cost of node i is defined as the minimum power necessary for i to establish a path to destination. Wang et al. [21] proposed a localized algorithm that first constructs a Gabriel graph from the given unit disk graph and then reduces the total transmit power by allowing each node individually excises some replaceable of links. Wattenhofer et al. [22] followed an opposite approach compared with the method in reference 21. Based on directional information, nodes increase their transmit power until they find at least one neighboring node in each direction. Here, transmit power control is based on local information, while guaranteeing global connectivity.

11.5.3 Power Management

Mechanism. In typical wireless networks, terminals have to be powered on at all time to detect possible signals that target them. However, this "always on" results in significant unwanted energy consumption, For example, measurements show that idle listening consumes 50–100% of the energy required for receiving [69]. Therefore, idle listening should be avoided as much as possible. Ideally, the wireless radio is powered on only when there are data waiting to be transmitted or received. Otherwise, the radio is powered off and the node is in the sleep state. In this way, the wasted energy for idle mode is minimized.

The energy wasted in idle mode is mainly due to the overhearing in wireless channel. If the radio is powered on in idle mode, the ongoing transmissions in the neighborhood would be overheard by this node, which frequently happens in ad hoc networks. In this case, energy consumed for packets is not directed to them, thereby causing a waste of energy. A large amount of energy can be consumed unnecessarily is this case. For example, If a transmitter i has n neighbors, then the total energy consumed for an m-packet transmission is $m \times (E_t + n \times E_r)$, where E_t is the energy required for transmitting one packet, and E_r is the energy required for receiving one packet, and an amount of $m \times (n - 1) \times E_r$ energy is wasted in this case.

Based on this observation, a new kind of energy conservation mechanism is proposed in which some nodes are allowed to stay in the sleep state whenever they are not scheduled to transmit or receive. Obviously, this approach can reduce energy consumption, thereby prolonging the battery life of nodes in the ad hoc network. There schemes can be classified into three categories: scheduled, on-demand, and asynchronous power managements. In a scheduled scheme [24, 25], nodes switch to the low-power sleeping state periodically. Because all the participating nodes have to synchronize their clocks, this mechanism is most appropriate for single-hop networks, but is not well-suited in multihop ad hoc networks. In on-demand schemes [27], nodes in the sleeping state are wakened by requests from neighboring nodes. Therefore, the power management is very simple. In asynchronous schemes [26], each node follows its own wake-up schedule in idle mode, as long as the wake-up intervals among neighboring nodes overlap. Thus, this type of scheme does not require time synchronization, and it is more suitable than a scheduled scheme for multihop ad hoc networks.

Related Protocols. Geographic adaptive fidelity (GAF) [24] is an energy-aware location-based routing algorithm. It forms a virtual grid of covered area. Each node uses its GPS-indicated location to associate itself with a point in the virtual grid. Nodes associated with the same point on the grid are considered equivalent in terms of the cost of routing. Such equivalence is exploited in keeping some nodes located in a particular grid area in the sleeping state in order to save energy. Thus, GAF can substantially increase the network lifetime if the number of nodes increases.

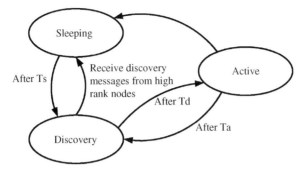

Figure 11.9. State transitions in GAF.

Nodes change states from sleep to active in turn so that the load is balanced. Three states are defined in GAF: the *discovery* state for determining the neighbors in the grid, the *active* state reflecting participation in routing, and the *sleep* state when the radio is turned off. The transition of states is depicted in Figure 11.9. The duration of the sleeping state is application-dependent, and related parameters are tuned accordingly during the routing process. In order to handle to mobility, each node in the grid estimates its leaving time of the grid and sends this to its neighbors. The sleeping neighbors adjust their sleeping time accordingly in order to keep the routing fidelity. Before the leaving time of the active node expires, sleeping nodes wake up and one of them becomes active. In this way, GAF keeps the network connected by keeping a representative node always in the active state for each region on its virtual grid. Simulation results show that the GAF performs at least as well as a normal ad hoc routing protocol in terms of latency and packet losses while increasing the lifetime of the network by saving energy.

Span [25] is an energy-saving technique for multihop ad hoc networks without significantly diminishing the capacity or connectivity of the network. Span adaptively elects "coordinators" from all nodes in the network. These coordinators stay awake continuously and perform multihop packet routing within the ad hoc network, while other nodes remain in power-saving mode and periodically check if they should wake up and become a coordinator. In this way, energy is saved because nodes will stay in power-saving state periodically, and network connectivity is still maintained.

A noncoordinator node determines if it should become a coordinator or not periodically, and the eligibility rule is that if two neighbors of a noncoordinator node cannot reach each other either directly or via one or two coordinators, the node should become a coordinator. Announcement contention is resolved by deferring the transmission of this message with a randomized backoff delay. If announcements from other nodes are received during this period, eligibility is reevaluated. The value of delay takes into account the number of neighboring nodes and the remaining energy, and it uses a randomized "slotting and damping" method reminiscent of various multicast protocols [70, 71]. The function for the delay is of the form

$$\text{delay} = \left(\left(1 - \frac{E_r}{E_m}\right) + \left(1 - \frac{C_i}{\left(\frac{N_i}{2}\right)}\right) + R\right) \times N_i \times T$$

where N_i is the number of neighboring nodes, C_i is the number of additional pairs of neighboring nodes that would be connected if node i were to become a coordinator, E_r and E_m are the remaining energy and initial energy of node i, respectively, and R is a random value uniformly generated from the interval [0, 1]. The intuition is that a noncoordinator node with more normalized remaining energy, which can make more pairs of nodes connected, is preferred to become a coordinator. In this way, the number of coordinators can be reduced. Each coordinator periodically checks if it should withdraw as a coordinator. When every pair of its neighbors can reach each other either directly or via some other coordinators, it transits to the power-saving mode. However, when the duration of service exceeds some threshold, the coordinator would withdraw whenever any pair of nodes in the neighborhood can reach other with other neighboring nodes, even though these nodes are not coordinators. Simulation results show that Span can save energy by a factor of 3.5 (or more) over the 802.11 MAC protocol.

Zhang et al. [26] present a systematic approach to asynchronous power management in ad hoc networks. The optimal wake-up schedule design is formulated as a block design problem in combinatorics. A wake-up schedule function (WSF) of a node v is defined as $f_v(x) = \sum_{i=0}^{T-1} a_i x^i$, where T is the length of the schedule, $a_i = 0$ or $1, \forall i \in [0, T-1]$, and x is the place holder. Thus, $k_v = f_v(1)$ is the number of slots in which node v is scheduled to be awake every T slots. The optimal asynchronous wake-up problem can be formulated as follows: Given a fixed value of T, minimize k_v/T subject to the constraint that the schedules of two neighboring nodes should overlap by at least m slots. A neighbor discovery and schedule bookkeeping protocol is proposed. All the nodes in the network choose the same WSF to schedule its own active and sleeping slots. Neighboring discovery is performed at the beginning of an active slot, and a node transmits a beacon message piggybacking its own id and wake-up schedule. Thus, nodes can detect neighboring nodes in the power-saving states. The schedule bookkeeping protocol stores the schedule and clock shifts of neighboring nodes; thus a node can infer its neighbors' wake-up schedule, and this can facilitate the data communication. The sender will transmit a packet to the receiver only when it infers that the receiver is in its own active slot.

Zheng et al. [27] propose an on-demand power management framework for ad hoc networks. In this framework, power management decisions are driven by data transmission in the network, and connectivity is only maintained between pairs of sources and destinations along the routes of data communication. This reduces energy consumption while maintaining effective communication. Specifically, transitions from the power-saving mode to the active mode are triggered by communication events in the network, and transitions from active mode to power-saving mode are determined by a soft state timer, called the *keep-alive timer*. Initially, all nodes are in power-saving modes. Whenever a routing message or a data packet is received, the timer is set or refreshed to the maximum of what is left for the current timer and the value associated with the received message. Thus, the timer is maintained on a per-node basis. When the timer expires, the node switched to the active state. Simulation results show that a reduction of energy consumption near 50% can be achieved compared to networks without power management.

11.5.4 Open Issues

Traditional minimum-energy routing protocols are proved to be inefficient in ad hoc networks where nodes are of equal importance to the operation of the network; hence energy-efficient routing protocols that maximize the system lifetime of network are desired. Most previous works that target this problem first formulate it as an optimization problem and solve it with some optimization techniques, or propose some heuristics. These algorithms typically require a large amount of message passing. However, the energy consumption of control overhead is always ignored, and whether these proposals still work in real network is an open issue. Moreover, most algorithms assume that the energy consumed for transmitting one packet along a link is fixed. This assumption does not hold if the underlying MAC layer employs power control on a per-packet basis. Maximizing network lifetime in this situation is open, and many works can be done for this issue.

The main concern of the power-efficient topology control algorithm proposed in recent years is to reduce the transmission range of nodes while maintaining moderate connectivity. However, because traffic in ad hoc networks is typically dynamic, a load-sensitive power-efficient topology control may further reduce the total energy consumption in ad hoc networks and prolong the network lifetime.

Power management can save a significant amount of energy for nodes in ad hoc networks because nodes are scheduled to switch to the sleeping state when they are in the idle mode. However, finding an optimal schedule is still open. For example, the following questions need to be answered in the optimal schedule: when to switch to sleep mode, how long to stay in sleeping state, how a node can send or receive signaling message when in sleeping state, and so on. Moreover, power management involves many challenging problems for the upper layers. For example, since there always exists a subset of nodes in any region, traditional broadcasting and multipath routing protocols would not work efficiently in such environment. Thus, either of these routing protocols are redesigned to adjust to such a situation, or the power management scheme is carefully designed take into account these considerations. Another interesting problem is the tradeoff between energy conservation by power management and network performance. Because the network can only make use of a subset of nodes at any time, packet delay is inevitably increased. How to strike a balance between them in power management is a critical issue to the application of power management in ad hoc networks.

11.6 CONCLUSIONS

Since power control has a significant impact on the functionalities in layers of the protocol stack in wireless ad hoc networks, it is a prototypical example of cross-layer design for wireless networks. Power control can reduce transmit power level (and hence gain energy conservation) and performance (e.g., throughput, delay) improvement by alleviating the congestion of wireless medium. There are many approaches to obtain the above benefits with power control. In this survey, we first distill some design principles that can serve as a guide to the design of the power control problem based on the impacts of power control on different layers. Then we address the power control problem in a single layer. The basic idea is intuitively given, and also several algorithms are introduced. Then we address the power control problem in a systematic approach, motivated by the principle that cross-layer design should be tackled holistically. Finally we consider the power control problem in a totally different perspective. We study various methods combined with power control to improve energy efficiency of wireless ad hoc networks. Power control is a good strategy to improve the network performance and conserve energy if joined with other approaches. Though there are many works that follow this methodology, research is still needed to further exploit its benefits, and some interesting and critical open issues are given in each section.

11.7 EXERCISES

1. Briefly explain why power control can affect the congestion level of the wireless medium, and propose methods to alleviate the congestion level with power control under different load conditions.

2. Describe the differences between contention-based medium access control schemes and interference-limited medium access control schemes.

3. In the so-called BASIC schemes, there are two methods to compute the minimum necessary transmit power for data packet delivery. Compare the pros and cons of both methods.

4. BASIC schemes are designed to conserve energy of wireless devices. However, in some wireless ad hoc network scenarios, BASIC schemes cannot achieve this goal and may even increase the energy consumption instead. Explain why this can happen, and illustrate with one such scenario.

5. Explain why PCM can reduce the potential collisions caused by BASIC schemes.

6. Give a scenario where the hidden terminal problem still exists in POWMAC schemes.

7. Explain why power control can affect functionalities at the network layer.

8. In the COMPOW scheme, nodes choose a common power level. However, this may decrease the performance of wireless ad hoc networks under certain scenarios. Illustrate this behavior with an example.

9. Briefly explain the interaction between power control and congestion control in JOCP algorithm.

10. List the metrics used for energy-aware routing algorithms.

REFERENCES

1. V. Kawadia and P. R. Kumar. Principles and protocols for power control in wireless ad hoc networks. *IEEE Journal on Selected Areas in Communications*, **23**(1):65–75, 2005.
2. S. Agarwal, S. Krishnamurthy, R. H. Katz, and S. K. Dao. Distributed power control in ad-hoc wireless networks. In *Proceedings of PIMRC*, 2001.
3. J. Gomez, A. T. Campbell, M. Naghshineh, and C. Bisdikian. Conserving transmission power in wireless ad hoc networks. In *Proceedings of IEEE ICNP*, 2001.
4. P. Karn. MACA—A New Channel Access Method for Packet Radio. In *Proceedings of 9th ARRL Computer Networking Conference*, 1990.
5. M. B. Pursley, H. B. Russell, and J. S. Wysocarski. Energy-efficient transmission and routing protocols for wireless multiple-hop networks and spread-spectrum radios. In *Proceedings of EUROCOMM*, 2000, pp. 1–5.
6. E.-S. Jung and N. H. Vaidya. A power control MAC protocol for ad hoc networks. In *Proceedings of ACM MobiCom*, 2002, pp. 36–47.
7. E.-S. Jung and N. H. Vaidya. A power control MAC protocol for ad hoc networks. *Wireless Networks*, **11**(1–2):55–66, 2005.
8. J. P. Monks, V. Bharghavan, and W.-M. W. Hwu. A power controlled multiple access protocol for wireless packet networks. In *Proceedings of IEEE INFOCOM*, Vol. 1, 2001, pp. 219–228.
9. J. Gomez, A. T. Campbell, M. Naghshineh, and C. Bisdikian. PARO: Supporting dynamic power controlled routing in wireless ad hoc networks. *ACM/Kluwer Journal on Wireless Networks*, **9**:443–460, 2003.
10. M. Chiang. Balancing transport and physical layers in wireless multihop networks: Jointly optimal congestion control and power control. *IEEE Jounal on Selected Areas in Communications*, **23**(1):104–116, 2005.
11. M. Johansson and L. Xiao. Scheduling, routing and power allocation for fairness in wireless networks. In *Proceedings of IEEE VTC*, Vol. 3, 2004, pp. 1355–1360.
12. B. Radunovic and J.-Y Le Boudec. Joint scheduling, power control and routing in symmetric, one-dimensional, multi-hop wireless networks. In *Proceedings IEEE WiOpt*, 2003.
13. B. Radunovic and J.-Y. Le Boudec. Power control is not required for wireless networks in the linear regime. In *Proceedings of Sixth IEEE International Symposium on World of Wireless Mobile and Multimedia Networks (WoWMoM)*, 2005, pp. 417–427.
14. R. L. Cruz and A. V. Santhanam. Optimal routing, link scheduling and power control in multihop wireless networks. In *Proceedings of IEEE INFOCOM*, 2003, pp. 702–711.
15. R. Bhatia and M. Kodialam. On power efficient communication over multi-hop wireless networks: Joint routing, scheduling and power control. In *Proceedings of IEEE INFOCOM*, 2004.
16. J.-H. Chang and L. Tassiulas. Energy conserving routing in wireless ad hoc networks. In *Proceedings of IEEE INFOCOM*, 2000, pp. 22–31.
17. J.-H. Chang and L. Tassiulas. Maximum lifetime routing in wireless sensor networks. *IEEE/ACM Transactions on Networking*, **12**(4):pp. 609–619, 2004.
18. G. Zussman and A. Segall. Energy efficient routing in ad hoc disaster recovery networks. In *Proc. IEEE INFOCOM*, 2003, pp. 682–690.
19. V. Rodoplu and T. H. Meng. Minimum energy mobile wireless networks. *IEEE Journal on Selected Areas in Communications*, **17**(8):1333–1344, 1999.

20. L. Li and P. Sinna. Throughput and energy efficiency in topology-controlled multi-hop wireless sensor networks. In *Proceedings ACM WSNA*, 2003, pp. 132–140.
21. S.-C. Wang, D. S. L. Wei, and S.-Y. Kuo. A topology control algorithm for constructing power efficient wireless ad hoc networks. In *Proceedings of IEEE GLOBECOM*, 2003, pp. 1290–1295.
22. R. Wattenhofer, L. Li, R. Bahl, and Y.-M. Wang. Distributed topology control for power efficient operation in multihop wireless ad hoc networks. In *Proceedings IEEE INFOCOM*, 2001, pp. 1388–1397.
23. R. Ramanathan and R. Rosales-Hain. Topology control of multihop wireless networks using transmit power assignment. In *Proceedings of IEEE INFOCOM*, 2000.
24. Y. Xu, J. S. Heidemann, and D. Estrin. Geography-informed energy conservation for ad hoc routing. In *Proceedings of ACM MobiCom*, 2001, pp. 70–84.
25. B. Chen, K. Jamieson, H. Balakrishnan, and R. Morris. Span: an energy-efficient coordination algorithms for topology maintenance in ad hoc wireless networks. In *Proceedings of ACM MobiCom*, 2001, pp. 85–96.
26. R. Zhang, J. Hou, and L. Sha. Asynchronous wakeup for ad hoc networks. In *Proceedings of ACM MobiHoc*, 2003, pp. 35–45.
27. R. Zheng, J. C. Hou, and L. Sha. Asynchronous wakeup for ad hoc networks. In *Proceedings of ACM MobiHoc*, 2003.
28. J. Macker and M. Corson. Mobile ad-hoc networking and the IETF. *ACM Mobile Computing and Communications Review*, **2**:9–14, 1998.
29. C. E. Perkins. Ad hoc on demand distance vector (AODV) routing. Internet draft, draft-ietf-manet-aodv-04.txt, October 1999.
30. D. B. Johnson, D. A. Maltz, and J. Broch. *The Dynamic Source Routing Protocol for Multihop Wireless Ad Hoc Networks, Ad Hoc Networking*, Addison-Wesley, Reading, MA, 2001, Chapter 5, pp. 139–172.
31. V. Kawadia and P. R. Kumar. A Cautionary perspective on cross-layer design. *IEEE Wireless Communications*, **12**(1):3–11, 2005.
32. P. Gupta and P. R. Kumar. The capacity of wireless networks. *IEEE Transactions on Information Theory*, **46**(2):388–404, 2000.
33. S. Narayanaswamy, V. Kawadia, R. S. Sreenivas, and P. R. Kumar. Power control in ad hoc networks: Theory, architecture, algorithms and implementation of the COMPOW protocol. In *Proceedings of European Wireless Conference*, 2002, pp. 156–162.
34. S. J. Park and R. Sivakumar. Load-sensitive transmission power control in wireless ad-hoc networks. In *Proceedings of IEEE GLOBECOM*, 2002, pp. 42–46.
35. Part11: Wireless LAN medium access control (MAC) and physical layer (PHY) specifications." International Standard ISO/IEC 8802-11; ANSI/IEEE Standard 802.11.
36. L. Jia, X. Liu, G. Noubir, and R. Rajaaraman. Transmission power control for ad hoc wireless networks: throughput, energy and fairness. In *Proceedings of IEEE WCNC*, Vol. 1, 2005, pp. 619–625.
37. A. Muqattash and M. Krunz. Power controlled dual channel (PCDC) medium access protocol for wireless ad hoc networks. In *Proceedings of IEEE INFOCOM*, Vol. 1, 2003, pp. 470–480.
38. A. Muqattash and M. Krunz. POWMAC: A single-channel power-control protocol for throughput enhancement in wireless ad hoc networks. *IEEE Journal on Selected Areas in Communications*, **23**:1067–1084, 2005.

39. S.-L. Wu, Y.-C. Tseng, and J.-P. Sheu. Intelligent medium access for mobile ad hoc networks with busy tones and power control. *IEEE Journal on Selected Areas in Communications*. **18**(9):1647–1657, 2000.
40. T. ElBatt and A. Ephremides. Joint scheduling and power control for wireless ad hoc networks. *IEEE Transactions on Wireless Communications*, **3**:74–85, 2004.
41. U. C. Kozat, I. Koutsopoulos, and L. Tassiulas. A framework for cross-layer design of energy-efficient communication with QoS provisioning in multi-hop wireless networks. In *Proceedings of IEEE INFOCOM*, 2004, pp. 1446–1456.
42. A. Behzad, I. Rubin, and A. Mojibi-Yazdi. Distributed power controlled medium access control for ad-hoc wireless networks. In *Proceedings of IEEE 18th Annual Workshop on Computer Communications*, 2003, pp. 47–53.
43. A. Behzad and I. Rubin. Multiple access protocol for power-controlled wireless access nets. *IEEE Transactions on Mobile Computing*, **3**(4):307–316, 2004.
44. J. C. Fang and R. R. Rao. An integrated and distributed scheduling and power control algorithm for maximizing network utility for wireless multihop networks. In *Proceedings of IEEE MILCOM*, 2003, pp. 1011–1017.
45. M. J. Neely, E. Modiano, and C. E. Rohrs. Dynamic power allocation and routing for time-varying wireless networks. *IEEE Journal on Selected Areas in Communications*, **23**(1):89–103, 2005.
46. O. Savas, M. Alanyali, and B. Yener. Joint route and power assignment in asynchronous multi-hop wireless networks. In *Proceedings MedHocNet*, 2004.
47. V. Kawadia and P. R. Kumar. Power control and clustering in ad hoc networks. In *Proceedings of IEEE INFOCOM*, 2003, pp. 459–469.
48. J. Gomez and A. T. Campbell. A case for variable-range transmission power control in wireless multihop networks. In *Proceedings of IEEE INFOCOM*, 2004, pp. 1425–1436.
49. F. P. Kelly, A. Maulloo, and D. Tan. Rate control for communication networks: Shadow prices, proportional fairness and stability. *Journal of Operations Research Society*, **49**(3):237–252, 1998.
50. S. Boyd and L. Vandenberghe. *Convex Optimization*, Cambridge University Press, 2004.
51. M. Chiang. To layer or not to layer: Balancing transport and physical layers in wireless multihop networks. In *Proceedings of IEEE INFOCOM*, 2004, pp. 2525–2536.
52. L. S. Brakmo and L. L. Peterson. TCP Vegas: End to end congestion avoidence on a global internet. *IEEE Journal on Selected Areas in Communications*, **13**(8):1465–1480, 1995.
53. E.-S. Jung and N. H. Vaidya. Power aware routing using power control in ad hoc networks. Department of Computer Science, Texas A&M University, College Station, TX, Technical Report, February 2005.
54. T. Nandagopal, T. Kim, X. Gao, and V. Bharghavan. Achieving MAC layer fairness in wireless packet networks. In *Proceeding of ACM MobiCom*, 2000.
55. B. Radunovic and J.-Y. Le Boudec. Optimal power control, scheduling and routing in UWB networks. *IEEE Journal on Selected Areas in Communications*, **22**(7):1252–1270, 2004.
56. Y. Nesterov and A. Nemirovskii. Inter-point polynomial algorithms in convex programming. In *SIAM Studies in Applied Mathematics*, 1994.
57. M. Minoux. *Mathematical Programming: Theory and Algorithms*, John Wiley & Sons, New York, 1986.

58. Y. Li and A. Ephermides. Joint scheduling, power control and routing algorithms for ad-hoc wireless networks. In *Proceedings of the 38th Hawaii International Conference on System Sciences*, 2005.
59. S. Singh, M. Woo, and C. S. Raghavendra. Power-aware routing in mobile ad hoc networks. In *Proceedings of ACM MobiCom*, 1998, pp. 181–190.
60. R. Duke, C. D. Rais, K.-Y. Wang, and S. K. Tripathi. Signal stability based adaptive routing (SSA) for ad hoc mobile networks. In *Proceedings of IEEE Personal Communications*, 1997, pp. 36–45.
61. C.-K. Toh. Maximum battery life routing to support ubiquitous mobile computing in wireless ad hoc networks. *IEEE Communication Magazine*, **June**:138–147, 2001.
62. A. Misra and S. Banerjee. MRPC: Maximizing network lifetime for reliable routing in wireless environments. In *Proceedings of IEEE WCNC*, 2002, pp. 800–806.
63. Q. Li, J. Aslam, and D. Rus. Online power-aware routing in wireless ad hoc networks. In *Proceedings of ACM SIGMOBILE*, 2001, pp. 97–107.
64. C. Gao and R. Jantti. A reactive power-aware on-demand routing protocol for wireless ad hoc networks. In *Proceedings IEEE VTC*, 2003, pp. 2171–2175.
65. P. Bergamo, A. Giovanardi, A. Travasoni, D. Maniezzo, G. Mazzini and M. Zorzi. Distributed power control for energy efficient routing in ad hoc networks. *Wireless Networks*. **10**(1):29–42, 2004.
66. S. Doshi, S. Bhandare, and T. X. Brown. An on-demand minimum energy routing protocol for a wireless ad hoc networks. In *Proceedings of ACM Mobile Computing and Communications Review*, 2002, pp. 50–66.
67. T. A. ElBett, S. V. Krishnamurthy, D. Connors, and S. Dao. Power management for throughput enhancement in wireless ad hoc networks. In *Proceedings of IEEE ICC*, 2000, pp. 1506–1513.
68. M. T. Haiiaghayi, N. Immorlica, and V. S. Mirrokni. Power optimization in fault-tolerant topology control algorithms for wireless multi-hop networks. In *Proceedings of ACM MobiCom*, 2003, pp. 300–312.
69. M. Stemm and R. H. Katz. Measuring and reducing energy consumption of network interfaces in hand-held devices. *IEEE Transactions on Communications*, **E80-B**(8):1125–1131, 1997.
70. G. Chesson. XTP protocol engine design. In *Proceedings of the IFIP WG6.1/6.4 Workshop*, May 1989.
71. S. Floyd, V. Jacobson, S. McCanne, C. G. Liu, and L. Zhang. A reliable multicast framework for lightweight sessions and applications level framing. In *Proceedings of ACM SIGCOMM*, 1995, pp. 342–356.

… **CHAPTER 12**

Power Saving in Solar-Powered WLAN Mesh Networks

AMIR A. SAYEGH, MOHAMMED N. SMADI, and TERENCE D. TODD

Department of Electrical and Computer Engineering, McMaster University, Hamilton, Ontario, L8S 4K1, Canada

12.1 INTRODUCTION

WLAN mesh networks have emerged as a cost-effective way of deploying outdoor coverage in metro-area Wi-Fi hotzones. In a WLAN mesh, multihop relaying is used to reduce the infrastructure cost of providing wired network connections to each WLAN mesh node. Protocols for these networks are currently being standardized under IEEE 802.11s, which will further reduce costs due to standard chipsets and other components.

One of the major costs of certain outdoor mesh deployments is that of providing nodes with continuous electrical power connections. In some cases, node powering is readily available; but in many others, it is expensive or practically impossible. An alternative to a fixed power connection is to operate the mesh nodes using a sustainable solar- powered design. In a solar-powered network, node resource allocation involves assigning solar panels and battery sizes to each mesh node. This assignment must incorporate the geographic location where the node is to be located, so as to account for its solar insolation history. Resource assignment is normally done using a target load profile for the node, which represents the workload for which it is being configured. In solar-powered mesh nodes the cost of the solar panel and battery can sometimes be a significant fraction of the total node cost. For this reason, node power consumption can strongly affect the end cost of the deployed network.

WLAN mesh networks typically carry traffic between IEEE 802.11 user stations and wired Internet connections. Such a network is usually provisioned for worse-case traffic conditions, and for this reason the time-averaged traffic load is always far lower than the worse case. For this reason, there is a potential for infrastructure

---

Algorithms and Protocols for Wireless and Mobile Ad Hoc Networks, Edited by Azzedine Boukerche
Copyright © 2009 by John Wiley & Sons Inc.

power saving that is inherent in the operation of the network. Unfortunately, IEEE 802.11 does not include native mechanisms that would allow a mesh AP to achieve this power saving, since IEEE 802.11 requires that all mesh APs be active at all times. This is an unfortunate impediment to the development of an energy-sustainable WLAN infrastructure.

In this chapter we will discuss the design and resource allocation for solar-powered IEEE 802.11 WLAN mesh networks. The potential for mesh node power saving is considered using both best-effort and real-time connection-oriented traffic scenarios. The presented results will compare various design alternatives including different infrastructure power-saving and non-power-saving options. Cases will be included using existing IEEE 802.11 standard assumptions and will also consider the case where modifications are made to the standard so that mesh AP power saving is possible. The results will also give a strong motivation for including access point power saving in outdoor WLAN mesh networks.

12.2 SOLAR-POWERED WLAN MESH NETWORKS

Figure 12.1 shows an example of a WLAN mesh network. In this network, solar mesh APs (SMAPs) are mesh node access points that provide both end-user station Wi-Fi coverage and also perform traffic relaying. This relaying is done to backhaul traffic so that all coverage areas are interconnected. In the figure, SMAP 2 provides IEEE 802.11 coverage for the two mobile stations shown on frequency f_4, and it also implements traffic relaying on f_2. Solar mesh points (SMPs) provide relaying services only; that is, they do not communicate directly with end-user stations. In

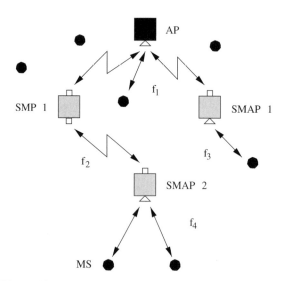

Figure 12.1. Solar-powered WLAN mesh network example.

Figure 12.1, for example, SMP 1 relays traffic between the wired AP and SMAP 2 but does not provide coverage for any end-user MSs.

One of the major costs of certain WLAN mesh deployments is that of providing mesh APs and mesh points with electrical power and wired network connections. This is especially true in Wi-Fi hotzones, where coverage is provided over extended outdoor areas. Although power can sometimes be supplied through power over Ethernet (PoE), such a solution requires a wired network connection, which is often very expensive. For the past several years, the SolarMESH network has been under development and has been undergoing deployment trials at McMaster University [1]. In SolarMESH, the WLAN mesh nodes are solar-powered, and they can be deployed quickly and inexpensively for certain outdoor Wi-Fi installations. This network has been carrying live traffic since 2003 and has helped to validate the use of solar power for this type of application. A rooftop picture of a SolarMESH Version II node is shown in Figure 12.2.

In an SMAP or SMP, energy is generated in the solar panel, and it is consumed by the SMAP/SMP electronics including its wireless network interface(s). A battery is needed to ensure continuous operation of the node over periods of time when there is insufficient solar insolation. Figure 12.3 shows a simplified block diagram of a solar-powered WLAN mesh node. The node includes a solar panel, a battery, and a charge controller that provides battery protection from over- and undercharging. The solar panel and battery are connected to the SMAP/SMP circuitry through the charge controller.

A discrete time energy flow model governs the operation of the node, where $\mathcal{E}_{panel}(k)$ is the energy produced in the solar panel between times $[(k-1)\Delta$ and $k\Delta]$, where Δ is the time-step length considered (typically 1 hour). The solar panel

Figure 12.2. SolarMESH Version II mesh AP.

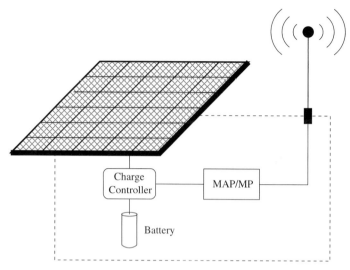

Figure 12.3. Basic SMAP components.

size is given by S_{panel}, and it is rated in watts at peak solar insolation. $\mathcal{B}(k)$ is defined as the residual battery energy stored at time $k\Delta$, and $\mathbf{B}_{max}$ is defined to be the total battery capacity. The load energy demand is given by $\mathcal{L}(k)$ over the time duration $[(k-1)\Delta, k\Delta]$, and therefore the energy balance equation can be written as [2]

$$\mathcal{B}(k) = \min\{\max[\mathcal{B}(k-1) + \mathcal{E}_{panel}(k) - \mathcal{L}(k), \mathbf{B}_{outage}], \mathbf{B}_{max}\} \quad (12.1)$$

where $\mathbf{B}_{outage}$ is the maximum allowed depth of battery discharge, based on safety and battery life considerations [3]. When $\mathcal{B}(k) < \mathbf{B}_{outage}$, the charge controller will disconnect the MAP/MP load and the node will experience a radio outage. Note that Eq. (12.1) can easily be modified to include non-ideal effects such as battery temperature dependence.

Since the total received solar energy and the traffic related workload are stochastic processes, this results in a level of uncertainty that must be addressed when the node is provisioned. The performance of an SMAP can be measured in terms of the frequency of system outage, P_{Out}. Using public solar insolation data for a particular geographic location [4, 5] and the energy flow model, a worst-case design can be performed. This is done by assuming that $\mathcal{L}(k) = \mathbf{P}_{APmax}\Delta$ for all k, where $\mathbf{P}_{APmax}$ is the peak power consumption of the SMAP. Using Eq. (12.1), a discrete time simulation of the node is done over the solar insolation history of the desired location. Using this method, it is straightforward to produce a set of solar panel versus battery capacity contours for a constant outage probability, P_{Out}. In order to complete this process, a solar energy conversion model is required that translates available solar insolation records into solar panel inputs for a given geographic location. A brief overview of the conversion model used is given in the Appendix.

An example output from such a simulation is shown in Figure 12.4 for two example locations (i.e., Toronto, Ontario, Canada, and Phoenix, Arizona), assuming a constant single-radio SMAP power consumption of 1 W. The graphs show plots of solar panel size versus battery capacity for curves of constant outage probability. Comparing the two geographic locations, the differences in node configuration are seen to be very large. In the more temperate case (Toronto), provisioning a solar node for negligible outage (i.e., $< 10^{-4}$) requires about a 22-W solar panel, which alone would be a significant fraction of the total node cost. In the Phoenix case the same performance can be obtained with about an 8-W panel. This will obviously result in a significant node cost difference between the two locations.

For each outage probability in Figure 12.4, there is a continuum of feasible battery/panel combinations. In this chapter we assume that the panel/battery selection is done on the basis of minimum cost. A simple cost model is developed that relates the power consumption, the target outage probability, and the overall battery and solar panel costs. Using the approximate linear relationship between cost and both solar panel and battery size (which is usually the case), lines of constant cost and slope can be superimposed on Figure 12.4. The optimum panel and battery cost corresponds to the line that intersects the desired outage curve tangentially.

We now consider the minimum cost of various design options. For comparison we will express these costs in Canadian dollars (CAD) using recent single unit retail costs. We assume that the variable component of the cost function will be made up of the costs of the battery and the solar panel. This component relies on the target outage probability, P_{Out}, and the power consumption profile associated with the node. In this example we will assume an average power consumption given by α watts. In this case the optimal cost can be approximately described, for $P_{Out} = 0.1$, for the city of Toronto, that is,

$$\text{Cost}_{\text{Solar}} \simeq (110\alpha + 37) \qquad (12.2)$$

The minimum panel/battery cost is a linear function of the power loading of the node. This is because the cost per Ah and W of batteries and panels, respectively, is roughly linear. Figure 12.5 shows contour plots of the design α (in watts) versus P_{Out}. This figure is useful when finding the cost optimal design for a given P_{Out} and α. For example, if the design budget is limited to be less than 200 CAD, then P_{Out} is less than 0.052 for loads varying from 0.675 W to 1 W.

Let us now consider the minimum cost comparison for three different North American cities representing three varying solar situations, that is, Toronto, Ontario, Canada, Phoenix, Arizona, and Yellowknife NWT. The normalized minimum cost for these cases is shown in Tables 12.1, 12.2 and 12.3. These values are normalized to one CAD per Ah and assume a 2:1 ratio of panel to battery cost. By examining the table, we can see that the total cost per watt follows a linear trend for a given P_{Out}. For Toronto a total cost of 55.32 achieves $P_{Out} = 0.0001$, while for Phoenix and Yellowknife the total cost is 22.62 and 113.21, respectively. We can also see that Phoenix needs about half the panel/battery cost investment per node while Yellowknife requires about double the Toronto case.

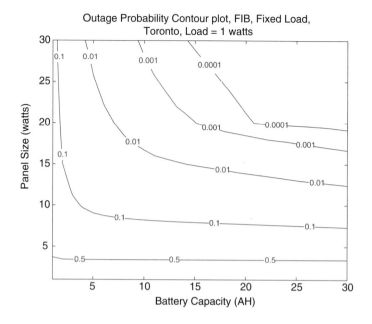

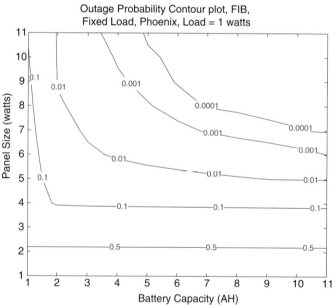

Figure 12.4. Outage example for Toronto and Phoenix, 1-W SMAP loading. FIB: full initial battery charge.

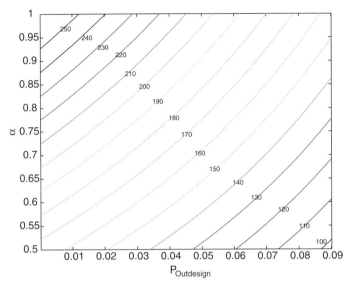

Figure 12.5. Cost contour for α versus P_{Out}.

TABLE 12.1. Cost (Normalized) for Different Outage Probabilities and Load Powers for City of Toronto

P_{Out}	1 W	2 W	3 W	4 W
10^{-1}	22	43.8	65.62	87.5
10^{-2}	40.46	80.88	121.25	161.58
10^{-3}	52	103.74	155.46	207.32
10^{-4}	55.32	110.47	165.46	220.45

TABLE 12.2. Cost (Normalized) Values for Different Outage Probabilities and Load Powers for City of Phoenix

P_{Out}	1 W	2 W	3 W	4 W
10^{-1}	9.74	18.5	27.6	36.72
10^{-2}	15.34	30	45.18	60.22
10^{-3}	20.64	40.7	60.86	80.95
10^{-4}	22.62	44.2	66	88.25

TABLE 12.3. Cost (Normalized) Values for Different Outage Probabilities and Load Powers for City of Yellowknife

P_{Out}	1 W	2 W	3 W	4 W
10^{-1}	36.54	73	109.47	146
10^{-2}	76.25	152.32	228.33	304.5
10^{-3}	104.43	208.85	313.16	417.7
10^{-4}	113.21	226.56	339.7	453.12

Obviously, P_{Out} is a decreasing function of cost for a fixed outage target. Let us first consider the temperate climate case exemplified by Toronto. For a 4-W load in order to reduce P_{Out} from 0.1 to 0.0001, an increase of investment from 87.5 to 220 is needed. If we consider a location with an abundance of solar insolation such as Phoenix, the cost increases from 36.72 to 88.25 (240%). For an extreme case such as Yellowknife, the investment must increase from 146 to 453.12 (300%). Careful examination of the results shows that the ratios are generally the same. On the other hand, what is not clear is that in order to reduce the outage from 0.001 to 0.0001, only a small increase in resources is needed. This holds for all three cities. In fact, the cost per watt needed to achieve a given outage is almost constant. For the temperate case, Toronto, increasing the cost from 207.32 to 220.45 gives this effect, which represents an increase of only 6%. For Phoenix, one needs a 9% increase; for Yellowknife, 8.5%.

12.3 PROTOCOL POWER SAVING IN SOLAR-POWERED WLAN MESH NETWORKS

From our previous discussion, it is clear that reducing mesh node power consumption can lead to significant cost savings. For this reason, SMAPs/SMPs should be designed so that power consumption is as low as possible. Unfortunately, there are limits to what can be achieved without any protocol-based mechanism for access point power saving. Using protocol-based power saving, SMAPs/SMPs could reduce their power consumption significantly, and in this case the achievable levels would be constrained only by satisfying user QoS requirements. Unfortunately, the current IEEE 802.11 standard only provides for mobile station power saving.

In IEEE 802.11 power saving the stations can be in one of two states: awake and doze. When the station's radio components are fully powered, this is defined as the awake state. In this state the station is capable of transmitting and receiving frames, as well as listening on the channel. In the doze state, however, the station is free to perform power saving (PS). In this case, the station is operating at reduced functionality and is not able to perform normal radio functions.

When a station is in doze mode the AP is responsible for buffering incoming frames for that station. The frames are then transmitted when the station is available in the awake state. A power-saving station listens to the AP beacons; and if the AP is holding frames for a given station, it indicates this in the traffic indication map (TIM), which is transmitted with the AP beacons. The station can then poll the AP (using PS-poll frames) to retrieve its buffered packets. Broadcast and multicast frames are handled using a delivery TIM (DTIM), and they are sent immediately after the beacons that contain the DTIM.

In IEEE 802.11e, an automatic power-save delivery (APSD) scheme is defined that allows the transmission of downlink frames synchronously based on periodic service intervals defined by the hybrid coordinator (HC). Upon the arrival of a new session destined to an APSD-enabled station, fixed positions in the superframe are scheduled by the HC for that station. This allows the APSD station to switch from the doze state into the awake state to receive or poll the HC for its buffered traffic. On the other hand, if the HC detects that the station is in the awake state outside the scheduled service periods, it can communicate with the APSD-enabled station asynchronously.

Since the IEEE 802.11 standard does not provide a mechanism for placing APs into power-save mode, the only available schemes are proprietary. Two such mechanisms have recently been proposed in reference 6 and 7, based on extensions to the IEEE 802.11 standard. In the first scheme, the AP makes use of fixed sleep and awake periods in order to perform power saving. The second approach extends the first by introducing a dynamic mechanism for updating the sizes of the sleep and awake periods. In this case the AP adaptively modifies its sleep schedule to support best effort traffic. The scheme is based on extensions to IEEE 802.11e [8] where an AP that supports power saving includes a network allocation map (NAM) in its beacon broadcasts. The NAM specifies periods of time within the superframe when it is unavailable, and during these periods the AP may conserve power. An example of this is shown in Figure 12.6, where a single inter-beacon period is given. The upper timeline shows channel activity, while the lower timeline shows the NAM times that are carried in the SMAP beacons. In this example, we show multiple HCCA periods and we assume that they are repeated every 20 ms. The NAMs are advertised by the SMAP so that power saving can occur when the channel is not needed.

Figure 12.7 shows another example for a best-effort traffic case. The SMAP has defined a power-saving period as shown which spans half of the inter-beacon period. MSs with best-effort traffic are thus prohibited from using this time. In reference 6, a movable boundary is defined which allows the SMAP to dynamically adjust the offered capacity to stations with best-effort traffic (as shown in Figure 12.7). An algorithm for dynamically updating the NAMs was proposed in reference 6. Both cases depicted in Figures 12.6 and 12.7 (connection-oriented and best-effort) can be integrated into the same SMAP superframe.

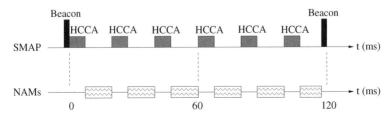

Figure 12.6. Power-saving MAP example with HCCA service periods.

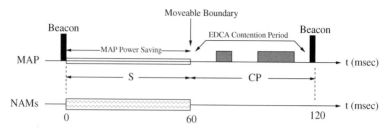

Figure 12.7. Best-effort MAP power saving with movable boundary.

12.4 POWER SAVING FOR SMAPs/SMPs

Cost-effective solar provisioning can be done based on average energy usage profiles due to solar panel and battery smoothing effects [9]. Using this approach, resource assignment results clearly show the potential for cost savings when SMAPs and SMPs are designed to a target usage which is less than peak power consumption. In this section we discuss the power saving that is available in both best-effort (BE) and typical connection-oriented (CO) traffic scenarios. In Section 12.4.1 we review a movable boundary mechanism that an SMAP can use to implement power saving for BE traffic using a modified version of IEEE 802.11. In Section 12.4.2 we discuss the SMAP power saving that is inherent in certain connection- oriented traffic such as VoIP.

12.4.1 Power-Saving AP Support for Best-Effort Traffic

Power-saving SMAPs require a mechanism for dynamically adjusting the capacity offered to MSs in accordance with current best-effort traffic demands. When there is very little traffic to be carried, the SMAP should be free to conserve power, and when loading increases, the level of power saving should decrease accordingly. When relaying traffic between two SMPs, the nodes can easily exchange backlog information and arrange for power saving. However, on the WLAN coverage radio it is much more difficult for an SMAP to determine the current state of mobile station backlog. In this section, we describe the work presented in reference 6, which includes a simple method for accomplishing this based on measuring channel utilization during best-effort time periods.

As previously discussed, the SMAP beacons advertise periods of time to the MSs (i.e., NAMs) during which the SMAP is unavailable and performing power saving. Figure 12.7 shows an SMAP superframe consisting of a power-saving sleep (i.e., **S**) subframe followed by a contention period (i.e., **CP**), which are separated by a movable boundary. For convenience the intervals of two adjacent **CP** and **S** subframes will be normalized to 1. During the **CP** time interval the SMAP updates the movable boundary between the two subframes. $t_{CP}(i)$ and $t_S(i)$ are defined to be the durations of the **CP** and **S** subframe intervals in the ith instance of this subframe, where $t_{CP}(i) + t_S(i) = 1$. For superframe i we define an error signal to be [6]

$$e(i) = (U(i) - \mathbf{U}_{th})t_{CP}(i) \qquad (12.3)$$

where $U(i)$ is the achieved **CP** usage and $\mathbf{U}_{th}$ is a target usage based on a normalized utilization threshold that has been set, $\mathbf{U}_{th}$. In reference 6 an update mechanism based on least mean square (LMS) adaptive control was used to dynamically update the boundary between **CP** and **S** subframes. According to the LMS optimization criterion, we want to minimize $(e(i))^2$, and it can be shown that this leads to

$$\frac{\partial}{\partial t_{CP}(i)} e(i)^2 = 2(U(i) - \mathbf{U}_{th}) t_{CP}(i) \frac{\partial e(i)}{\partial t_{CP}(i)} \quad (12.4)$$

Due to the difficulty in evaluating the partial derivative in this term, it is incorporated into a single constant denoted by $\mathcal{K}$ [10]. The value of $\mathcal{K}$ is then appropriately tuned for the situation being considered. Using this result and assuming that we do not let $t_{CP}(i)$ drop below a minimum value $t_{\min}$, the following update equation can be used:

$$t_{CP}(i+1) = \max \{\min(t_{CP}(i)[1 + \mathcal{K}(U(i) - \mathbf{U}_{th})], 1), t_{\min}\}. \quad (12.5)$$

The choice of $\mathcal{K}$ captures the tradeoff between responsiveness and steady-state error. In reference 6 the range of possible settings for U_{th} was characterized. Each SMAP has N_s-associated MSs, each fed by a Poisson packet arrival process. The default simulation parameters used are shown in Table 12.4.

In Figures 12.8 and 12.9 we show the tradeoff between packet mean delay and power consumption for an SMAP that runs the adaptive protocol in comparison to another SMAP that is using a fixed boundary for the contention period. The fixed boundary scheme was originally proposed in reference 7. In the fixed boundary scheme we show the curves corresponding to 50% and 75% of the SMAP time being spent in the **CP** interval. The capacity of the SMAP in these cases is mandated by the fixed boundary. It can be seen that the power consumption stays relatively flat, with minor variations due to the different times spent in transmit and receive modes. The power consumption of the SMAP associated with a 75% fixed boundary is 50% higher than that of the 50% fixed boundary case. Therefore, it can be seen that the power consumption in the fixed boundary case scales proportionally to the activity interval.

TABLE 12.4. Default Simulation Parameters

Parameter	Value
Superframe interval, T_{SF}	100 ms
WLAN transmission rate	11 Mbps
Number of data stations per AP, N_s	20
Data packet payload	200 bytes
Power consumption in LISTEN/RECEIVE mode	500 mW
Power consumption in TRANSMIT mode	750 mW
Power consumption in DOZE mode	2 mW
Minimum duration of **CP**-subframe, $t_{\min}$	10 ms

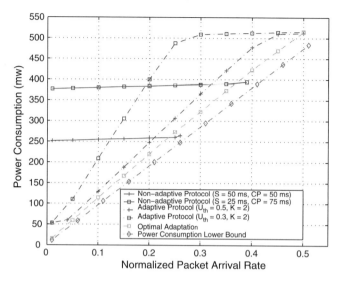

Figure 12.8. Algorithm performance-power consumption for SMAP.

Compared with the fixed subframe case, the adaptive protocol adapts the **CP**-subframe duration based on traffic loading. In Figure 12.8 we compare the power and delay performance for the 0.3 and 0.5 U_{th} cases. It can be seen that the case corresponding to U_{th} set to 0.5 has a significantly better SMAP power consumption for almost all the interarrival rates considered when compared to the case of U_{th} set to 0.3. This is true because larger values of normalized utilization ($U(i)$) are needed before the algorithm responds by increasing the **CP**-boundary. In Figure 12.8 we

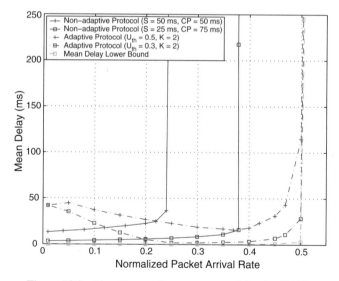

Figure 12.9. Algorithm performance-mean delay for SMAP.

also compare the performance of the adaptive protocol to optimal adaptation and to a lower bound. This curve was obtained by assuming that the SMAP instantly assumes a low-power sleep mode whenever there is no packet backlog, and it immediately reawakens when a backlog appears. It can be seen that the power consumption curve of the adaptive protocol closely resembles that of the optimal adaptation when $\mathbf{U}_{th}$ is set to 0.5. We also find the lower bound on power performance by summing up the components required for the successful transmission of packets at a particular packet arrival rate while ignoring the overhead due to packet collision.

In Figure 12.9 the curves show the mean delay performance of our adaptive protocol for the two cases selected of $\mathbf{U}_{th}$. When $\mathbf{U}_{th}$ is set to 0.5, more delay is incurred than when $\mathbf{U}_{th}$ is set to 0.3. This difference in mean delay is due to the less reactive SMAP driving the mean delay up in the 0.5 case compared with 0.3. However, in many cases this mean delay increase is probably not very significant for best-effort traffic. It is clear that the selection of $\mathbf{U}_{th}$ and $\mathcal{K}$ allows the SMAP to make a tradeoff between mean station delay and SMAP power consumption. Also we can explain the decrease in mean delay as arrival rate increases when implementing the adaptive protocol by noting that the responsiveness of the SMAP increases as loading goes up. Finally, we also present a mean delay lower bound that is obtained by making the SMAP always available for packet transmission, and we can observe that the 0.3 $\mathbf{U}_{th}$ curve approximates this bound very closely for medium and high loading.

It is clear from the above results that a simple control scheme can obtain excellent best-effort traffic performance and track the system backlog sufficiently well to achieve good power saving for the SMAP. Figures 12.8 and 12.9 clearly show that when traffic drops to low values (such as during nighttime hours in an outdoor hotspot), the SMAP power consumption can be significantly reduced. When traffic conditions are heavy, the SMAP operates as a fully powered AP.

12.4.2 SMAP/SMP Power Saving in Connection-Oriented Traffic

To achieve reasonable connection-oriented traffic performance, the SMAP/SMP node traffic load should not exceed certain limits; otherwise, unacceptable blocking rates will occur. In this section we study the SMAP/SMP operation using VoIP as an example of connection-oriented traffic. The parameters at the SMAPs are configured based on a target call blocking probability. Our objective is to show that significant cost reductions in provisioning the SMAPs can be obtained if potential power savings due to idle periods along the timeline are exploited.

Radio power consumption at the SMAP consists of coverage radio power consumption and relay radio power consumption. The relay radio protocols can be proprietary and capable of power saving. However, a significant potential for power saving is made possible by controlling the utilization of the coverage radio as well.

The analysis presented in reference 11 quantifies the capacity of an IEEE 802.11e HCCA access point carrying VoIP traffic. Table 12.5 summarizes the number of concurrent calls that the access point is capable of supporting using different codecs. This view of the access point capacity permits us to use the Erlang B formula to plot the call blocking probability as a function of the call arrival rate, λ. In Figure 12.10

TABLE 12.5. VoIP Capacities for IEEE 802.11e HCCA

Codec	802.11b 11 Mbps	802.11b 5.5 Mbps	802.11a 54 Mbps	802.11a 36 Mbps	802.11a 18 Mbps
G.711	16.1	12.7	105.6	89.3	60.9
G.726	17.8	14.8	120.7	106.1	77.9
G.729	19.3	17	135.3	123.6	98.3

we show the call blocking probability for different codecs as a function of the traffic arrival rate assuming an exponentially distributed holding time with a mean of 180 s for the 11-Mbps data rate. Throughout the following we assume the IEEE 802.11b air interface, although similar conclusions result from using IEEE 802.11a/g.

Figure 12.11 plots the utilization versus the call arrival rate. In practical telephony, designers aim at provisioning resources that result in a 1% maximum new call blocking probability. From Figure 12.10 we see that the call arrival rates corresponding to 1% blocking probability are 0.049, 0.053, and 0.062 calls per second for the G.711, G.726, and G.729, respectively. At these arrival rates, all the codecs have a similar utilization in the 55% region. This observation has strong implications on the potential for SMAP power saving. Even at unacceptably high blocking probabilities, the system continues to be underutilized; for example, at a blocking probability of 10% the system utilization is close to 75% for the G.711 codec, affirming the feasibility of power saving as a means of cost reduction when provisioning the system.

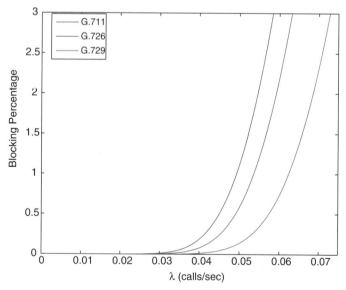

Figure 12.10. Blocking percentage versus call arrival rate for IEEE 802.11b HCCA at 11 Mbps.

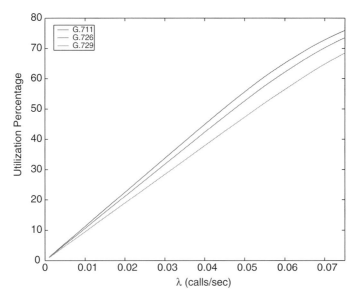

Figure 12.11. Utilization percentage versus call arrival rate for IEEE 802.11b HCCA at 11 Mbps.

To give concrete examples of potential cost savings as a result of capturing system underutilization, we define a power model for the access point. We assume that an 802.11e air interface is used over relay and coverage radios. The timeline is divided between connection-oriented (CO) traffic that spans a total superframe fraction, ρ, and best-effort (BE) traffic that occupies the remaining time, assuming a normalized superframe time. We assume that a SMAP has a single coverage radio used for mobile station access and a single relay radio used for back-hauling traffic. The total power consumption at the node consists of three components. The first is the power consumed by the host processor running on the node, we refer to this as P_H. The second component is the power consumed by the coverage radio P_c, and the third component is the power consumed by the relay radio P_r.

Most host processors are capable of operating in sleep mode when the network interfaces are idle. Therefore, in our power model we assume that the host processor only consumes significant power if there is traffic being processed by the coverage or relay radios. We capture this effect by

$$P_H = \max\{\min(U_c + U_r, 1)P_{H\max}, P_{H\min}\}, \tag{12.6}$$

where U_c and U_r represent the total utilization of the coverage and relay radios, respectively. $P_{H\max}$ represents power consumption at the host processor when its operating at maximum capacity. $P_{H\min}$ represents the power consumption of the host processor when no traffic is carried. The above-mentioned relation assumes a worse-case temporal correlation between U_c and U_r.

Next we consider the power consumption of the coverage radio. As a worse case we assume that the coverage radio is fully awake during the $1-\rho$ time interval when it is servicing best effort traffic. During the remaining ρ interval the coverage radio services CO traffic. The deterministic nature of VoIP traffic allows the coverage radio to sleep during idle intervals using HCCA NAMs such as those shown in Figure 12.6. The power consumption of these activities on the coverage radio's timeline is captured using

$$P_c = P_{BE} + P_{CO} \tag{12.7}$$

$$P_{BE} = (1 - \rho) \times P_{\text{Max}} \tag{12.8}$$

$$P_{CO} = U_{CO} \times P_{\text{Max}} + (\rho - U_{CO}) \times P_{\text{Sleep}} \tag{12.9}$$

where P_{Max} represents the average of the maximum transmit and receive power, and P_{Sleep} represents the power dissipated when the radio is in sleep mode. The relay radios can use any proprietary power saving protocol since they do not have to interact directly with legacy MPs, and all relay traffic is regarded as connection-oriented. Therefore we can describe the relay radio's power consumption as

$$P_R = U_r \times P_{\text{Max}} + (1 - U_r) \times P_{\text{Sleep}} \tag{12.10}$$

In Table 12.6 we summarize the values of the parameters assumed in our simulations. We assume that the best effort and the VoIP signaling traffic are negligible.

For the coverage radio there are two extreme cases to consider. First is the case when the relay radio does not carry any traffic, $U_r = 0$ and $P_H = \max\{U_c \times P_{H\max}, P_{H\min}\}$. The power consumed by the relay radio will be P_{Sleep} which is equal to 2 mW. The coverage radio consumption will depend on U_{CO} which varies from 0 to ρ. Since we neglect any signaling or best effort traffic by assuming $U_{BE} = 0$, then $P_c = (U_{CO}) \times P_{\text{Max}} + (1 - U_{CO}) \times P_{\text{Sleep}}$. Therefore, P_c varies between 0.752 W ($U_{CO} = \rho$) and 0.002 W ($U_{CO} = 0$). P_H varies between 1 and 0.02 W respectively. Thus the total power consumed by the node varies between 0.024W and 1.752W. In the second case, the relay radio is fully utilized, $P_r = 0.75$ W, $P_H = 1$ W and P_c varies between 0.002W and 0.75W, therefore, the total power will vary between 1.752W and 2.5W.

Figure 12.12 shows P_{Total} vs U_{CO} for several different relay radio consumption scenarios varying from $U_r = 0$ to 100%. It can be seen that as the utilization of both

TABLE 12.6. Parameters

Parameter	Value
P_{Max}	0.75 W
P_{Sleep}	2 mW
$P_{H\max}$	1 W
$P_{H\min}$	20 mW
ρ	1

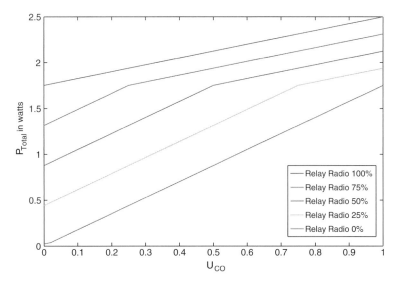

Figure 12.12. P_{Total} versus U_{CO}.

radios increases, the power consumption also increases. When the relay radio is fully utilized, the consumption varies between 1.752-2.5 W with a slope of 0.75 and for the case when the relay radio does not carry any traffic, the consumption varies between 0.024 to 1.752 W with a slope of 1.73. When the relay radio is under-utilized, the power consumption depends more strongly on the utilization of the CO traffic, this is clear in the variation of the slope from 0.75 to 1.73, an increase of more than 2.3 times. For example, when the relay radio utilization is 50% the power consumption increases rapidly with a slope of 1.7 up to $U_{CO} = 0.5$. After that point, the slope of the power consumption curve decreases to 0.9 which is a reduction of almost 50%. For example, when $U_{CO} = 0.5$, the power consumption is approximately 1.75 W. The reason for the change in slope is that when the sum of the utilization of both radios is less than $P_{H \max}$ the power consumption increases linearly, however when they exceed it, the slope of the power consumption curve remains linear but changes due to the fact that the host processor power is constant. This applies for all cases except when $U_r = 1$ or 0.

From this we can reach the conclusion that when the medium is less utilized, the power consumption is reduced rapidly and is not affected by the overheads; this is especially true when the relay radio is carrying less traffic. Therefore, there are two elements of motivation for reducing the medium utilization: satisfying the QoS guarantees as previously discussed, and the potential for power consumption reduction. If the system designer has perfect knowledge of the utilization of the relay radio, the power saving potential can be properly evaluated. For example, if the node is placed in a location within the mesh network where it is relaying large amounts of traffic for other users and it has several relay radios, then the potential for power saving is less and the designer may decide to forgo power-saving protocols on the coverage radio. This will also depend on the ratio of the coverage radios to the relay radios.

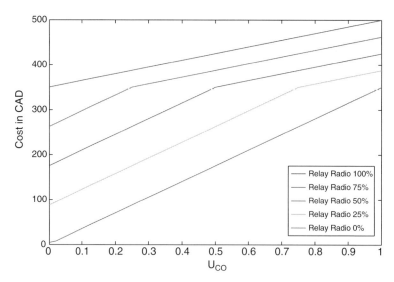

Figure 12.13. Cost versus U_{CO} for $P_{\text{Outdesign}} = 0.0001$.

As previously shown, cost can be approximated as a linear function of power consumption for a given outage probability. Figure 12.13 shows the cost in CAD versus P_{CO} for $P_{\text{Out}} = 0.0001$ for a single node. We can see that when the relay radio is off, the cost is heavily dependent on the expected value of the arrival rate. We can see that the cost varies between 35.8 and 350 CAD. We can also see that employing power saving at the SMAP will help significantly reduce the cost of the nodes, especially when accurate traffic forecasting is available to the designer based on traffic demands and the network topology.

We now integrate the previous results in order to estimate the potential of power saving for WLAN mesh networks. To facilitate the evaluation of the results, Figure 12.14 shows the power consumption versus the call arrival rate for different relay radio utilization levels. It is interesting to observe the effect of the relay radio utilization on the dependency of power consumption to the call arrival rate. We can observe this by comparing the two cases when the relay radio is on and when it is off. First, if we consider the no relay case, the power increases from approximately 0 W to 1.25 W while the arrival rate varies from 0 to 0.075 calls per second. The curve is a straight line with slope 16.67, while for the case when the relay radio is on, the power increases from 1.75 to 2.25 W again in a straight line with slope 6.7. Therefore, we notice that the slope has increased 2.5 times, signifying that the dependency on the call arrival rate increases when the relay radio is sleeping or has minimal traffic.

From our discussion and from the results shown in this section, we can see that limiting the call arrival rate in order to satisfy QoS constraints also provides a significant opportunity for power saving, which in turn leads to network cost savings.

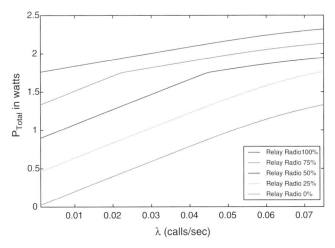

Figure 12.14. P_{Total} versus λ.

12.5 CONCLUSIONS

In this chapter we have discussed the design of solar-powered WLAN mesh networks. We first considered the basic solar-powered node components needed for energy sustainable operation. In order to provide this sustainability, the solar panel and battery have to be carefully provisioned, taking into account the solar insolation history of the node's intended location. Sizing of these components is important since they are often a significant fraction of the node cost. In order to properly provision a node, its target power consumption must be known in advance, which may be difficult if the node's power consumption is traffic-dependent.

Solar-powered node cost is often a strong function of the solar panel and battery resources needed by the node. Protocol power saving for mesh APs, however, is not defined under any existing IEEE standard. In this chapter we have shown the value that such power saving would provide. Movable boundary mechanisms are available which would allow a node to achieve power saving and still offer good best-effort traffic performance. Similarly, mesh AP power saving can be achieved for connection-oriented traffic using minor modifications to the IEEE 802.11 standard. The advantages of this were clearly shown in our results.

12.6 APPENDIX 1: SOLAR INSOLATION CONVERSION FOR SIMULATION OF PV SYSTEMS

This appendix provides an overview of how solar radiation data can be modeled and converted in order to simulate a photovoltaic (PV) system. Current commercial simulators take as an input the current geographic location and the required load profile, and they then perform discrete event simulations in order to predict different outage events and battery and PV sizing necessary to provide sustainable operation of the system. The simulator must take into account the inefficiencies caused by losses in the charge controller and other components in the system including the battery temperature efficiency; for most battery families, as the ambient temperature decreases, the charge carriers within the battery become more sluggish and the battery capacity decreases significantly.

Publicly available meteorological data are only available for horizontal and fully tracking (direct normal) components and cannot be used directly for a fixed planar solar panel. For this reason it is necessary to develop a conversion methodology to compute the energy incident on the panel using the available datasets. The total radiation received by a panel consists of three terms: the direct component, I_c, which is the radiation coming directly from the sun; the diffuse component, D_c (due mainly to sky diffraction), which consists of non-direct-radiation components; finally the ground-reflected component, R_c, which is often neglected.

Calculation of the direct component can be performed using well-known trigonometric calculations and is straightforward. On the other hand, the calculation of the diffuse component estimation requires a more complex conversion model, and several have appeared in the literature. Before examining the diffuse component estimation methodologies, let us first discuss the data available in North American records [4, 5]. Generally, five different solar radiation data fields are available. They are briefly discussed below.

Extraterrestrial Horizontal Radiation. This is the amount of solar radiation received on a horizontal surface at the top of the atmosphere. It is also known as "top-of-atmosphere" (TOA) irradiance and is the amount of global horizontal radiation that a location on earth would receive if there were no atmosphere. This number is used as the reference amount against which actual solar energy measurements are compared.

Extraterrestrial Direct Normal Radiation. This reading is the level of solar radiation received on a surface normal to the sun at the top of the atmosphere.

The previous two fields are deterministic and can be calculated using the sun–earth distance and position equations. The rest of the fields are random processes. In general, due to the motion of the earth around the sun and its rotation, solar insolation experiences cyclic changes over a year, and these variations are deterministic to a large extent. However, complex weather processes such as humidity, temperature, air pressure, and cloud type affect the received insolation. The following data is typically included.

Global Horizontal Radiation. This is the total amount of direct and diffuse solar radiation received on a horizontal surface at ground level where the measurements are taken.

Direct Normal Radiation. This is the amount of solar radiation received directly from the sun at the measurement point within a field of view centered on the sun.

Diffuse Horizontal Radiation, (D_h). This is the level of solar radiation received from the sky (excluding the solar disk) on a horizontal surface.

The conversion models attempt to translate the above readings into the actual total solar energy impinging on a tilted plane solar panel. This is done by defining various solar angles that are used in the conversion. In addition to the solar radiation intensity, the angle of incidence determines the energy received by the panel, which is a function of the time of day and the day of the year. There are several methodologies available in the literature; some are based on the assumptions that the diffuse radiation is measured by the weather station, while others assume that it is not. References 9 and 12 provide methodologies that can be used to develop a simulator. In reference 9 the following methodology is described; in this case the data fields available are assumed to be available from US and Canadian weather stations and the diffuse horizontal radiation is available. In the first phase, the direct component is estimated based on the zenith angle and the tilt angle of the plane and the hour angle and solar declination angle. The sky clearness is then estimated and discretized based on the Perez equations 13, which depend on the zenith angle, the direct component, and the horizontal diffuse radiation. Next, the sky brightness is estimated using the procedure in reference 13. Finally the diffuse component on a tilted plane is calculated based on the Perez equations. In addition to the above computations, it is important to make adjustments for temperature effects, inaccuracies in measurements, and losses in the electrical control circuitry.

12.7 EXERCISES

1. In the text it was stated that one criterion for choosing the battery and panel size is minimum cost. Can you suggest other criteria? What is the effect of varying the cost model?

2. In the solar panel size versus battery capacity graphs, discuss the values of the asymptotes of these curves; that is, What happens when the solar panel or battery becomes very large?

3. If you were designing a controller to prevent node outages and also to minimize the energy withheld from the node, what would be the control variable? Develop a theoretical framework for the time evolution of the system.

4. What modifications would you need to incorporate into the analysis if you were considering other sources of energy? For example, if you were incorporating a wind turbine into the system, what would be the effect on the required battery and panel sizes?

5. Discuss other possible criteria for updating the moveable boundary defined for the best-effort traffic case. Design a media access control protocol that achieves the same effect as the proposed mechanism.

6. In the text we assumed an Erlang B model with discrete channel capacity. Can you think of any refinements to this model? What would happen if the mesh MAC was contention-based?

REFERENCES

1. The SolarMESH Network. http://owl.mcmaster.ca/~todd/SolarMESH/, McMaster University, Hamilton, Ontario, Canada., 2004.
2. F. M. Safie. Probabilistic modeling of solar power systems. In *Reliability and Maintainability Symposium, 1989. Proceedings, Annual*, 1989, pp. 425–430.
3. L. Narvarte and E. Lorenzo. On the usefulness of stand-alone PV sizing methods. *Progress in Photovoltaics: Research and Applications*, **8**:391–409, 2000.
4. National Solar Radiation Data Base. http://rredc.nrel.gov/solar/, National Renewable Energy Laboratory (NREL), U.S. Department of Energy, 2004.
5. National Climate Data and Information Archive. http://www.climate.weatheroffice.ec.gc.ca/, The Meteorological Service of Canada, 2004.
6. Y. Li, T. D. Todd, and D. Zhao. Access point power saving in solar/battery powered IEEE 802.11 ESS mesh networks. In *The Second International Conference on Quality of Service in Heterogeneous Wired/Wireless Networks (QShine)*, August 2005.
7. F. Zhang, T. D. Todd, D. Zhao, and V. Kezys. Power saving access points for IEEE 802.11 Wireless network infrastructure. In *IEEE Wireless Communications and Networking Conference 2004, WCNC '2004*, March 2004.
8. IEEE Standards Department. *Part 11: Wireless Medium Access Control (MAC) and Physical Laye r (PHY) Specifications: Medium Access Control (MAC) Quality of Service (QoS) Enhancements*, IEEE Press, New York, 2005.
9. A. Farbod. Design and resource allocation for solar-powered ESS mesh networks. M.S. thesis, McMaster University, West, Hamilton, Ontario, Canada, August 2005.
10. Y. Chung, J. Kim, and C.-C. J. Kuo. Network friendly video streaming via adaptive LMS bandwidth control. In *International Symposium on Optical Science, Engineering, and Instrumentation, San Diego, CA*, July 1998.
11. T. Todd and A. Kholaif. WLAN VoIP capacity using an adaptive voice packetization server. In *Internal technical Report-Wireless Networking Lab, McMaster University*, 2006.
12. M. Shahidehpour, M. Marwali, and M. Daneshdoost. Probabilistic production costing for photovoltaic-utility systems with battery storage. *IEEE Transactions on Energy Conversion*, **12**(2):175–180, 1997.
13. R. Perez, P. Ineichen, R. Seals, J. Michalsky, and R. Stewart. Modeling daylight availability and irradiance components from direct and global irradiance. *Solar Energy*, vol. **44**:271–289, 1990.

CHAPTER 13

Reputation-and-Trust-Based Systems for Ad Hoc Networks

AVINASH SRINIVASAN

Department of Mathematics, Computer Science, and Statistics, Bloomsburg University, Bloomsburg, PA 17815

JOSHUA TEITELBAUM

Microsoft, Redmond, Seattle 98052

JIE WU and MIHAELA CARDEI

Department of Computer Science and Engineering, Florida Atlantic University, Boca Raton, FL 33431

HUIGANG LIANG

Department of ITOM, College of Business, Florida Atlantic University, Ft. Lauderdale, FL 33301

13.1 INTRODUCTION

Reputation and trust are two very useful tools that are used to facilitate decision making in diverse fields from an ancient fish market to state-of-the-art e-commerce. Reputation is the opinion of one entity about another. In an absolute context, it is the trustworthiness of an entity [1]. Trust, on the other hand, is the expectation of one entity about the actions of another [2]. For over three decades, formal studies have been done on how reputation and trust can affect decision-making abilities in uncertain conditions. Only recently has trust and reputation been adapted to wireless communication networks. Trust is a multidimensional entity that, if effectively modeled, can resolve many problems in wireless communication networks.

Wireless communication networks, mobile ad hoc networks (MANETs), and wireless sensor networks (WSNs), in particular, have undergone tremendous technological advances over the last few years. With this development comes the risk of newer threats and challenges, along with the responsibility of ensuring the safety, security, and integrity of information communication over these networks. MANETs, due to

---

Algorithms and Protocols for Wireless and Mobile Ad Hoc Networks, Edited by Azzedine Boukerche
Copyright © 2009 by John Wiley & Sons Inc.

the individualized nature of the member nodes, are particularly vulnerable to selfish behavior. Because each node labors under a energy constraint, there is incentive for a node to be programmed to selfishly guard its resource, leading it to behave in a manner that is harmful to the network as a whole.

WSNs, on the other hand, are open to unique problems due to their usual operation in unattended and hostile areas. Since sensor networks are deployed with thousands of sensors to monitor even a small area, it becomes imperative to produce sensors at very low costs. This inevitably dilutes the tamper-resistant property of sensors. The aforementioned unique situation of WSNs leaves the nodes in the network open to physical capture by an adversary. Because each node has valid cryptographic key information, standard cryptographic security can be bypassed. The adversary has as much knowledge about the system as was available to the node itself, and it is therefore able to reprogram the captured node in such a way as to cause maximum damage to the system.

These problems can be resolved by modeling MANETs and WSNs as reputation and trust-based systems. As in real life, we tend to believe and interact only with people who we see as having a good reputation. Reputation can be defined as a person's history of behavior, and it can be positive, negative, or a mix of both. Based on this reputation, trust is built. Trust can be seen as the expectation that a person will act in a certain way. For example, if a person has a reputation for not getting jobs done, then people will not trust that this person will get a job done in the future. Based on this, people will not assign critical jobs to him, since they believe there is a good chance that he will not get the job done.

Similarly, nodes in MANETs and WSNs can make reputation-and-trust-guided decisions—for example, in choosing relay nodes for forwarding packets for other nodes, or for accepting location information from beacon nodes as in reference 26. This not only provides MANETs and WSNs with the capability of informed decision making, but also provides them security in the face of internal attacks where cryptographic security gives way. The way in which a system discovers, records, and utilizes reputation to form trust, and uses trust to influence behavior, is referred to as a reputation-and-trust-based system. This chapter is dedicated to providing the reader with a complete understanding of reputation-and-trust-based systems from the wireless communication perspective.

The rest of this chapter is organized as follows. In Section 13.2, we will give an in-depth background of reputation and trust from social perspective. Then in Section 13.3, we will present reputation and trust from network perspective and discuss various challenges in modeling networks as reputation-and-trust-based systems. We move on to present the reputation-and-trust-based systems and discussing their goals, properties, initialization, and classification in Section 13.4. Then in Section 13.5, we will discuss the components of reputation and trust-based systems in detail. In Section 13.6 we will present the current reputation and trust-based systems that are in use. We have presented the current open problems in reputation-and-trust-based systems from the network perspective in Section 13.7. Finally in Section 13.8 we conclude this chapter.

13.2 SOCIAL PERSPECTIVE

In this section we provide an in depth discussion on trust and uncertainty from a societal perspective and use e-commerce in the cyber space to illustrate various concepts. Trust and uncertainty have played a very critical role in the marketplace, modeling both consumer and seller behavior and providing a great deal of insight into meaningful transactions. We first define trust and uncertainty in a general framework exploring their cause–effect relationship. Then we discuss the various trust antecedents. Finally, we end the section with a brief discussion on information asymmetry and opportunistic behavior.

13.2.1 Trust and Uncertainty

Trust has been widely recognized as an important factor affecting consumer behavior, especially in the e-commerce context where uncertainty abounds. Trust is necessary only when there is uncertainty. Transaction economics research demonstrates that uncertainty increases transaction cost and decreases acceptance of online shopping [3]. The Internet environment is uncertain, and consumers' perceived uncertainty deters their adoption of online shopping [4]. Trust contributes to e-commerce success by helping consumers overcome uncertainty.

Trust is complex and multidimensional in nature. Multiple research streams in the past have shed light on various antecedents of trust. Major antecedents of trust include calculus-based trust, knowledge-based trust, and institution-based trust [5–7].

Uncertainty originates from two sources: *information asymmetry* and *opportunism*. The former refers to the fact that either party may not have access to all of the information it needs. The latter indicates that different goals exist between transacting partners, and both parties tend to behave opportunistically to serve their self-interests.

13.2.2 Trust Beliefs and Trust Intention

Trust means that the trustor believes in, and is willing to depend on, the trustee [6]. Based on the theory of reasoned action, McKnight et al. [8] breaks the high-level trust concept into two constructs: trusting beliefs and trusting intention. Trusting beliefs are multidimensional, representing one's beliefs that the trustee is likely to behave in a way that is benevolent, competent, honest, or predictable in a situation. According to McKnight et al. [5], three trusting beliefs appeared most frequently in trust research: *competence*, *benevolence*, and *integrity*. Trusting intention is the extent to which one is willing to depend on the other person in a given situation. Stewart [9] explained that there are several intended actions that represent trusting intentions such as the intent to continue a relationship, the intent to pursue long-term orientation toward future goals, and the intent to make a purchase. Theory of reasoned action supports the proposition that "positive beliefs regarding an action have a positive effect on intentions to perform that action" [9]. Trusting beliefs have been found to have a significant positive influence on trusting intention [5].

Trust is a critical aspect of e-commerce. Online purchase renders a customer vulnerable in many ways due to the lack of proven guarantees that an e-vendor will not behave opportunistically. The Internet is a complex social environment, which still lacks effective regulation. When a social environment cannot be regulated through rules and customs, people adopt trust as a central social complexity reduction strategy. Therefore, online customers have to trust an e-vendor from which they purchase; otherwise, the social complexity will cause them to avoid purchasing [7].

13.2.3 Calculus-Based Trust Antecedents

In economic transactions, parties develop trust in a calculative manner [10]. To make a calculus-based trust choice, one party rationally calculates the costs and benefits of other party's cheating or cooperating in the transaction. Trust develops if the probability of that party performing an action that is beneficial or at least not detrimental to the first party is high [10].

Consumers can obtain pieces of information regarding an e-vendor's trustworthiness through a variety of channels. Calculus-based trust develops because of credible information regarding the intentions or competence of the trustee. The credible information, such as reputation and certification, can signal that the trustee's claims of trustworthiness are true. Theoretical evidence has shown that calculus-based trust can be a powerful form of trust to facilitate electronic transactions [11].

13.2.4 Knowledge-Based Trust Antecedents

Previous research proposes that trust develops as a result of the aggregation of trust related knowledge by the involved parties [12]. This knowledge is accumulated either first-hand (based on an interaction history) or second-hand. One of the knowledge-based antecedents for trust tested by previous researches is familiarity, an understanding of what, why, where, and when others do what they do [7]. Familiarity emerges as a result of one's learning gained from previous interactions and experiences. It reduces environmental uncertainty by imposing a structure. For example, consumers who are familiar with the site map and the ordering process of a website are more likely to trust the website. In general, familiarity with the situation and various parties involved is found to build trust in business relationships [8]. Past research has found that familiarity with an e-vendor positively influences trust in that e-vendor [13].

13.2.5 Institution-Based Trust Antecedent

Institution-based trust means that one believes the necessary impersonal structures are in place to enable one to act in anticipation of a successful future endeavor [8]. Such trust reflects the security one feels about a situation because of guarantees, safety nets, or other structures. The concept of institution-based trust comes from sociology, which deals with the structures (e.g., legal protections) that make an environment feel trustworthy [5]. Two types of institution-based trust have been discussed in the literature: situational normality and structural assurance [8]. Situational normality is an assessment that the success is likely because the situation appears to be normal or favorable. In the context of the Internet, for example, high situational normality would mean that the Internet environment is appropriate, well-ordered, and favorable for doing personal business [5]. Structural assurance is an assessment that the success is likely because safeguard conditions such as legal recourse, guarantees, and regulations are in place [7, 8, 14]. Prior research has found that institutional-based antecedents positively affect trust in e-vendors [7].

13.2.6 Uncertainty

Uncertainty refers to the degree to which an individual or organization cannot anticipate or accurately predict the environment [15]. Prior research has demonstrated that uncertainty increases transaction cost and decreases acceptance of online purchasing [3]. Uncertainty regarding whether trading parties intend to and will act appropriately is the source of transaction risk that erodes exchange relationships and increases transaction cost [6]. Transaction risks can result from the impersonal nature of the electronic environment. These risks are rooted in two types of uncertainty: about the identity of online trading parties or about the product quality [11]. In the cyberspace that lacks security, a dishonest seller can easily masquerade as an honest one to attract a credulous buyer into a fraudulent transaction. In addition, the lack of information about the true quality of the product or service prior to actual purchase makes the buyer more uncertain. One objective of trust building is to reduce the trustor's perceived uncertainty so that transaction cost is lowered and a long-term exchange relationship sustains [16]. Prior studies have stressed the important role of trust in reducing risk or uncertainty in Internet shopping [13]. It has been found that trust mitigates opportunism [17] and information asymmetry [11] in uncertain contexts.

13.2.7 Information Asymmetry

Information asymmetry is defined as the difference between the information possessed by buyers and sellers [11]. Due to information asymmetry, it is difficult and costly for buyers to ascertain the attributes of products and services before purchase. Necessary information regarding quality of products or services may be incomplete or not readily available. Information asymmetry is a problem for Internet shopping due to the physical distance between buyers and sellers. Two sets of agent problems result from information asymmetry: adverse selection and moral hazard [18]. Adverse selection problems take place when the buyer is not capable of knowing the seller's characteristics or the contingencies under which the seller operates. Moral hazard problems occur after a contract is signed, but the seller may misbehave because the buyer is unable to observe its behavior. Marketing researchers have observed that most buyer–seller relationships are characteristic of information asymmetry [19]. When consumers cannot be adequately informed to make a judgment, they are likely to be subject to moral hazard and adverse selection problems and perceive a high degree of uncertainty.

13.2.8 Opportunistic Behavior

Opportunistic behavior is prevalent in exchange relationships. In the online buyer–seller relationship, the seller may behave opportunistically by trying to meet its own goals without considering the consumer's benefits. Examples of opportunistic behavior could include misrepresentation of the true quality of a product or service, incomplete disclosure of information, actual quality cheating, contract default, or failure to acknowledge warranties [19]. Some researchers argue that trust can be defined as the expectation that an exchange partner will not engage in opportunistic behavior, and one of the consequences of trust is to reduce perceived uncertainty associated with opportunistic behavior [16].

13.3 WIRELESS COMMUNICATION NETWORK PERSPECTIVE

This general perspective of reputation and trust has been applied to the various means and places in so-called "cyberspace" where people interact. The most obvious is the realm of e-commerce, which has already been discussed in detail. Other extensions of this style of research in the electronic media include e-mail, electronic social networks such as Friend-of-a-Friend and Friendster, Peer-to-Peer (P2P) systems, and Google's PageRank algorithm [20].

We move from this general perspective, and we examine how these broad strokes of research can aid security in a wireless networking environment. Here, as in the more general domain, we find elements of information asymmetry and opportunism. Just as in e-commerce, nodes in MANETs and WSNs have no way of gathering information about nodes beyond their sensing range and therefore have a great deal of uncertainty. Also, in systems involving asymmetrical duties and designs, some nodes may have different capabilities, allowing them to distribute information that cannot be checked by other nearby nodes.

In addition, the digital, artificial nature of a wireless network both poses new problems and simplifies others. The human brain is a far more complicated and less understood machine than a wireless node. The node is created with certain capabilities, and it is limited by such. Whereas a human can imagine and innovate new methods of opportunistic behavior, a node can only perform its programming. But, on the other hand, humans have an evolved sense of reputation and trust that does not require logic to function. We, as designers, must attempt to build reputation and trust networks without the benefit of thousands of years of evolution.

In the following sections, we first give a background on MANETs and WSNs and then discuss the challenges in modeling them as reputation-and-trust-based systems. At the end of this section, we present the various types of node misbehavior.

13.3.1 Background

A MANET is a self-configuring system of mobile nodes connected by wireless links. In a MANET, the nodes are free to move randomly, changing the network's topology rapidly and unpredictably. MANETs are decentralized, and therefore all network activities are carried out by nodes themselves. Each node is both (a) an end-system and (b) a relay node to forward packets for other nodes. Such a network may operate as a stand-alone network or may be part of a larger network like the Internet. Since MANETs do not require any fixed infrastructure, they are highly preferred for connecting mobile devices quickly and spontaneously in emergency situations like rescue operations, disaster relief efforts, or other military operations. MANETs can either be managed by an organization that enforces access control or they can be open to any participant that is located closely enough. The later scenario poses greater security threats than the former. In MANETs, more often than not, nodes are individuals and do not have any common interests. It may be advantageous for individual nodes not to cooperate. Hence, they need some kind of incentive and motivation to cooperate. The noncooperative behavior of a node could be due to selfish intention (for example, to save power) or completely malicious intention (for example, to launch attacks such as denial-of-service).

A WSN is a network of hundreds and thousands of small, low-power, low-cost devices called sensors. The core application of WSNs is to detect and report events. WSNs have found critical applications in military and civilian life, including robotic landmine detection, battlefield surveillance, environmental monitoring, wildfire detection, and traffic regulation. They have been invaluable in saving lives, be it a soldier's life on the battlefield or a civilian's life in areas of high risk of natural calamities. In WSNs, all the sensors belong to a single group or entity and work toward the same goal, unlike in MANETs. Also, individual sensors have little value unless they work in cooperation with other sensors. Hence, there is an inherent motivation for nodes in WSNs to be cooperative, and so incentive is less of a concern.

However, since WSNs are often deployed in unattended territories that can often be hostile, they are subject to physical capture by enemies. The obvious solution is to make the nodes themselves tamperproof. The difficulty with this solution is that full tamper-proofing makes each individual node prohibitively expensive, and even then does not prevent a truly determined adversary from eventually cracking the node. Since many nodes are often required to cover an area, each node must be economically affordable. As such, simple tamperproofing is not a viable solution. Hence, sensors can be modified to misbehave and disrupt the entire network. This allows the adversary to access the cryptographic material held by the captured node and allow the adversary to launch attacks from within the system as an insider, bypassing encryption and password security systems. Even though cryptography can provide integrity, confidentiality, and authentication, it fails in the face of insider attacks. This necessitates a system that can cope with such internal attacks.

13.3.2 Node Misbehavior

The intentional noncooperative behavior of a node, as identified in reference 21, is mainly caused by two types of misbehavior: *selfish behavior*—for example, nodes that want to save power, CPU cycles, and memory—and *malicious behavior*, which is not primarily concerned with power or any other savings but interested in attacking and damaging the network. When the misbehavior of a node manifests as selfishness, the system can still cope with it because this misbehavior can always be predicted. A selfish node will always behave in a way that maximizes its benefits; and as such, incentive can be used to ensure that cooperation is always the most beneficial option. However, when the misbehavior manifests as maliciousness, it is hard for the system to cope with it, since a malicious node always attempts to maximize the damage caused to the system, even at the cost of its own benefit. As such, the only method of dealing with such a node is detection and ejection from the network.

Malicious misbehavior in packet forwarding can be generally divided into two types of misbehavior: forwarding and routing. Some common forwarding misbehavior are packet dropping, modification, fabrication, timing attacks, and silent route change. Packet dropping, modification, and fabrication are self-explanatory. Timing misbehavior is an attack in which a malicious node delays packet forwarding to ensure that packets expire their Time-to-Live (TTL), so that it is not immediately obvious that it is causing the problem. Silent route changes is an attack in which a malicious node forwards a packet through a different route than it was intended to go through. A malicious node with routing misbehavior attacks during the route discovery phase. Three common attacks of this type are black hole, gray hole, and worm hole. A black-hole attack is one in which a malicious node claims to have the shortest path and then when asked to forward will drop the received packets. In a gray-hole attack, which is a variation of the black-hole attack, the malicious node selectively drops some packets. A worm-hole attack, also known as tunneling, is an attack in which the malicious node sends packets from one part of the network to another part of the network, where they are replayed.

The selfish behavior of a node, as shown in Figure 13.1, can be generally classified as either *self-exclusion* or *nonforwarding*. The self-exclusion misbehavior is one in which a selfish node does not participate when a route discovery protocol is

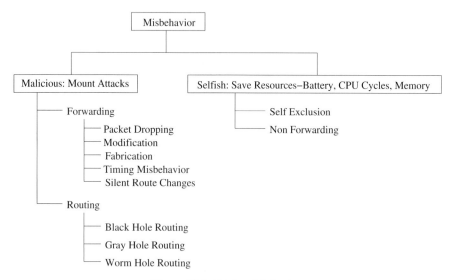

Figure 13.1. Node misbehavior.

executed. This ensures that the node is excluded from the routing list of other nodes. This benefits a selfish node by helping it save its power because it is not required to forward packets for other nodes. A reputation model is an effective way to thwart the intentions of such selfish nodes. Since a node does not forward packets for other nodes in the networks, it is denied any cooperation by other nodes. So, it is in the best interest of a selfish node to be cooperative. On the other hand, the nonforwarding misbehavior is one in which a selfish node fully participates in route discovery phase but refuses to forward the packets for other nodes at a later time. This selfish behavior of a node is functionally indistinguishable from a malicious packet dropping attack.

Since reputation-based systems can cope with any kind of observable misbehavior, they are useful in protecting a system. Reputation-and-trust-based systems enable nodes to make informed decisions on prospective transaction partners. Researchers have been steadily making efforts to successfully model WSNs and MANETs as reputation-and-trust-based systems. Adapting reputation-and-trust-based systems to WSNs presents greater challenges than MANETs and Peer-to-Peer (P2P) systems due to their energy constraints. CORE [22], CONFIDANT [23], RFSN [2], DRBTS [1], KeyNote [24], and RT Framework [25] are a few successful attempts. However, RFSN and DRBRTS are the only works so far focusing on WSNs. The others concentrate on MANETs and P2P networks.

13.4 REPUTATION-AND-TRUST-BASED SYSTEMS

Reputation-and-trust-based systems have now been used for over half a decade for Internet, e-commerce, and P2P systems [26–29]. More recently, efforts have been made to model MANETs and WSNs as reputation-and-trust-based systems [21, 23, 30, 31]. In this section we will discuss the various aspects of reputation-and-trust-based systems. To begin with, we present the goals of a reputation-and-trust-based system. Then, we discuss the properties of reputation-and-trust-based systems including the properties of reputation and trust metrics. Then we discuss various types of system initialization followed by different classifications of reputation-and-trust-based systems. Finally, we end the section with a discussion of the pros and cons of reputation and trust-based systems.

13.4.1 System Goals

The main goals of reputation-and-trust-based systems, as identified in reference 28, after adapting them to wireless communication networks, are as follows:

- Provide information that allows nodes to distinguish between trustworthy and nontrustworthy nodes.
- Encourage nodes to be trustworthy.
- Discourage participation of nodes that are untrustworthy.

In addition, we have identified two more goals of a reputation-and-trust-based system from a wireless communication network perspective. The first goal is to cope with any kind of observable misbehavior. The second goal is to minimize the damage caused by insider attacks.

13.4.2 Properties

To operate effectively, reputation-and-trust-based systems require at least three properties, as identified in reference 26:

- Long-lived entities that inspire an expectation of future interaction.
- The capture and distribution of feedback about current interactions (such information must be visible in the future).
- Use of feedback to guide trust decisions.

The trust metric itself has the following properties:

- *Asymmetric*: Trust is not symmetric; that is, if node A trusts node B, then it is not necessarily true that node B also trusts node A.
- *Transitive*: Trust is transitive; that is, if node A trusts node B and node B trusts node C, then node A trusts node C.
- *Reflexive*: Trust is reflexive; that is, a node always trusts itself.

13.4.3 Initialization

Reputation-and-trust-based systems can be initialized in one of the following ways: (1) All nodes in the network are considered to be trustworthy. Every node trusts every other node in the network. The reputation of the nodes is decreased with every bad encounter. (2) Every node is considered to be untrustworthy and no node in the network trusts any other node. The reputation of nodes with such an initialization system is increased with every good encounter. (3) Every node in the network is neither considered trustworthy nor untrustworthy. They all take a neutral reputation value to begin with. Then with every good or bad behavior, the reputation value is increased or decreased, respectively.

13.4.4 Classification

We have recognized that reputation-and-trust-based systems can be broadly classified into the following groups.

1. Observation
    a. **First-hand**: The system uses direct observation or its own experience to update reputation.

b. **Second-hand**: The system uses information provided by peers in its neighborhood.

Most systems proposed so far use both first-hand and second-hand information to update reputation. This allows the system to make use of the experience of its neighbors to form its opinions. Some systems choose not to use both types of information. In systems that use only first-hand information, a node's reputation value of another node is not influenced by others. This makes the system completely robust against rumor spreading. OCEAN and *pathrater* [30] are two such systems that make use of only first-hand information. Only one system so far, DRBTS, has the unique situation where a certain type of node uses only second-hand information. In this system, a node does not have any first-hand information to evaluate the truthfulness of the informers. One way to deal with this situation is to use a simple majority principle.

2. Information Symmetry

   a. **Symmetric**: All nodes in the network have access to the same amount of information—that is, both first-hand and second-hand. When making a decision, no node has more information than any other node.

   b. **Asymmetric**: All nodes do not have access to the same amount of information. For example, in DRBTS [1], sensor nodes do not have first-hand information. Thus, in the decision-making stage, the sensor nodes are at a disadvantage.

3. Centralization

   a. **Centralized**: One central entity maintains the reputation of all the nodes in the network—for example, like in eBay or YAHOO auctions. Such a reputation system can cause both a security and information bottleneck.

   b. **Distributed**: Each node maintains the reputation information of all the nodes it cares about. In this kind of a reputation system, there could be issues concerning the consistency of reputation values at different nodes; that is, there may not be a consistent local view. Each node can have either local or global information.

      - **Local**: Nodes have reputation information only about nodes in their neighborhood. This is the most reasonable option, particularly for static sensor networks, since nodes interact only with their immediate neighbors. This mitigates memory overhead to a large extent.

      - **Global**: Nodes have reputation information of all the nodes in the network. This is suitable for networks with lots of node mobility. Even after moving to a new location, nodes are not completely alienated and have a reasonable understanding of their new neighborhood. Unfortunately, this leads to a large overhead and can lead to scalability problems.

13.4.5 Pros and Cons

Reputation-and-trust-based systems are one of the best solutions for dealing with selfish misbehavior. They are also very robust solutions to curtail insider attacks.

Reputation-and-trust-based systems are, for the most part, self-maintaining. Unfortunately, along with the added overhead, both in computation and communication, they also add another dimension of security consideration. Now, an adversary has another vector to attack the system with, namely, the reputation system itself. It is difficult to properly design the reputation system in order to make it robust enough to survive such attacks, but we will examine that tradeoff a little later in the chapter.

13.5 COMPONENTS

This section is dedicated completely to an in-depth discussion of the various components of reputation-and-trust-based systems. We have identified four important components in a reputation-and-trust-based system: information gathering, information sharing, information modeling, and decision making. In the rest of this section we will discuss each of these components in detail, presenting examples of real-world systems whenever appropriate.

13.5.1 Information Gathering

Information gathering is the process by which a node collects information about the nodes it cares about. This component of the reputation-and-trust-based systems is concerned only with first-hand information. First-hand information is the information gathered by a node purely based on its observation and experience. However, according to CONFIDANT [23], first-hand information can be further classified into *personal experience* and *direct observation*. Personal experience of a node refers to the information it gathers through one-to-one interaction with its neighboring nodes. On the other hand, direct observation refers to the information a node gathers by observing the interactions among its neighboring nodes. CONFIDANT is currently the only system to make this distinction.

Most reputation-and-trust-based systems make use of a component called *watchdog* [30] to monitor their neighborhood and gather information based on promiscuous observation. Hence, first-hand information is confined to the wireless sensing range of a node. However, the watchdog system is not very effective in the case of directional antennas, limiting its broad application, especially in very secure situations. This limitation has received very little study currently.

13.5.2 Information Sharing

This component of reputation-and-trust-based systems is concerned with dissemination of first-hand information gathered by nodes. There is an inherent tradeoff between efficiency in using second-hand information and robustness against false ratings. Using second-hand information has many benefits. The reputation of nodes builds up more quickly, due to the ability of nodes to learn from each others' mistakes. No information in the system goes unused. Using second-hand information has an additional benefit in that, over time, a consistent local view will stabilize.

However, sharing information comes at a price. It makes the system vulnerable to false report attacks. This vulnerability can be mitigated by adopting a strategy of limited information sharing—that is, sharing either only *positive information* or *negative information*.

The difficulty with the solution is that while sharing only positive information limits the vulnerability of the system to false praise attacks, it has its own drawbacks. When only positive information is shared, not all the information in the system is used, since nodes cannot share their bad experiences. This is particularly detrimental since learning from ones own experience in this scenario comes at a very high price. Also, colluding malicious nodes can extend each other's survival time through false praise reports. CORE [21] suffers from this attack.

Similarly, sharing only negative information prevents the false praise attack mentioned above, but has its own drawbacks. Not all the information in the system is used, since nodes cannot share their good experiences. More importantly, malicious nodes can launch a bad-mouth attack on benign nodes either individually or in collusion with other malicious nodes. CONFIDANT [23] suffers from this attack.

Yet another way of avoiding the negative consequences of information sharing is to not share information at all. OCEAN [32] is one such model that builds reputation purely based on its own observation. Such systems, though they are completely robust against rumor spreading, have some serious drawbacks. The time required for the system to build a reputation is increased dramatically; and it takes longer for a reputation to fall, allowing malicious nodes to stay in the system longer.

Systems like DRBTS [1] and RFSN [2] share both positive and negative information. The negative effects of information sharing, as discussed above, can be mitigated by appropriately incorporating first-hand and second-hand information into the reputation metric. Using different weighting functions for different information is one efficient technique.

With respect to information sharing, we have identified three ways in which information can be shared among nodes: *friends list*, *blacklist*, and *reputation table*. A friends list shares only positive information, a blacklist shares only negative information, while a reputation table shares both positive and negative information.

For sharing information, three important issues have to be addressed: dissemination frequency, dissemination locality, and dissemination content.

Dissemination frequency can be classified into two types:

- *Proactive Dissemination.* In proactive dissemination, nodes publish information during each dissemination interval. Even if there has been no change to the reputation values that a node stores, it still publishes the reputation values. This method is suitable for dense networks to prevent congestion since they have to wait till the dissemination interval to publish.
- *Reactive Dissemination.* In reactive dissemination, nodes publish only when there is a predefined amount of change to the reputation values they store or when an event of interest occurs. This method mitigates the communication overhead in situations where there is not frequent updates to the reputation values. However, reactive dissemination may cause congestion in dense networks with

high network activity, or cause the system to stall if network activity is especially low.

In both of these types of information dissemination, the communication overhead can be mitigated to a large extent by piggybacking the information with other network traffic. For example, in CORE the reputation information is piggybacked on the reply message, and in DRBTS it is piggybacked on the location information dispatch.

Dissemination locality can be classified into two types:

- *Local*: In local dissemination, the information is published within the neighborhood. It could be through either a local broadcast, a multicast, or a unicast. In DRBTS [1], the information is published in the neighborhood through a local broadcast. This enables all the beacon nodes to update their reputation table accordingly. Other models may choose to unicast or multicast, depending on the application domain and security requirements.
- *Global*: In global dissemination, the information is sent to nodes outside the range of the publishing node. Like local dissemination, global dissemination could use one of the three publishing techniques: broadcast, multicast, or unicast. For networks with node mobility, global dissemination is preferred because this gives nodes a reasonable understanding of the new location they are moving to.

Content of information disseminated can be classified into two types:

- *Raw*. The information published by a node is its first-hand information only. It does not reflect the total reputation value, which includes the second-hand information of other nodes in the neighborhood. Disseminating raw information prevents information from looping back to the originator.
- *Processed*. The information published by a node is its overall opinion of the nodes in its reputation tables. This includes information provided to the node by others in the neighborhood.

13.5.3 Information Modeling

This component of a reputation-and-trust-based system deals with combining the first-hand and second-hand information meaningfully into a metric. It also deals with maintaining and updating this metric. Some models choose to use a single metric, reputation, like CORE and DRBTS [1, 21], while a different model like RFSN may choose to use two separate metrics, reputation and trust, to model the information [2]. While most models make use of both first-hand and second-hand information in updating reputation and/or trust, some models like OCEAN [32] may use just first-hand information. The first-hand information can be directly incorporated into the reputation metric without much processing.

However, this is not the case with second-hand information. The node providing the second-hand information could be malicious, and the ratings it provides could be spurious. Hence it is necessary to use some means of validating the credibility of the reporting node. One method is to use a deviation test like the one in references 34 and 35. If the reporting node qualifies the deviation test, then it is treated as trustworthy and its information is incorporated to reflect in the reputation of the reported node. However, different models may choose to use a different strategy in accepting the second-hand information, depending on the application domain and security requirements. For instance, the model in reference uses Dempster–Shafer belief theory [35] and discounting belief principle [36] to incorporate second-hand information. However, Beta distribution has been the most popular among researchers in reputation-and-trust-based systems. It was first used by Josang and Ismail [37]. Since then, many researchers have used the beta distribution including Ganeriwal and Srivastava [2] and Buchegger and Boudec [33]. Beta distribution is the simplest among the various distribution models, (namely, Poisson, binomial, or Gaussian) that can be used for representation of reputation. This is mainly because of the fact that it is indexed by only two parameters.

An important issue in maintaining and updating reputation is how past and current information is weighted. Different models tend to weight them differently, each with a different rationale. Models like CORE tend to give more weight to the past observations with the argument that a more recent sporadic misbehavior should have minimal influence on a node's reputation that has been built over a long period of time. This helps benign nodes that are selfish due to genuinely critical battery conditions. Also, occasionally, nodes may temporarily misbehave due to technical problems such as link failure, in which case it is justified not to punish them severely.

On the other hand, models like RFSN tend to give more weight to recent observations than the past, with the argument that a node has to contribute and cooperate on a continuous basis to survive. This is known as aging, where old observations have little influence on a node's current reputation value. This forces nodes to be cooperative at all times. Otherwise, a malicious node will build its reputation over a long period and then start misbehaving by taking advantage of its accumulated goodness. The higher the reputation a malicious node builds, the longer it can misbehave before it can be detected and excluded. However, there is a drawback in adopting the aging technique. In periods of low network activity, a benign node gets penalized. This can be resolved by generating network traffic in regions and periods of low network activity using mobile nodes. DRBTS resolves this problem with beacon nodes being capable of generating network traffic on a need basis.

13.5.4 Decision Making

This component of a reputation-and-trust-based system is responsible for taking all the decisions. The decisions made by this component are based on the information provided by the information modeling component. The basic decision is a binary decision, on who to trust and who not to. The actual decision could translate to be one of cooperate/don't cooperate, forward/don't forward, and so on, based on the function being monitored by the system.

The decision of this component varies along with the reputation and trust values in the information modeling component. The decision can vary from trust to no-trust, wherein a node that was trusted so far will no longer be trusted after its reputation and trust values fall below a predetermined threshold. Similarly, it can vary from no-trust to trust, wherein a node that was initially not trusted will be trusted soon after its reputation and trust values exceed a predetermined threshold.

13.6 EXAMPLES OF REPUTATION AND TRUST-BASED MODELS

In this section we will review various reputation and trust-based models that have been proposed for MANETs and WSNs over the last few years. For each model, we will review the working principle, in light of the discussions presented in the previous sections.

13.6.1 CORE

CORE stands for "a COllaborative REputation mechanism to enforce node cooperation in mobile ad hoc networks." This model was proposed by Michiardi and Molva [22] to enforce node cooperation in MANETs based on a collaborative monitoring technique. CORE is a distributed, symmetric reputation model that uses both first-hand and second-hand information for updating reputation values. It uses bidirectional communication symmetry and dynamic source routing (DSR) protocol for routing. CORE also assumes wireless interfaces that support promiscuous mode operation.

In CORE, nodes have been modeled as members of a community and have to contribute on a continuing basis to remain trusted; otherwise, their reputation will degrade until eventually they are excluded from the network. The reputation is formed and updated along time. CORE uses three types of reputation—namely, *subjective reputation*, *indirect reputation*, and *functional reputation*—and addresses only the selfish behavior problem. CORE assigns more weight to the past observations than to the current observations. This ensures that a more recent sporadic misbehavior of a node has minimum influence on the evaluation of overall reputation value of that node. CORE has two types of protocol entity, namely, a requestor and a provider.

- *Requestor*: Refers to a network entity asking for the execution of a function f. A requestor may have one or more providers within its transmission range.
- *Provider*: Refers to any entity supposed to correctly execute the function f.

In CORE, nodes store the reputation values in a reputation table (RT), with one RT for each function. Each entry in the RT corresponds to a node and consists of four entries: unique ID, recent subjective reputation, recent indirect reputation, and composite reputation for a predefined function. Each node is also equipped with a watchdog mechanism for promiscuous observation. RTs are updated in two situations: during the request phase and during the reply phase.

Information Gathering. The reputation of a node computed from first-hand information is referred to as subjective reputation . It is calculated directly from a node's observation. CORE does not differentiate between interaction and observation for subjective reputation unlike CONFIDANT [23]. The subjective reputation is calculated only for the neighboring nodes—that is, nodes in the transmission range of the subject. The subjective reputation is updated only during the request phase. If a provider does not cooperate with a requestor's request, then a negative value is assigned to the rating factor σ of that observation, and consequently the reputation of the provider will decrease. The reputation value varies between -1 and 1. New nodes, when they enter the network, are also assigned a neutral reputation value of 0 since enough observations are not available to make an assessment of their reputation.

Information Sharing. CORE uses indirect reputation—that is, second-hand information—to model MANETs as complex societies. Thus, a node's impression of another node is influenced by the impression of other members of the society. However, there is a restriction imposed by CORE on the type of information propagated by the nodes. Only positive information can be exchanged. As we have stated before, this prevents bad-mouthing attacks on benign nodes. It is assumed that each reply message consists of a list of nodes that cooperated. Hence indirect reputation will be updated only during the reply phase.

Information Modeling. CORE uses functional reputation to test the trustworthiness of a node with respect to different functions. Functional reputation is the combined value of subjective and indirect reputation for different functions. Different applications may assign different weight to routing and similarly to various other functions such as packet forwarding, and so on.

The authors argue that reputation is compositional; that is, the overall opinion on a node that belongs to the network is obtained as a result of the combination of different types of evaluation. Thus, the global reputation for each node is obtained by combining the three types of reputation.

Finally, reputation values that are positive are decremented along time to ensure that nodes cooperate and contribute on a continuing basis. This prevents a node from initially building up a good reputation by being very cooperative and contributive but then abusing the system later.

Decision Making. When a node has to make a decision on whether or not to execute a function for a requestor, it checks the reputation value of that node. If the reputation value is positive, then it is a well-behaved entity. On the other hand, if the reputation value of the requestor is negative, then it is tagged as a misbehaving entity and denied the service. A misbehaving entity is denied service unless it cooperates and ameliorates its reputation to a positive value. Reputation is hard to build, because reputation gets decreased every time the watchdog detects a noncooperative behavior, and it also gets decremented over time to prevent malicious nodes from building reputation and then attacking the system resources.

Discussions. Giving greater weight to the past does enable a malicious node to misbehave temporarily if it has accumulated a high reputation value. Since only positive information is shared for indirect reputation updates, CORE prevents false accusation attacks, confining the vulnerability of the system to only false praise. The authors argue that a misbehaving node has no advantage by giving false praise to other unknown entities. This is true only so long as malicious nodes are not colluding. When malicious nodes start collaborating, then they can help prolong the survival time of one another through false praise. However, the effect of false praise is mitigated in CORE to some extent by coupling the information dissemination to reply messages. Also, since only positive information is shared, the possibility of retaliation is prevented.

There is an inherent problem in combining the reputation values for various functions into a single global value. This potentially helps a malicious node to hide its misbehavior with respect to certain functions by behaving cooperatively with respect to the remaining functions. The incentive for a node to misbehave with respect to a particular function is to save scarce resource. The node may choose to not cooperate for functions that consume lots of memory and/or power and choose to cooperate for functions that don't require as much memory and/or power. Nonetheless, functional reputation is a very nice feature of CORE that can be used to exclude nodes from functions for which their reputation value is below the threshold and include them for functions for which their reputation value is above the threshold.

CORE also ensures that disadvantaged nodes that are inherently selfish due to their critical energy conditions are not excluded from the network using the same criteria as for malicious nodes. Hence, an accurate evaluation of the reputation value is performed that takes into account sporadic misbehavior. Therefore, CORE minimizes false detection of a nodes misbehavior.

13.6.2 CONFIDANT

CONFIDANT stands for "cooperation of nodes—fairness in dynamic ad hoc networks." This model was proposed by Buchegger and Le Boudec [23] to make misbehavior unattractive in MANETs based on selective altruism and utilitarianism. CONFIDANT is a distributed, symmetric reputation model that uses both first-hand and second-hand information for updating reputation values. It aims at detecting and isolating misbehaving nodes, thus making it unattractive to deny cooperation. In CONFIDANT, the Dynamic Source Routing (DSR) protocol has been used for routing and it assumes a promiscuous mode operation. CONFIDANT also assumes that no tamper-proof hardware is required for itself, since a malicious node neither knows its reputation entries in other nodes nor has access to other nodes to modify their values. CONFIDANT is inspired by *The Selfish Gene* by Dawkins [23], which states reciprocal altruism is beneficial for every ecological system when favors are returned simultaneously because of instant gratification. The benefit of behaving well is not so obvious in the case where there is a delay between granting a favor and the repayment. The CONFIDANT protocol consists of four components at each node: monitor, trust manager, reputation system, and path manager.

Information Gathering. The monitor helps each node in passively observing their 1-hop neighborhood. Nodes can detect deviations by the next node on the source route. By keeping a copy of a packet while listening to the transmission of the next node, any content change can also be detected. The monitor registers these deviations from normal behavior and as soon as a given bad behavior occurs, it is reported to the reputation system. The monitor also forwards the received ALARMS to the Trust Manager for evaluation.

Information Sharing. The Trust Manager handles all the incoming and outgoing ALARM messages. Incoming ALARMs can originate from any node. Therefore, the source of an ALARM has to be checked for trustworthiness before triggering a reaction. This decision is made by looking at the trust level of the reporting node. CONFIDANT has provisions for several partially trusted nodes to send ALARMs which will be considered as an ALARM from a single fully trusted node. The outgoing ALARMs are generated by the node itself after having experienced, observed, or received a report of malicious behavior. The recipients of these ALARM messages are called friends, which are maintained in a friends list by each node.

The trust manager consists of the three components: *alarm table*, *trust table*, and *friends list*. An alarm table contains information about received alarms. A trust table manages trust levels for nodes to determine the trustworthiness of an incoming alarm. A friends list has all friends a node has to which it will send alarms when a malicious behavior is observed. The Trust Manager is also responsible for providing or accepting routing information, accepting a node as part of a route, and taking part in a route originated by some other node.

Information Modeling. The Reputation System manages a table consisting of entries for nodes and their rating. Ratings are changed only when there is sufficient evidence of malicious behavior that has occurred at least a threshold number of times to rule out coincidences. The rating is then changed according to a rate function that assigns the greatest weight for personal experience, a smaller weight for observations in the neighborhood, and an even smaller weight to reported experience. Buchegger and Le Boudec state that the rationale behind this weighting scheme is that nodes trust their own experiences and observations more than those of other nodes. Then, the reputation entry for the misbehaving node is updated accordingly. If the entry for any node in the table falls below a predetermined threshold, then the Path Manager is summoned.

Decision Making. The *Path Manager* is the component that is the decision maker. It is responsible for path re-ranking according to the security metric. It deletes paths containing misbehaving nodes and is also responsible for taking necessary actions upon receiving a request for a route from a misbehaving node.

Discussions. In CONFIDANT, only negative information is exchanged between nodes and the authors argue that it is justified because malicious behavior will ideally be the exception not the normal behavior. However, since only negative information is exchanged, the system is vulnerable to false accusation of benign nodes by malicious nodes. Unlike in CORE, even without collusion, malicious nodes benefit by falsely accusing benign nodes. With collusion of malicious nodes, this problem can explode beyond control. However, false praise attacks are prevented in CONFIDANT because no positive information is exchanged. This eliminates the possibility of malicious nodes colluding to boost the survival time of one another. Also, since negative information is shared between nodes, an adversary gets to know his situation and accordingly change his strategy. This may not be desirable. Sharing negative information in the open also introduces fear of retaliation which may force nodes to withhold their true findings.

In CONFIDANT, despite designing a complete reputation system, the authors have not explained how the actual reputation is computed and how it is updated using experienced, observed, and reported information. Also, in CONFIDANT, nodes that are excluded will recover after a certain timeout. This gives malicious nodes a chance to reenter the system and attack repeatedly unless they are revoked after a threshold number of reentries. Failed nodes in CONFIDANT are treated like any other malicious node that is both good and bad. It is good, because any node that is uncooperative should be punished, but it is bad because a genuine node may not be able to cooperate due to a temporary problem, and punishment may make it even worse. Also, the authors have not provided any evidence to support their rationale behind differentiating first-hand information as personal experience and direct observation and assigning them different weights.

13.6.3 Improved CONFIDANT—Robust Reputation System

Buchegger and Le Boudec [34] presented an improved version of CONFIDANT. They called this "A Robust Reputation System" (RRS). The RRS was an improvement on CONFIDANT, introducing a Bayesian framework with Beta distribution to update reputation. Unlike in CONFIDANT, the RRS uses both positive and negative reputation values in the second-hand information.

The RRS uses two different metrics: reputation and trust. The reputation metric is used to classify other nodes as normal/misbehaving while the trust metric is used to classify other nodes as trustworthy/untrustworthy. Whenever second-hand information is received from a node, the information is subjected to a deviation test. If the incoming reputation information does not deviate too much from the receiving node's opinion, then the information is accepted and integrated. Since the reporting node sent information the receiver sees as valid, the reporting node's trust rating is increased. On the other hand, if the reputation report deviates past the threshold, then the reporting node's trust value is lowered. The receiving node also decides whether to integrate the deviating information or not, depending on whether the reporting node is trusted or untrusted, respectively. To use a real-world example, if a trusted friend tells us something that is counter to our experience, we give the benefit of the doubt, but too many such deviations will cause us to lose our trust in the friend.

In RRS, only fresh information is exchanged. Unlike CORE, RRS gives more weight to current behavior than the past. The authors argue that, if more weight is given to past behavior, then a malicious node can choose to be good initially until it builds a high reputation and trust value and then chooses to misbehave. By assigning more weight to current behavior, the malicious node is forced to cooperate on a continuing basis to survive in the network.

13.6.4 RFSN

RFSN stands for Reputation-based Framework for Sensor Networks. This model was proposed by Ganeriwal and Srivastava [2]. RFSN is a distributed, symmetric reputation-based model that uses both first-hand and second-hand information for updating reputation values. In RFSN, nodes maintain the reputation and trust values for only nodes in their neighborhood. The authors argue that there exists no sensor network application in which a node will require prior reputation knowledge about a node many hops away. The objective of RFSN is to create a community of trustworthy sensor nodes. RFSN was the first reputation and-trust-based model designed and developed exclusively for sensor networks. RFSN distinguishes between trust and reputation and uses two different metrics.

From Figure 13.2, it is clear that the first-hand information from the watchdog mechanism and second-hand information are combined to get the reputation value of a node. Then the trust level of a node is determined from its reputation. Finally, based on the trust value, the node's behavior toward the node in question is determined. If the trust value is above a certain threshold, then the behavior is "cooperate"; otherwise, it is "don't cooperate."

Information Gathering. RFSN, like many other systems, uses a watchdog mechanism for first-hand information. In RFSN, the watchdog mechanism consists of different modules, typically one for each function it wants to monitor. The higher the number of modules, the greater the resource requirements. RFSN maintains the reputation as a probabilistic distribution which gives it complete freedom, unlike a discrete value. The authors argue that reputation can only be used to statistically predict the future behavior of other nodes and it cannot deterministically define the action performed by them. The reputation of all nodes that node *i* interacts with is maintained in a reputation table.

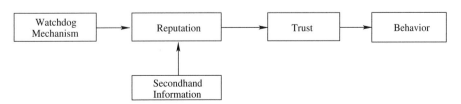

Figure 13.2. Architectural design of RFSN [2].

The direct reputation, $(R_{ij})_D$, is updated using the direct observations, that is, the output of the watchdog mechanism.

Information Sharing. In RFSN, nodes share their findings with fellow nodes. However, they are allowed to share only positive information. RFSN gives higher weight to second-hand information from nodes with higher reputation, which is reasonable and fairly novel. The weight assigned by node i to second-hand information from a node k is a function of the reputation of node k maintained by node i. Like many other reputation-and-trust-based systems, RFSN also makes use of the popular Beta distribution model.

Information Modeling. Assume node i has established some reputation value, R_{ij}, for node j. Let the two nodes have $r + s$ more interactions with r cooperative and s noncooperative interactions, respectively. The reputation value is now updated as follows.

$$R_{ij} = \text{Beta}(\alpha_j^{\text{new}} + 1, \beta_j^{\text{new}} + 1) \tag{13.1}$$

$$\alpha_j^{\text{new}} = (w_{\text{age}} * \alpha_j) + r \tag{13.2}$$

$$\beta_j^{\text{new}} = (w_{\text{age}} * \beta_j) + s \tag{13.3}$$

RFSN, unlike CORE, gives more weight to recent observations. This is done by using w_{age}, which is range bound in the space (0,1). This is for updating reputation value using direct observation.

To update the reputation value using second-hand information, RFSN maps the problem into the Dempster–Shafer belief theory[1] [35] domain. Then, the problem is solved using the belief discounting theory [36]. Using this theory, the reputation of the reporting node is automatically taken into account in the calculation of the reputation of the reported node. Hence, a separate deviation test is not necessary. A node with higher reputation gets higher weight than a node with lower reputation. Then, the trust level of a node is determined using its reputation value. It is determined as the statistically expected value of the reputation.

Decision Making. Finally, in the decision-making stage, node i has to make a decision on whether or not to cooperate with node j. The decision of node i is referred

---

[1]The Dempster–Shafer theory is a mathematical theory of evidence used to combine different pieces of information to calculate the probability of an event. It is a generalization of the Bayesian theory of subjective probability.

to as its behavior B_{ij} and is a binary value: {cooperate, don't cooperate}. Node i uses T_{ij} to make a decision as follows:

$$B_{ij} = \begin{cases} \text{cooperate} & \forall T_{ij} \geq B_{ij} \\ \text{don't cooperate} & \forall T_{ij} < B_{ij} \end{cases} \quad (13.4)$$

Discussions. Like CONFIDANT, RFSN also treats misbehaving and faulty nodes the same way. The rationale is that a node that is uncooperative has to be excluded irrespective of the cause of uncooperative behavior. In RFSN, nodes are allowed to exchange only good reputation information. Also, in RFSN, only direct reputation information is propagated. This prevents the information from looping back to the initiating node. However, this increases the memory overhead slightly since a separate data structure has to be maintained for direct reputation.

13.6.5 DRBTS

DRBTS stands for "distributed reputation and trust-based beacon trust system." This model was proposed by Srinivasan, Teitelbaum, and Wu recently to solve a special case in location-beacon sensor networks. DRBTS is a distributed model that makes use of both first-hand and second-hand information. It has two types of nodes: beacon node (BN) and sensor node (SN). It is symmetric from the BN perspective but asymmetric from the SN perspective. This is because beacon nodes are capable of determining their location and must pass this information to the sensor nodes. The difficulty is that without knowledge of their own location, a sensor node has no way of telling if a beacon node is lying to it. DRBTS enables sensor nodes to exclude location information from malicious beacon nodes on the fly by using a simple majority principle. DRBTS addresses the malicious misbehavior of beacon nodes.

Information Gathering. In DRBTS, information gathering is addressed from two different perspectives: the sensor node perspective and the beacon node perspective. From a beacon node perspective, DRBTS uses a watchdog for neighborhood watch. When an SN sends a broadcast asking for location information, each BN will respond with its location and the reputation values for each of its neighbors. The watchdog packet overhears the responses of the neighboring beacon nodes. It then determines its location using the reported location of each BN in turn, and then it compares the value against its true location. If the difference is within a certain margin of error, then the corresponding BN is considered benign, and its reputation increases. If the difference is greater than the margin of error, then that BN is considered malicious and its reputation is decreased. From a sensor node perspective, there is no first-hand information gathered by sensor nodes through direct observations. They rely completely on the second-hand information passed to them from nearby beacon nodes during the location request stage.

DRBTS also includes a method by which BNs can send out location requests disguised as SNs, in case of low network activity. However, unlike CONFIDANT, DRBTS does not differentiate first-hand information into personal experience and direct observation.

Information Sharing. DRBTS does make use of second-hand information to update the reputation of its neighboring nodes. However, information sharing is only with respect to BNs. SNs do not share any information since they do not collect any by virtue of their own observation of their neighborhood. In DRBTS, nodes are allowed to share both positive and negative reputation information. This is allowed to ensure a quick learning time.

Information Modeling. Let BN j respond to a SN's request. Then BN i, in the range of j updates its reputation entry of j using the direct observation as follows.

$$R_{ki}^{New} = \mu_1 \times R_{ki}^{Current} + (1 - \mu_1) \times \tau \tag{13.5}$$

where $\tau = 1$ if the location was deemed to be truthful and $\tau = 0$ otherwise, and μ_1 is a weight factor.

To use second-hand information, assume B j is reporting about BN k to BN i. Now BN i first performs a deviation test to check if the information provided by BN j is compatible.

$$|R_{ji}^{Current} - R_{ki}^{Current}| \leq d \tag{13.6}$$

If the above test is positive, then information provided is considered to be compatible and the entry R_{ik} is updated as follows.

$$R_{j,i}^{New} = \mu_2 \times R_{j,i}^{Current} + (1 - \mu_2) \times R_{k,i}^{Current} \tag{13.7}$$

But, if the deviation test in equation 13.6 is negative, then j is considered to be lying and its reputation is updated as follows.

$$R_{j,k}^{New} = \mu_3 \times R_{j,k}^{Current} \tag{13.8}$$

Equation (13.8) ensures that nodes that lie are punished so that such misbehavior can be discouraged.

Decision Making. Decisions are made from the sensor node's perspective. After sending out a location request, an SN waits until a predetermined timeout. A BN has to reply before the timeout with its location information and its reputation ratings for its neighbors. Then, the SN, using the reputation ratings of all the responding BNs, tabulates the number of positive and negative votes for each BN in its range. Finally, when the SN has to compute its location, it considers the location information only from BNs with positive votes greater than negative votes. The remaining location information is discarded.

Discussions. DRBTS addresses the malicious misbehavior of beacon nodes. This unique problem that this system solves, though very important to a specific branch of WSNs, is not encountered very frequently. However, the idea can be easily extended to other problem domains.

13.7 OPEN PROBLEMS

Though lots of research has been done in this field, reputation-and-trust-based systems are still in their incubation phase when it comes to MANETs and WSNs. There are some open problems that have been identified which need to be resolved sooner, rather than later. One such problem is the bootstrap problem. It takes a while to build reputation and trust among nodes in a network. Minimizing this startup period is still an open issue. Using all the available information does help in building reputation and trust among nodes quickly; but as examined earlier, it makes the system vulnerable to false report attacks. Also, in systems like CORE where nodes have to contribute on a continued basis to survive, periods and regions of low network activity pose new problems. Aging will deteriorate the reputation of even benign nodes since there is no interaction.

Another important problem that needs to be addressed is the intelligent adversary strategies. An intelligent adversary can manifest the degree of his misbehavior such that he can evade the detection system. A game-theoretic approach to resolve this problem may be a worthy investigation.

A system like CORE uses functional reputation to monitor the behavior of nodes with respect to different functions. However, CORE unifies the reputation of a node for various functions into a global reputation value of that node. This may not be very effective since it will allow an adversary to cover his misbehavior with respect to one function by being well-behaved for other functions. No literature so far has addressed the benefits of using functional reputation values independently. It may be beneficial to exclude nodes for a particular function if it is known for misbehaving with respect to that function rather than judging him with respect to over-all behavior. For example, a specific lawyer might be known to be a great person to have on your side if you need a case dismissed, while being known to cheat at cards. Knowing this, you would hire him to argue a case, but perhaps not invite him to the weekly poker game.

Finally, a scheme needs to be developed to motivate nodes to publish their ratings honestly. This is particularly important for MANETs since nodes often don't belong to the same entity. However, in WSNs, since nodes tend to belong to some overarching system, there is an inherent motivation for nodes to participate honestly in information exchange. Market schemes seem to show some promise in similar problems in routing in P2P and MANETs, and inspiration might be drawn from these fields.

13.8 CONCLUSION

Reputation and trust are two very important tools that have been used since the beginning to facilitate decision making in diverse fields from an ancient fish market to state-of-the-art e-commerce. This chapter has provided a detailed understanding of reputation-and-trust-based systems both from a societal and a wireless communication networks' perspective. We have examined all aspects of reputation-and-trust-based systems including their goals, properties, initialization, and classifications. Also, we have provided an in-depth discussion of the components of reputation and trust-based systems. A comprehensive review of research works focusing on adapting reputation-and-trust-based systems to MANETs and WSNs has been presented, along with an in-depth discussion of their pros and cons. We have also presented some open problems that are being researched even now.

ACKNOWLEDGEMENTS

This work was supported in part by NSF grants ANI 0073736, EIA 0130806, CCR 0329741, CNS 0422762, CNS 0434533, and CNS 0531410.

13.9 EXERCISES

1. Discuss any two areas of application of reputation-and-trust-based systems with respect to ad hoc and sensor networks other than those discussed in this chapter.
2. How can the bootstrap problem in reputation-and-trust-based systems be efficiently addressed?
3. Information asymmetry plays a critical role. Discuss one situation in detail, other than those presented in this chapter where information asymmetry plays a vital role.
4. In all the systems discussed in this chapter, the reputation-and-trust metric have been considered to be either black or white; that is, an entity has either a good reputation or a bad reputation, or an entity is either trusted or not trusted. However, in real life there is an element of uncertainty. Discuss a reputation-and-trust-based model that incorporates uncertainty into the reputation-and-trust metric.

5. Can you think of any real-life metric other than reputation and trust onto which mobile ad-hoc and sensor network security problems can be mapped and solved? Discuss in detail.

REFERENCES

1. A. Srinivasan, J. Teitelbaum and J. Wu. DRBTS: Distributed reputation-based beacon trust system. In *The 2nd IEEE International Symposium on Dependable, Autonomic and Secure Computing (DASC '06)*, Indianapolis, USA, 2006.
2. S. Ganeriwal and M. Srivastava. Reputation-based framework for high integrity sensor networks. In *Proceedings of the 2nd ACM Workshop on Security of Ad Hoc and Sensor Networks (SASN '04)*, October 2004, pp. 66–77.
3. T.-P. Liang and J.-S. Huang. An empirical study on consumer acceptance of products in electronic markets: A transaction cost model. *Decision Support Systems*, **24**:29–43, 1998.
4. H. Liang, Y. Xue, K. Laosethakul, and S. J. Lloyd. Information systems and health care: Trust, uncertainty, and online prescription filling. *Communications of AIS*, **15**:41–60, 2005.
5. D. H. McKnight, V. Choudhury, and C. Kacmar. Developing and validating trust measures for e-commerce: An integrating typology. *Information Systems Research*, **13**:334–359, 2002.
6. D. M. Rousseau, S. B. Sitkin, R. S. Burt, and C. Camerer. Not so different after all: A cross-discipline view of trust. *Academy of Management Review*, **23**:393–404, 1998.
7. D. Gefen, E. Karahanna, and D. W. Straub. Trust and TAM in online shopping: an integrated model. *MIS Quarterly*, **27**:51–90, 2003.
8. D. H. McKnight, L. L. Cummings, and N. L. Chervany. Initial trust formation in new organization relationships. *Acadamy of Management Review* **23**:473–490, 1998.
9. K. J. Stewart. Trust transfer on the world wide web. *Organization Science*, **14**:5–17, 2003.
10. P. Dasgupta. Trust as a commodity. In D. G. Gamretta, Editor, *Trust*, Basil Blackwell, New York, pp. 49–72, 1988.
11. S. Ba and P. A. Pavlou. Evidence of the effect of trust building technology in electronic markets: Price premiums and buyer behavior. *MIS Quarterly*, **26**:243–268, 2002.
12. R. J. Lewicki and B. B. Bunker. Trust in relationships: A model of trust development and decline. In J. Z. Rubin, editor, *Conflict, Cooperation and Justice*, Jossey-Bass, San Francisco, 1995, pp. 133–173.
13. D. Gefen. E-commerce: the role of familiarity and trust. *Omega*, **28**:725–737, 2000.
14. S. P. Shapiro. The social control of impersonal trust. *American Journal of Sociology*, **93**:623–658, 1987.
15. J. Pfeffer and G. R. Salancik. The external control of organizations: A resource dependence perspective. New York: Harper Row, 1978.
16. S. Ganesan. Determinants of long-term orientation in buyer–seller relationships. *Journal of Marketing*, **58**:1–19, 1994.
17. P. M. Doney and J. P. Cannon. An examination of the nature of trust in buyer–seller relationships. *Journal of Marketing*, **61**:35–51, 1997.
18. P. R. Nayyar. Information asymmetries: A source of competitive advantage for diversified service firms. *Strategic Management Journal*, **11**:513–519, 1990.

19. D. P. Mishra, J. B. Heide, and S. G. Cort. Information asymmetry and levels of agency relationships. *Journal of Marketing Research*, **35**:277–295, 1998.
20. J. Golbeck and J. Hendler. Inferring binary trust relationships in web-based social networks. *ACM Transactions on Internet Technology*, **6**(4):497–529, 2006.
21. P. Michiardi and R. Molva. Simulation-based analysis of security exposures in mobile ad hoc networks. *European Wireless Conference*, 2002.
22. P. Michiardi and R. Molva. CORE: A COllaborative REputation mechanism to enforce node cooperation in mobile ad hoc networks. *Communication and Multimedia Security*, September, 2002.
23. S. Buchegger and J.-Y. Le Boudec. Performance Analysis of the CONFIDANT Protocol (Cooperation Of Nodes- Fairness In Dynamic Ad-hoc NeT-works). *Proceedings of MobiHoc 2002*, Lausanne, CH, June 2002.
24. M. Blaze, J. Feigenbaum, J. Ioannidis, and A. Keromytis. RFC2704. The KeyNote Trust Management System Version 2, 1999.
25. N. Li, J. Mitchell, and W. Winsborough. Design of a role-based trust management framework. In *Proceedings of the IEEE Symposium on Security and Privacy*, Oakland, 2002.
26. P. Resnick, R. Zeckhauser, E. Friedman, and K. Kuwabara. Reputation systems. *Communications of the ACM*, **43**(12):4548, 2000.
27. C. Dellarocas. Immunizing online reputation reporting systems against unfair ratings and discriminatory behavior. In *Proceedings of the ACM Conference on Electronic Commerce*, pp. 150–157, 2000.
28. P. Resnick and R. Zeckhauser. Trust among strangers in internet transactions: Empirical analysis of ebays reputation system. Working paper for the *NBER Workshop on Empirical Studies of Electronic Commerce*, 2001.
29. K. Aberer and Z. Despotovic. Managing trust in a peer-2-peer information system. In *Proceedings of the Ninth International Conference on Information and Knowledge Management* (CIKM 2001), 2001.
30. S. Marti, T. J. Giuli, K. Lai, and M. Baker. Mitigating routing misbehaviour in mobile ad hoc networks. In *Proceedings of the Sixth Annual International Conference on Mobile Computing and Networking (MobiCom 2000)*.
31. S. Buchegger and J.-Y. Le Boudec. The effect of rumor spreading in reputation systems in mobile ad-hoc networks. *Wiopt03*, Sofia- Antipolis, March 2003.
32. S. Bansal and M. Baker. Observation-based cooperation enforcement in ad hoc networks. http://arxiv.org/pdf/cs.NI/0307012, July 2003.
33. S. Buchegger and J.-Y. Le Boudec. A robust reputation system for peer-to-peer and mobile ad-hoc networks. In *Proceedings of P2PEcon 2004*, Harvard University, Cambridge MA, June 2004.
34. R. Jurca and B. Faltings. An incentive compatible reputation mechanism. In *Proceedings of the IEEE Conference on E-Commerce*, Newport Beach, CA, June 24–27, 2003.
35. G. Shafer. *A Mathematical Theory of Evidence*, Princeton University Press, Princeton, NJ, 1976.
36. A. Jsang. A logic for uncertain probabilities. *International Journal of Uncertainty, Fuzziness and Knowledge-Based Systems*, **9**(3):279–311, June 2001.
37. A. Josang and R. Ismail. The beta reputation system. In *Proceedings of the 15th Bled Electronic Commerce Conference*, Bled, Slovenia, June 2002.

38. S. Buchegger and J.-Y. Le Boudec. Self-policing mobile ad-hoc networks by reputation systems. *IEEE Communications Magazine*, July 2005.
39. J. Mundinger, J.-Y. Le Boudec. Analysis of a Reputation System for Mobile Ad-Hoc Networks with Liars. In *Proceedings of The 3rd International Symposium on Modeling and Optimization*, Trento, Italy, April 2005.

CHAPTER 14

Vehicular Ad Hoc Networks: An Emerging Technology Toward Safe and Efficient Transportation

MAEN M. ARTIMY, WILLIAM ROBERTSON, and WILLIAM J. PHILLIPS

Department of Engineering Mathematics and Internetworking, faculty of Engineering, Dalhousie University, Halifax, Nova Scotia, Canada B3H 3J5

14.1 INTRODUCTION

Research on mobile ad hoc networks (MANETs) has covered vast ground in the past decade. The majority of this research has focused on networks where nodes are assumed to be mobile devices carried by people or vehicles with a range of applications from controlling military units to e-commerce. Recently, a number of projects around the world have initiated research with the aim of investigating the use of ad hoc networks as a communication technology for vehicle-specific applications within the wider concept of intelligent transportation systems (ITS).

ITS include the application of computers, communications, sensor, and control technologies and management strategies in an integrated manner to improve the functioning of the transportation systems. These systems provide traveler and real-time information to increase the safety and efficiency of the ground transportation network. Interest in ITS and similar activities has started in North America, Europe, and Japan since the 1970s. The interest in ITS represented a shift from an early trend of constructing more and larger highways as a solution to epidemic traffic congestion, increased accident rates and their consequences on human life, lost productivity, increased air pollution, and expensive repairs. Such a trend is no longer seen as a viable option due to its high financial, social, and environmental costs. The new trend is to utilize the existing infrastructure efficiently via new technologies that allow the transportation system to respond dynamically to traffic congestion and improve safety.

Intervehicle communication (IVC) is usually viewed as a critical element of the ITS architecture. Many ITS projects employ short- to medium-range wireless technology

---

Algorithms and Protocols for Wireless and Mobile Ad Hoc Networks, Edited by Azzedine Boukerche
Copyright © 2009 by John Wiley & Sons Inc.

to build Vehicle-to-Vehicle (V2V) and Vehicle-to-Roadside (V2R) communication networks. These networks collect and disseminate time-critical information that helps improve the safety and increase the efficiency on the transportation infrastructure. With the current advances of MANETs, it is apparent that such technology should be suitable for IVC applications. Vehicular ad hoc networks (VANETs) applications will include on-board active safety systems leveraging V2V or V2R networking. These systems may assist drivers in avoiding collisions or dangerous road conditions. Other applications include adaptive cruise control, real-time traffic congestion and routing information, high-speed tolling, and mobile infotainment.

Establishing reliable, scalable, and secure VANETs poses an extraordinary engineering challenge. However, direct vehicle communication based on VANETs is the most promising technology to support on-board safety systems due to high availability, high flexibility of connection, high data rates, and low latency for the transfer of information between vehicles. Furthermore, direct IVC can exploit the geographic relevance of data and can be deployed with low costs [1, 2].

Despite many similarities to MANET, VANETs have some advantages over general ad hoc networks that include the ability to assemble large computational resources, plentiful power sources, and constrained mobility patterns. As such, research in VANETs have followed significantly different directions. This chapter provides a literature survey that covers several research directions in VANETs with more emphasis on the topics related to connectivity, transmission range assignment, and vehicle mobility simulation, which are among the research interests of the authors.

The remainder of this chapter is structured as follows. Section 14.2 lists the characteristics that distinguish VANETs from traditional ad hoc networks. Section 14.3 provides a summary of the enabling technologies that are expected to support VANETs and IVC. Section 14.4 describes the challenge of maintaining connectivity in VANETs. The section focuses on the analytical estimates of the required common transmission range in the cases of homogeneous and nonhomogeneous distribution of vehicles. Section 14.5 introduces an algorithm that adjusts a vehicle's transmission range based on local vehicle density. The use of this algorithm eliminates the need for a static transmission range that can have some adverse effects on the network performance in certain conditions. Section 14.6 introduces the routing schemes that are suitable for vehicular networks. Section 14.7 classifies the applications of ITS and VANETs into four categories and summarizes the challenges and common approaches in each category. Section 14.8 provides a brief introduction to traffic flow theory in order to emphasize the highly dynamic nature of VANETs. The section introduces also some traffic simulation models that are used in VANETs research. The chapter is concluded in Section 14.9 with a summary and final remarks. Section 14.9 lists some exercises that the reader can use as a guide to further study.

14.2 VANET CHARACTERISTICS

VANETs exhibit some unique characteristics that differentiate them from other types of ad hoc network. These characteristics are [3]:

1. Vehicles mobility is restricted to one-dimensional road geometry. Considering that current standards of wireless ad hoc networks have a limited coverage of a few hundred meters, from the perspective of these networks, vehicles on long highways move in one dimension. Moreover, in urban centers, buildings and similar obstacles limit the propagation of the communication signal and split the network topology.
2. Factors such as road configuration, traffic laws, safety limits, and physical limits affect the mobility of vehicles. Drivers' behavior and interaction with each other also contribute to the vehicle mobility pattern. Simulating vehicle traffic is a complex task that has been the focus of study for its applications in transportation engineering.
3. Vehicle mobility creates a highly dynamic topology. In heavy traffic congestion, vehicles may be within a couple of meters proximity of one another, whereas on a sparsely populated road the distance may be hundreds of meters. Traffic condition may change rapidly between congested and sparse due to traffic jams, accidents, and road constraints.
4. VANTs are potentially large-scale networks.
5. Research shows that the end-to-end connectivity is limited to short range communications in one-dimensional networks. These networks remain almost surely divided into an infinite number of partitions.
6. Vehicles can provide more resources than other types of mobile networks such as large batteries, antennas, and processing power. Therefore, conserving such resources in VANETs is not a major concern.

14.3 ENABLING TECHNOLOGIES

Two approaches exist in developing wireless technologies for VANETs. The first starts from existing wireless local area networks (WLAN) technologies such as IEEE 802.11. The second approach depends on the modification of 3G cellular technologies to VANET environment.

14.3.1 IEEE 802.11

The IEEE 802.11 standard [4] specifies the physical (PHY) and the medium access control (MAC) layers of a wireless local area network (WLAN) and offers the same interface to higher layers as other IEEE 802.x LAN standards. The physical layer can be either infrared or spread-spectrum radio transmission. Other features also include the support of power management, the handling of hidden nodes, license-free operation in the 2.4-GHz industrial, scientific, medical (ISM) band, and data rates of 1 or 2 Mbps. The standard does not support routing or exchange of topology information.

The IEEE 802.11 standard is the most commonly used PHY/MAC protocol in the research and demonstration of MANET and VANET due to the standard's support of infrastructure or ad hoc networks. There are three basic access mechanisms defined for IEEE 802.11. The first is a version of CSMA/CA. The second uses RTS/CTS handshake signals to avoid hidden terminal problem. The third mechanism uses polling to provide time-bounded service.

14.3.2 DSRC

The direct short-range communication (DSRC) standard [5] describes a MAC and PHY specifications for wireless connectivity. This standard is based on IEEE Standard 802.11a [4] in the 5-GHz band and is meant to be used in high-speed vehicle environments. In fact, the PHY layer of DSRC is adapted from IEEE 802.11a PHY based on orthogonal frequency division multiplex (OFDM) technology. Moreover, the MAC layer of DSRC is very similar to the IEEE 802.11 MAC based on the CSMA/CA protocol with some minor modifications.

The DSRC standard utilizes the 75-MHz spectrum, which is allocated by the United States Federal Communications Commission to provide wireless V2R or V2V communications over short distances. The standard specifies communications that occur over line-of-sight distances of less than 1000 m between roadside units and mostly high-speed, but occasionally stopped and slow-moving, vehicles or between high-speed vehicles. It is expected, with the support of industry and government organizations, that the DSRC system will be the first wide-scale VANET in North America.

14.3.3 UTRA TDD

The FleetNet and CarTALK projects have selected UMTS terrestrial radio access time division duplex (UTRA TDD) as a potential candidate for the air interface that can be used for their VANET applications [6–8]. The UTRA TDD offers connectionless and connection-oriented services as well as control over quality of service (QoS) through flexible assignment of radio resources and asymmetric data flows. Other features include: availability of the unlicensed frequency band at 2010–2020 MHz in Europe; large transmission range; support of high vehicle velocities; and high data rates of up to 2 Mbps.

To adapt UTRA TDD to ad hoc operation, decentralized schemes for synchronization, power control, and resource management are needed to replace the centralized schemes. Coarse time synchronization for time multiplexing is achieved using the global positioning system (GPS). For fine synchronization, an additional synchronization sequence is transmitted along with every data burst. A combination of frequency, timeslot, and code provides transmission capacity for one data connection, which can be a unicast, anycast, multicast, or broadcast connection. Only one station is allowed to transmit in one timeslot. Up to 16 stations can be simultaneously reached by one station. This restriction eliminates the need for power control.

UTRA TDD reserves a number of timeslots for high-priority services. The remaining slots can be dynamically assigned and temporarily reserved by different stations for lower priority services using the Reservation ALOHA (R-ALOHA) scheme. Potential collisions are reduced by reserving a small transmit capacity for a circuit-switched broadcast channel that is primarily used for signaling purposes but may also be used for transmitting small amounts of user data.

14.4 VANET CONNECTIVITY

VANETs and MANETs are faced with the nontrivial task of maintaining connectivity so that a vehicle (a mobile node) may establish a single hop or multihop communication link to any other vehicle in the network. The connectivity of the network is affected by factors that include transmitter power, environmental conditions, obstacles, and mobility. Increasing the transmission range of mobile transceivers may solve the connectivity problem. This solution results in more neighbors being reached and fewer hops being traveled by packets. However, more nodes have to share the medium, causing more contentions, collisions, and delays that reduce capacity. A short transmission range, on the other hand, increases the capacity of the network by allowing simultaneous transmissions in different geographic locations (due to frequency reuse) at the expense of connectivity. This problem is complicated by the fact that the network topology is constantly changing [9].

Extensive research is dedicated to determine the minimum transmission range (MTR) in MANETs. The MTR corresponds to the minimum common value of the nodes' transmitting range that produces a connected communication graph. The motivation behind this research is the difficulty, in many situations, to adjust the nodes' transmitting range dynamically, making the design of a network with a static transmitting range a feasible option. It is also known that setting the nodes' transmitting range to the minimum value minimizes energy consumption while maximizing network capacity [10].

14.4.1 Homogeneous MTR in One-Dimensional Networks

If all node positions in a network are known, the MTR is determined by the length of the longest edge of the Euclidean Minimum Spanning Tree (MST) in a geometric graph composed of all nodes [11]. In the most realistic scenarios, node positions are unknown but it can be assumed that the nodes are distributed according to some probability distribution in the network. In the latter case, it is necessary to estimate the MTR that provides connectivity with a probability that converges to 1 as the number of nodes or the network size increases.

In the probabilistic approaches to determine the MTR, authors have relied on the theory of geometric random graphs (GRG) to represent wireless ad hoc networks [12]. A network can be modeled by a graph $G(V, r_c)$ in which two nodes are connected if the Euclidean distance between them is no more than r_c. Most studies consider how the transmission range is related to the number of nodes n, dispersed according to a uniform or Poisson distribution in a fixed area (or line). The continuum percolation theory and the occupancy theory have also been used in the probabilistic analysis of ad hoc networks.

In the case of one-dimensional models, Piret [13] discussed the coverage problem to find that the lower bound of the transmission range, r_{cover}, for nodes located according to the Poisson distribution in a line of length L, with density k, is $r_{cover} = \ln(Lk)/(2k)$. The author shows that, if $r_c = m\, r_{cover}$, then the connectivity, $Q()$, among nodes approaches 1 (i.e., $\lim_{L\to\infty} Q(r_c, L) = 1$) when $m > 2$, where m is a constant.

Santi and Blough [14] provide tighter bounds on MTR using occupancy theory. Their primary result shows that when nodes are distributed uniformly over a line of length L, the network is connected if

$$r_c n \in \Theta(L \ln(L)), \qquad r_c \gg 1 \tag{14.1}$$

Dousse et al. [15] approach the connectivity problem in both pure ad hoc and hybrid networks. They conclude that connectivity is limited to short-range communications in one-dimensional and strip networks (two-dimensional networks of finite width and infinite length), because the network remains divided (with high probability) into an infinite number of bounded clusters. Since VANETs in highway environment can be represented as a one-dimensional or strip network, it can be concluded that it is not practical to maintain connectivity in the entire network that may stretch for tens of kilometers. Instead a VANET should tolerate a certain level of partitioning. A relationship derived by Cheng and Robertazzi [16] predicts that the expected number of broadcasts needed to disseminate a message before a gap is encountered in a one-dimensional network increases exponentially relative to the product of transmission range and the node density.

To investigate the effect of mobility, Sánchez et al. [17] present an algorithm to calculate the MTR required to achieve (with high probability) full network connectivity. The main empirical results show that the MTR decreases as $\sqrt{\ln(n)/n}$ and, when considering mobility, the range has little dependence on the mobility model. The results were obtained through simulations using Random Waypoint (RWP), random direction, and Brownian-like mobility models.

Santi and Blough [14, 18] also used simulations to investigate the relationship between the MTR in stationary and mobile networks. They consider RWP and Brownian-like mobility models to analyze different MTR values in various connectivity requirements. The simulation results show that, due to mobility, the MTR has to be increased relative to the stationary case to ensure the network connectivity during 100% of the simulation time. Furthermore, the simulation results show that the transmitting range can be reduced considerably if the connectivity is maintained during 90% of the simulation time.

The paper by Füßler et al. [19] focuses on vehicular networks. In the context of comparing various routing strategies, simulations are used to find the effect of the transmission range on the number of network partitions and provide an estimate of the transmission range that minimizes the partitions. It should be noted that the simulations in reference 19 are limited to free-flow traffic of low density.

14.4.2 Transmission Range in Nonhomogeneous Vehicle Traffic

Most of the work presented in Section 14.4.1 assumes that nodes are distributed homogeneously according to uniform or Poisson distributions within the network space. Jost and Nagel's analysis and simulations [20] show that vehicles in free-flow conditions are distributed along the road according to uniform distribution while in congested traffic, the vehicle distribution in nonhomogeneous. It is also shown in Section 14.4.1 that the work of Santi and Blough [14] applies to an environment similar to free-flow vehicle traffic in a single-lane road. Based on these two principles and the assumption that vehicle mobility is described by the Nagle and Schreckenberg (NaSch) model [21], we have derived an estimate for the lower-bound of the MTR for VANETs in highways by considering the non-homogeneous distribution of vehicles along the highway in congested traffic [22].

A congested traffic system can be characterized as patches of traffic jams separated by jam-free regions [23], as shown in the top half of Figure 14.1. In such a network, the worst-case MTR value depends only on the traffic of the free-flow region since the MTR in the jam region always has a lower value. Suppose that all free-flow regions can be consolidated in one side of the road and all congested regions are consolidated in the other side, as illustrated in bottom half of Figure 14.1. The critical range within this free-flow section, r_{c,L_F}, has a lower-bound given by Eq. (14.1). Therefore,

$$r_{c,L_F} \geq \frac{\ln(L_F)}{k_F} \qquad (14.2)$$

where L_F is the length of the free-flow section of the road and k_F is the density of the traffic in the same section.

In a highway such as the one shown in Figure 14.1, the flow of vehicles escaping the traffic jam, q_F, is a free-flow traffic that is virtually independent of the vehicle density. The value of q_F is determined by the average waiting time for the first vehicle in the

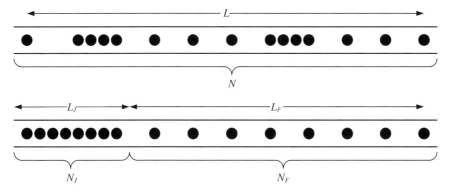

Figure 14.1. Congested traffic in a highway section. (Top) Patches of traffic jams. (Bottom) Consolidation of the free-flow and congested sections of the road.

traffic jam to move [24]. This outflow is the maximum flow achievable in the road [23]. Therefore, given that the speed in free-flow traffic is u_f, and from the fundamental relation (14.8) (see Section 14.8.1), the maximum density in the free-flow section of the road is

$$k_F = \frac{q_F}{u_f} \qquad (14.3)$$

We use basic mathematical manipulations to determine the distance and the density of vehicles in each region of Figure 14.1 as a function of the total length and density (omitted here). Also, by knowing that the MTR in the road must be equal to or higher than that of the free-flow region, we obtain

$$r_{c,L} \geq r_{c,L_F} \qquad (14.4)$$

Using (14.2) and (14.4), the MTR, $r_{c,L}$, for a VANET that spans the entire road length is given,

$$r_c(k) = \begin{cases} \dfrac{\ln(L)}{k}, & k \leq k_c \\ \dfrac{\ln(L)}{k_F} + \dfrac{1}{k_F}\ln\left(\dfrac{k_{\text{jam}} - k}{k_{\text{jam}} - k_F}\right), & k > k_c \end{cases} \qquad (14.5)$$

The range, r_c, in (14.5) depends only on the length, L, of the road section and the vehicle density within the section, k. The traffic jam density, k_{jam}, is a constant. The value of k_F is obtained from (14.3). The density, k_c, is the critical density that separates the free-flow traffic from congested traffic (see Section 14.8.1).

Figure 14.2 shows the plot of Eq. (14.5). The values of k_c and k_F are determined for different values of u_f. Equation (14.5) indicates that the nonhomogeneous distribution of vehicles results in a considerable increase in the MTR. At the speed of $u_f = 135$ km/hr, the rise in MTR reaches more than six times the value obtained from (14.1).

14.5 DYNAMIC TRANSMISSION RANGE ASSIGNMENT

The Dynamic Transmission Range Assignment (DTRA) algorithm proposed in references 22 and 25, employs information about the local vehicle density estimate and local traffic condition to set a vehicle's transmission range dynamically. In the DTRA algorithm, no information about the neighboring nodes is collected and no message exchange is required. A vehicle can determine its transmission range based on its own mobility pattern, which provides hints about the local traffic density. Unlike protocols that depend on overhearing of data and control traffic [26–28] to determine the transmitting power level, the DTRA depends on estimation of vehicle density. As

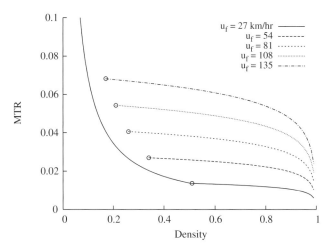

Figure 14.2. The lower bound of MTR in a highway of finite length and various values of free-flow speed, u_f.

a result, the protocol is transparent to data communication protocols and can be used in conjunction with existing protocols.

As an alternative to using a high transmission power to maintain connectivity in all traffic conditions, estimation of traffic density provides the necessary means to develop a power control algorithm that sets a vehicle's transmission range dynamically as traffic conditions change. The DTRA algorithm, shown in Algorithm 1, uses the ratio, T_s/T_t, as an order parameter to distinguish between the free-flow and the congested traffic conditions. This parameter represents the fraction of time during which a vehicle remains motionless T_s, within a short time window T_t. In line 6, the time ratio is also used to estimate the local traffic density, K [22, 25].

The algorithm sets the transmission range to its maximum level in free-flow traffic, which is characterized by low vehicle density. Assuming that a vehicle occupy a finite length, the maximum transmission range, $r_{\max}$, needed to maintain VANET connectivity in a single-lane highway of length, L, is

$$r_{\max} = L\left(1 - \frac{k}{k_{\text{jam}}}\right) + \frac{1}{k_{\text{jam}}} \tag{14.6}$$

where k is the vehicle density and k_{jam} is the density of a traffic jam. Equation (14.6) is obtained by assuming that all but one vehicle are packed at a distance of $1/k_{\text{jam}}$ from each other at one side of the road while the remaining vehicle is located at the opposite side.

The experiments in reference 25 show that the highest required transmission range is much lower than $r_{\max}$ for most of the density range. The upper bound, r, can be

ALGORITHM 1. Dynamic Transmission Range Algorithm

Input: α, λ'		▷ constants
Input: MR		▷ maximum transmission range
1:	**function** CALCULATEDYNAMICRANGE(T_s/T_t)	
2:	$\quad f_s \leftarrow T_s/T_t$	
3:	$\quad$ **if** $f_s = 0$ **then**	▷ free-flow traffic
4:	$\quad\quad$ TR $\leftarrow$ MR	
5:	$\quad$ **else**	▷ congested traffic
6:	$\quad\quad K \leftarrow \left[(1-f_s)/\lambda' + 1\right]^{-1}$	
7:	$\quad\quad t_1 \leftarrow$ MR $\times (1-K)$	
8:	$\quad\quad t_2 \leftarrow \sqrt{\text{MR} \times \log(\text{MR})/K} + \alpha \times \text{MR}$	
9:	$\quad\quad$ TR $\leftarrow \min(t_1, t_2)$	
10:	$\quad$ **end if**	
11:	$\quad$ **return** TR	▷ dynamic transmission range
12:	**end function**	

approximated by

$$r \leq \sqrt{L \ln(L)/k} + \alpha L \qquad (14.7)$$

where $0 \leq \alpha \leq 0.25$. The results of Artimy et al. [25] show also that the minimum of Eqs. (14.6) and (14.7) represents the upper-bound MTR needed to maintain connectivity in VANETs in single- or multilane highways. These results are used in Algorithm 1 to set the transmission range in congested traffic. Figure 14.3 shows the

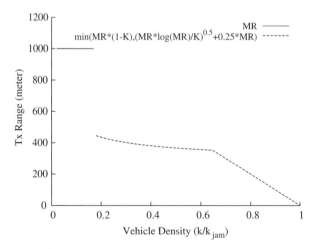

Figure 14.3. Values of transmission range selected by the DTRA algorithm versus estimated vehicle density.

transmission range values returned by the DTRA algorithm in the entire range of vehicle density, assuming that the maximum range MR = L = 1000 meters.

14.6 ROUTING

Factors such as the vast number of nodes that lack inherent organization, as well as frequent topological changes, make routing a challenging task in a VANET. MANETs' topology-based routing protocols [29] are ineffective in VANETs either due to slow reaction to the rapid change in the topology, in the case of proactive protocols, or due to the explicit route establishment phase that render reactive protocols impractical for most of the ITS applications that have low latency requirement. Moreover, for many ITS applications, the target node identity is a priori unknown, which rules out unicast as the main delivery mechanism.

A routing approach for a VANET should, therefore, show a high degree of adaptability with respect to the dynamics of the network, cope with scalability, and employ the broadcast nature of wireless networks. The MANET-style packet-forwarding protocols may still be used for relatively delay-tolerant applications such as Internet services. Since the cars of the future are likely to be equipped with an on-board GPS receiver and have access to digital maps, routing in VANET will be achieved using a position-based approach.

In contrast to topology-based routing, position-based routing relies on vehicles' geographic position to forward data packets. Depending on the packet forwarding strategy, position-based routing can be classified into geocast routing and position-based routing approaches. The basic idea of geocast routing is to flood data packets within a restricted geographic region or direction range. Position-based routing forwards the packet to only one neighbor at a time, which is successively closer to the packet's destination.

14.6.1 Geocast Routing

Unlike the traditional multicast schemes, the geocast group is implicitly defined as the set of nodes within a geographically specified region. If a node resides within the region at a given time, it will automatically become a member of the corresponding geocast group at that time. In highly dynamic networks, many of the existing tree-based multicast protocols cannot keep the tree structure up-to-date. Therefore, multicast flooding may be considered as an alternative approach for MANET [30].

The location-based multicast protocols flood a message within the "multicast region." If the message source is outside the mulicast region, the message must traverse a "forwarding zone." The algorithms in reference 30 define the forwarding zone based on distance information only. The forwarding zone can be implicitly determined by the interest of the participating nodes in the distributed information. For instance, only vehicles heading toward a dangerous situation will be interested in the warning message. In reference 31, the message is forwarded to a region if the interest rate (in the message) within the region is higher than a certain threshold. Depending on the size of the message and anticipated overhead, the interest rate is determined prior to sending the message or when the message reaches the boundaries of the forwarding zone.

The term "zone of relevance" is used in the Role-Based Multicast (RBM) protocol [32] to identify the region where vehicles are likely to be interested in the message. The multicast group is defined implicitly by location, speed, driving direction, and time. The message is forwarded from the source to all vehicles within the zone of relevance. To avoid overwhelming the communication channel, a vehicle defers the rebroadcast of the received message by a time delay that is inversely proportional to its distance from the source. The vehicle then may rebroadcast the message only when no other vehicle broadcasts the message or when new vehicles move into its vicinity. The Inter-Vehicle Geocast (IVG) protocol [33] works on similar principles but avoids the routing overhead caused by maintaining a set of neighbors in the RBM.

14.6.2 Position-Based Routing

In the position-based routing, each host makes packet forwarding decisions based on the location of itself, its neighboring hosts, and the destination. Therefore, no global knowledge of the network is required in order to forward packets. This approach avoids the overhead of maintaining information about the dynamic network topology, but requires location services for distributing and querying position information among nodes [2].

In position-based routing, data packets are often forwarded to the neighbor that is geographically closest to the destination until the destination is reached using a strategy called "greedy forwarding" [34]. However, the assumption that physically close hosts are also close in the network topology may not hold true in VANETs, where the underlying road infrastructure and other obstacles creates permanent topology holes that restrict the propagation of the data messages [35].

One proposed routing scheme, called Geographic Source Routing (GSR), requires the support of on-board navigation systems to provide digital maps and makes use of Reactive Location Service (RLS) in order to learn the current position of a desired communication partner [2]. With this information at hand, the sending node can compute a path to the destination as a sequence of road junctions that the packet has to traverse in order to reach the destination. Forwarding a packet between two successive junctions is done on the basis of greedy forwarding since no obstacles should block the way.

The Spatially Aware Routing (SAR) protocol [35] can predict permanent topology holes caused by spatial constraints and avoid them beforehand. SAR makes use of relevant spatial information, like the road network topology, to generate a simple graph-based spatial model. Based on the spatial model, a source node can predict a permanent topology hole. The sender then selects a Geographic Source Route to avoid these holes in packet forwarding. To construct a spatial model, the relevant spatial information has to be extracted from available Geographic Information Systems (GIS), such as digital road maps used in vehicle navigation systems [36].

Access to traditional Internet applications will continue to rely on unicast routing protocols such as Ad Hoc On-Demand Distance Vector (AODV) and Dynamic Source Routing (DSR). However, routes created by these protocols may be susceptible to frequent breaks due to the dynamic nature of mobility involved. The adverse effects of the frequent route breaks on routing performance can be mitigated by either (a) incorporating mobility prediction techniques into the routing protocol to estimate the route lifetime or (b) maintaining multiple routes between the source and destination nodes. The former approach is followed in the Prediction-based AODV (PRAODV) protocol where the lifetime of a route between two nodes is estimated by the nodes forwarding the route-reply packet. This lifetime estimate is used to request a new route just before the expiration of the current one [37]. In the PRAODV-M protocol the predicted lifetime is used as a metric to select the route with the longest lifetime instead of the AODV's minimum hop count metric [37]. The multiple-route approach is followed in the Opportunity-Driven Multiple Access (ODMA) protocol [38].

14.7 APPLICATIONS

The ITS initiatives include many proposed applications for V2V and V2R communications. Due to the current advances in wireless technology and the extensive research in mobile ad hoc networks, VANETs are advocated as a supporting technology for many of these applications. Applications that rely on IVC technologies, including VANETs, can be classified into four categories:

- Safety-related applications
- Automated highways and cooperative driving
- Local traffic information systems
- IP-based applications

14.7.1 Safety-Related Applications

Human drivers rely on visual signals (e.g., brake light) for emergency warnings. The slow driver reaction time, which typically ranges from 0.7 to 1.5 s, and the dependency on line-of-sight propagation of the warning signals result in large delay in propagating road emergency warnings and may result in unavoidable collisions [39, 40]. Emerging communication technologies, such as VANETs, can reduce the delay in propagating emergency warnings significantly. Using IVC, a vehicle that recognizes a dangerous situation can report it instantly to neighboring vehicles to facilitate a fast reaction to the situation. The dangerous situation can be recognized by on-board sensors that detect events such as the deployment of airbags (as a result of a collision), loss of tire traction, or sudden application of brakes.

There is a considerable amount of research in the safety-related applications due to the interest in improving the public safety on transportation systems. Examples of the safety-related applications include proximity warning, road obstacle warning, and intersection collision warning. The objective of these applications is to use IVC to collect surrounding vehicle locations and dynamics and warn the driver when a collision is likely.

There are two approaches to collision warning systems (CWS) [41]. In passive approaches, all vehicles must maintain accurate knowledge of all vehicle positions. The risk of a collision is assessed using algorithms that analyze the gathered data (position, speed, acceleration, and direction) [42]. In active approaches, warning packets are sent only when emergency events occur. The emergency event is detected using on-board sensors. It is likely that a practical system will combine both approaches.

The unreliable nature of wireless communication and the fast changing group of affected vehicles create challenges in satisfying the strict latency and reliability constraints in CWS. As a result, CWS rely on repeated broadcast as a delivery mechanism for warning messages [39, 43]. Unlike unicast communication where reliability is enhanced by receiver's feedback (e.g., via RTS/CTS handshake), it is difficult to identify all receivers of a broadcast message and obtain their feedback in a highly mobile network. To enhance probability of successful reception, without receiver feedback, several copies of each message are transmitted without acknowledgment in combination with the CSMA mechanism of IEEE 802.11 [40, 44, 45].

A priority access mechanism can also be used to improve the probability of successful reception in a saturated medium. Moreover, the priority mechanism allows a quicker access to the medium compared to nonprioritized mechanisms [40, 46, 47].

Due to the broadcast nature of Emergency Warning Message (EWM) transmissions, it is not possible to control the rate of transmission based on channel feedback. Instead, application-specific properties are used for EWM congestion control [39]. In the Vehicular Collision Warning Communication (VCWC) [39], a control policy is proposed to reduce the transmission rate of the warning messages as more vehicles react to the initial broadcast by starting their own broadcast. Another proposed method in reference 39 eliminates redundant emergency warning messages by exploiting the natural chain effect of emergency events.

14.7.2 Automated Highways and Cooperative Driving

This type of application is concerned with the automation of some driving functions in order to increase driving safety and improve the capacity of highways. Among the applications considered are assisted/automated takeover and lane merge, platooning, automatic cruise control, and emergency vehicles announcement. These applications are responsible for the exchange of the drivers' intentions, signals, and data on varying relative positions [48]. With the help of an infrastructure that provides V2R communication, other applications can be included such as hidden driveway warning, electronic road signs, intersection collision warning, railroad crossing warning, work zone warning, highway merge assistance, and automated driving.

Among the early projects in IVC are those that investigated automated car platooning. The principal motivation for this application is increasing highway capacity. Essentially, the capacity can be increased substantially by reducing the spacing between vehicles traveling at high speed. Since drivers' reaction time is too slow to achieve this goal safely, vehicles must operate under automatic control [49].

Notable examples of automated platooning are the California PATH project [50], the European Chauffeur project [51], and the Japanese DEMO 2000 [52]. In these projects, the vehicles were able to perform several tasks, such as lane merge, lane split, and avoid an obstacle. The IVC played an essential role in the vehicle control because the necessary data for the platooning had to be exchanged among vehicles over the communication links. Safe platooning imposes the most severe conditions on the IVC. Thus it requires a high-speed wireless communication network capable of delivering reliable and fast messages [48].

One of the challenges facing safety and cooperative driving applications is the coexistence of equipped and nonequipped vehicles, which may affect the function of these systems. Since it is not realistic to expect that all vehicles are equipped at the early stages of this technology, a candidate solution may be the fusion with other systems. For instance, collision avoidance at intersections and similar applications could be realized with the assistance of V2R communications [53]. Another solution may rely on an Automated Highway System Architecture (AHS) [50].

14.7.3 Local Traffic Information Systems

Established traffic information systems have been traditionally based on centralized architectures. Roadside sensors deliver traffic data to a central unit where the information is processed. The traffic information is then disseminated to drivers via radio broadcast or on demand via cellular phone [54]. By utilizing on-board sensors, the GPS, and digital maps, a powerful traffic information system can be realized using a VANET that can be rapidly deployed without the need for much of the expensive infrastructure that is required in existing traffic and travel information systems [55].

Information systems that make use of IVC face two major challenges: required market penetration and scalability. With the initial low market penetration of equipped vehicles, the effectiveness of these system is limited. However, it is argued that an effective system can be created even if only 1–3% of all vehicles are equipped with an IVC system [54]. Scalability becomes an issue once a higher market penetration is reached due to data overload conditions.

Traffic information is most likely to influence drivers' decisions when it relates to the immediate area or an area they are likely to enter. Therefore, the general approach to the scalability challenge is to increase the level of abstraction of the collected data as the information is propagated further from the source. Fore instance, periodical data reports are forwarded, unmodified, over a fixed number of hops, and then only particular results from the on-board traffic analysis is forwarded further [56]. Similarly, traffic information will only be transmitted to a vehicle if that information is relevant in terms of the vehicle's likely route [55].

In the system proposed in reference 54, scalability is achieved by dividing each road into segments of limited size (application dependent). Each vehicle monitors the locally observed traffic situation using its own sensors and by recurrently receiving data packets with detailed information from other vehicles. A traffic situation analysis is performed in each individual vehicle, and the aggregated information is transmitted locally using 1-hop broadcast. The per-segment information is forwarded by an application layer that selects which information is to be sent to the relevant segments.

14.7.4 IP-Based Applications

Applications in this category deal mainly with passenger comfort and entertainment. Examples of these applications include online games and promotional broadcast from commercial vehicles or stationary gateways (e.g., restaurants). Many of these applications offer the traditional IP-based services (e.g., email, web access); thus, they require a fixed infrastructure to provide connectivity to the Internet.

The FleetNet project has done an extensive amount of work on an architecture that consists of roadside installed Internet Gateways (IGWs) and proxy servers. The IGWs provide temporary access to Internet services for passing vehicles through the proxy server that ensures protocols interoperability [57]. Several difficulties must be addressed in order to realize such an approach, which includes (a) the discovery of the IGWs and (b) the handover of connections from one gateway to the next [58].

While classic service discovery protocols may not be efficient in VANETs due to the highly dynamic topology, a proactive approach allows the IGWs to advertise their presence to vehicles. The multihop capability of the ad hoc routing protocol increases the communication range significantly and reduces the cost of the system by lowering the number of IGWs. A fuzzy logic can be used by vehicles to choose the most suitable IGW among many that can be accessed at the same time, either directly or over multiple hops [58].

Using the IVC paradigm for resource discovery can provide better information authenticity, accuracy, and reliability than a fixed infrastructure, especially for real-time information. In an opportunistic approach to resource discovery, a vehicle either senses the resources or obtains new resources from its exchanges with encountered vehicles. The relevance of the resources is assessed based on its proximity to the original source and its age [59].

Longer-range wireless technologies, such as cellular networks, may provide Internet access to VANETs. To avoid the high cost of this architecture, some vehicles can act as gateways for other vehicles. These gateways must have the capability to communicate with the nearest base station and with other vehicles. Other vehicles can access the Internet by routing packets thought the gateways [37].

14.8 VEHICLE MOBILITY

Given the dominating effect of mobility on the performance of MANETs, most related studies involve a discussion of some node mobility model. In the absence of real-life mobility traces, the results of these studies depend heavily on making the appropriate choice of a mobility model [60, 61]. Most studies of MANET protocols focus on node mobility in various environments where mobile nodes change their speed and direction more or less randomly. Vehicle mobility, on the other hand, is more restricted since vehicle traffic typically follows the road and vehicles interact with each other by accelerating, decelerating, and changing lanes. For these reasons, general ad hoc mobility models such as RWP are less useful in VANETs. Instead, many studies of VANETs use complex vehicle traffic simulators to generate vehicle movement traces. The studies in references 8, 19, and 62, for example, show that realistic representation of vehicle traffic is necessary to evaluate intervehicle ad hoc networks in highway environments.

Section 14.8.1 provides an introduction to traffic flow theory in order to describe the relationships between vehicle flow, speed, and density and their effect on the movement and distribution of vehicles. In Section 14.8.2, a survey of the various approaches to vehicle mobility simulation in the literature is provided. Section 14.8.3 introduces the vehicle traffic simulator, *RoadSim*.

14.8.1 Introduction to Traffic Flow Theory

Traffic flow theories explore relationships among three main quantities; vehicle density, flow, and speed. The flow q measures the number of vehicles that pass an observer per unit time. The density k represents the number of vehicles per unit distance. The speed, u, is the distance a vehicle travels per unit time. The units of these quantities are usually expressed as veh/h/lane, veh/km/lane, and km/h, respectively. In general, traffic streams are not uniform, but vary over both space and time. Therefore, the quantities q, k, and u are meaningful only as averages or as samples of random variables [63]. The three quantities are related by the so-called Fundamental Traffic Flow relationship [64],

$$q = u \times k \tag{14.8}$$

Several theories attempt to define the relationships among each pair of variables in (14.8), but no single theory provides the complete picture. The following subsections provide a brief introduction to the principles most relevant to the scope of this chapter.

Speed–Density Relationship. Car-following models [65] provided early means to describe the speed–density relationship. These models apply to single-lane, dense traffic with no overtaking allowed. They also assume that each driver reacts in some specific fashion to a stimulus from the vehicle(s) ahead or behind. The models do not apply in low densities where interactions between vehicles disappear.

A model proposed by Pipes [66] belongs to a class of car-following models that assume that drivers maintain constant time gap between vehicles. The Pipes' model results in the following speed–density relation [64, 66]:

$$u = \lambda \left(\frac{1}{k} - \frac{1}{k_{\text{jam}}} \right) \tag{14.9}$$

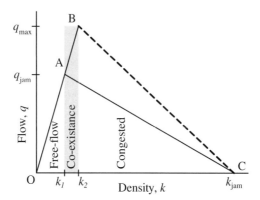

Figure 14.4. Fundamental diagram of road traffic.

where λ measures the sensitivity of the vehicle interaction and k_{jam} is the maximum vehicle density at traffic jam.

The Fundamental Diagram of Road Traffic. The Fundamental Diagram of Road Traffic describes the flow–density relationship. A typical q–k relationship follows the general shape of Figure 14.4. The figure shows that the flow is zero when there are no cars on the road, and also when there is a complete traffic jam at maximum density, k_{jam}. The shape of Figure 14.4 suggests that there are two q–k regimes. The left branch (OA) of the relationship represents free-flow traffic at densities below k_1. In the free-flow phase, interactions between vehicles are rare because of the low density. As a result, vehicles can travel at free-flow speed u_f, determined by the slope of the left branch, $u_f = q/k$.

At densities above k_2, traffic become congested and hindered by traffic jams. The q–k relationship of the congested traffic is represented by the right branch (AC). Empirical evidence and traffic simulations suggest that traffic may also be in a co-existence phase. In densities between k_1 and k_2, drivers may accept shorter headway (time gap) between vehicles, thus achieving the maximum flow, q_m, at the critical density, k_2. High fluctuations of speed in this traffic phase may cause a breakdown in the flow and drive the traffic into the congested phase.

At densities higher than k_1, speed can be expressed as a function of density as in (14.9). By substituting (14.9) in (14.8) and accounting for low-density traffic, a piecewise linear q–k relationship can represent the OAC triangle of Figure 14.4 [64]:

$$q = \min\left(u_f k, q_1\left(1 - \frac{k}{k_{jam}}\right)\right) \qquad (14.10)$$

where $q_1 = q(k_1) \approx \lambda$.

Traffic Jams and Phase Transition. Traffic jams are caused by bottlenecks at red lights, road constructions, access ramps, or similar constraints. Vehicle traffic in

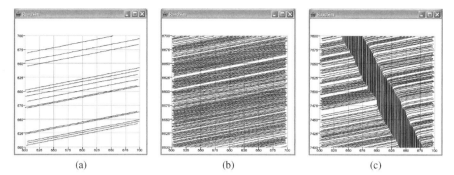

Figure 14.5. Space–time diagram illustrating (a) free-flow traffic, (b) traffic approaching critical density, and (c) a traffic jam in a 200-m highway segment (time progresses upward).

these scenarios can be described by a standard queuing system where the jam at the bottleneck will grow spatially backward if the arrival rate (inflow) is greater than the service rate (outflow) of the bottleneck. The spatial growth can be described also by the theory of kinematic waves [67].

It is also suggested that waves of traffic jams may occur without the presence of obvious constraints but mainly due to fluctuation (noise) in speed. These fluctuations may be caused by bumps, curves, lapses of attention, and different engine capabilities. In moderate traffic flow, a noise of high amplitude may cause the traffic to become unstable and create traffic jams [68]. This type of traffic jam can be described by the theory of kinematic waves, and it has been studied extensively using Cellular Automata (CA) models [20, 69].

The transition from free-flow traffic to congested traffic can be described with the help of the space–time diagram of Figure 14.5. A space–time diagram shows vehicle positions as they would appear in a series of aerial photographs taken at fixed time intervals of a road section and lined-up vertically according to their time index. A vehicle moving at constant speed will appear as a diagonal line in the space–time diagram, while a stationary vehicle will appear as a vertical line.

Figure 14.5a shows the traffic flow at density lower than k_1 (in Figure 14.4) where a vehicle can travel freely without influence from other vehicles. Figure 14.5b shows traffic whose density is in the range $[k_1, k_2]$. At this density, drivers may adapt to the dense traffic by slowing down and maintaining a minimum safety distance. However, high fluctuations in speed by a lead vehicle may break the flow and create traffic jams. A traffic jam appears in Figure 14.5c as a cluster of vertical lines. Vehicle that are not caught in jams are traveling in free-flow traffic. Traffic jams grow as density increases and merge with other traffic jams. Eventually, the entire road section is occupied with one wide traffic jam.

14.8.2 Traffic Simulators in VANETs Research

In some research related to VANETs, simulation of vehicle movement is considered a special type of a general mobility model such as the RWP model, perhaps with some extensions [35, 70, 71]. Such mobility models cannot produce complex vehicle maneuvers such as car-following, accelerating, decelerating, or passing, nor can they produce global phenomena such as traffic jams. For this reason, many studies of VANETs rely on vehicular mobility models or traffic simulators to generate vehicle movement traces.

Traffic simulation models can be classified as either microscopic, mesoscopic, or macroscopic. Microscopic models are models that continuously or discretely describe the state of individual vehicles, which may include speed and locations and their interactions. Macroscopic models ignore the detailed description of each vehicle and provide an aggregate view of traffic flow information such as speed, flow, and density. Mesoscopic models are models that have aspects of both macroscopic and microscopic models [72].

Models of VANET simulations can be classified into two categories. In the two-stage model of Figure 14.6a, a vehicle traffic simulator (VTS) generates vehicle movement traces in the form of vehicle locations updated at regular time intervals. These movement traces are fed in the second stage to a wireless network simulator (WNS) to carry out the simulation of communication protocols independently (e.g., 3, 73, 74). This approach takes advantage of many existing traffic simulators that were designed originally for the purpose of studying vehicle traffic in transportation networks and have the ability to generate realistic vehicle movements. Examples of the simulators used in this category include CORSIM [3, 62, 73, 74], FARSI [19], SHIFT [45], and Videlio [2]. Moreover, traffic simulators based on CA models are used for FleetNet simulations [54, 75]. Many other simulators that were developed in support of ITS applications are surveyed in reference 76.

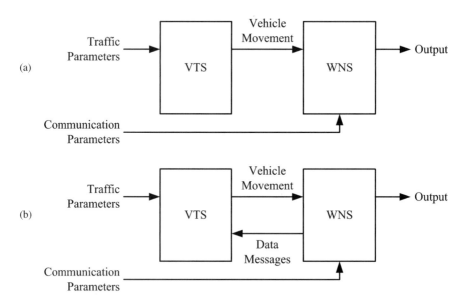

Figure 14.6. VANET Simulation Models: (a) 2-stage model and (b) integrated model.

The second category of VANET simulations integrates the mobility and the communication components. The model shown in Figure 14.6b provides a feedback path from the WNS to the VTS components in the simulator in order to allow the simulation of vehicles' reaction to communication messages. The model is needed to study some of the applications discussed in Section 14.7, in which vehicles are expected to react to messages generated by information, safety, or cooperative driving applications by, say, changing speed or choosing an alternative route. The simulators that are reviewed here are still in the early stages of development and may lack the feedback loop between the WNS and VTS components, but they were developed specifically to support VANET research.

Among the recently proposed simulators is GrooveSim [71]. In simulations using GrooveSim, vehicles can travel on real city street maps using one of four mobility models. The communication components of the simulator include routing and transport protocols. GrooveSim, however, does not support the car-following model or multilane traffic. Dense traffic can only be created by manipulating the input parameters of one of the mobility models, and not by interactions between the vehicles. Moreover, the simulator does not account for effects buildings and other obstacles have on IVC.

The simulator described in reference 77 also uses real street maps, but vehicles move according to the STRAW model. STRAW combines a car-following model, which sets the vehicle speed in relation to the speed of the preceding vehicle, with additional rules to deal with intersections and traffic controls (stop sign and traffic lights). The simulator does not support lane-change and it does not include buildings and obstacles.

To simulate vehicle interactions at intersections, Dogan et al. [78] developed an IVC simulator that consists of VTS and WNS components. In order to evaluate an intersection collision warning system, the VTS supports vehicles of different sizes (cars, buses, trucks, and motorcycles) and various speed capabilities. The vehicle dynamics are similar to a car-following model. The WNS supports IEEE 802.11x [4] and DOLPHIN [79] as MAC protocols.

14.8.3 *RoadSim*

We developed a traffic microsimulator, *RoadSim* [22, 80], to generate vehicle movement on highways. *RoadSim* implements the NaSch model [21], which generates complex vehicles movement patterns by specifying a few simple rules. The rules used in the NaSch model create behavior similar to car-following models with a constant headway. As a result, the speed–density and flow-density relationships are described by equations (14.9) and (14.10), respectively.

In *RoadSim*, more rules are added for additional components such as intersections, traffic controls, and multiple lanes. Vehicle maneuvers associated with these components cannot be created easily with car-following models. Moreover, it is possible, using *RoadSim*, to create any grid of intersections and roads (with multiple lanes and two directions) in order to study VANETs in urban environments. *RoadSim* includes several extensions to support VANET research. Among these extensions are the components responsible for creating communication graphs among nodes based on their transmission range. *RoadSim* can also translate node movements and transmission range into ns-2 [81] scripts and generate connection scenarios for ns-2 simulations.

14.9 CONCLUDING REMARKS

Current advances in the field of wireless ad hoc networks show that intervehicle communication (IVC) based on vehicular ad hoc networks (VANET) is a feasible approach that has a competitive edge over other technologies such as cellular networks with respect to several aspects: low latency for emergency warnings, high reliability due to link redundancy, and low cost of usage due to the lack of dependency on infrastructure and the use of unlicensed frequency hands. These features are ideal for the applications envisioned for the Intelligent Transportation Systems (ITS). VANETs can support vehicle-to-vehicle (V2V) and vehicle-to-roadside (V2R) communication, which will enable vehicular safety applications including collision warning as well as other ITS applications like local traffic information for routing, high-speed tolling, and mobile infotainment.

Some of the properties that distinguish VANETs from general ad hoc networks can affect their design significantly. It is expected, at early stages of market introduction, that only a small fraction of vehicles will support communication capabilities. Such conditions will have a significant impact on information propagation due to network partitioning. Therefore, message dissemination have to rely on data buffering, repeated broadcast, and opportunistic forwarding.

Also due to the partitioned, highly dynamic nature of these networks, it is difficult to maintain a global view of the network. Moreover, in many ITS applications, the destination vehicles are not known in advance. Thus, routing based on geographic multicast is more desirable, where only knowledge about immediate neighbors is retained. The target area for message propagation is application-dependent and is determined by the level of interest in the message.

VANETs also exhibit a rapid phase change from free-flow to congested networks due to traffic jams. Extensive research has focused on the impact of vehicle density on network connectivity, transmission range, capacity, and dissemination of messages by assuming a homogeneous distribution of vehicles. Yet, little has been done to investigate the effect of nonhomogeneous distribution of vehicles, due to traffic jams and road constraints, on these factors.

The VANET research community is increasingly aware of the limitations of the traditional mobility models. These limitations are manifested in the lack of representing interactions among vehicles and defining a relationship between node density and speed. There are several recent attempts to develop realistic vehicle mobility models based on car-following models or use off-the-shelf traffic simulators to generate realistic vehicle movement for VANET research. Since many application of VANETs aim to change the drivers' behavior in response to traffic-related messages, an ideal VANET simulator should include a mobility model that responds to such messages.

14.10 EXERCISES

1. Derive Eq. (14.5).

2. Equation (14.5) was derived with the assumption that all vehicles are equipped with wireless transceivers and are willing to participate in the network. Remove this restriction by assuming that only a fraction of vehicles participate in the network and derive lower-bound MTR in terms of the total vehicle density.

3. Algorithm 1 relies on estimating the vehicle local density as shown in reference 25. Study the reference and provide a list of proposed enhancements to existing vehicular ad hoc protocols using local density estimates.

4. Discuss the parameters that have to be considered for routing in a VANET.

5. Figure 14.4 might be used to render a policy for the transmission power of a wireless unit. Identify the variables and any derivations from those variables which will lead to a power control strategy for VANET units.

6. Common ad hoc mobility models, such as the Random Waypoint model, do not have a provision for node density. What are the conditions in which it is possible to use these mobility models to generate vehicle movements?

7. The NaSch model [20, 21] generates an accurate traffic behavior at the macroscopic level. However, the model's fixed car size, low vehicle speed resolution, and simplistic acceleration model reduce its accuracy at the microscopic level. Identify the model's limitations and suggest ways to enhance the model to reflect more accurate microscopic car movement.

REFERENCES

1. I. Chisalita and N. Shahmehri. Vehicular communication—a candidate technology for traffic safety. In *Proceedings of 2004 IEEE International Conference on Systems, Man and Cybernetics*, Vol. 4, 2004, pp. 3903–3908.
2. C. Lochert, H. Hartenstein, J. Tian, H. Füßler, D. Hermann, and M. Mauve. A routing strategy for vehicular ad hoc networks in city environments. In *Proceedings of IEEE Intelligent Vehicles Symposium (IV' 03)*, 2003, pp. 156–161.
3. H. Wu, R. Fujimoto, R. Guensler, and M. Hunter. MDDV: a mobility-centric data dissemination algorithm for vehicular networks. In *Proceedings of the 1st ACM International Workshop on Vehicular Ad Hoc Networks (VANET' 04)*, Philadelphia, ACM Press, New York, 2004, pp. 47–56.
4. IEEE STD 802.11-1997 information technology—telecommunications and information exchange between systems-local and metropolitan area networks-specific requirements—part 11: Wireless lan medium access control (mac) and physical layer (phy) specifications, 1997.

5. ASTM E2213-03. Standard specification for telecommunications and information exchange between roadside and vehicle systems—5 Ghz band dedicated short range communications (dsrc) medium access control (mac) and physical layer (phy) specifications, September 2003.
6. M. Lott, R. Halfmann, E. Schultz, and M. Radimirsch. Medium access and radio resource management for ad hoc networks based on UTRA TDD. In *Proceedings of the 2nd ACM International Symposium on Mobile Ad Hoc Networking & Computing (MobiHoc '01)*, Long Beach, CA, ACM Press, New York, 2001, pp. 76–86.
7. W. Franz, H. Hartenstein, and B. Bochow. Internet on the road via inter-vehicle communications. In *GI Workshop, Communication over Wireless LANs*, Vienna, Austria, September 2001.
8. M. Rudack, M. Meincke, K. Jobmann, and M. Lott. On traffic dynamical aspects of inter vehicle communications (IVC). In *Proceedings of the 58th IEEE Semiannual Vehicular Technology Conference (VTC'03-Fall)*, Vol. 5, 2003, pp. 3368–3372.
9. T. Issariyakul, E. Hossain, and D. I. Kim, Medium access control protocols for wireless mobile ad hoc networks: issues and approaches. *Wireless Communications and Mobile Computing*, **3**(8):935–958, 2003.
10. P. Santi. Topology control in wireless ad hoc and sensor networks. *ACM Computer Surveys*, **37**(2): pp. 164–194, 2005.
11. M. D. Penrose. The longest edge of the random minimal spanning tree. *The Annals of Applied Probability*, **7**(2):340–361, 1997.
12. J. Díaz, M. D. Penrose, J. Petit, and M. Serna, Convergence theorems for some layout measures on random lattice and random geometric graphs. *Combinatorics, Probability and Computing*, **9**(6):489–511, 2000.
13. P. Piret. On the connectivity of radio networks. *IEEE Transactions on Information Theory*, **37**(5):1490–1492, 1991.
14. P. Santi and D. M. Blough. An evaluation of connectivity in mobile wireless ad hoc networks. In *Proceedings of the International Conference on Dependable Systems and Networks*, June 2002, pp. 89–98.
15. O. Dousse, P. Thiran, and M. Hasler. Connectivity in ad-hoc and hybrid networks. In *Proceedings of the Twenty-First Annual Joint Conference of the IEEE Computer and Communications Societies (INFOCOM '02)*, Vol. 2, 2002, pp. 1079–1088.
16. Y. C. Cheng and T. G. Robertazzi. Critical connectivity phenomena in multihop radio models. *IEEE Transactions on Communications*, **37**(7):770–777, 1989.
17. M. Sánchez, P. Manzoni, and Z. J. Haas. Determination of critical transmission range in ad-hoc networks. In *Proceedings of Multiaccess, Mobility and Teletraffic for Wireless Communications Conference*, Venice, Italy, October 1999.
18. P. Santi and D. M. Blough. The critical transmitting range for connectivity in sparse wireless ad hoc networks. *IEEE Transactions on Mobile Computing*, **2**(1):25–39, 2003.
19. H. Füßler, M. Mauve, H. Hartenstein, D. Vollmer, and M. Käsemann. A comparison of routing strategies in vehicular ad-hoc networks. In *Reihe Informatik*, March 2002.
20. D. Jost and K. Nagel. Probabilistic traffic flow breakdown in stochastic car following models. *Transportation Research Record*, **1852**, pp. 152–158, 2003.
21. K. Nagel and M. Schreckenberg. A cellular automaton model for freeway traffic. *Journal de Physique I France*, **2**:2221–2229, 1992.

22. M. M. Artimy. Modelling of Transmission Range in Vehicular Ad Hoc Networks. Ph.D. thesis, Dalhousie University, 2006.
23. K. Nagel and M. Paczuski. Emergent traffic jams. *Physical Review E*, **51**: 2909–2918, 1995.
24. R. Barlovic, L. Santen, A. Schadschneider, and M. Schreckenberg. Metastable states in cellular automata for traffic flow. *The European Physical Journal B—Condensed Matter*, **5**(3):793–800, 1998.
25. M. M. Artimy, W. Robertson, and W. J. Phillips. Assignment of dynamic transmission range based on estimation of vehicle density. In *Proceedings of the 2nd ACM International Workshop on Vehicular Ad Hoc Networks (VANET '05)*, Cologne, Germany, ACM Press, New York, 2005, pp. 40–48.
26. R. Ramanathan and R. Rosales-Hain. Topology control of multihop wireless networks using transmit power adjustment. In *Proceedings of the Nineteenth Annual Joint Conference of the IEEE Computer and Communications Societies (INFOCOM '00)*, Vol. 2, 2000, pp. 404–413.
27. J. Liu and B. Li. MobileGrid: capacity-aware topology control in mobile ad hoc networks. In *Proceedings of the Eleventh International Conference on Computer Communications and Networks*, 2002, pp. 570–574.
28. E. C. Arvelo. Open-loop power control based on estimations of packet error rate in a Bluetooth radio. In *Proceedings IEEE Wireless Communications and Networking (WCNC '03)*, 2003, Vol. 3, pp. 1465–1469.
29. E. M. Royer and C.-K. Toh. A review of current routing protocols for ad hoc mobile wireless networks. *IEEE Personal Communications*, **6**(2):46–55, 1999.
30. Y.-B. Ko and N. H. Vaidya. Geocasting in mobile ad hoc networks: location-based multicast algorithms. In *Proceedings of the Second IEEE Workshop on Mobile Computing Systems and Applications (WMCSA '99)*, 1999, pp. 101–110.
31. T. Kosch, C. Schwingenschlogl, and L. Ai. Information dissemination in multihop inter-vehicle networks. In *Proceedings of the IEEE 5th International Conference on Intelligent Transportation Systems (ITSC '02)*, 2002, pp. 685–690.
32. L. Briesemeister and G. Hommel. Role-based multicast in highly mobile but sparsely connected ad hoc networks. In *Proceedings of the First Annual Workshop on Mobile and Ad Hoc Networking and Computing*, August 2000, pp. 45–50.
33. A. Bachir and A. Benslimane. A multicast protocol in ad hoc networks inter-vehicle geocast. In *The 57th IEEE Semiannual Vehicular Technology Conference (VTC'03-Spring)*, Vol. 4, 2003, pp. 2456–2460.
34. B. Karp and H. T. Kung. GPSR: Greedy perimeter stateless routing for wireless networks. In *Proceedings of the 6th annual international conference on Mobile computing and networking (MobiCom '00)*. ACM Press, New York, 2000, pp. 243–254.
35. J. Tian, L. Han, and K. Rothermel. Spatially aware packet routing for mobile ad hoc inter-vehicle radio networks. In *Proceedings of the IEEE Intelligent Transportation Systems*, Vol. 2, October 2003, pp. 1546–1551.
36. V. Dumitrescu and J. Guo. Context assisted routing protocols for inter-vehicle wireless communication. In *Proceedings of the 2005 IEEE Intelligent Vehicles Symposium*, 2005, pp. 594–600.
37. V. Namboodiri, M. Agarwal, and L. Gao, A study on the feasibility of mobile gateways for vehicular ad-hoc networks. In *VANET '04: Proceedings of the 1st ACM International*

Workshop on Vehicular Ad Hoc Networks, Philadelphia, ACM Press, New York, 2004, pp. 66–75.
38. B.-W. Chuang, J.-H. Tarng, J. Lin, and C. Wang. System development and performance investigation of mobile ad-hoc networks in vehicular environments. In *Proceedings of the 2005 IEEE Intelligent Vehicles Symposium*, 2005, pp. 302–307.
39. X. Yang, L. Liu, N. H. Vaidya, and F. Zhao. A vehicle-to-vehicle communication protocol for cooperative collision warning. In *Proceedings of the First Annual International Conference on Mobile and Ubiquitous Systems: Networking and Services (MOBIQUITOUS 2004)*, 2004, pp. 114–123.
40. S. Biswas, R. Tatchikou, and F. Dion. Vehicle-to-vehicle wireless communication protocols for enhancing highway traffic safety. *Communications Magazine, IEEE*, **44**(1):74–82, 2006.
41. H. Alshaer and E. Horlait. An optimized adaptive broadcast scheme for inter-vehicle communication. In *Proceedings of the IEEE 61st Vehicular Technology Conference (VTC 2005-Spring)*, Vol. 5, 2005, pp. 2840–2844.
42. J. Ueki, J. Mori, Y. Nakamura, Y. Horii, and H. Okada. Development of vehicular-collision avoidance support system by inter-vehicle communications—vcass. In *Proceedings of the IEEE 59th Vehicular Technology Conference (VTC-Spring 2004)*, Vol. 5, 2004, pp. 2940–2945.
43. F. Giudici, E. Pagani, and G. Paolo Rossi. Adaptive retransmission policy for reliable warning diffusion in vehicular networks. In *Proceedings of the Third Annual Conference on Wireless On-Demand Network Systems and Services (WONS 2006)*. January 18–20 2006, pp. 213–218, INRIA.
44. Q. Xu, R. Segupta, D. Jiang, and D. Chrysler. Design and analysis of highway safety communication protocol in 5.9 GHz dedicated short range communication spectrum. In *Poceedings of the 57th IEEE Semiannual Vehicular Technology Conference (VTC'03-Spring)*, Vol. 4, April 2003, pp. 2451–2455.
45. Q. Xu, T. Mak, J. Ko, and R. Sengupta. Vehicle-to-vehicle safety messaging in DSRC. In *Proceedings of the 1st ACM International Workshop on Vehicular Ad Hoc Networks (VANET'04)*, Philadelphia, ACM Press, New York, 2004, pp. 19–28.
46. U. Ozguner, F. Ozguner, M. Fitz, O. Takeshita, K. Redmill, W. Zhu, and A. Dogan. Inter-vehicle communication: recent developments at ohio state university. In *Proceedings of the IEEE Intelligent Vehicle Symposium*, 2002, Vol. 2, pp. 570–575.
47. M. Torrent-Moreno, D. Jiang, and H. Hartenstein. Broadcast reception rates and effects of priority access in 802.11-based vehicular ad-hoc networks. In *VANET '04: Proceedings of the 1st ACM International Workshop on Vehicular Ad Hoc Networks*, Philadelphia, ACM Press, New York, 2004, pp. 10–18.
48. M. Aoki and H. Fujii. Inter-vehicle communication: technical issues on vehicle control application. *IEEE Communications Magazine*, **34**(10):90–93, 1996.
49. P. Varaiya. Smart cars on smart roads: problems of control. *IEEE Transactions on Automatic Control*, **38**(2):195–207, 1993.
50. R. Horowitz and P. Varaiya. Control design of an automated highway system. *Proceedings of the IEEE*, **88**(7):913–925, 2000.
51. O. Gehring and H. Fritz Practical results of a longitudinal control concept for truck platooning with vehicle to vehicle communication. In *Proceedings of the IEEE Conference on Intelligent Transportation System (ITSC 97)*, 1997, pp. 117–122.

REFERENCES

52. S. Kato, S. Tsugawa, K. Tokuda, T. Matsui, and H. Fujii. Vehicle control algorithms for cooperative driving with automated vehicles and intervehicle communications. *IEEE Transactions on Intelligent Transportation Systems*, **3**(3):155–161, 2002.
53. S. Kato, N. Minobe, and S. Tsugawa. Applications of inter-vehicle communications to driver assistance system. *JSAE Review*, **24**(1):9–15, 2003.
54. L. Wischhof, A. Ebner, and H. Rohling. Information dissemination in self-organizing inter-vehicle networks. *IEEE Transactions on Intelligent Transportation Systems*, **6**(1):90–101, 2005.
55. M. Thomas, E. Peytchev, and D. Al-Dabass. Auto-sensing and distribution of traffic information in vehicular ad hoc networks. In *Proceedings of the United Kingdom Simulation Society Conference*, St Catherine's College, Oxford, March 2004, pp. 124–128.
56. A. Ebner and H. Rohling. A self-organized radio network for automotive applications. In *Proceedings 8th World Congress on Intelligent Transportation Systems (ITS '01)*, Sydney, Australia, October 2001.
57. M. Bechler, W. J. Franz, and L. Wolf, Mobile internet access in fleetnet. In *Verteilten Systemen KiVS 2003*, Leipzig, Germany, February 2003.
58. M. Bechler, L. Wolf, O. Storz, and W. J. Franz. Efficient discovery of internet gateways in future vehicular communication systems. In *Proceedings of the 57th IEEE Semiannual Vehicular Technology Conference (VTC 2003-Spring)*, Vol. 2, 2003, pp. 965–969.
59. Bo Xu, A. Ouksel, and O. Wolfson. Opportunistic resource exchange in inter-vehicle ad-hoc networks. In *Proceedings of the 2004 IEEE International Conference on Mobile Data Management*, 2004, pp. 4–12.
60. T. Camp, J. Boleng, and V. Davis. A survey of mobility models for ad hoc network research. *Wireless Communications and Mobile Computing*, **2**(5):483–502, 2002.
61. C. Bettstetter. Mobility modeling in wireless networks: categorization, smooth movement, and border effects. *SIGMOBILE Mobile Computer Communication Reviews*, **5**(3):55–66, 2001.
62. Z. D. Chen, H. T. Kung, and D. Vlah. Ad hoc relay wireless networks over moving vehicles on highways. In *Proceedings of the 2nd ACM International Symposium on Mobile Ad Hoc Networking & Computing (MobiHoc'01)*, Long Beach, CA, 2001, pp. 247–250, ACM Press.
63. F. L. Hall. Traffic stream characteristics. In N. H. Gartner, C. Messer, and A. K. Rathi, editors, *Traffic Flow Theory: A State of the Art Report—Revised Monograph on Traffic Flow Theory*, Oak Ridge National Laboratory, Oak Ridge, TN, 1997, Chapter 2.
64. R. Haberman. *Mathematical Models: Mechanical Vibrations, Population Dynamics, and Traffic Flow: An Introduction to Applied Mathematics*, Prentice-Hall, Englewood Cliffs, NJ, 1977.
65. R. W. Rothery. Car-following models. In N. H. Gartner, C. Messer, and A. K. Rathi, editors, *Traffic Flow Theory: A State of the Art Report—Revised Monograph on Traffic Flow Theory*, Oak Ridge National Laboratory, Oak Ridge, TN, 1997, Chapter 4.
66. L. A. Pipes. An operational analysis of traffic dynamics. *Journal of Applied Physics*, **24**(3):274–281, 1953.
67. R. Kuhne and P. Michalopoulos. Continuum flow models. In N. H. Gartner, C. Messer, and A. K. Rathi, editors, *Traffic Flow Theory: A State of the Art Report—Revised Monograph on Traffic Flow Theory*, Oak Ridge National Laboratory, Oak Ridge, TN, 1997, Chapter 5.

68. T. Nagatani. The physics of traffic jams. *Reports on Progress in Physics*, **65**(9):1331–1386, 2002.
69. K. Nagel, P. Wagner, and R. Woesler. Still flowing: Approaches to traffic flow and traffic jam modeling. *Operations Research*, **51**(5):681–710, 2003.
70. J. Tian, J. Hahner, C. Becker, I. Stepanov, and K. Rothermel. Graph-based mobility model for mobile ad hoc network simulation. In *Proceedings of the 35th Annual Simulation Symposium*, 2002, pp. 337–344.
71. R. Mangharam, D. S. Weller, D. D. Stancil, R. Rajkumar, and J. S. Parikh. GrooveSim: A topography-accurate simulator for geographic routing in vehicular networks. In *Proceedings of the 2nd ACM International Workshop on Vehicular Ad Hoc Networks (VANET'05)*, Cologne, Germany, ACM Press, New York, 2005, pp. 59–68.
72. E. Lieberman and A. K. Rathi. Traffic simulation. In N. H. Gartner, C. Messer, and A. K. Rathi, editors, *Traffic Flow Theory: A State of the Art Report—Revised Monograph on Traffic Flow Theory*, Oak Ridge National Laboratory, Oak Ridge, TN, 1997, Chapter 10.
73. J. Yin, T. ElBatt, G. Yeung, B. Ryu, S. Habermas, H. Krishnan, and T. Talty. Performance evaluation of safety applications over dsrc vehicular ad hoc networks. In *Proceedings of the First ACM Workshop on Vehicular Ad Hoc Networks*, Philadelphia, ACM Press, New York, 2004, pp. 1–9.
74. J. J. Blum, A. Eskandarian, and L. J. Hoffman. Challenges of intervehicle ad hoc networks. *IEEE Transactions on Intelligent Transportation Systems*, **5**(4):347–351, 2004.
75. H. Hartenstein, B. Bochow, A. Ebner, M. Lott, M. Radimirsch, and D. Vollmer. Position-aware ad hoc wireless networks for inter-vehicle communications: The FleetNet project. In *Proceedings of the 2nd ACM International Symposium on Mobile Ad Hoc Networking & Computing (MobiHoc '01)*, Long Beach, CA, October 2001.
76. S. Adams Boxill and L. Yu. An evaluation of traffic simulation models for supporting ITS development. Technical Report SWUTC/00/167602-1, Center for Transportation Training and Research, Texas Southern University, October 2000.
77. D. R. Choffnes and F. E. Bustamante. An integrated mobility and traffic model for vehicular wireless networks. In *Proc. The 2nd ACM International Workshop on Vehicular Ad Hoc Networks (VANET '05)*, Cologne, Germany, ACM Press, New York, 2005, pp. 69–78.
78. A. Dogan, G. Korkmaz, Y. Liu, F. Ozguner, U. Ozguner, K. Redmill, O. Takeshita, and K. Tokuda. Evaluation of intersection collision warning system using an inter-vehicle communication simulator. In *Proceedings of the 7th International IEEE Conference on Intelligent Transportation Systems*, 2004, pp. 1103–1108.
79. K. Tokuda, M. Akiyama, and H. Fujii. Dolphin for inter-vehicle communications system. In *Proceedings of the IEEE Intelligent Vehicles Symposium (IV 2000)*, 2000, pp. 504–509.
80. M. M. Artimy, W. Robertson, and W. J. Phillips. Vehicle traffic microsimulator for ad hoc networks research. In *Proceedings of the International Workshop on Wireless Ad-Hoc Networks (IWWAN'04)*, Oulu, Finland, May 2004, pp. 105–109.
81. Anonymous. The network simulator—ns-2. February 28 2006.

■■■■■ CHAPTER 15

Performance Issues in Vehicular Ad Hoc Networks

MARIA KIHL

Department of Electrical and Information Technology, Lund University, Sweden

MIHAIL L. SICHITIU

Department of Electrical and Computer Engineering, North Carolina State University, Raleigh, NC 27695

15.1 INTRODUCTION

Ad hoc networks are wireless networks without a fixed infrastructure, which are usually assembled on a temporary basis to serve a specific deployment such as emergency rescue or battlefield communication. They are especially suitable for scenarios where the deployment of an infrastructure is either not feasible or not cost effective. The differentiating feature of an ad hoc network is that the functionality normally assigned to infrastructure components, such as access points, switches, and routers, needs to be achieved by the regular nodes participating in the network. For most cases, there is an assumption that the participating nodes are mobile, do not have a guaranteed uptime, and have limited energy resources.

In infrastructure-based wireless networks, such as cellular networks or WiFi, the wireless connection goes only one hop to the access point or the base station; the remainder of the routing happens in the wired domain. At most, the decision that needs to be made is either (a) which base station a mobile node should talk to or (b) how it should handle the transfer from one station to another during movement. Routing in the wired domain was long considered a mature field, where trusted and reliable solutions exist. The topology of the infrastructure, its bandwidth, and its routing and switching resources are provisioned to provide a good fit with the expected traffic.

In ad hoc networks, however, routing becomes a significant concern, because it needs to be handled by ordinary nodes that have neither specialized equipment nor a fixed, privileged position in the network. Thus, the introduction of ad hoc networks

---

Algorithms and Protocols for Wireless and Mobile Ad Hoc Networks, Edited by Azzedine Boukerche
Copyright © 2009 by John Wiley & Sons Inc.

signaled a resurgent interest in routing through the challenges posed by the mobility of the nodes, their limited energy resources, their heterogeneity (which under some conditions can lead to asymmetric connections), and many other issues. These challenges were answered with a large number of routing algorithms, and ad hoc routing remains an active and dynamically evolving research area. Ad hoc routing algorithms are serving as a source of ideas and techniques to related technologies such as wireless sensor networks and mesh networks.

In this chapter we survey the field of wireless ad hoc routing. While we attempted to include most of the influential algorithms, our survey cannot be exhaustive: The number of proposed algorithms and variations exceeds 1000. However, we strived to represent most of the research directions in ad hoc routing, thus giving the reader an introduction to the issues, the challenges, and the opportunities offered by the field.

15.2 VANETs VERSUS MANETs

A VANET is a special type of mobile ad hoc network (MANET) [12, 26]. In this section we present some characteristics that differentiate a class of VANETs from the more general MANETs.

- *Embedded System.* All VANETs are basically embedded systems. The communication software is implemented into a vehicle, which may already have support for either tele- or data communication. Therefore, it will be necessary to incorporate the VANET protocols into the existing communication technologies that a car will have access to. This may be facilitated with a gateway architecture, which allows for several access network technologies [28].
- *Infinite Energy Supply.* One problem in many MANETs—and, in particular, wireless sensor networks—is that the node's energy is supplied by batteries with a finite capacity. This is, however, not a problem in VANETs. A vehicle can be assumed to have infinite battery capacity, and therefore the protocols do not have to be energy-efficient. The unlimited power supply can be used to improve the performance of the protocols—for example, by using topology control [13].
- *Rate of Link Changes.* A common assumption in MANETs is that nodes have reasonably long-lived links. Most routing protocols for MANETs assume that links are sufficiently long-lived to allow for route discoveries and periods of stability between link changes. In contrast, in VANETs, the nodes (i.e., the vehicles) move along the roads with high velocities. The end result is that the links between neighboring nodes are short-lived (in the range of seconds [51]). Buildings on the side of the road will also negatively affect connectivity, and the communication range is significantly smaller in urban areas than in rural areas [36]. These two issues will affect the communication channel models, and new models need to be developed for VANETs. Furthermore, the common mobility models used for MANETs (e.g., the random waypoint model) do not apply to VANETs.

- *Localization.* One advantage of VANETs is that, owing to their GPS receiver, all nodes always know their (quasi-exact) position and time. Also, the navigation system likely has detailed road maps, with roads that are divided into segments, each with a unique identifier. The information from the GPS and road maps can be used to, for example, improve the performance of the routing protocols [22].
- *Applications.* In a MANET, the major application considered is data transmission between two nodes (unicast transmission between two nodes that are aware of each others IP addresses). However, the envisioned major applications in a VANET will be very different from the above scenario. Most VANET applications require data to be transmitted to a certain *area* instead of being transferred to a certain node. This area is usually called the zone of relevance (ZOR) [31]. In other words, this is a scenario with multicasting to a certain geographical area. Furthermore, a node will most probably not be aware of other nodes' IP addresses. A VANET is continuously changing, and detailed information about each node cannot be transferred to all other nodes. Instead, location-based routing will likely have to be used.

15.3 PHYSICAL AND LINK LAYER CONSIDERATIONS

The physical layer (PHY) is literally the base of the networking stack. It directly limits the performance of the entire stack, and its characteristics have to be taken into account when designing the upper-layer protocols. Since wireless communications are inherently broadcasted at the physical layer, the data link control (DLC) necessarily includes a medium access control (MAC) protocol. The main difficulties in designing an efficient MAC protocol stem from the lack of a natural central coordination point and the inherent dynamics of VANETs.

15.3.1 The Physical Layer

At the physical layer, most solutions developed for wireless cellular systems (especially G3) are likely to work reasonably well because the two environments are similar. Cellular systems are single-hop wireless systems based on extensive infrastructure (base stations). The base stations coordinate the medium access, which usually consists of a combination of FDMA, TDMA, and CDMA. The most common modulations employed are frequency and phase modulation. Several research groups (e.g., Fleetnet [17]) proposed to use cellular technology for VANETs. The main differences involve the relative speeds of the sender and receiver and the propagation characteristics resulting from antenna placement. The expected capacity and link duration are also different.

While in cellular systems the base stations are always stationary, in a VANET environment both ends of a wireless link may move toward each other at highway speeds, thus considerably increasing the rate of change in the channel. Thus, in order to maintain the same characteristics, as in a cellular environment, the equalization and modulation has to be more robust. It was pointed out [46] that this increased rate of change may render OFDM modulations designed for wireless local area networks (WLANs) (e.g., 802.11a) unsuitable for VANETs (at least at large packet sizes). This observation is especially relevant, because the current standardization work focuses on a physical layer derived from IEEE 802.11a.

The second important difference between the PHY of VANETs and cellular systems regards the antenna placement. In cellular systems, the antenna of the base stations is usually placed at carefully planned locations and at a considerable distance from the ground. This placement ensures a reasonably uniform and predictable signal propagation in most of the cell covered by that base station (there are always exceptions, especially close to the edge of the cell). This placement allows for long-lived semipredictable links: In most cases a mobile phone has ample time to observe the decline in signal strength and initiate a hand-off before the link is broken. In VANETs, since the antennas of both the transmitter and the receiver are placed lower, the links tend to be far less predictable than in a cell environment. This characteristic prompts the development of opportunistic higher layers that should take advantage of a good link while it lasts without counting on its longevity.

Finally, the links of VANET systems are often assumed to be much shorter (a few hundreds of meters in DSRC) in comparison to cellular systems (that can have cells of tens of kilometers). This choice is also prompted by the assumed high data rates needed by VANET applications: The reduced range of transmissions in VANETs translate directly in an increased spatial reuse (i.e., allow for many noninterfering simultaneous transmissions), thus increasing the capacity of the system. However, this reduced link range coupled with the high-speeds of VANETs also contribute to the short-lived nature of VANET links.

15.3.2 The Data Link Control Layer

The data link control layer is traditionally responsible for efficient (and, perhaps, reliable) communications over one hop. In this section we present several MAC layers that have been proposed for VANETs. On one hand, there are MAC layers designed for other purposes and used "as is" or with minor modifications. On the other hand, since none of these protocols have been optimized for VANETs, they tend to be suboptimal; therefore there has been a significant amount of research geared toward developing new MAC protocols specifically for VANETs.

DSRC and 802.11p. By far the most promising technology is based on the popular IEEE 802.11 standard. The recently published Dedicated Short-Range Communications (DSRC) standard [14] is heavily based on the IEEE 802.11a [24]. Interestingly, IEEE formed task group P working at a protocol named Wireless Access for the Vehicular Environment (WAVE). While the final specifications of 802.11p are not finalized at this time, it is likely that it will be similar to DSRC and, by transitivity, to 802.11a.

Thus, 802.11 is on its way to become the VANET standard at the MAC layer. Its reliability, throughput, and range characteristics are a good match for most applications. When communicating to roadside access points, the infrastructure mode can be used, while for pure VANETs the ad hoc mode of 802.11 will be employed. Many researchers used 802.11 for their testbeds primarily for convenience and the good quality-to-price ratio of the hardware.

However, to use the unicast capability of 802.11, a node has to know the MAC address of the destination. To this end, either an ARP type of mechanism or a new neighbor discovery scheme has to be employed. While in ad hoc mode, 802.11 nodes periodically broadcast beacon packets (without guarantees on the timing and the time to the next beacon). However, these beacons are typically not visible beyond the network card (network driver at the best) and, thus, are not useful for a neighbor discovery mechanism.

However, for the MAC layer broadcast, no such state about the neighbors is needed, thus reducing the associated overhead. Therefore, several routing protocols—for example, flooding, diffusion, and geocasting (see Section 15.4)—assume an 802.11-like broadcast protocol at the MAC layer.

Bluetooth (802.15.1). Bluetooth [8, 25] (or IEEE 802.15.1) is a spread-spectrum, low-data-rate protocol for short-range communications. Bluetooth is, by design, very cheap to implement and it is already available in several high-end vehicles, primarily serving to the cell phone to the vehicle's audio system. Given its ready availability, some researchers considered it for VANET applications.

The protocol requires a pico-cell architecture, essentially one-hop clusters with a master node and up to seven active slaves. Similar to the case of cellular networks, the master coordinates the activity of the pico-cell. These characteristics make Bluetooth less than ideal as a MAC layer for VANETs. One of the significant hurdles for using Bluetooth in VANETs is the formation and maintenance of pico-cells in the highly dynamic VANET environment [41, 49]. Furthermore, multihop communications require highly complex management for the formation of scatternets (a multihop structure supported by the standard).

Dedicated Schemes. Before DSRC was standardized, many research groups proposed to use existing MAC protocols, or designed their own specifically for VANET communications. One of the simplest MAC protocols, ALOHA [1], foregoes the need for coordination among the base stations. In essence, every node transmits whenever it has to a packet without delay. Not surprisingly, the throughput of the system is dismal at best (at most 18% under the assumption of Poisson traffic and infinite number of users).

Slotted ALOHA (S-ALOHA) has been shown to double the efficiency (under the same assumptions) simply by allowing the nodes to transmit only in well-defined timeslots. This scheme obviously requires reasonably good time-synchronization between the nodes in order to obtain an agreement on the slot boundaries. The time-synchronization problem in multihop wireless networks is relatively well studied, and several solutions for static networks have been proposed. However, an external means of synchronization (e.g., GPS) may solve this problem in a trivial and scalable manner. Once the time synchronization problem is solved, S-ALOHA does not require any coordination among the nodes. Finally, reservation ALOHA (R-ALOHA) uses small slots for contention; and if the contention is successful, it sends data in the successful slots. R-ALOHA, in addition to time synchronization, requires a significant coordination between the nodes.

Other Technologies. Interestingly enough, third-generation cellular technologies have also been proposed for VANET communications despite the apparent mismatch between the design of the protocol and the application. Indeed, cellular communication protocols have been designed to operate in a single-hop, infrastructure-based environment. The coordination and management is, naturally, performed by the base stations. In contrast, VANETs are multihop and infrastructureless. Despite these considerations, a group of researchers, as part of the Fleetnet project [17], developed a modified version of UTRA-TDD suitable for ad hoc communications. Not surprisingly, the main issues faced were the tight time synchronization and coordination necessary to make the scheme work.

Code Division Multiple Access (CDMA) communications have the remarkable property that a receiver can successfully decode two simultaneous transmissions at the same time. Thus, the nodes have no need to coordinate their transmissions in order to avoid collisions. However, a dynamic and fully distributed code assignment scheme has to be developed. A very interesting idea for code assignment (a similar idea was also used for slot assignment in S-ALOHA and R-ALOHA systems [16, 38, 39]) exploits the fact that (unless there is a collision) no two vehicles occupy the same position at the same time. Thus, based on GPS information on each vehicle's position (augmented with lane information), each vehicle can transmit using the code corresponding to its location [30]. An orthogonal code assignment may ensure efficient code reuse in neighboring road segments. An issue often overlooked in systems proposing the use of CDMA communications is the power control. In order to be able to decode packets from two or more simultaneous transmitters, the received power of all these transmissions should be equal (at least to a good approximation). In long-lived, connection-oriented communications in cellular systems, this power control is achieved through constant feedback. In VANET environments with volatile links, it is unclear how this power control can be practically implemented.

Final Words. In conclusion, the design of a MAC layer for VANETs has to solve a fundamental tussle. On one hand, the lack of coordination among nodes inevitably results in collisions and overall poor efficiency (e.g., ALOHA). On the other hand, too much coordination involves keeping track of state information (e.g., clustering, availability of neighbors, time synchronization, reservations, power control, and slot and code assignments) and the dissemination (or discovery) of this information. In a highly dynamic VANET maintaining this state, information is bound to result in a large overhead. It is very likely that reasonably good solutions will have to find a suitable compromise between the two extremes. DSRC (in essence, a CSMA/CA scheme) may very well be such a compromise.

15.4 ROUTING LAYER

Many VANET applications require that data are transmitted in a multihop fashion, thus requiring a routing protocol. In this section we discuss how routing can be performed in a VANET, and we also describe some proposed routing algorithms.

15.4.1 Routing Constraints

In a VANET, similar to a MANET, data have to be forwarded in multiple hops from a sender to one or several receivers. The objective of the routing protocol is to find an optimal path between the sender and the receiver. However, the nature of a VANET imposes the following three constraints for a routing protocol:

1. Short-lived links.
2. Lack of global network configuration.
3. Lack of detailed knowledge about a node's neighbors.

The first issue is due to the mobility of the vehicles. Studies have shown that the lifetime of a link between two nodes in a VANET is in the range of seconds [51]. The second issue stems from the fact that a VANET always is in a transient phase, since vehicles are continuously joining and departing. The third issue is caused by both mobility and security aspects. Each node in a VANET is a vehicle owned by a different person, in contrast to general MANETs, where it is commonly assumed that all nodes have the same owner (or belong to the same organization, e.g., US DoD). Therefore, a driver may not want to share detailed information that can be used for identification purposes. A routing protocol for VANETs must be able to work under these constraints.

15.4.2 Geographical Addressing

All traditional networks, both wired and wireless, use fixed addresses, which means that a node has a certain identifier that never changes. Two well-known examples are the MAC addresses used in the IEEE 802 standards and the IP-addresses used by the Internet protocols. However, for many VANET applications, it is not practical to use fixed addresses. First, it may be difficult for a node to know the address of another node due to the third constraint mentioned above. Second, in many applications the intended destination is not a particular node, but rather a group of nodes that happen to be at a certain location.

Therefore, *geographical addresses* may be used in a VANET. In the simplest case a geographical address is a unique coordinate. Data should be forwarded to that coordinate and the node in that location will be the destination node. This communication is not only uncommon, but it also assumes that the source node is able to know the exact location of the destination at the time it receives the packet. In the following discussion we always assume that a node can find its exact location by using GPS.

However, usually the intended destination is all nodes that happen to be in a certain area, called *Zone of Relevance* (ZOR) [31]. The ZOR is usually specified as a rectangular or circular range rather than a single coordinate. In VANETs, the ZOR can be enhanced by other attributes so that a subset of vehicles that should receive the message is selected. Example of such attributes are:

- The direction of movement of the vehicle
- The road identifier (e.g., number, name, etc.)
- The type of vehicle (e.g., trucks, 18 wheelers, etc.)
- Some physical characteristics (e.g., taller than, weighing more than or with a speed higher than)
- Some characteristic of the driver (e.g., beginner, professional, etc.)

15.4.3 Geocasting in VANETs

"Traditional" routing protocols for MANETs—for example, AODV [43, 44]—were developed for unicast with fixed addresses. In VANETs, protocols for multicast with geographical addresses are more suitable. *Geocast* is a class of routing protocols that fits that description [32]. The main objective of a Geocast protocol is to forward data from a sender to all nodes in a ZOR. The protocol should work even without any previous knowledge about the nodes that are in the ZOR. An example of Geocast is shown in Figure 15.1. Here, vehicles move along a highway. Vehicle A wants to send data (e.g., information about an emergency stop) to all vehicles within a ZOR behind itself, as shown in Figure 15.1. Vehicle B is not within the ZOR and should, therefore, not receive the data.

The Basics. Geocast is usually implemented as a controlled flooding mechanism. A sender starts with setting the ZOR and then broadcasting the packet to all its neighbors. A node that receives a geocast packet checks if it is within the ZOR. If the node is

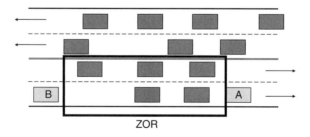

Figure 15.1. Geocasting example.

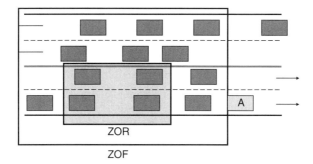

Figure 15.2. Zone of forwarding (ZOF).

within the ZOR and has not received the packet before, it sends the packet to the higher layer (e.g., transport) and it also broadcasts it.

It is clear that geocasting with a full flooding mechanism will not work well in VANETs. First, a VANET may have no practical boundaries (although, realistically, the road network is bounded by the seas). Full flooding within a VANET will most likely cause severe network congestion. Second, it is clearly not optimal to flood a message in the entire network, especially if the ZOR is significantly smaller than the entire network. Since the intended destinations are within a specific ZOR, the packet should only be forwarded *toward* that region.

Therefore, a geocasting protocol specifies not only a ZOR but also a zone of forwarding (ZOF). When a node in the ZOF receives a geocast packet for the first time, it only forwards the packet (as opposed to the nodes in the ZOR that also accept the packet). The forwarding zone must include at least the target area and a path between the sender and the target area. An example is shown in Figure 15.2. Here, the ZOR is specified as the shaded area. The ZOF is specified as shown, which means that all nodes within this region will help to forward the packet toward the ZOR.

Geocasting Protocols for VANETs. Several geocast protocols have been proposed for general MANETs. Good surveys are presented in references 32 and 35. Because the geocasting mechanism fits the routing requirements for VANETs very well, the concept has been studied since the beginning of 1990s [31]. However, so far, most geocasting protocols proposed for VANETs often assume knowledge about neighbors. Below, we describe a geocasting protocol for VANETs that works under all constraints listed above.

Geocast with Distance-Based Delay of Forwarding. In references 6 and 9 a geocasting protocol that avoids network congestion is proposed for VANETs. The basic idea is that when a node receives a new packet and is within the ZOF, it will forward the packet after a certain delay. The delay is calculated so that nodes close to the sender will wait a longer time.

The source node is assumed to set a ZOR that includes itself. This is likely a common requirement for most VANET applications. For example, in a collision avoidance application, the ZOR is the region right *behind* the sender vehicle. In a traffic monitoring application, the ZOR is probably the region *around* the source vehicle. In references 6 and 9 the ZOF is not defined as a region, but rather as a maximum number of hops, *MaxHops*. A node that receives a packet for the first time will forward the packet if the current number of hops is less than *MaxHops*. The routing protocol assumes that the link layer uses IEEE 802.11 with CSMA.

When a node receives a packet and determines that it should forward it, it first determines a waiting time, *WT*. This waiting time depends on the distance, d, between this node and the previous sender node. The objective of this waiting time is to let nodes further away from the previous sender forward the packet with a higher priority than nodes close to the sender. Note that this is a wireless environment, and many (but possibly not all) nodes within range will receive the broadcast transmission. In this way, a broadcast storm (packet collisions due to simultaneous forwarding) is avoided, and, in addition, the forwarding is optimized (the number of broadcasting nodes is minimized).

The waiting time is determined as

$$WT = WT_{\max} - \frac{WT_{\max}}{R} \cdot \hat{d} \qquad (15.1)$$

where R is the transmission range, $WT_{\max}$ is the maximum waiting time, and $\hat{d} = \min\{d, R\}$. This algorithm is rather simple and can be improved, although at an increase in complexity. Variants of the algorithm above have been proposed in references 2, 3, and 5.

15.4.4 Unicast Protocols for VANETs

Some VANET applications require unicast routing. For example, some envisioned comfort applications, as on-board games and file transfer, will likely need unicast routing with fixed addresses. Papers proposing unicast protocols for VANETs usually assume that a vehicle (e.g., the client) somehow "magically" knows the address of the intended destination (e.g., the server).

Cluster-Based Routing on Highways. Cluster-based routing has been considered very practical for VANETs because vehicles driving on a highway may form clusters in a natural way. Therefore, many papers have proposed cluster-based routing protocols for VANETs. We will here describe one proposal [48] that uses a simple algorithm to perform cluster-based unicast routing in a highway scenario.

The network is divided into clusters, as in Figure 15.3. A node can be one of the following roles (also called status): *cluster head*, *Gateway*, or *Member*. Each cluster has exactly one cluster head. A gateway is a node that belongs to at least two clusters. All the other nodes are members of a cluster. In Figure 15.3, clusters are depicted

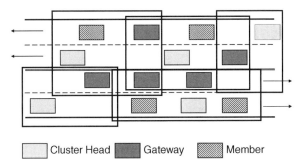

Figure 15.3. Cluster-based routing.

as rectangular, however in reality the clusters can have any shape depending on the particular network topology.

All nodes periodically broadcasts hello messages containing their status, address, and location. Each time a node receives a hello message, it stores the included information in a so-called *neighbor table*. When a node first switches on, it listens for hello packets. If it receives any hello message from a cluster head, the node stores this information in its table, changes is status to member, and replies to the hello message. If no cluster head is found, the node changes its state to cluster head and broadcasts this in a hello message.

If a node belonging to one cluster receives a hello message from another cluster head, it stores this information, changes its status to gateway, and broadcasts this new information. A gateway node always belongs to at least two clusters. If a gateway looses contact with one of its cluster heads, it will change its status to member of this cluster.

All information about neighbors and routes is time-stamped with an expiration time. If a node is not receiving updated information about its cluster head before the expiration time, it will change its status to cluster head. Thus, all cluster heads have information about its cluster members and about which nodes are gateways to other clusters. The objective with this partitioning of the network is to minimize the number of retransmissions. In this setup, only cluster heads and gateways will forward packets in the network.

When a node wants to transmit data to another node, it proceeds as follows: If the destination node is included in the sender's neighbor table, the sender directly sends the packet to the destination. Otherwise, it broadcasts a location request (LREQ) packet. When a cluster head receives a new LREQ packet, it first checks if the destination node is a member of a cluster. If so, the cluster head sends a location reply (LREP) back to the source node with information about the destination's location. If the destination node is not a member of the cluster, the Cluster head rebroadcasts the LREQ packet. Gateways that receives a LREQ packet from one of its cluster heads will rebroadcast the packet to the next cluster.

In this way, LREQ packets will be flooded by cluster heads and gateways throughout the network until they reach the cluster head of the destination. This cluster head will send back a LREP packet to the sender's cluster Head. The LREP packet will be routed the shortest way back, by using the information about the source node's location. When the LREP packet reaches the source node, it will transmit its data packet. All cluster heads and gateways on the way will have information stored in their tables about the next hop.

Geographic Forwarding in City Environments. In cities, buildings can cause multipath fading and shadowing, thus resulting in radio transmission failures. The shortest (geographically speaking) path between two nodes may not be the best path (in terms of reliability, delay, throughput, etc.). However, the fact that vehicles drive on streets, with intersections, can be used by the routing protocol. All vehicles belonging to a VANET are assumed to have access to a navigation systems, with detailed map data. Roads are divided into segments, each with a unique identifier. Intersections can also be identified.

In reference 33 a routing protocol, called Geographic Source Routing (GSR), is proposed for unicast with fixed addresses in city environments. It uses the fact that packets can be routed through intersections on the way to their destination. A node is assumed to discover the position of its destination by using a location service [37]. Using this information, the sender can compute a path, consisting of a sequence of intersections that the packet should traverse in order to reach the destination. This is performed by using the map data from the navigation system and, for example, the Dijkstra shortest path algorithm. The path is placed in the packet header or is computed by each intermediate node. The packet is forwarded to each intersection using greedy forwarding [29]. Forwarding to intersections should work without major problems, since no buildings are in the way for the radio transmission (the roads are assumed straight).

15.4.5 Diffusion Mechanisms

Several papers focusing on traffic monitoring and management applications employ a unique dissemination procedure, called a *diffusion mechanism*. In these approaches the data forwarding is performed at the application layer that collects information from other vehicles, aggregates it, and then broadcasts it at regular intervals. The result is that the information is "diffused" in the network, with each vehicle having more accurate information on the state of the nearby traffic. Note that the diffusion mechanism is not a proper routing protocol, since it is performed at the application layer. However, we have chosen to describe diffusion in the routing section since its objective is the same as routing, that is, forwarding data in the network.

Segment-Oriented Data Abstraction and Dissemination (SODAD). In reference 52 a diffusion mechanism for traffic management applications, called SODAD, is described. The objective of the mechanism is to distribute information about road segments to all vehicles in the area. It is assumed that vehicles close to a specific road segment need to get fresher information than vehicles further away. The application in each node aggregates the data and sends out updated information at certain time intervals. All data are always transmitted via single-hop broadcast.

It is assumed that all nodes have access to a detailed digital map, where the roads are divided into segments, each with a unique (globally known) identifier. A node maintains a table, called the knowledge base, with aggregated information about each road segment it has received information about. Denote by $s_{n,i}$ the aggregated information about road segment i at node n. If N information values $d_1, d_2, ..., d_N$ have been received at node n for a segment i, $s_{n,i}$ is calculated as

$$s_{n,i} = a(d_1, d_2, ..., d_N) \tag{15.2}$$

where $a(\cdot)$ is an aggregation function. The aggregation function depends on the application. If, for example, the information values correspond to vehicle speeds, the aggregation function could calculate the moving average or minimum speed, so that the vehicle can detect traffic congestion. Every time an aggregated information value, $s_{n,i}$, is updated, a time stamp, $t_{n,i}$, is set to the current global time. The tuple $(s_{n,i}, t_{n,i})$ describes the information available for segment i at node n.

The node recurrently sends broadcast messages consisting of parts of its knowledge base. It is up to each application to decide how much information should be included in a broadcast message. Information about more important road segments (e.g., segments close to the vehicle) should be broadcasted more often.

Assume that node n receives a message, m, with information for S road segments with identifiers $i_1, ..., i_S$. The message consists of the tuples $(s_{m,k}, t_{m,k})$, where $k = i_1, ..., i_S$. The information value $s_{n,k}$ is updated with $s_{m,k}$ by using the aggregation function if $t_{m,k} > t_{n,k}$.

This means that information regarding a road segment will be forwarded to all nodes in the area. A node further away from a certain road segment will receive information aggregated by several nodes, and it can thereby get a updated information on the situation on that road segment. A node close to a road segment will often get fresh information broadcasted by its neighbors.

15.5 TRANSPORT AND SECURITY ISSUES

15.5.1 End-to-End QoS

The transport layer is typically responsible with providing end-to-end services—for example, reliability, flow control, and congestion control. A problem in mobile multihop networks is the high probability of packet loss due to noisy links. It is well known that regular flavors of TCP are working very poorly in MANETs. There are also numerous papers proposing improvements for TCP in MANET scenarios. However, so far, very few papers discuss transport layer issues in VANETs.

The end-to-end QoS is crucial for many VANET applications: without end-to-end and delay guarantees, many of the potential "killer" applications, (e.g., public safety applications) will not be feasible. For other applications, end-to-end delivery guarantees will be of paramount importance (e.g., for toll-road payments).

Many VANET applications need congestion control mechanisms to avoid network overload and unbounded delays. For example, in a collision warning applications, it was shown [58], that, in the absence of a congestion control mechanism, the vital information may encounter unacceptable delays.

15.5.2 Security and Privacy Issues

If and when VANETs will become pervasive, they will probably be the largest open-access ad hoc networks in existence. The right balance between security and privacy of this network will be of utmost importance for its long-term success. VANETs have challenges similar to those of MANETs (especially the lack of a single coordination point). For a good overview of data security issues in MANETs, see reference 57. In this section we will discuss security and privacy issues that occur specifically in VANETs (see reference 19 for more details).

Electronic License Plates. Currently, every vehicle is registered with its national or regional authority and has allocated a unique identifier. However, for electronic identification and tracking of vehicles, *electronic license plates* must be used. In a simple scenario, a vehicle could provide its identity by periodically broadcasting a beacon when the vehicle's engine is on. In the United States, electronic license plates are currently evaluated in several states under the Heavy Vehicle Electronic License Plate Program (HELP) [50].

The electronic license plate could be used in a number of scenarios—for example, to facilitate the payment of road tolls. It could of course also be used by authorities—for example, the police—to find criminals or drivers who flee from the scene of an accident.

Authentication. Before two vehicles can exchange data, their identity and validity must be certified to avoid misuse and hacker attacks. Therefore, the authentication of vehicles will be a necessary part of a security architecture for VANETs.

Authentication is usually performed with public key technology and digital certificates. Since VANETs will most probably be deployed world-wide, the authentication process of vehicles will be similar to the authentication of credit cards or mobile phones. Registration authorities will need to implement a public key infrastructure (PKI). Also, each vehicle must be provided with a private/public key pair, a shared symmetric key, and a digital certificate of its identity and public key.

However, in VANETs, new problems arise. For example, some VANET applications have very stringent real-time requirements. One example is emergency warning systems. When a colliding vehicle transmits an emergency warning message, this message must be forwarded to all nodes in the ZOR rapidly. A time-consuming authentication scheme is at odds with the real-time requirements of this application.

Privacy. The privacy issue in a VANET will be an obstacle that must be solved before VANETs can be socially accepted in the community. The fact that a vehicle may be tracked by authorities may not be harder to get accepted than the possible tracking of mobile phones. However, since VANET applications require vehicle-to-vehicle communication, the privacy issues will also include the possibility that drivers may obtain detailed location and identity information about other drivers.

Therefore, one important part of a security architectures for VANETs will be development of privacy-preserving protocols. These protocols could be based on an anonymity scheme, relying on temporary pseudonyms [19]. Only authorities should be able to translate a vehicle's pseudonym to its real identity.

Location Verification. All vehicles participating in a VANET are assumed to get their exact location from GPS. Furthermore, all position-based protocols assumes that all nodes cooperate in determining and reporting their locations or distances. However, all GPS-based systems are vulnerable to several different attacks, where intruders may report a wrong location or produce fake GPS signals. Therefore, there must be procedures for location verification in a security architecture for VANETs.

To ensure that vehicles report correct locations, it is of course important that the GPS receivers in vehicles participating in a VANET are tamper-proof. However, this will not solve the problem of potentially fake GPS signals.

Location verification can also be performed with so called distance-bounding protocols, see reference 4. With distance-bounding protocols, it is possible for one vehicle to determine the upper bound on the physical distance to another vehicle. The main problem with the existing distance-bounding protocols is that they rely on roadside infrastructure. In this scheme, base stations along the road, controlled by a certified authority, are used in the protocols in order to verify a vehicle's position in two or three dimensions.

15.6 PERFORMANCE MODELING AND EVALUATION ISSUES

Numerous papers have evaluated MANET protocols both theoretically and with simulations. However, since the behavior of nodes and channels in a VANET is so different from a MANET, we believe that it is important to know how this will affect the performance modeling and evaluation of the protocols. In this section we will present and discuss some of these issues.

15.6.1 Mobility Models

One of the basic features of a MANET (and VANET) is that nodes move according to some mobility model. It is well known that the mobility models can significantly affect the results of the evaluation (quantitatively as well as qualitatively [e.g., 27]). For MANETs the random way-point model (RWP) is, by far, the most popular mobility model [56]. Several others (e.g., random walk [11] and random group [21]) are less common.

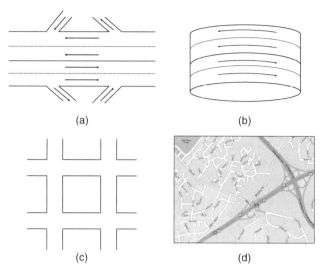

Figure 15.4. Road models: (a) straight road (highway), (b) circular road, (c) road grid, (d) real(istic) road.

In a VANET, nodes can only move along streets, prompting the need for a *road model* is needed. Another important aspect in a VANET is that nodes cannot move independently of each other; instead they have to move in lanes according to fairly well established *traffic models*.

Road Models. In the literature, two basic road models are often considered: the straight highway and the road grid.

The *straight highway* (shown in Figure 15.4a) is a reasonable model for highways outside cities. In this model, vehicles move in lanes, in either one or two directions. The model can include ramps and exits where vehicles can enter or leave the highway. The main characteristic of a straight highway model is that messages transmitted between vehicles only move along one dimension.

A special case of the straight highway is the *circular road* model depicted in Figure 15.4b. In this model, vehicles also move in one or several lanes similar to the straight highway. However, in this case, the road is a closed loop, where no vehicles can enter or leave the road.

The *road grid* (Fig. 15.4c) is fairly representative for urban roads and town centers, where the straight highway model is not applicable. The road grid can also incorporate traffic lights. In this model, traffic moves in two (or even three) dimensions. The road grid is more complicated to implement than the straight highway, and it is therefore mostly used in very specific scenarios. As an alternative to a road grid model, a real *road map* (e.g., Fig. 15.4d) can be used.

PERFORMANCE MODELING AND EVALUATION ISSUES   **449**

Traffic Models. Traffic modeling is a well-known research area in civil engineering. It is important to model vehicular traffic during the design phase of new roads and intersections. There are a number of traffic models that accurately mimic vehicle traffic, and a good survey focused at VANETs can be found in reference 20.

With respect to the movement of the vehicles, two distinct strategies are common. In the first model, each vehicle, upon entering an intersection, randomly makes a decision on whether to go straight or turn left or right. This model is somewhat similar to Brownian motion mobility models. In a second model, each vehicle randomly chooses a destination somewhere else on the map and attempts to go there on the shortest possible route. Once it reaches the destination, it pauses and then chooses another random destination. This model is close in spirit with the commonly used RWP model encountered in MANETs.

One important aspect in a traffic model is the *driver behavior*. The reaction of the drivers in different situations will affect, for example, the traffic throughput. A driver must decide when to overtake a slow-moving vehicle, when to change lanes on a multilane highway, and when to slow down or accelerate. The driver behavior will affect, indirectly, the performance of a VANET and must be taken into account in the modeling process.

15.6.2 Simulation Issues

Due to the cost and difficulties involved in deploying large vehicular testbeds, the vast majority of the proposed VANET protocols and systems are evaluated via simulations or with small testbeds. Usually, the simulation of a VANET system includes two stages.

In the first stage, the vehicle movements are determined, usually using a *traffic simulator*—for example, CORSIM [42]. The input of the traffic simulator includes the road model, scenario parameters (including maximum speed, rates of vehicle arrivals and departures, etc.). The output from the traffic simulator is a trace file where every vehicle's location is determined at every time instant for the entire simulation time.

In the second stage, the trace file is used as input in a *network simulator*—for example, ns-2 [40]. Each vehicle becomes a node in a MANET with the trace file specifying the movements of each node.

However, there is a clear need for integrated traffic and network simulators, to evaluate of the performance of VANETs. To begin with, it is cumbersome to first use a traffic simulator (e.g., CORSIM) to simulate the vehicle movements, and then a network simulator (e.g., ns-2), to simulate the network behavior. Second, it can be argued that the movement of the vehicles should change as a function of the information they receive from other vehicles. For example, if a vehicle receives information about traffic congestion or an accident on a road segment, it could re-route and go another way.

To our knowledge, there has so far only been one attempt to build a completely integrated traffic and network simulator for VANETs. The simulator Jist/SWANS [7] is a Java-based network simulator for MANETs that has an integrated street mobility model STRAW [10]. However, significant more work is needed for this simulator to gain wide acceptance in the community.

15.6.3 Communication Channel Models

Accurate models for the communication channel are a prerequisite for meaningful simulation results. Simulators for MANETs usually implement a circular propagation model, with no obstacles.

There are two features of VANET that are not incorporated in the commonly used channel models for MANETs. First, vehicles will move with, sometimes, very high velocity. The channel models used today usually assume no or very low mobility. In reference 55 the physical layer of DSRC was modeled in detail and used in the simulations. The results showed high bit error rates due to the high velocity of the vehicles. Second, the roadside obstacles will affect transmission footprint. In a city scenario, buildings will block the radio waves and cause shorter transmission ranges than in a rural highway scenario. In reference 18, experiments showed that the radio range for IEEE 802.11 was much smaller in cities than on highways.

15.7 OPEN ISSUES

Despite considerable advances in VANET protocol design, the commercial deployment of VANETs is still several years ahead. Many issues remain to be addressed; in this section we will discuss some of them.

15.7.1 Access Network

DSRC and the future 802.11p has been proposed as the access protocol of choice. However, several challenges remain both in the physical as well as in the MAC layers.

In the physical layer, only very few papers [46] considered the effects of the very high mobility on an 802.11a-like OFDM modulation scheme. While the typical issues have been thoroughly studied for the WLAN case of stationary users (or users with low mobility), VANETs are substantially different. Furthermore, since OFDM is not used in any cellular telephony standard, not even estimates of how this modulation will behave in a high-speed environment are available.

At the MAC layer, there are several issues to be resolved. To begin with, the MAC layer of 802.11 supports two distinct modes: infrastructure mode and ad hoc mode. In infrastructure mode, a central coordination point manages a single cell of the associated station. This mode fits well for roadside-to-vehicle communications. However, in ad hoc mode, the design of the MAC protocol assumes that all nodes share a single collision domain—that is, that every node is capable of communicating directly with any other node (in one hop). This is clearly not the case for VANETs, and it has been shown that several coordination mechanisms of 802.11 (e.g., time synchronization and power savings) fail in a multi hop environment.

The second research focus at the MAC layer is in configuring the parameters of the MAC protocol, such as inter frame spacings, initial values, evolution of the collision window, and so on. These values have been fine-tuned for common WLAN situations. Similar to the case of using 802.11 for long links, we can expect the default 802.11 values to be grossly different from optimal.

15.7.2 Routing Protocols

There are several issues yet to be resolved at the network layer. Perhaps the first problem to be resolved (as it shapes the rest of the protocol stack) is the addressing mode in VANETs. While several addressing modes have been developed for MANETs, only fixed address and geographical addressing seem to be good matches for VANET applications. Furthermore, some applications fit a fixed addressing model very well, while others require geographical addressing. Using both modes requires a mapping between these to fundamentally different modes of addressing. While, for MANETs, there exist scalable and efficient location services [e.g., 59], their mobility assumptions do not apply to VANETs.

The vast majority of MANET routing protocols have been designed for fixed addressing modes. Furthermore, those protocols (implicitly) assume that the network is in quasi-equilibrium and only needs to repair a few links at the time. In contrast, in VANETs, the links are so dynamic that an on-demand protocol such as AODV would get the routes broken even before they are formed [54]. Thus, a new routing protocol capable of handling both a mixture of addressing modes and highly dynamic links should be developed. The new routing protocol will have to support position-based multicast (similar to geocasting), because many VANET applications require support for such a service.

Finally, a host of QoS-related issues, from guarantees for delays to priorities and fairness, should receive support from the networking layer.

15.7.3 Transport Protocols

The role of the transport layer is to provide a comprehensive and complete set of services to the application layer. Surprisingly little work has been done in this area. This lack of contributions can be explained by the lack of a clear networking layer interface to the transport layer. In other words, it is unclear on what network layer services can the transport layer be designed.

There are several services that traditionally are associated with the transport protocol. Multiplexing (for multiple application support) and preserving the order of the packets should pose no special difficulties given port-like and sequence numbers similar to TCP. However, reliability is far more difficult to implement. Especially for geographical multicasting applications, where the number and identity of the recipients is not known at the time a packet is sent, it may be very well impossible to ensure that all the vehicles in the ZOR received the packet. Similarly, we believe that it will be very difficult (if not impossible) to provide delivery guarantees given the highly dynamic and unreliable lower layers.

Congestion control is a particularly useful service that can be implemented either at the network or at the transport layer. It is well-known that a network may perform extremely poorly in the absence of a congestion control mechanism. The problem of congestion control in wireless networks is especially difficult. Most congestion control mechanisms (e.g., the one in TCP, or the ones designed for ATM-ABR) have been designed with the assumption of a chain of links and routers (or switches) stretching between two end systems. In multihop wireless networks, there are no proper "links," because each wireless transmission does not affect the congestion only at its intended receiver but to all nearby receivers as well. Therefore, an appropriate congestion control mechanism should monitor and react to congestion not only on the "path" taken by the packets, but also around it. The problem is considerably more difficult for multicast applications.

15.7.4 Security Architecture

It is clear that in the Internet security was an afterthought. The initial network design (despite being funded by DARPA!) was focused on practical issues, from correctness to efficiency. After all the "technical" challenges (e.g., addressing, efficient routing, reliable transport, congestion control) have been solved and after the network succeeded commercially, the issue of security arose. Numerous stop-gap measures for various security aspects have since been developed. However, all the current activity and media attention point to the shortcomings of the current solution (after all, how much coverage did routing get in the media lately?).

In contrast, since VANET protocols are in their nascent phase, it *should* be much easier to incorporate strong security mechanisms into their core. Ideally, security will be a non-issue by the time VANETs become a commercial success. The security architecture will have to walk the fine line between security and privacy. On one hand, consumers are often extremely concerned with privacy issues, and these issues can easily prevent a mass adoption of the technology. On the other hand, security issues and the associated liability are the main concern of the companies commercializing the technology. Further complicating the matter, several layers of back doors should be accessible by law-enforcement agencies.

We believe that, at this early stage in the development of VANETs, a comprehensive and flexible security architecture providing a complete set of services is a very worthy goal.

Finally, there is a long list of nontechnical issues related to security aspects in VANETs—for example, how to enforce a "recommended" (or optimal) speed, whether to penalize a narrowly avoided accident, or who can track who's vehicle under what conditions.

15.8 EXERCISES

1. Guess how many car-accident deaths occur in your country every year and how many millions of gallons (liters) are consumed by the vehicles every year. Then find the exact numbers on the Internet or at your local library.

2. Discuss the advantages and disadvantages of a VANET solution for a traffic monitoring application. Compare it with a system based on sensors embedded in the highway and visual signs informing us about the current traffic conditions (such systems are deployed in Chicago and in many places in Japan).

3. Enumerate at least three applications of VANETs. Compare the advantages and disadvantages of a VANET solution to those applications with existing solutions.

4. Design and describe a slotted-TDMA-like solution (similar to GSM) for MAC access in VANETs. Obviously, more than one solution can be presented here. Focus on coordination techniques in the absence of a central coordination point.

5. Geographical addressing and fixed addressing can be mapped to each other (much like the way that IP addresses are resolved into MAC addresses by ARP or domain names into IP addresses by DNS). Design and describe a networking service that given a set of geographical addresses would convert them into IP addresses and the other way around.

6. Elaborate on several reasons why TCP is a poor choice for a transport protocol for VANETs.

7. Give examples of other wireless communication systems with essentially infinite energy supplies.

8. Consider the following two congestion control schemes for VANETs: The first one is based on local information with each node backing off more before transmitting upon detecting a high link utilization in its area. The second is one based on end-to-end (either explicit or implicit) feedback (like TCP). Discuss advantages and disadvantages of each of these strategies in a VANET environment. Which one do you believe would be more suitable for VANETs, and why?

9. Consider a traffic monitoring application that generates B bytes of data for each vehicle (e.g., position, time, speed) periodically with a period P (s). Assume a transmission range T_R (m), a header length H for every packet, and a transmission data rate R. Further assume a vehicle density D (vehicles/m) and a range of monitoring M (m) (i.e., how far behind the data needs to be sent). How much data traffic (bits per second) has to be forwarded by each vehicle if no data aggregation is assumed? Make and state any reasonable assumptions you need to solve this problem. Assume some reasonable values for the variables defined above and compute the traffic.

10. Repeat the problem above if you assume that each vehicle aggregates all data it receives over an entire sampling period P and it then forwards it as a single (larger) packet.

REFERENCES

1. N. Abramson. The aloha system—another alternative for computer communication. In *AFIPS Conference Proceedings*, Vol. 37, Montvale, NJ, AFIPS Press, Montvale, NJ, 1970, pp. 281–285.
2. H. Alshaer and E. Horlait. An optimized adaptive broadcast scheme for inter-vehicle communicatiorn. In *Proceedings of the 61st IEEE Vehicular Technology Conference*, 2005, pp. 2840–2844.
3. A. Bachir and A. Benslimane. A multicast protocol in ad hoc networks inter-vehicle geocast. In *Proceedings of the 57th IEEE Vehicular Technology Conference*, Jeju, Korea, 2003.
4. S. Brands and D. Chaum. Distance-bounded protocols. *Lecture Notes in Computer Science*, 1440, 1998.
5. A. Benslimane. Optimized dissemination of alarm messages in vehicular ad-hoc networks (VANET). In *Proceedings of the 7th IEEE International Conference on High Speed Networks and Multimedia Communications*, 2004, pp. 665–666.
6. L. Briesemeister and G. Hommel. Role-based multicast in highly mobile but sparsely connected ad hoc networks. In *Proceedings of the IEEE First Annual Workshop on Mobile and Ad Hoc Networking and Computing (MobiHoc '00)*, 2000, pp. 45–50.
7. R. Barr, Z. J. Haas, and R. van Renesse. JIST: An efficient approach to simulation using virtual machines. *Software - Practice and experience*, **35**:539–576, 2004.
8. Bluetooth specifications version 1.1. http://www.bluetooth.org/spec/.
9. L. Briesemeister, L. Schäfers, and G. Hommel. Disseminating messages among highly mobile hosts based on inter-vehicle communication. In *Proceedings of the IEEE Intelligent Vehicle Symposium*, 2000, pp. 522–527.
10. D. Choffnes and F. Bustamante. An integrated mobility and traffic model for vehicular wireless networks. In *Proceedings of the Second ACM International Workshop on Vehicular Ad Hoc Networks*, Cologne, Germany, 2005.
11. T. Camp, J. Boleng, and V. Davies. A survey of mobility models for ad hoc network research. *Wireless Communications & Mobile Computing (WCMC): Special issue on Mobile Ad Hoc Networking: Research,Trends and Applications*, **2**(5):483–502, 2002.
12. I. Chlamtac, M. Conti, and J. Liu. Mobile ad hoc networking: imperatives and challenges. *Ad Hoc Networks*, **1**(1):13–64, 2003.
13. G. Caizzone, P. Giacomazzi, L. Musumeci, and G. Verticale. A power control algorithm with high channel availability for vehicular ad hoc networks. In *Proceedings of the IEEE International Conference on Communications*, 2005, pp. 3171–3176.
14. Standard specification for telecomunications and information exchange between roadside and vehicle systems - 5GHz band dedicated short range communications (DSRC) medium access control (MAC) and physical layer (PHY). ASTM E2213-03, September 2003.
15. W. Enkelmann. Fleetnet—applications for inter-vehicle communication. In *Proceedings of the IEEE Intelligent Vehicles Symposium*, 2003, pp. 162–167.

16. K. Fujimura and T. Hasegawa. Performance comparisons of inter-vehicle communication networks—including the modified V-PEACE scheme proposed. In *Proceedings of 2001 IEEE Intelligent Transportation Systems Conference*, 2001, pp. 968–972.
17. Fleetnet website. http://www.et2.tu-harburg.de/fleetnet.
18. Y. Günter and H.P Grossmann. Usage of wireless lan for inter-vehicle communication. In *Proceedings of the 8th IEEE International Conference on Intelligent Transportation Systems*, 2005, pp. 296–301.
19. J. P. Hubaux, S. Capkun, and J. Luo. The security and privacy of smart vehicles. *IEEE Security & Privacy Magazine*, **2**:49–55, 2004.
20. Y. Hörmann, H. P. Grossmann, W. H. Khalifa, M. Salah, and O. H. Karam. Simulator for inter-vehicle communication based on traffic modeling. In *Proceedings of the IEEE Intelligent Vehicles Symposium*, 2004, pp. 99–104.
21. X. Hong, M. Gerla, G. Pei, and C. Chiang. A group mobility model for ad hoc wireless networks. . In *Proceedings of the ACM/IEEE MSWIN'99*, Seattle, WA, August 1999.
22. A. H. Ho, Y. H. Ho, and K. A. Hua. A connectionless approach to mobile ad hoc networks in street environments. In *Proceedings of the 2005 IEEE Intelligent Vehicles Symposium*, 2005, pp. 575–582.
23. B. Hofmann-Wellenhof, H. Lichtenegger, and J. Collins. *Global Positioning System: Theory and Practice*. 4th edition, Springer-Verlag, 1997.
24. IEEE. Wireless LAN medium access control (MAC) and physical layer (PHY) specification: High-speed physical layer in the 5 Ghz band. IEEE Standard 802.11a, September 1999.
25. IEEE. Part 15.1: Wireless medium access control (MAC) and physical layer (PHY) specifications for wireless personal area networks (WPANs). IEEE Std. 802.15.1, June 2002.
26. Internet Engineering Task Force. MANET—mobile ad hoc working group. http://www.ietf.org/html.charters/manet-charter.html.
27. A. Jardosh, E. M. Belding-Royer, K. Almeroth, and S. Suri. Towards realistic mobility models for mobile ad hoc networks. In *Proceedings of the 9th Annual ACM/IEEE International Conference on Mobile Computing and Networking (MobiCom '03)*, San Diego, CA, September 2003.
28. W. Kellerer, C. Bettstetter, C. Schwingenschlögl, P. Sties, and H. Vögel. (Auto) mobile communication in a heterogeneous and converged world. *IEEE Personal Communications*, **8**(6):41–47, 2001.
29. B. Karp and H. T. Kung. Gpsr: Greedy perimeter stateless routing for wireless networks. In *Proceedings of the 6th annual ACM/iEEE International Conference on Mobile Computing and Networking*, 2000, pp. 243–254.
30. S. Katragadda, G. Murthy, R. Rao, M. Kumer, and R. Sachin. A decentralized location-based channel access protocol for inter-vehicle communication. In *Proceedings of the 57th IEEE Vehicular Technology Conference*, 2003, pp. 1831–1835.
31. W. Kremer. Realistic simulation of a broadcast protocol for a inter vehicle communication system (IVCS). In *Proceedings of the 41st IEEE Vehicular Technology Conference*, 1991, pp. 624–629.
32. Y. Ko and N. H Vaidya. Flooding-based geocasting protocols for mobile ad hoc networks. *Mobile Networks and Applications*, **7**(6):471–480, 2002.

33. C. Lochert, H. Hartenstein, J. Tian, H. Füssler, D. Hermann, and M. Mauve. A routing strategy for vehicular ad hoc networks in city environments. In *Proceedings of the IEEE Intelligent Vehicles Symposium*, 2003, pp. 156–161.
34. J. Li, J. Jannotti, D. De Couto, D. Karger, and R. Morris. A scalable location service for geographic ad-hoc routing. In *Proceedings of the 6th ACM International Conference on Mobile Computing and Networking (MobiCom '00)*, August 2000, pp. 120–130.
35. C. Maihöfer. A survey of geocast routing protocols. *IEEE Communications Surveys & Tutorials*, **6**:32–42, 2004.
36. M. Möske, H. Füssler, H. Hartenstein, and W. Franz. Performance measurements of a vehicular ad hoc network. In *Proceedings of the 59th IEEE Vehicular Technology Conference*, 2004, pp. 2016–2020.
37. M. Mauve, J. Widmer, and H. Hartenstein. A survey on position-based routing in mobile ad hoc networks. *IEEE Network*, **15**(6):30–39, 2001.
38. T. Nagaosa and T. Hasegawa. A new scheme of nearby vehicles' positions recognition and inter-vehicle communication without signal collision - V-PEACE scheme. In *Proceedings of the 51st IEEE Vehicular Technology Conference*, 2000, pp. 1616–1620.
39. T. Nagaosa, Y. Kobayashi, K. Mori, and H. Kobayashi. An advanced CSMA inter-vehicle communication system using packet transmission timing decided by the vehicle position. In *Proceedings of the 2004 IEEE Intelligent Vehicles Symposium*, 2004, pp. 111–114.
40. ns-2 home page. http://www.isi.edu/nsnam/ns/.
41. T. Ohyama, S. Nakabayashi, Y. Shiraki, and K. Tokuda. A study of real-time and autonomous decentralized DSRC system for inter-vehicle communications. In *Proceedings of the 3rd IEEE International Conference on Intelligent Transportation Systems*, 2000, pp. 190–195.
42. L. Owen, Y. Zhang, L. Rao, and G. McHale. Traffic flow simulation using CORSIM. In *Proceedings of the 2000 Winter Simulation Conference*, 2000, pp. 1143–1147.
43. C. Perkins, E. Belding-Royer, and S. Das. Ad hoc on-demand distance vector (AODV) routing. RFC 3561, July 2003.
44. C. Perkins. Ad-hoc on-demand distance vector routing. In *Proceedings of MILCOM*, November 1997.
45. C. Plenge. The performance of medium access protocols for inter-vehicle communication systems. In *Proceedings of Mobile Kommunikation*, 1995, pp. 189–196.
46. S. Sibecas, C. A. Corral, S. Emami, and G. Stratis. On the suitability of 802.11a/RA for high-mobility DSRC. In *Proceedings of the 55th IEEE Vehicular Technology Conference*, 2002, pp. 229–234.
47. S. E. Shladover, C. A. Desoer, J. K. Hedrick, M. Tomizuka, J. Walrand, W. Zhang, D. H. McMahon, H. Peng, S. Sheikholeslam, and N. McKeown. Automatic vehicle control developments in the PATH program. *IEEE Transactions on Vehicular Technology*, **40**(1):114–130, 1991.
48. R. A. Santos, A. Edwards, R. M. Edwards, and N. L. Seed. Performance evaluation of routing protocols in vehicular ad-hoc networks. *International Journal of Ad Hoc and Ubiquitous Computing*, **1**:80–91, 2005.
49. H. Sawant, J. Tan, Q. Yang, and Q. Wang. Using Bluetooth and sensor networks for intelligent transportation systems. In *Proceedings of the 7th IEEE International Conference on Intelligent Transportation Systems*, 2004, pp. 767–772.

50. C. M. Walton. The heavy vehicle electronic license plate program and crescent demonstration project. *IEEE Transactions on Vehicular Technology*, **40**:147–151, 1991.
51. S.Y. Wang. The effects of wireless transmission range on path lifetime in vehicle-formed mobile ad hoc networks on highways. In *Proceedings of the IEEE International Conference on Communications*, 2005, pp. 3177–3181.
52. L. Wischhof, A. Ebner, and H. Rohling. Information dissemination in self-organizing inter-vehicle networks. *IEEE Transactions on Intelligent Transportation Systems*, **6**(1):90–101, 2005.
53. M. Williams. PROMETHEUS-the European research programme for optimising the road transport system in europe. In *Proceedings of the IEE Colloquium on 'Driver Information' (Digest No.127)*, 1988, pp. 1–9.
54. S. Y. Wang, C. C. Lin, Y. W. Hwang, K. C. Tao, and C. L. Chou. A practical routing protocol for vehicle-formed mobile ad hoc networks on the roads. In *Proceedings of the 8th IEEE International Conference on Intelligent Transportation Systems*, 2005, pp. 161–166.
55. J. Yin, T. ElBatt, G. Yeung, B. Ryu, S. Habermas, H. Krishnan, and T. Talty. Performance evaluation of safety applications over DSRC vehicular ad hoc networks. In *Proceedings of the first ACM workshop on Vehicular ad hoc networks*, 2004, pp. 1–9.
56. J. Yoon, M. Liu, and B. Noble. Random waypoint considered harmful. In *Proceedings of INFOCOM 2003*, April 2003.
57. H. Yang, H. Luo, F. Ye, S. Lu, and L. Zhang. Security in mobile ad hoc networks: Challenges and solutions. *IEEE Wireless Communications*, **11**:38–47, 2004.
58. X. Yang, J. Liu, F. Zhao, and N. H. Vaidya. A vehicle-to-vehicle communication protocol for cooperative collision warning. In *Proceedings of the First Annual International Conference on Mobile and Ubiquitous Systems: Networking and Services*, 2004, pp. 1–14.

CHAPTER 16

Cluster Interconnection in 802.15.4 Beacon-Enabled Networks

JELENA MIŠIĆ AND RANJITH UDAYSHANKAR

Department of Computer Science, University of Manitoba, Winnipeg, Manitoba, R3T 2N2, Canada

16.1 INTRODUCTION

The recent IEEE 802.15.4 standard [1] for low-rate wireless personal area networks supports small, cheap, energy-efficient devices operating on battery power that require little infrastructure to operate [2, 3]. It is considered as enabling technology for home networks and wireless sensor networks. IEEE 802.15.4 networks can appear in star topology where all communications are routed through the PAN coordinator or in peer-to-peer topology where nodes can communicate with each other directly while the PAN coordinator is still needed for cluster management [1]. Networks with peer-to-peer topology have homogeneous nodes with the same initial energy, computational resources, and link capacity, while in star topology the PAN coordinator can have higher energy, a greater number of computational resources, and potentially higher capacity of intercoordinator links than ordinary nodes in the cluster.

In the recent period, several evaluations related to the performance of IEEE 802.15.4 networks either in peer-to-peer or in cluster topology have been conducted, and results have been reported in references 4–13. The choice of topology for IEEE 802.15.4 standard is still an open question. However, it seems that the choice of topology is an issue of tradeoff between the node simplicity and homogeneity versus the duration of network lifetime. For sensor networks covering large geographic areas, it is difficult to replace sensor's batteries when they are exhausted; therefore, when nodes close to the sink die, the whole network is unavailable. It was shown in reference 14 that in the homogeneous network case, nodes close to the sink die first because their batteries are exhausted due to excessive packet relaying. The concept of power heterogeneity enhanced with link heterogeneity was further considered in

---

Algorithms and Protocols for Wireless and Mobile Ad Hoc Networks, Edited by Azzedine Boukerche
Copyright © 2009 by John Wiley & Sons Inc.

reference 15, and it was proven that modest number of nodes with higher power can provide a five-fold increase of network lifetime. For this reason we choose cluster with star topology as the basic network building block and explore the ways to achieve efficient cluster interconnection in order to implement larger networks.

In this chapter, we consider two 802.15.4 clusters operating in beacon-enabled, slotted CSMA-CA mode; the clusters will be referred to as the source and sink cluster. The clusters can be interconnected in two ways:

- In master–slave (MS) fashion, with the coordinator node of the source cluster acting as the bridge.
- In slave–slave (SS) fashion where bridge is ordinary node in both clusters.

The bridge periodically visits the sink cluster in order to deliver the data gathered from the sensor nodes in the source cluster. Bridge visits are made possible by the existence of active and inactive parts of the superframe; that is, the bridge visits the sink cluster during the inactive period of the source cluster superframe. The bridge delivers its data either by competing with other nodes in the sink cluster using the CSMA-CA access mode or by using the guaranteed timeslots (GTS) allocated by the sink cluster coordinator. Also, standard [1] allows transmissions in the CSMA part of the superframe to be either (a) acknowledged, giving base for the reliable MAC with retransmissions, or nonacknowledged, which has applications in sensor networks where packets' content is correlated. In this chapter, we discuss and compare the performance of MS- and SS-based bridges. At this point we don't consider power management algorithms explicitly since we target applications of both (a) personal area networks where reliable MAC is needed and (b) sensor networks that can demand either reliable or nonreliable MAC. However, our framework is open for power management algorithms either through control of inactive superframe part or through individual power control for each node.

The rest of the chapter is organized as follows. In Section 16.2 we review the properties of 802.15.4 beacon-enabled MAC related to the operation of bridges. In Section 16.3, we present the details of master–slave (MS) bridge operation. In Section 16.4 the operation of slave-slave (SS) bridge is described. In Section 16.5, we discuss comparative performance issues of MS bridges and SS bridges. Finally, Section 16.6 concludes the chapter.

16.2 BASIC PROPERTIES OF IEEE STD 802.15.4 MAC

In beacon-enabled networks the PAN coordinator divides its channel time into superframes [1]. Each superframe begins with the transmission of a network beacon, followed by an active portion and an optional inactive portion, as shown in Figure 16.1. The coordinator interacts with its PAN during the active portion of the superframe, and may enter a low power mode during the inactive portion. The superframe duration, SD, is equivalent to the duration of the active portion of the superframe, which cannot exceed the beacon interval BI.

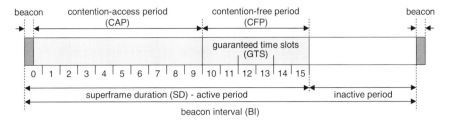

Figure 16.1. The composition of the superframe. (Adapted from reference 1.)

All communications in the cluster take place during the active portion of the superframe, the duration of which is referred to as the superframe duration SD. The superframe is divided into 16 slots of equal size. Each slot consists of $3 \cdot 2^{SO}$ backoff periods, which gives the shortest active superframe duration $aBaseSuperframeDuration$ of 48 backoff periods. In the ISM band, the duration of the backoff period is 10 bytes, giving the maximum data rate of 250 kbps. The duration of the active part of the superframe is $SD = aBaseSuperframeDuration \cdot 2^{SO}$ (expressed in backoff periods), where the parameter $SO = 0, \ldots, 14$ is known as $macSuperframeOrder$. The time interval between successive beacons is $BI = aBaseSuperframeDuration * 2^{BO}$, where BO denotes the so-called $macBeaconOrder$. The duration of the inactive period of the superframe can easily be determined as $I = aBaseSuperframeDuration * (2^{BO} - 2^{SO})$. While the default access mode in beacon-enabled operation is slotted CSMA-CA, some slots may optionally be reserved for certain devices (GTS). Any device can request a GTS (one for uplink and/or one for downlink transmission), but the actual allocation is ultimately the responsibility of the cluster coordinator. The structure of the superframe is shown in Figure 16.1.

Data transfers from a node to PAN coordinator are synchronized with beacons as shown in Figure 16.2a and are done using slotted CSMA-CA access described below. Data transfers from the coordinator are more complex and are first announced by the coordinator which transmits the list of nodes which have the pending downlink packets. The device periodically listens to the network beacon; and if a packet is pending, transmits a MAC command requesting the data. MAC command frames are very short.

The PAN coordinator acknowledges the successful reception of the data request by transmitting an acknowledgement frame. After the acknowledgement the node turns on its receivers for period of $aMaxFrameResponseTime$, which is equal to 1220 symbols and the PAN coordinator has to transmit the pending frame within that period.

Downlink transmission can be achieved without slotted CSMA-CA only if the coordinator's MAC layer can start transmission of the data frame between $aTurnaroundTime$ (12 symbols) and $aTurnaroundTime + aUnitBackoffPeriod$ and there is time remaining in the Contention Access Part for the packet appropriate interframe spacing and acknowledgment. If this is not possible, transmission is done using slotted CSMA-CA. We consider that downlink transmissions without CSMA-CA mechanism are detrimental for network performance since they will cause

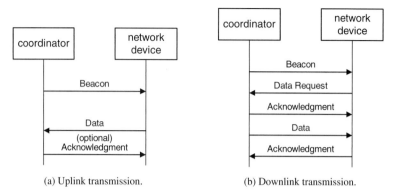

Figure 16.2. (a) Uplink and (b) downlink data transfers in beacon-enabled PAN.

additional collisions and we assume that all transmissions will be achieved using slotted CSMA-CA.

If the transmission was correctly received within the time limit, the node will acknowledge it. If not, the whole process of announcement through the beacon has to be repeated. The standard allows for informing the device that there are more frames waiting at the PAN coordinator's queue by using the frame pending subfield of the data frame received from the coordinator (a *more* bit). If *more* bit is set to 1, the device still has frames with the coordinator and it has to send new data request frame in order to retrieve it. The cycle of downlink transmission is shown in Figure 16.2b. According to the standard, maximally seven devices can be advertised in the beacon. In this work we assume that PAN will advertise nodes in round-robin fashion in the case if it has more than seven pending downlink packets.

The active portion of each superframe is divided into equally sized slots; the beacon is transmitted at the beginning of slot 0, and the contention access period (CAP) of the active portion starts immediately after the beacon. In each slot, the channel access mechanism is contention-based, using the CSMA-CA access mechanism (more details are given below). A device must complete all of its contention-based transactions within the contention access period (CAP) of the current superframe.

Within the timeslots of the active portion of the superframe, the PAN coordinator may reserve slots to allow dedicated access to some devices. These slots are referred to as guaranteed timeslots (GTS), and together they comprise the so-called contention-free period (CFP). In this work we do not consider the impact of the GTS, although their presence will clearly decrease the usable bandwidth of the PAN for other devices.

The basic time unit of the MAC protocol is the duration of the so-called backoff period. Access to the channel can occur only at the boundary of the backoff period. The actual duration of the backoff period depends on the frequency band in which the 802.15.4 WPAN is operating. Namely, the standard allows the PAN to use either one of three frequency bands: 868–868.6 MHz, 902–928 MHz and 2400–2483.5 MHz. In the two lower-frequency bands, BPSK modulation is used, giving the data rate of 20 kbps and 40 kbps, respectively. Each data bit represents one modulation symbol

BASIC PROPERTIES OF IEEE STD 802.15.4 MAC

TABLE 16.1. Timing Structure of the Slotted Mode MAC Protocol

Type of Time Period	Duration	MAC Constant
Modulation symbol	1 Data bit in 860-MHz and 915-MHz bands, 4 data bits in 2.4-GHz band	N/A
Unit backoff period	20 symbols	*aUnitBackoffPeriod*
Basic superframe slot (SO = 0)	Three unit backoff periods (60 symbols)	*aBaseSlotDuration*
Basic superframe length (SO = 0)	16 basic superframe slots (960 symbols)	*aBaseSuperframeDuration* = *NumSuperframeSlots* · *aBaseSlotDuration*
(Extended) superframe duration SD	*aBaseSuperframeDuration*·2^{SO}	*macSuperframeOrder*, SO
Beacon interval BI	*aBaseSuperframeDuration*·2^{BO}	*macBeaconOrder*, BO
Maximal time to wait for downlink transmission	1220 symbols	*aMaxFrameResponseTime*
Rx-to-Tx or Tx-to-Rx maximum turnaround time	12 symbols	*aTurnaroundTime*
Timeout value to wait for the acknowledgement	54 symbols	*macAckWaitDuration*

Note: The values of both *BO* and *SO* must be less than 15 in the beacon enabled mode.

that is further spread with the chipping sequence. In the third band, the O-QPSK modulation is used before spreading; in this case, four data bits comprise one modulation symbol that is further spread with the 32-bit spreading sequence. Table 16.1 summarizes the basic timing relationships in the MAC sublayer. Note that the constants and attributes of the MAC sublayer, as defined by the standard, are written in italics. Constants have a general prefix of "a" (e.g., *aUnitBackoffPeriod*), while attributes have a general prefix of "mac" (e.g., *macMinBE*).

16.2.1 CSMA-CA Algorithm

This algorithm is comprised of downlink data transmission, uplink data transmission, and uplink request transmission states. As is the case with other contention-based access control schemes, transmission will be attempted only when the medium is clear, but withheld if there is channel activity or when contention occurs. The CSMA-CA protocol, shown as a flowchart in Figure 16.3, is invoked when a packet is ready to be transmitted. In this algorithm, three variables are maintained for each packet:

1. *NB* is the number of times the algorithm was required to experience backoff due to the unavailability of the medium during channel assessment.
2. *CW* is the contention window—that is, the number of backoff periods that need to be clear of channel activity before the packet transmission can begin.

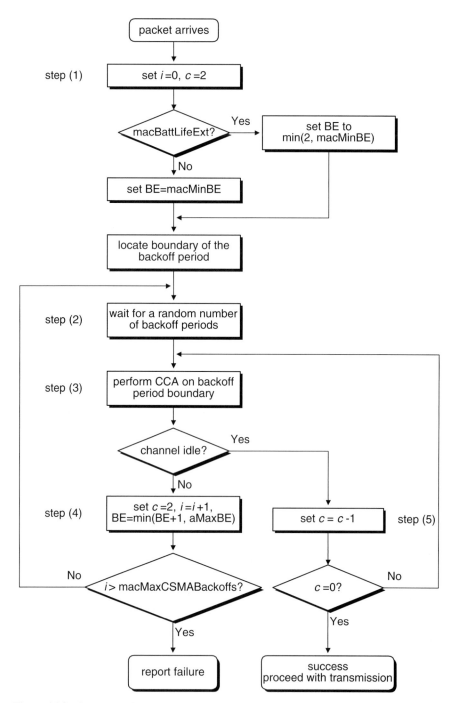

Figure 16.3. Operation of the slotted CSMA-CA MAC algorithm in the beacon-enabled mode. (Adapted from reference 1.)

3. *BE* is the backoff exponent that is related to the number of backoff periods a device should wait before attempting to assess the channel (see below for a detailed explanation).

In step 1, the algorithm begins by setting *NB* to zero and *CW* to 2. If the device operates on battery power (as determined by the attribute *macBattLifeExt*), *BE* is set to 2 or to the constant *macMinBE*, whichever is less; otherwise, it is set to *macMinBE* (the default value of which is 3). The algorithm then locates the boundary of the next backoff period.

In step (2), the algorithm attempts to avoid collisions by generating random waiting time in the range $0, \ldots, 2^{BE} - 1$ backoff periods. When the wait period is over, the MAC sub-layer needs to perform *CW* clear channel assessment (CCA) procedures, transmit the frame, and optionally wait for acknowledgment. The time to wait for an acknowledgment, *macAckWaitDuration*, is equivalent to 54 or 120 symbols, depending on the currently selected physical channel. If the remaining time within the CAP area of the current superframe is sufficiently long to accommodate all of these, the MAC sublayer will proceed with step (3) and perform the first CCA to see whether the medium is idle. If the remaining time is not sufficient, the MAC sublayer will pause until the next superframe.

If the channel is busy, the values of *NB* and *BE* are increased by one (but *BE* cannot exceed *macMaxBE*, the default value of which is 5), while *CW* is reset to 2; this is step (4) in the flowchart. If the number of retries is below or equal to *macMaxCSMABackoffs* (the default value of which is 5), the algorithm returns to step (2), otherwise the algorithm terminates with a channel access failure status. Failure will be reported to the higher protocol layers, which can then decide whether to reattempt the transmission as a new packet or not.

If the channel is idle, step (5), the value of *CW* is decreased by one, and the channel is assessed again. When *CW* becomes zero, the transmission of the packet may begin, provided that the remaining number of backoff periods in the current superframe suffices to handle both the packet and the subsequent acknowledgment. If this is not the case, the standard requires that the transmission is deferred until the beginning of the next superframe.

Note that the backoff unit boundaries of every device should be aligned with the superframe slot boundaries of the PAN coordinator; that is, the start of first backoff unit of each device is aligned with the start of the beacon transmission. The MAC sublayer should also ensure that the PHY layer starts all of its transmissions on the boundary of a backoff unit.

16.3 MASTER–SLAVE BRIDGING ALGORITHM

We consider two interconnected clusters operating in the ISM band around 2.4 GHz. Each cluster operates in a different frequency sub-band so that intercluster interference is avoided (standard prescribes 16 distinct cluster channels). We assume that both clusters operate in beacon-enabled CSMA-CA mode control of their respective cluster (PAN) coordinators. In each cluster, the channel time is divided into superframes that are bounded by beacon transmissions from the coordinator [1]. For clarity, variables pertaining to the source and sink cluster will be labeled with subscripts *src* and *snk*, respectively, while the variables linked to the bridge will have the subscript *bri*.

During the inactive portion of the superframe shown in Figure 16.2, any device may enter a low-power mode or perform other functions, including the interconnection function. This facilitates the creation of larger networks through bridging, with the cluster coordinator of the source cluster acting as the bridge. When the active part of the superframe is completed in the source cluster, its cluster coordinator/bridge switches to the sink cluster. The bridge has stored the packet(s) that need to be delivered to the sink cluster coordinator (which acts as the network sink), and it waits for the beacon so that it can deliver its data to the sink cluster coordinator.

In case the bridge has been allocated GTS access, it will wait until its slot arrives and then transmit the data without any backoff countdown; otherwise, it will execute the CSMA-CA transmission procedure just like any other node in sink cluster. In the latter case, should the bridge be unable to transmit its data when the (active portion of the) superframe in the sink cluster ends, it will freeze its backoff counter and leave the sink cluster. The bridge will resume the backoff countdown upon returning to the sink cluster for the next superframe. Also, if the bridge's buffer becomes empty before the end of the sink's active superframe part, the bridge will immediately return to the source cluster and wait for the time to transmit the beacon denoting the beginning of the next superframe, and the source cluster continues to operate. Bridge operation in both access modes is presented in Figure 16.4; in the discussions that follow, we will refer to those modes as the CSMA-CA and GTS mode, respectively.

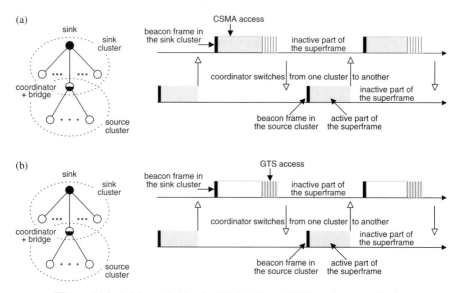

Figure 16.4. Bridge switching in CSMA-CA and GTS mode, respectively.

We will assume that all the traffic from the source cluster occurs in the uplink direction and that the bridge actually delivers it to the sink cluster coordinator. The sink cluster has some local traffic as well. (In more complex networks, the sink cluster may contain several bridges, each with its own "source" cluster.) This is a reasonable assumption in sensor networks, where most, if not all, of the traffic will be directed toward the network sink. All ordinary nodes in either cluster use the CSMA-CA access mode.

16.3.1 Queuing Model of MS Bridge Exchange

Let us now consider the source cluster and calculate the amount of traffic that reaches the sink cluster. The source cluster contains n ordinary sensor nodes that have the packet arrival rate of λ_i during inactive and active parts of the superframe. Coordinator in the source cluster also functions as the bridge. During the active part of the superframe, nodes send packets to the bridge. However, not all generated packets make it to the transmission medium because of packet blocking at the source nodes. Namely, ordinary sensor node is assumed to have finite input buffer with the capacity of L packets; once the buffer is full, packets will be simply dropped. This poses few problems in sensing applications because redundant information is available from other nodes. The coordinator/bridge has a finite buffer as well; this buffer has the capacity of L_{bri} packets. In case this buffer is full, new packets will not be admitted and the coordinator/bridge will not send the acknowledgment. Let us denote the blocking probabilities at ordinary nodes in the source cluster and at the bridge as Pb_{src} and Pb_{bri}, respectively.

Nonacknowledged Transfer. In the case of nonacknowledged transmission, the cluster coordinator does not acknowledge the packets which are successfully received and stored in its buffer. Without acknowledgments, packets transmitted toward the bridge which are lost due to the noise at the physical layer, collisions, or blocking will not be retransmitted.

The amount of traffic admitted in the source cluster is $n\lambda_i(1 - Pb_{src})$ and the total packet arrival rate offered to the bridge is $\lambda_{bri} = n\lambda_i(1 - Pb_{src})\gamma_{src}\delta_{src}$, where γ_{src} denotes probability that no collision has occurred for a particular packet in the source cluster and δ_{src} denotes the probability that packet is not corrupted by the noise. Given the bit error rate of the physical medium and the total packet length, δ_{src} can be calculated as the probability that none of the bits in the packet is corrupted by the noise. The graphical representation of the queuing, blocking, and bridging between the two clusters is shown in Figure 16.5a.

Acknowledged Transfer. The purpose of acknowledged transfer is to achieve reliable packet transfer at the MAC layer. All successful packet transmissions which are received by the respective cluster coordinator and placed in its buffer are acknowledged. However, when the incoming packet reaches the coordinator/bridge when

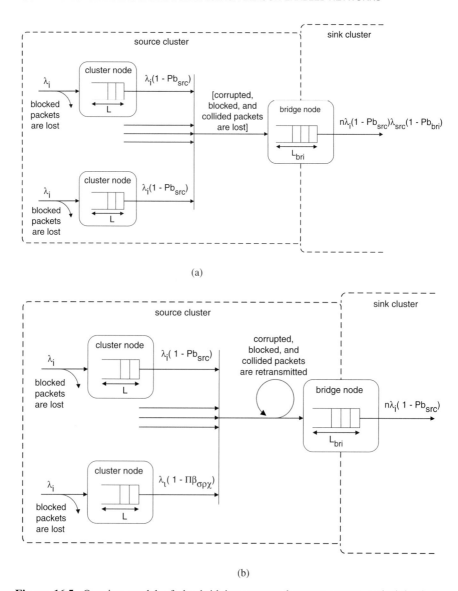

Figure 16.5. Queuing model of the bridging process between source and sink cluster. (a) Nonacknowledged transfer. (b) Acknowledged transfer.

its buffer is full, it will not send the acknowledgment even if the transmission was successful. The lack of acknowledgment may also be due to a collision or noise at the physical layer. If the acknowledgment is not received within the time prescribed by the standard [1], the sending node will repeat the transmission. In our model, the ordinary node will repeat the packet transmission until it receives the acknowledgment. (The standard prescribes the maximum number of transmission re-tries, but in that case the final reliability of transmission has to be achieved at higher protocol layers, which is equivalent to our approach.)

In this case, we can say that traffic blocked by the bridge "stays" in the network and contributes to an increase in traffic, as well as the number of collisions, in the source cluster. The graphical representation of the queuing, blocking, and bridging between the two clusters is shown in Figure 16.5b.

The amount of traffic admitted in the source cluster is $n\lambda_i(1 - Pb_{src})$. Since the transfer is reliable, the total arrival rate offered to the bridge satisfies the following equality $n\lambda_i(1 - Pb_{src}) = \lambda_{bri}(1 - Pb_{bri})$, which gives the offered bridge packet arrival rate as

$$\lambda_{bri} = \frac{n\lambda_i(1 - Pb_{src})}{1 - Pb_{bri}} \quad (16.1)$$

Since the number of nodes n is relatively large and events of packet blocking by the bridge, collisions, and corruptions by noise are noncorrelated, we will assume that packet arrival process to the bridge is Poisson with average rate λ_{bri}.

16.4 *SLAVE*–SLAVE BRIDGING ALGORITHM

Slave–Slave bridge is an ordinary node in both the source and sink cluster. It visits both clusters in time-division basis as shown in Figure 16.6.

Upon leaving one cluster, bridge has to wait for the beacon of new cluster in order to start communication. This synchronization time is also shown in Figure 16.7, where t indicates duration of the superframe and points 1 and 2 indicate start and end of the bridge's synchronization time.

However, SS bridge operation is more complex than MS bridge operation. The reason for this is complex downlink communication where each downlink packet has first to be advertised in the beacon, requested by the bridge node by sending a request packet; request has to be acknowledged by the coordinator and finally the downlink transmission can commence. The request packet is sent in CSMA-CA mode and can collide with other uplink data packets. The downlink data packet is also sent in

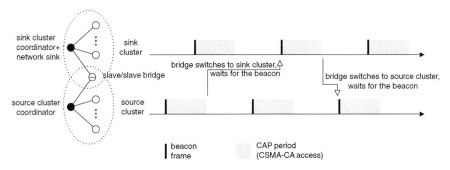

Figure 16.6. Bridge switching between the source and sink cluster.

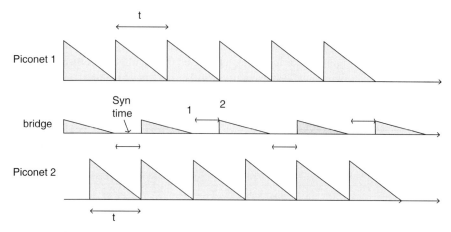

Figure 16.7. Timing of the SS bridge's presence in source cluster and in sink cluster.

CSMA-CA mode and can collide with the other packets. The standard also prescribes that downlink transmission has to be completed within 61 backoff periods; and if it comes later due to many backoff attempts, it will not be acknowledged by the bridge. The diagram, which shows interaction among all the nodes involved in the operation of source and sink clusters is shown in Figure 16.8. Note that this figure indicates operations in active superframe parts from Figures 16.4a and 16.4b. As mentioned in Section 16.3, clusters operate in different frequency bands and intercluster interference is not present.

From the discussions presented above, the following states can be identified for the source or sink PAN coordinator node:

1. The source or sink coordinator may be transmitting the beacon.
2. The source coordinator may be listening to its nodes and receiving data packets from ordinary nodes or request packets from the bridge node. The sink coordinator only receives data packets from the bridge and its local nodes.
3. The source coordinator may be transmitting the downlink data packet as a result of the previously received request packet. As soon as downlink transmission is finished, the source coordinator switches to the listening mode.

Similarly, ordinary or bridge node in the source cluster can be in one of the following states:

1. The ordinary node may be transmitting an uplink data packet.
2. The bridge node may be transmitting an uplink request packet.
3. The bridge node may be in an uplink request synchronization state, which is a virtual state that lasts from the moment of new downlink packet arrival at the coordinator (or the failure of the previous downlink reception) up to

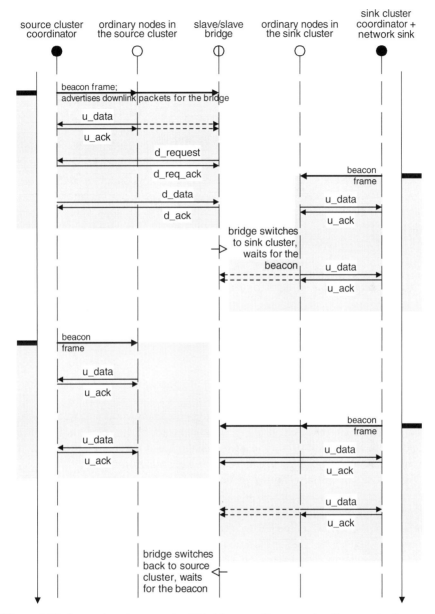

Figure 16.8. Interactions between the SS bridge, coordinator, and ordinary nodes in source cluster and in sink cluster.

the beginning of the CSMA-CA procedure for the uplink request. Note that the arrivals of downlink packets at the coordinator follow the Poisson process, whereas the corresponding announcements in the beacon (from which the target node finds out about those packets) do not.

4. The bridge node may be waiting for a downlink packet.
5. The ordinary or bridge node may also be in an idle state, without any downlink or uplink transmission pending or in progress.

In sink cluster, ordinary or bridge node may be transmitting uplink data packets to the coordinator.

16.4.1 Queuing Model of SS Bridge Exchange

Let us now consider the source WPAN cluster and calculate the amount of traffic that reaches the sink WPAN cluster. The source WPAN cluster contains n ordinary sensor nodes that have the packet arrival rate of λ_i packets per second during inactive and active time of the superframe. During the active period of the superframe nodes, send packets to the coordinator and further route the packets to the bridge during its stay. However, not all generated packets make it to the transmission medium because of packet blocking at the source nodes, which we discussed in the case of MS bridge. Moreover, the coordinator and the bridge also have a finite buffer; these buffers have the capacity of L_c and L_b packets. If the buffer is full, a new packet will not be admitted and the node (coordinator or the bridge) will not send the acknowledgment. Let us denote the blocking probabilities at ordinary nodes in the source cluster, at the coordinator, and at the bridge as follow: Pb_{src}, Pb_c, and Pb_b, respectively.

Nonacknowledged Transfer. In the case of nonacknowledged transmission, the WPAN cluster coordinator does not acknowledge the packets that are successfully received and stored in its buffer. Without acknowledgment, uplink packets which are lost due to the noise at the physical layer, collisions, or blocking will not be retransmitted. However, request packets sent from the bridge as a result of the coordinator's advertisement of downlink packets have to be acknowledged.

The amount of traffic admitted in the source WPAN cluster is $n\lambda_i(1 - Pb_{src})$ and the successful packet arrival rate offered to the coordinator is $\lambda_c = n\lambda_i(1 - Pb_{src})\gamma_{src}\delta_{src}$, where γ_{src} denotes the probability that no collision has occurred for a particular packet in the source cluster and δ_{src} denotes the probability that a packet is not corrupted by the noise. Furthermore, the successful packet arrival rate offered to the bridge is $\lambda_b = \lambda_c(1 - Pb_c)\gamma_c\delta_{src}$, where γ_c denotes the probability that no collision has occurred for a particular packet from a coordinator in the source cluster.

Given the bit error rate of the physical medium and the total packet length, δ_{src} is calculated as the probability that none of the bits in the packet is corrupted by the noise. The graphical representation of the queuing, blocking and bridging between the two WPAN clusters is shown in Figure 16.9.

Acknowledged Transfer. Acknowledged transfer within 802.15.4 beacon-enabled MAC was discussed in Section 16.3.1, and in this section we only discuss its

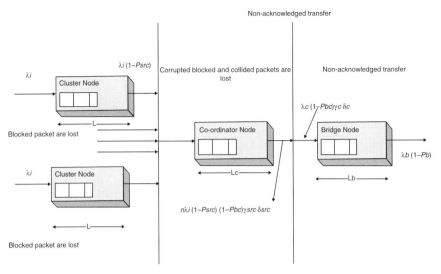

Figure 16.9. Queuing model of the SS bridging process without packet acknowledgments.

implications on the operation of SS bridge. The queuing model of bridging between the two WPAN clusters using the SS bridge is shown in Figure 16.10.

Since the transfer is reliable, the total data arrival rate offered to the coordinator satisfies the following equality: $n\lambda_i(1 - Pb_{src}) = \lambda_c(1 - Pb_c)$, which gives the

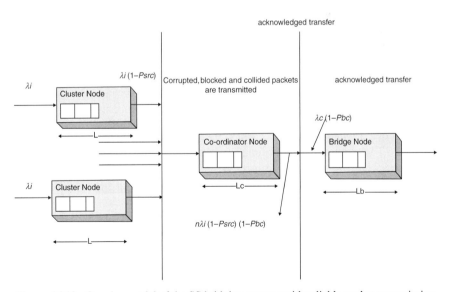

Figure 16.10. Queuing model of the SS bridging process with reliable packet transmission.

offered data packet arrival rate to the coordinator as

$$\lambda_c = \frac{n\lambda_i(1 - Pb_{src})}{(1 - Pb_c)}$$

Similarly, the offered data packet arrival rate to the bridge satisfies the following equality $\lambda_c = \lambda_b(1 - Pb_b)$, which gives the offered bridge packet arrival rate as

$$\lambda_b = \frac{n\lambda_i(1 - Pb_{src})}{(1 - Pb_c)(1 - Pb_b)}$$

Since the number of nodes n is relatively large and the events of packet blocking by the bridge, collisions, and corruptions by noise are noncorrelated, we will assume that the packet arrival process to the coordinator is Poisson with average rate λ_c.

16.5 COMPARATIVE PERFORMANCE EVALUATION

In this section we compare performance of MS and SS bridge under CSMA-CA acknowledged bridge access. Master–slave bridges were evaluated through analytical processing using Maple 10 from Maplesoft. In the evaluation environment for slave–slave bridges, we used a simulator built using the Artifex simulation engine by RSoft Design, Inc. [16]. The network under evaluation consists of one source cluster and one sink cluster. Clusters operate on different channels in the ISM band with a raw data rate of 250 kbps, which means that *aUnitBackoffPeriod* has 10 bytes and *aBaseSlotDuration* has 30 bytes and a bit error rate of $BER = 10^{-4}$. Both source and sink cluster had n ordinary nodes, with n varying between 5 and 30. The packet arrival rate to each ordinary node was varied between 30 packets per minute and 3 packets per second for an MS bridge. For an SS bridge packet the arrival rate was varied between 80 packets per minute and 280 packets per minute.

The data packet size was fixed at 3 backoff periods, and the request packet size was 2 backoff periods. Ordinary nodes had buffers that can hold $L = 3$ packets, while the bridge buffer capacity was $L_{bri} = 6$ packets.

The superframe size in both clusters was controlled with $SO = 0$, $BO = 1$; as the value of *aNumSuperframeSlots* is 16, the *aBaseSuperframeDuration* is exactly 480 bytes. The minimum and maximum values of the backoff exponent, *macMinBE* and *aMaxBE*, were set to 3 and 5, respectively, while the maximum number of backoff attempts was 5.

For both MS and SS bridge, the bridge residence time in source and sink cluster was equal to the active superframe time (48 backoff periods). During residence in the sink cluster, the bridge was trying to deliver as many packets as possible. If the bridge's buffer was emptied before the end of superframe, the bridge has returned to the source cluster. If the bridge was in the process of backoff countdown when the end of the active superframe part occurred, it has frozen the backoff counter and resumed it upon the next visit to the sink cluster.

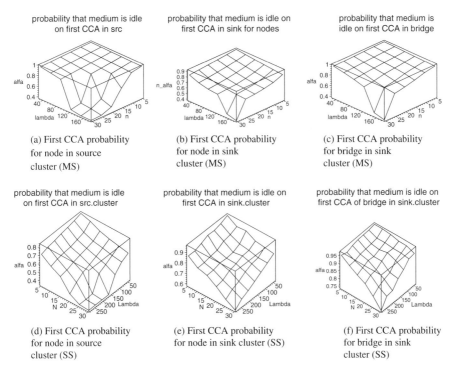

Figure 16.11. Probability that the medium is idle on first CCA. Top row shows the results for the MS bridge; bottom row shows results for the SS bridge.

Figure 16.11 shows the probability that the medium is idle at first Clear Channel Assessment (CCA) for source nodes, sink nodes, and bridge. By comparing Figures 16.11a and 16.11d, we observe that nodes in the source cluster will observe less activity on first CCA with the presence of SS bridge. Therefore, the source cluster will a enter saturation condition where this probability is low (around 0.4) and flat later when it operates with an SS bridge. This is because a large portion of packets in the source cluster with an SS bridge are request packets from the bridge for which backoff countdown is started immediately after the beacon with backoff window equal to 8. Therefore, many request packets from the bridge will choose a small backoff value like 0 or 1 and pass the first CCA successfully (first two backoff periods after the beacon are idle). By the same token, the SS bridge will start backoff countdown immediately after joining the sink cluster and sense the idle channel if it gets a small backoff value. Similar reasons explain Figure 16.12 also. Milder probability that second CCA is successful is just a result of the fact that bridge's request packets will obtain small backoff values and test the medium in the second or third backoff period after the beacon. Also, when the bridge returns to the sink cluster after being emptied in the previous visit, it will start the backoff count immediately after the beacon and will very likely sense that the medium is idle after the beacon. However, the

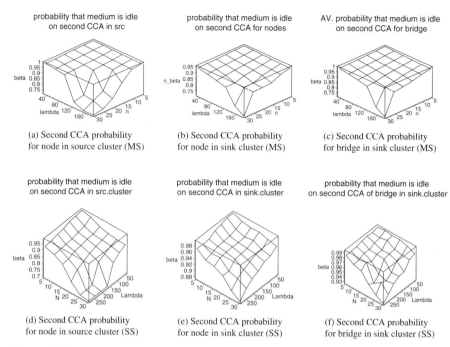

Figure 16.12. Probability that the medium is idle on the second CCA. Top row shows results for the MS bridge; bottom row shows results for the SS bridge.

price for somehow unrealistic feeling about the activity on the medium in the source cluster is high level of collisions, which is shown in Figure 16.13. Indeed, many packets from the bridge which pass the first and second CCAs collide with each other. However, data packets from the previous superframe which did not have enough room to conduct two CCAs, send packets, and receive acknowledgment will have to wait for the next beacon, conduct two CCAs, and transmit. Therefore, request packets from the SS bridge will collide with many delayed packets from the previous superframe, which results in a worse success probability for the source cluster with an SS bridge than with an MS bridge. A large number of collisions of request packets will in turn result in repeated advertisements of downlink transmissions in the beacon, and the flow of downlink packets to the bridge will slow down until it finally stops. Therefore, relatively good success probabilities for the SS bridge in the sink cluster are a result of very few packets which are transferred to the sink cluster and transmitted successfully due to lower congestion in the medium. All these observations are nicely confirmed with throughput values shown in Figure 16.14. By comparing Figures 16.14a and 16.14d, we see that throughput in the source cluster reaches a higher maximum value for the SS bridge, but this is just due to the throughput of request packets. The bridge throughput in the sink cluster is actually higher for the MS bridge as shown in Figures 16.14c and 16.14f.

COMPARATIVE PERFORMANCE EVALUATION 477

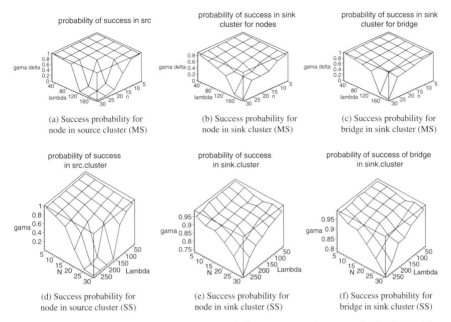

Figure 16.13. Probability of successful transmission. Top row shows results for MS bridge, bottom row shows results for SS bridge.

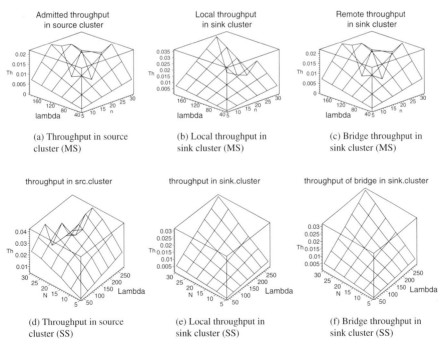

Figure 16.14. Throughput. Top row shows results for MS bridge; bottom row shows results for SS bridge.

16.6 CONCLUSION

In this chapter we have described design and performance issues of cluster interconnection for beacon-enabled 802.15.4 clusters. Our discussion shows that there are pros and cons for both approaches. The SS bridge removes the task of bridging from the WPAN coordinator, but it generates more traffic in the source cluster. The MS bridge efficiently uses the inactive superframe period where all nodes sleep and utilizes uplink data transmissions, but it becomes a single point of failure and a target for security attacks.

16.7 EXERCISES

1. What are advantages and disadvantages of star-based topology over peer-to-peer topology?
2. What is major performance problem with downlink transmissions in IEEE 802.15.4 beacon-enabled networks?
3. What are the main differences between master–slave and slave–slave bridges in IEEE 802.15.4 beacon-enabled networks?
4. What are the pros and cons of CSMA-CA bridge access over GTS-based bridge access in the remote cluster?
5. Do you think that slave–slave bridging would have better performance if in the source the cluster the packets are delivered to the bridge using GTS channel? Why or why not?
6. Discuss the change of the performance of the cluster interconnection if the data transfers were nonacknowledged. In which applications can nonacknowledged transmissions be tolerated?

REFERENCES

1. Standard for part 15.4: Wireless MAC and PHY specifications for low rate WPAN. IEEE Std 802.15.4, IEEE, New York, October 2003.
2. J. A. Gutiérrez, E. H. Callaway, Jr., and R. L. Barrett, Jr. *Low-Rate Wireless Personal Area Networks*, IEEE Press, New York, 2004.
3. Edgar H. Callaway, Jr. *Wireless Sensor Networks, Architecture and Protocols*, Auerbach Publications, Boca Raton, FL, 2004.
4. G. Lu, B. Krishnamachari, and C. Raghavendra. Performance evaluation of the IEEE 802.15.4 MAC for low-rate low-power wireless networks. In *Proceedings Workshop on Energy-Efficient Wireless Communications and Networks EWCN '04*, Phoenix, AZ, April 2004.

5. J.-S. Lee. An experiment on performance study of IEEE 802.15.4 wireless networks. In *10th IEEE Conference on Emerging Technologies and Factory Automation ETFA 2005*, Catania, Italy, September 2005, Vol. 2, pp. 451–458.
6. B. Bougard, F. Catthoor, D. C. Daly, A. Chandrakasan, and W. Dehaene. Energy efficiency of the IEEE 802.15.4 standard in dense wireless microsensor networks: Modeling and improvement perspectives. In *DATE '05: Proceedings of the Conference on Design, Automation and Test in Europe*, Vol. 1, Munich, Germany, March 2005, pp. 196–201.
7. I. Howitt, R. Neto, J. Wang, and J. M. Conrad. Extended energy model for the low rate WPAN. In *Proceedings IEEE International Conference on Mobile Ad-hoc and Sensor Systems MASS2005*, Washington, DC, November 2005, pp. 315–322.
8. T. H. Kim and S. Choi. Priority-based delay mitigation for event-monitoring IEEE 802.15.4 LR-WPANs. *IEEE Communication Letters*, **10**(3):213–215, March 2006.
9. N. F. Timmons and W. G. Scanlon. Analysis of the performance of IEEE 802.15.4 for medical sensor body area networking. In *First Annual IEEE Communications Society Conference on Sensor and Ad Hoc Communications and Networks SECON 2004*, Santa Clara, CA, October 2004, pp. 16–24.
10. A.-C. Pang and H.-W. Tseng. Dynamic backoff for wireless personal networks. In *IEEE GLOBECOM 2004*, Vol. 3, Dallas, TX, December 2004, pp. 1580–1584.
11. M. Neugebauer, J. Plonnigs, and K. Kabitzsch. A new beacon order adaptation algorithm for IEEE 802.15.4 networks. In *Proceeedings of the Second European Workshop on Wireless Sensor Networks*, Istanbul, Turkey, February 2005, pp. 302–311.
12. J. Mišić, S. Shafi, and V. B. Mišić. The impact of MAC parameters on the performance of 802.15.4 PAN. *Ad Hoc Networks*, **3**(5):509–528, 2005.
13. J. Mišić, S. Shafi, and V. B. Mišić. Performance of beacon enabled ieee 802.15.4 cluster with downlink and uplink traffic. *IEEE Transactions on Parallel and Distributed Systems*, **17**(4):361–377, 2006.
14. V. P. Mhatre, C. Rosenberg, D. Kofman, R. Mazumdar, and N. Shroff. A minimum cost heterogeneous sensor network with a lifetime constraint. *IEEE Transactions on Mobile Computing*, **4**(1):4–15, 2005.
15. M. Yarvis, N. Kushalnagar, H. Singh, A. Rangarajan, Y. Liu, and S. Singh. Exploiting heterogeneity in sensor networks. in *Proceedings INFOCOM05*, Vol. 2, Miami, FL, March 2005, pp. 878–890.
16. RSoft Design, Inc. *Artifex v.4.4.2*, San Jose, CA, 2003.

INDEX

ABR algorithm, 137
Acceptable-power-to-send (APTS), 324
Access network, 450
Access window (AW), 325
ACK-reducing proposals, *See* ACK-thinning proposals
ACK-thinning techniques, 277, 285, 286, 287, 288, 296, 305, 308
 ACK-thinning proposals, 278
 aim, 287
 cross-layer approaches, 287
 end-to-end techniques, 288
 goals, 285
 hybrid techniques, 296
 importance, 285
 single-layer approaches, 287
ACK expiration timer, 289
ACK segments, 278
Acknowledgment (ACK), 253, 308
 effects, 308
 packet, 73
 segment, 283
Active delay control (ADC), 295
 method, 307
 technique, 306
Active queue management (AQM), 307
 mechanism, 295
 RED, 295
Ad hoc broadcast protocol (AHBP), 58–59, 64, 79
 protocol, 79, 81
Ad hoc mobility models, 421
 RWP, 421
Ad hoc multicast routing protocols, 166
Ad hoc network routing algorithm, 3, 6, 24, 25

goal of, 6
Ad hoc networking techniques, 88
Ad hoc networks, 5, 12, 15, 21, 22, 129, 375, 426
 performance issues, 434
 reputation/trust-based systems, 375
Ad hoc networks applications, 130
 automotive/PC interaction, 131
 conferencing, 130
 embedded computing applications, 130
 emergency services, 130
 home networking, 130
 personal area networks/bluetooth, 131
 sensor dust, 130
Ad hoc networks routing protocols, 132
Ad hoc on-demand distance vector (AODV), 7, 9, 134, 136, 417
 routing protocol, 8, 10, 134, 135, 299
Ad hoc on-demand multipath distance vector routing (AOMDV), 155–156
 protocol, 155
Ad hoc routing algorithms, 434
Ad hoc routing protocol, 133, 345
 schematic presentation, 133
Ad hoc satellite networks, 231
Ad hoc transport protocol (ATP), 267
 assisted congestion control, 267
 layer coordination, 267
 rate-based transmissions, 267
 receiver, 11
Ad hoc wireless networks, 142
Adaptive dynamic backbone construction algorithm (ADB), 167–169, 171, 182
 core connection process, 168
 core selection process, 168
 neighbor discovery process, 168

Algorithms and Protocols for Wireless and Mobile Ad Hoc Networks, Edited by Azzedine Boukerche
Copyright © 2009 John Wiley & Sons, Inc.

481

Adaptive dynamic backbone multicast (ADBM), 166, 177, 179, 182
  protocol, 178
Adaptive pacing (AP), 262, 263
Adaptive protocol, 365
Additive Gaussian white noise (AWGN) channel, 328, 338
Algebra tools, 30
Algorithms design, 3
Angle of arrival (AOA), 32, 42, 325
  technique, 32
Ant-colony-based routing algorithm (ARA) 139–140
Area delimiting methods, 34–36
Associativity-based routing (ABR), 137–138
Automated highway system architecture (AHS), 419
Automatic power-save delivery (APSD), 361
  scheme, 361
  station, 361
Automatic repeat request (ARQ) paradigm, 278

Backward ant (BANT), 140, 259
Backward ant route discovery phase, 141
Bandwidth estimation, 257, 260
  end-to-end proposals, 257
  link-layer proposals, 257
  split-connection proposals, 257
Bandwidth estimation, 259
Basic max-flow formulation, 205
Basic power control schemes, 322
Battlefield based communications, 24, 165
Beacon nodes, 397
Beacon-supported protocols, 24
Beacon vector routing (BVR), 25, 26
  protocol, 101
Beacon vectors, 26
Bellman-Ford routing algorithm, 142, 144
Bellman-Ford shortest path algorithm, 343
Best-effort (BE) traffic, 362, 367
  power-saving AP Support, 362
Best next-hop candidates, 232
Beta distribution model, 389
  binomial, 389
  Gaussian, 389
  Poisson, 389
Bipartite graph, 102, 106
  illustration, 106

Bit error probability (BER), 251, 257, 316
Block design problem, 346
Body area network (BAN, 2
Bootstrap nodes, 29
Bordercast resolution protocol (BRP), 146
Bounding box localization algorithm, 34, 35
Bounding box network, 35
BPSK modulation, 462
BRG nodes, 59
Bridging process, 468
  queuing model, 468
Broadband wireless multihop networks, 337
Broadcast query request packet, 137, 138
Broadcast relay gateway (BRG) nodes, 59
Broadcast robustness, 53
Broadcast storm, 51
Brownian-like mobility models, 410
Buffering capability and sequence information (TCP-BUS), 266
Butterfly network, 111

CacHing and multipath routing protocol (CHAMP), 154
California PATH project, 419
Canadian dollars (CAD), 357
Canadian weather stations, 373
Car-following model, 421, 425
Carrier sense multiple access (CSMA), 6, 40, 53, 321
  based schedulers, 209
  CA algorithm, 463, 464
  CA mechanism, 461
  CA mode, 460
  IEEE 802.11, 321
  protocol, 210, 211, 463
Cellular automata (CA) models, 423
Cellular networks, *See* WiFi
Channel error state, 262
Clear-to-send (CTS) packet, 321
Cluster-based routing, 442
  cluster head, 442
  gateway, 442
  member, 442
Clustered group multicast (CGM) 181
Cluster-head gateway switch routing (CGSR), 143–144
  protocol, 143
Clustering algorithms, 5, 23, 181
Cluster topologies, 66

Code division multiple access (CDMA)
  communications, 438
  code, 143, 144
  wireless networks, 337
Coding-based algorithms, 110, 242
Collaborative reputation mechanism to enforce node cooperation (CORE), 390
  decision making, 391
  functional reputation, 390
  indirect reputation, 390
  information modeling, 391
  model, 387, 389, 391, 399
  provider protocol, 390
  requestor protocol, 390
  subjective reputation, 390
Collision avoidance (CA), 40
  information, 323, 325
Collision avoidance reduce collisions technique, 53
Collision warning systems (CWS), 418
  active approaches, 418
  passive approaches, 418
Communication channel models, 450
Compass-selected routing (COMPASS) algorithm, 5
Complex weather processes, 372
  air pressure, 372
  humidity, 372
  temperature, 372
COMPOW power control scheme, 331
Computing resources, 2
  battery power, 2
  hardware, 2
  trust manager, 393
Congested traffic system, 411
Congestion avoidance, 253, 284
  phase, 254, 283
Congestion control, 267, 315
  algorithms, 251, 252, 333
  mechanism, 252, 255, 258, 261
Congestion indication event, 283, 284
Congestion window, 253, 258, 284, 290, 292, 294, 295, 301
Congestion window limit (CWL), 262
  scheme, 263
Connected dominating sets (CDS), 51, 167
  algorithm, 59, 60, 64
  approach, 59, 63
  protocols, 168

tree, 63, 74
Connection-oriented stream transport protocol, 10
Connection-oriented traffic, 362, 365, 367, 362
  scenarios, 362
  SMAP/SMP power saving, 365
  VoIP, 362
Connection request, 41
Connection states, 261
  classification, 261
Constant bit rate (CBR) traffic, 61
Constrained optimization problem, 326, 330, 335
Contention-based path selection (COPAS), 299
  method, 306
  routing path, 300
Continuum percolation theory, 409
Control-based approaches, 236
Conversion models, 373
Cooperation of nodes-fairness in dynamic ad hoc networks (CONFIDANT) model, 387, 392
  decision making, 393
  monitor, 393
  path manager, 393
  reputation system, 393
Coordinated routing illustration, 202
Coordinate system, 36
  geometry, 37
Core-assisted multicast protocol (CAMP), 180
Core connection process, 174–176
Core requested table (CRT), 169
Core selection process, 171–174
Cost-effective last-mile technology, 187
Coverage radio, 367, 368
  consumption, 368
Cross layer vs. single layer techniques, 306
Cryptographic key information, 376
Cyclic redundancy checksums (CRCs), 92

Data applications, 88
  data sharing, 88
  e-mails, 88
  messaging, 88
Data carousel, 100
Data communication engineer, 2

484 INDEX

Data communication protocols, 413
Data distribution systems, 88
Data Link Control Layer, 435, 436–438
　DSRC, 436
　schemes, 437
　technologies, 438
Data packets, 166, 189
Data retransmission buffer, 148
DATA segments, 278, 291
Decentralized MAC protocol, 286
Decoding algorithms, 90
Decoding process, 97, 103, 117
Decoding system, 90
Dedicated short-range communications
　　(DSRC), 436
Delay-free and delayed networks, 112
Delay tolerant networks (DTNs), 110, 119,
　　220, 221, 244
　DataMule, 220
　Interplanetary network (IPN), 220
　Village networks, 220
　Zebranet, 220
Delayed acknowledgments (DACKs)
　mechanism, 288, 289
　schematic presentation, 289
Dempster–Shafer belief theory, 389
Designator device (DD), 42
Destination-sequenced distance-vector
　　(DSDV), 7, 134, 136, 142, 144
　protocol, 7, 142
Differential destination multicast (DDM), 9,
　　10
Diffuse horizontal radiation, 373
Diffusion mechanisms, 444–445
Digital fountains, 93, 99, 107
Dijkstra shortest path algorithm, 444
Directed broadcast (DB), 75, 82
　DB transmissions, 75
Directed cyclic graph (DAG), 136
Direct normal radiation, 373
Discounting belief principle, 389
Dissemination frequency, 387
　proactive dissemination, 387
　reactive dissemination, 387
Dissemination locality, 388
　global dissemination, 388
　local dissemination, 388
Distance-bounding protocols, 447
Distance-minimizing routing algorithm, 25

Distance routing effect algorithm for
　　mobility (DREAM), 150–151
　routing protocol, 150, 151
Distance vector-hop (DV-hop) algorithm, 33
Distributed algorithm, 36
Distributed power control algorithm, 326,
　　328
Distributed reputation and trust-based
　　beacon trust system (DRBTS), 397–
　　399
　beacon node (BN), 397
　decision making, 398
　information gathering, 397
　information modeling, 398
　information sharing, 398
　sensor node (SN), 397
　systems, 387
Distributed source routing (DSR), 9, 134,
　　209
　protocol, 7, 300
　routes, 212
Double-covered broadcast (DCB), 52, 60,
　　64, 74
　algorithm, 83
Drop-least-encountered (DLE), 230
DV-hop algorithm, 33
DV-hop network, 33
　schematic presentation, 33
Dynamic adaptive ACK, 292, 293
　demonstration, 293
Dynamic core-based multicast routing
　　protocol (DCMP), 180
Dynamic delayed ACK (DDA), 290, 291
　schematic presentation, 291
Dynamic delaying window, 292
Dynamic network topology, 251
Dynamic networks, 226
Dynamic routing and power control (DRPC)
　　algorithm, 329
Dynamic routing protocol (DRP), 138
Dynamic Source Routing (DSR), 7, 133,
　　134, 390, 292, 417
Dynamic transmission range assignment
　　(DTRA) algorithm (412–415) 412

EDR metric, 197
Efficient communication, 119
Eifel algorithm, 264
Electronic license plates, 446

Electronic social networks, 380
  friend-of-a-friend, 380
  Google's Page rank algorithm, 380
  Peer-to-Peer (P2P) systems, 380
Explicit link failure notification (ELFN), 11
Emergency warning message (EWM) transmissions, 418
Enclosure graph, 343
Encoding algorithms, 89
Encoding/decoding processes, 95
  schematic presentation, 95
Encoding process, 90, 99, 112–116
Encoding techniques, 89
Encoding, transmitting, decoding processes, 90
  schematic presentation, 90
Encoding/decoding system, 91, 92
  success of, 91
End-to-end congestion control, 282
End-to-end connectivity, 88
End-to-end delay, 320
  processing delay, 320
  propagation delay, 320
  queueing delay, 320
  transmission delay, 320
End-to-end metrics, 230
End-to-end proposals, 252, 257
End-to-end QoS, 444–445
End-to-end semantics, 252
End-to-end vs. point-to-point techniques, 305–306
Energy-aware routing algorithms, 341, 342
Energy-aware routing protocols, 341–343
  metrics, 341
Energy-efficient routing algorithms, 320
Energy-saving technique, 345
Epidemic/partial routing-based approach, 226
Erasure codes, 92, 93
Error correcting codes, 92
Error-induced segment loss, 286
Error-prone transmission channels, 101
Error detection code, *See* Error correcting code
Estimated Data Rate (EDR), 197
Estimation-based erasure coding (EBEC), 243
Estimation based approach, 229–230
Euclidean distance, 29, 409

European Chauffeur project, 419
European Commission, 87
Exact receiver-based delay control (RDC), 307
Expected MAC latency (ELR), 198
Expected transmission count (ETX), 195, 200
Explicit congestion notification (ECN) congestion signal, 295
Explicit feedback algorithms, 72–73
  hyper-flooding, 74
  hyper-gossiping, 74
Explicit link failure notification (ELFN), 260, 261, 266
Explicit route disconnection notification (ERDN), 266
Explicit route successful notification (ERSN), 266
Exposed node problem, 256
Extending timeout values, 266
Extraterrestrial direct normal radiation, 372
Extraterrestrial horizontal radiation, 372

Fast-moving nodes, 166
Ferry-initiated MF (FIMF) scheme, 238
Finite field, 98
Fire signals, 1
  limitation of, 1
  mechanism, 1
Fire torches, 1
First-hand information, 386, 387
  direct observation, 386
  personal experience, 386
Fixed-wired networks, 12
Flat-routed ad hoc networks, 22
FleetNet project, 420
Flexible radio network (FRN), 297
Forward ant route discovery phase, 140
  distance, 141
  link cost tables, 141
  routing, 141
Forward RTO-Recovery (F-RTO), 264
Forward error correction (FEC), 99
Forwarding factor, 120
Forwarding group multicast protocol (FGMP), 10
Frank–Wolfe method, 338
Free-flow traffic, 423
  space-time diagram, 423

F-RTO, 265
FSR protocol, 148
Fundamental traffic flow relationship, 421
Fuzzy logic, 420

Gabriel graph, 343
GAF state transitions, 345
Galois field, *See* Finite field
Game-theoretic approach, 399
Gateway nodes, 23, 144
General-purpose processors, 93
Geocast protocol, 440
Geographic adaptive fidelity (GAF), 344
Geographic information systems (GIS), 416
Geographic source routing (GSR) 416, 444
Geometric coordinates, 28
Geometric random graphs (GRG) theory, 409
Global horizontal radiation, 372
Global positioning system (GPS), 30, 34, 44, 149, 170, 408
　based systems, 447
　receiver, 30
　signaling channel, 343
　system, 33
Greedy forwarding strategy, 416
Greedy hop-count-based routing, 190
Greedy perimeter stateless routing (GPSR), 152
Greedy routing scheme (GRS), 4
Grid topologies, 62
Grouping of nodes, *See* Ad hoc networks clustering
Guaranteed timeslots (GTS), 460, 462

Hamming distance, 91, 92
Heavy vehicle electronic license plate program (HELP), 446
High-layer erasure codes, 93
High-speed modems, 95
　ADSL, 95
　xDSL, 95
Host processor, 367
HW-implemented network adaptor, 92
Hybrid coordinator (HC), 361
Hybrid routing schemes, 145

IBRR protocol, 232
IEEE 802.11, 354, 360, 418

CSMA mechanism, 418
IEEE 802.11 transceivers, 285
IEEE 802.11 user stations, 353
IEEE 802.15.4 networks, 459
Implicit feedback algorithms, 73
Implicit feedback solutions, 73
Incoming/generated packets, 112
Incomplete decoding algorithm, 91
　code-word, 91
Industrial, scientific, medical (ISM) band, 407
Information Asymmetry, 379
　adverse selection problems, 379
　definition, 379
　moral hazard problems, 379
Information dissemination, 388
　types, 388
Information society technologies program, 87
Infrastructure-based wireless networks, 129
　cellular networks, 129
　WiFi, 433
Installed internet gateways (IGWs), 420
Institution-based trust antecedent, 378
　situational normality, 378
　structural assurance, 378
Intelligent transportation systems (ITS), 405, 426
　applications, 405
Interactive services, 88
　gaming, 88
　video streaming, 88
　VoIP, 88
Interarrival smoothing factor, 294
Interflow contention, 255, 256
Intermediate/relay nodes, 110
Intermittently connected networks (ICNs) 220
Internet research task force (IRTF), 221
Interpacket delay difference (IDD), 261
Interrogation-based relay routing (IBRR), 231
Inter-vehicle geocast (IVG) protocol, 416
Intervehicle communication (IVC), 405, 418, 419, 426
　challenges, 419
　paradigm, 420
　role, 419
　simulator, 425

Inter-vehicle geocast (IVG) protocol, 416
Intraflow contention, 255, 256
Intrazone routing protocol (IARP), 146
IP-based services, 420
  email, 420
  web access, 420
IP multicast, 88
Iterative algorithm, 338

Joint power control, 337
Joint routing, 330
Joint scheduling algorithm, 339
Jointly optimal congestion control and power control (JOCP) algorithm, 333

Keep-alive timer, 346
K-fault tolerance, 343

Landmark Ad Hoc Routing (LANMAR), 147–148, 147
LAR routing protocol, 149, 151
LCC algorithm, 143
Least cluster change (LCC), 143
Least mean square (LMS), 363
Location-guided tree construction algorithm (LGT), 10
Lightweight mobile routing (LMR), 136
Linear codes, 95
Linear network coding, 112
Link-layer proposals, 252, 257
Link-quality-aware routing, 188, 200, 201
  limitations of, 200
Link-quality-based routing, 191
  advantages of, 191
  environment, 194
  factors, 192
  impact of, 192
Link layer (435–436) 435
  considerations, 435
Link-quality-based routing, 191
Link quality indicator (LQI), 42
LMS optimization criterion, 363
Local core node, 166
Location-based algorithm (LBA), 58, 64, 67
Location-based multicast protocols, 150, 415
Location aided knowledge extraction routing for mobile Ad hoc networks (LAKER), 152–153
Location reply (LREP) packet, 443

Location request (LREQ) packet, 443
Logical link control (LLC) sublayers, 318
Long-range radio, 238
Longer-range wireless technologies, 420
  cellular networks, 420
Low-density areas of networks, 54
Low-energy-consuming devices, 87
Loop-free multipaths, *See* ROAM
Low-rate wireless personal area networks, 459
Lower and upper-bound constraints, 205
Lower energy consumption, 190
Lower energy expenditure, 190
LT code, 105, 106, 107, 108
Luby-transform codes, 105, 107

Macroscopic models, 424
Master-slave (MS) fashion, 460
Master-slave bridging algorithm, 465
Maximal independent contention set (MICS), 210
Maximal independent set (MIS), 210
MCF formulation, 206
Mean delay performance, 365
Medium Access Control (MAC) layer, 73, 317, 407
  approach, 329
  background, 53
  broadcast, 52, 61, 74
  contention-based, 322
  formulation, 213
  interactions, 213
  interference-based, 323
  IP, 179
  layer 152, 179, 436, 461
  mechanism, 287
  modeling, 209, 2122
  protocol, 11, 189, 190, 194, 210, 211, 212, 462
  scheduling-based, 326
  schemes, 322, 323, 326
  specific interactions, 210
  UDP headers, 179
Medium access control (MAC) protocol, 324, 435
Meets and visits (MV), 233
Mesh-based protocols, 166
Mesh networks, 187
Micrelnet network architecture, 42–47

Microscopic models, 424
Minimal dominating set (MDS), 5
Minimal estimated expected delay (MEED) routing, 234
Minimum connected dominating set (MCDS), 5, 56
Minimum-energy data transmissions, 89
Minimum energy routing (MER), 338
Minimum hop routing protocol (MHRP), 335, 341
Minimum spanning tree (MST), 237, 343, 409
Minimum transmission range (MTR), 409, 410, 412
Mobile-IP specification, 279
  scope, 279
Mobile (wireless) ad hoc networks (MANETs), 6, 7, 8, 9, 11, 12, 13, 14, 15, 21, 75, 131, 166, 219, 376, 380, 381, 383, 391, 400, 409
  algorithmic perspective, 3–15
  applications, 277
  data communication, 2
  definition, 279
  design algorithms, 2
  evaluation issues, 447
  mesh networks, 51
  multihop sensor, 51
  network characteristics, 251
  performance modeling, 447
  protocol design, 282, 451
  research, 67
  topology, 3
  transport layer protocol, 267
  types, 281
  design issues, 131–132
  transport layer protocols, 251
Mobility-adaptive multicast routing protocol, 166
Mobility model, 69, 447
  RWP, 423
Model-based routing (MBR), 235
Modeling for interference-aware routing, 204
Modems, 92
Modulo operations, 98
Most forward within radius (MFR), 4
Movement-based algorithm for Ad hoc networks (MORA), 153–154
Movement-based algorithm for Ad hoc networks protocol, 154
  D-MORA, 154
  UMORA, 154
Multi-bipartite graph, 102, 104
Multicast core-extraction distributed Ad hoc routing (MCEDAR), 181
Multicast packet forwarding, 178–179
Multicast protocol, *See* DDM
Multicast protocols, 94
Multicast route table (MRT), 10
Multicast routing protocol, 9, 165
Multicast support, 180
Multicast ad-hoc on-demand distance vector routing (MAODV), 9, 180
Multichannel radios, 24
  intersection of, 35
Multihop ad hoc network concept, *See* End-to-end connectivity
Multihop ad hoc networks, 89
Multihop ad hoc wireless technologies, 87
Multihop wireless communications, 165
Multihop wireless networks (MHWNs), 51, 187, 337
  background, 189
  introduction, 187
Multipath routing protocols, 154
Multipath routing schemes, 336
Multiple retransmissions, 94
Multipoint relays (MPR), 143
Multi-route algorithm (MURA), 239

Natural disaster, 2
  earthquake, 2
  flooding, 2
  hurricane, 2
  military personnel, 2
Near-optimal approximation algorithm, 338
Nearest forward progress (NFP), 4
Neighbor-table-based multipath routing (NTBR), 157–158
  data-driven, 158
  time-driven, 157
Neighbor discovery process, 169–171, 173
Network allocation map (NAM), 361
Network coding model, 120
Network coding scheme, 111
Network-interconnected devices, 87
Network-wide broadcast, 51, 83

algorithms, 51
Network coordinate system, 38
Network layer problem, 329
  power control, 329
Network lifetime 342, 344
Network lifetime routing, 341
Network nodes localization, 30
Network protocol data units (NPDUs) 142
Network simulator, 449
Networking protocols, 89
Node-initiated MF (NIMF) scheme, 238
Node misbehavior, 381
  malicious behavior, 381
  selfish behavior, 381
Node power levels, 341
  minimize variance, 341
Node selfish behavior, 382
  self-exclusion misbehavior, 382
  self-exclusion/nonforwarding, 382
Noise tolerance, 324
Nonhomogeneous vehicle traffic, 411–412
  transmission range, 411
Network-wide broadcast (NWB), 51
  packet, 64, 73
  protocols 52, 63, 67, 71, 74
  robustness, 75
  solutions, 73
  transmission, 66, 73, 75, 82
Network-wide broadcast algorithms, 52, 53, 55, 58, 61, 72, 73, 74, 76
  redundancy control, 52, 72
  robustness control, 52, 72
  overview, 54

Occupancy theory, 409, 410
OCEAN model, 387, 388
Off-the-shelf traffic simulators, 426
On-board safety systems, 406
On-demand multicast routing protocol (ODMRP), 10, 180
One-dimensional networks (409–410) 409
  homogeneous MTR, 409
One-source two-sink network, 110
Opportunity-driven multiple access (ODMA) protocol, 417
Optimized link-state routing (OLSR), 7, 9, 143
  protocols, 7, 299

Orthogonal frequency division multiplex (OFDM) technology, 408

Packet collision, 336
Packet delivery ratio, 179
Packet forwarding algorithms, 4–5
Packet identifier, 120
Packet interarrival times, 288
Packet loss ratio (PLR), 262
Packet out-of-order delivery ratio (POR), 261
Packet transmission, 316
Parent variable, 172
Partial topology knowledge forwarding (PTKF) algorithm, 5
Peer-to-peer systems, 383
  applications, 88
Perez equations, 373
Perimeter nodes, 28, 29
  broadcasts, 29
  criterion, 29
Personal area network (PAN), *See* Piconet network
  coordinator, 459, 462, 465, 470
  router, 40
Pervasive devices, 87
Photovoltaic (PV) systems simulation, 371–373
  solar insolation conversion, 371
Physical layer (PHY), 407, 435–436
  considerations, 435
Pico-cell architecture, 437
Piconet network, 131
Poisson distribution, 409–411
Power allocation, 330
Power attenuation factor, 323
Power-aware algorithms, 316
Power-aware routing protocol (PARO), 335
Power-based control algorithms, 6
Power-based interference graph, 327
Power control 316, 318, 320, 321, 332, 335, 339, 340, 347
  approaches, 339
  design principles, 316, 337
  energy-oriented perspective, 340
  impact, 318
  issues 339, 340
  MAC layer problem, 321, 334
  single-layer approach, 320

systematic approach, 335
transport layer problem, 332
Power control algorithm, 328, 334, 335, 339
flowchart, 328
Power-controlled dual channel (PCDC), 324, 339
Power controlled multiple access protocol (PCMA), 322–323
Power control problem, 326
Power-efficient topology, 341, 343
control 343–344
control algorithm, 347
Power management system, 344–347
mechanism, 344
protocols, 344
Power-saving MAP, 361
HCCA service, 361
Power-saving mode, 12
Power-saving protocols, 369
Power-saving station, 360
POWMAC protocol, 325
Proactive routing algorithms, 6
Proactive routing protocols, 299
AODV, 299
Probabilistic random walk, 68, 70
bar grapgh, 70
Probabilistic routing protocol using history of encounters and transitivity (PROPHET), 231
Probe-ACK packet, 196
Protocol model, 191
Pseudo-coordinates, *See* Beacon vectors
Pseudo-geometric routing, 27, 28
algorithm, 26
Pseudo-geometric space, 26
Public key infrastructure (PKI), 446

q-k relationship, 422
Quadratically constrained programming problem, 338
Quality of service (QoS), 408
constraints, 370
parameters, 4

Radio-location techniques, 32
Radio beacon, 24
Radio signals 30, 34
Radio wire MicrelNet, 42
Random access MAC protocols, 319

Random linear coding, 117
Random walk model, 68
probabilistic version, 68
Random way-point model (RWP), 67, 410, 447
Random, linear network coding, 112
Raptor codes, 108–109, 110
applications, 110
Reactive location service (RLS), 416
Reactive protocols, 8
Real-time tracking and monitoring, 45
applications, 45
Received-signal-strength (RSS), 42
Received signal strength indicator (RSSI) signal, 5
Receiver-based delay control (RDC), 295
Receiver sensitivity threshold, 189
Recursive encoding, 114
Reduced-function device (RFD), 40
Reed–Solomon codes, 95, 96, 99
Relative distance micro-discovery ad hoc routing (RDMAR), 148
Relay nodes, 111
schematic presentation, 111
Relay radio's power consumption, 368
Remote control toys, 46
Reputation-and-trust-based system, 376, 377, 378, 379, 380, 383–386, 389, 390, 392, 394, 399
calculus-based trust antecedents, 378
classification, 384–385
CORE, 390–392
codes, 101, 102
decision making, 389–390
information asymmetry, 379
information gathering, 386
information modeling, 388
information sharing, 386–388
initialization, 384
institution-based trust antecedent, 378
knowledge-based trust antecedents, 378
open problems, 399
opportunistic behavior, 380
properties, 384
social perspective, 377
system goals, 383–384
trust and uncertainty, 377
trust beliefs and trust intention, 377
uncertainty degree, 379

Reputation-based framework for sensor
    networks (RFSN), 395–397
  architectural design, 395
  decision making, 396
  information modeling, 396
  information sharing, 396
  model, 388
  systems, 387
Reputation-based model, 383, 395
Reputation table (RT), 390
Request-power-to-send (RPTS), 324
Request-to-send (RTS) packet, 189, 193, 321
Request to send/clear to send (RTS/CTS)
    exchange, 302
Reservation ALOHA (R-ALOHA) systems,
    438
Retransmission-based solutions, 94
Retransmission yimeout (RTO) 264, 283,
    305
  value, 254, 255, 266
Retransmitted packet loss, 256
Road models, 448
  circular road model, 448
  road grid, 448
  straight highway, 448
Road traffic, 422
  fundamental diagram, 422
Robust reputation system (RRS), 394–395
  metrics, 394
Role-based multicast (RBM) protocol, 416
Round-trip hop count (RTHC), 262
Round-trip time (RTT), 196, 253, 288, 295
Route-quality estimation, 199–200
Route-tree-based multicast algorithms, 10
Route failure determination, 140, 252, 255,
    260
Route maintenance temporally ordered
    routing algorithm, 137
  flow diagram of, 137
Route replies (RREPs), 299
  message, 8, 133, 299
  packet, 135
Route request (RREQ), 8, 299
  message, 8, 133, 156
  packet, 135, 156, 300
Routing algorithm, 342
  bandwidth, 6
  feature, 6
  power, 6

  QoS support, 6
  scalability, 6
Routing approach, 415–417
  geocast routing, 415–417
  position-based routing, 416–417
Routing layer, 439–445
  geographical addressing, 439–440
  protocol, 266
  routing constraints, 439
  VANETs geocasting, 440–442
Routing on-demand acyclic multipath
    (ROAM), 140, 141
  algorithm, 140
Routing protocol(s), 11, 21, 159, 220, 266,
    315, 329, 442, 451
  AODV, 220
  DSR, 11
  OLSR, 220

Satellite networks, 94
Satellite signals, 30
Satellite systems, 92
Scalable broadcast algorithm (SBA), 57, 59,
    64
Scalable location update-based routing
    protocol (SLURP), 148–149
Schedule bookkeeping protocol, 346
Scheduling algorithm, 326, 328
Scheduling and interference-aware (SIA),
    212
Second-hand information, 386, 387, 391,
    398
  benefits, 386
Security architecture, 452
Segment-oriented data abstraction and
    dissemination (SODAD), 444–445
Segment reduction effect, 286
Selective acknowledgment responses, 309
  effect, 309
Selective acknowledgment (SACK), 267,
    309
  information, 267
  scheme, 268
Selective additional rebroadcast (SAR), 75,
    79
  approaches, 79
  broadcast, 74, 79
  protocols, 80
Security-aware ad hoc routing technique, 13

Self-positioning algorithm (SPA), 36
Sender-based delay control (SDC), 307
Sender maximum segment size (SMSS), 253
Sensor network, 46
Sensor node(s), 30, 45
Sensor protocols for information via negotiation (SPIN), 108
Shannon coding system, 91, 92
Shared wireless infostation model (SWIM), 228
Short-term throughput (STT), 261
Shorter end-to-end delay, 190
Signal-strength-based technique, 32
Signal-to-noise ratio (SNR), 195, 319
Signal stability-based adaptive routing (SSBR), 138
Signal stability table, 138
Signal to interference and noise ratio (SINR), 203, 329, 335
Single-channel mobile phones, 24
Single-path routing schemes, 336
Slave-slave (SS) fashion, 460
  bridging algorithm, 469–472
Slave-slave (SS) bridge exchange, 472
  queuing model, 472
Slotted ALOHA (S-ALOHA) systems, 437, 438
Slotting/damping method, 345
Slow-start threshold, 253
Smart-networked toy application, 46
SMAP/SMP operation, 362, 365
  VoIP, 365
  power saving, 362
Soft state timer, *See* Keep-alive timer
Solar mesh ad hoc positioning system (SMAPs), 354, 356, 361, 363, 364, 365, 367, 370
  algorithm performance-mean delay, 364
  algorithm performance-power consumption, 364
  beacons, 362
  components, 356
  power consumption, 364
  power saving, 362
Solar mesh points (SMPs), 354
SolarMESH network, 355
SolarMESH version II node, 355
Solar-powered mesh nodes, 353
Solar-powered network, 353

Solar-powered node cost, 371
Solar-powered WLAN mesh networks 353, 354, 360
  power saving protocol, 353, 360
Source-initiated protocols, 132
Source-tree adaptive routing (STAR), 145
Space-time graph, 226
  illustration for, 226
Spatial contention effects, 310
  application layer, 310
  link layer, 310
  network layer, 310
  transport layer, 310
Spatially aware routing (SAR) protocol, 416
Speed-density relationship, 421
Split-connection proposals 252, 257
Split multipath routing (SMR), 156–157
  protocol, 156
Spurious retransmission detection, 263
Static ad hoc networks, 297
Static routing protocol (SRP), 138
Storage devices, 95
  barcodes, 95
  compact disk, 95
  DVD, 95
  tape, 95
Store-carry-forward (SCF), 222
  routing, 222
STRAW model, 425
SW-implemented codes, 92

Table-driven protocols, 142
Teleconference current mission scenarios, 165
Temporally ordered routing algorithm (TORA), 136–137
  AODV, 137
  DSR, 137
  LMR, 137
Time-evolving ad hoc networks, 220
  illustration of, 220
Time-evolving graph, 247
Time-to-live (TTL), 28, 382
Time difference of arrival (TDOA) method, 32
Time domain (TDMA), 144
Time of arrival (TOA) technique, 32, 33, 42
  estimatation, 32
Timeout tolerance factor, 294

Top-of-atmosphere (TOA), 372
Topology-ambivalent (flooding) approaches, 54
Topology-aware algorithms, 57
  properties of, 57
Topology-based reverse path forwarding (TBRPF) protocols, 7
Topology control algorithms, 5
Topology dissemination-based reverse path forwarding (TBRPF), 9
Topology-sensitive approaches, 56
Topology-sensitive CDS-based algorithm, 60
Topology-sensitive model, 58
Tornado code, 101–102, 104, 105
  efficiency of, 105
  history, 101
  properties of, 104
Traffic approaching critical density, 423
  space-time diagram, 423
Traffic demand oracle, 225
Traffic flow theory, 421
  introduction, 421
Traffic indication map (TIM), 360
Traffic microsimulator, 425
  RoadSim, 425–426
Traffic simulation models, 424, 449
  driver behavior, 449
Traffic simulator, 449
  CORSIM, 449
Transmission control protocol (TCP) 10, 251, 277, 282, 445
  ACK segments, 287
  acknowledgments, 107
  BUS, 265
  characteristics, 282–284
  congestion control mechanisms, 253
  DATA segments, 287
  Eifel, 264
  end-to-end semantics, 301
  enhancements characteristics, 268, 269–272
  goal, 297
  Jersey, 257, 259, 260
  NewReno agents, 295
  performance enhancement schemes, 257
  problems, 254
  protocol, 11
  RFC, 297
  RTO recovery, 265

  sfchemes, 257
  timestamp option, 265
  use, 252, 277
  Vegas, 257, 258
  Westwood, 257, 259
Transmission errors, 256
Transmission power control, 316, 319, 332
  algorithms, 319, 334
  problem, 315, 336
Transmission scheduling policy, 256
Transmission scheme, 100
Transmission time, 297
Transmitting packets, 3
Transport layer protocol, 251, 267, 451
  ad hoc transport protocol (ATP), 267
Traveling salesman problem (TSP), 239
Tree approach, 224
  illustration of, 224
Tree-based protocol, 166
Trust manager, 393
  alarm table, 393
  friends list, 393
  trust table, 393
Trust metric, 384
  properties, 384

Ultra-wide band (UWB), 2
UMTS terrestrial radio access time division duplex (UTRA TDD), 408
Uncertainty degree, 379
  types, 379
United States Federal Communications Commission, 408
United States Weather Stations, 373

Vandermonde matrices, 97
VANETS vs. MANETS, 434–435
Vector-hop algorithm, 30
Vehicle mobility, 420
Vehicle platoon formation, 46, 47
  maintenance, 47
Vehicle-to-roadside (V2R) communication networks, 406, 426
Vehicle-to-vehicle (V2V) communication networks, 406, 426
Vehicle traffic simulator (VTS), 424
Vehicle's transmission range 412, 413
Vehicular ad hoc network (VANET) 2, 14, 405, 406, 407, 409, 410, 415, 416, 417,

419, 421, 426, 433, 435, 438, 441, 442, 444, 446
  applications, 445
  automated highways and cooperative driving, 418–419
  characteristics, 406–407
  classification, 417
  communications, 438
  connectivity, 409–412
  direct short-range communication (DSRC) standard, 408
  emerging technology, 405
  geocasting protocols, 441
  IEEE 802.11 standard, 407–408
  IP-based applications, 420
  local traffic information systems, 419
  performance issues, 433
  privacy issues, 446
  protocols, 449, 452
  research community, 426
  safe/efficient transportation, 405
  safety-related applications, 417–418
  security issues, 446
  unicast protocols, 442–444
  UTRATDD, 408
Vehicular ad hoc network (VANET) protocol design, 450
  open issues, 450
Vehicular ad hoc network (VANET) simulation models, 424
  integrated model, 424
  2-stage model, 424
Vehicular ad hoc network (VANET) system, 436, 449
  stages, 449
Vehicular ad hoc network (VANET) research, 423–425
  traffic simulators, 423
Vehicular collision warning communication (VCWC), 418
Virtual coordinates, 28
Virtual dynamic backbone protocol (VDBP), 167
Visual communication system, 1
VoIP traffic, 368

Waiting time, 442
Wake-up schedule function (WSF), 346
WiFi, 129

Wireless access for the vehicular environment (WAVE), 436
Wireless ad hoc networks 110, 315, 319, 326, 339, 340
  power control survey, 315
Wireless cellular systems, 435
Wireless channel errors, 252, 255
Wireless communication network(s), 34, 53, 88, 375, 380
  background, 380–381
  node misbehavior, 381
  perspective, 380
Wireless error determination, 255, 260
Wireless links, *See* Modems
Wireless local area networks (WLAN), 407, 436
  802.11a, 436
  coverage radio, 362
  mesh deployments, 355
  mesh networks, 353
  mesh nodes, 355
  multihop relaying, 353
Wireless mesh routers, 24, 87
Wireless multihop ad hoc networks, 88, 343
Wireless multihop communications, 278
Wireless network, 87, 94, 129, 318
  protocol stack, 318
  architecture, 317
Wireless network simulator (WNS), 424
  CORSIM, 424
  FARSI, 424
  SHIFT, 424
  Videlio, 424
Wireless/mobile communications, 95
  cellular telephones, 95
  digital television
  high-speed modems, 95
  microwave links, 95
  satellite communications, 95
Wireless personal networks (WPAN), 39, 41, 462
Wireless routing protocol (WRP), 144
Wireless sensor networks (WSNs), 2, 101, 110, 375, 376, 380, 381, 383, 400
Wireless transmission, 189
Wizzy digital courier service, 221

ZigBee data broadcasting, 39
ZigBee device, 40

ZigBee network(s), 40, 42
  broadcasting, 42
  localization, 42
  ZigBee coordinator, 40
  ZigBee end device, 40
  ZigBee router, 40
ZigBee protocol 39, 40, 44
  architecture, 39
ZigBee routing, 39
Zone of forwarding (ZOF), 441
Zone of relevance (ZOR), 416, 440
Zone routing protocol (ZRP), 13, 145–146, 181
  architecture 146, 147

WILEY SERIES ON PARALLEL AND DISTRIBUTED COMPUTING
Series Editor: Albert Y. Zomaya

Parallel and Distributed Simulation Systems / Richard Fujimoto

Mobile Processing in Distributed and Open Environments / Peter Sapaty

Introduction to Parallel Algorithms / C. Xavier and S. S. Iyengar

Solutions to Parallel and Distributed Computing Problems: Lessons from Biological Sciences / Albert Y. Zomaya, Fikret Ercal, and Stephan Olariu (*Editors*)

Parallel and Distributed Computing: A Survey of Models, Paradigms, and Approaches / Claudia Leopold

Fundamentals of Distributed Object Systems: A CORBA Perspective / Zahir Tari and Omran Bukhres

Pipelined Processor Farms: Structured Design for Embedded Parallel Systems / Martin Fleury and Andrew Downton

Handbook of Wireless Networks and Mobile Computing / Ivan Stojmenović (*Editor*)

Internet-Based Workflow Management: Toward a Semantic Web / Dan C. Marinescu

Parallel Computing on Heterogeneous Networks / Alexey L. Lastovetsky

Performance Evaluation and Characteization of Parallel and Distributed Computing Tools / Salim Hariri and Manish Parashar

Distributed Computing: Fundamentals, Simulations and Advanced Topics, *Second Edition* / Hagit Attiya and Jennifer Welch

Smart Environments: Technology, Protocols, and Applications / Diane Cook and Sajal Das

Fundamentals of Computer Organization and Architecture / Mostafa Abd-El-Barr and Hesham El-Rewini

Advanced Computer Architecture and Parallel Processing / Hesham El-Rewini and Mostafa Abd-El-Barr

UPC: Distributed Shared Memory Programming / Tarek El-Ghazawi, William Carlson, Thomas Sterling, and Katherine Yelick

Handbook of Sensor Networks: Algorithms and Architectures / Ivan Stojmenović (*Editor*)

Parallel Metaheuristics: A New Class of Algorithms / Enrique Alba (*Editor*)

Design and Analysis of Distributed Algorithms / Nicola Santoro

Task Scheduling for Parallel Systems / Oliver Sinnen

Computing for Numerical Methods Using Visual C++ / Shaharuddin Salleh, Albert Y. Zomaya, and Sakhinah A. Bakar

Architecture-Independent Programming for Wireless Sensor Networks / Amol B. Bakshi and Viktor K. Prasanna

High-Performance Parallel Database Processing and Grid Databases / David Taniar, Clement Leung, Wenny Rahayu, and Sushant Goel

Algorithms and Protocols for Wireless and Mobile Ad Hoc Networks / Azzedine Boukerche (*Editor*)

Algorithms and Protocols for Wireless Sensor Networks / Azzedine Boukerche (*Editor*)